Catalogue of Risks

Dirk Proske

Catalogue of Risks

Natural, Technical, Social and Health Risks

 Springer

Dr.-Ing Dirk Proske, MSc.
University of Natural Resources and Applied
Life Sciences, Vienna
Institute of Mountain Risk Engineering
Peter-Jordan-Street 82
1190 Vienna
Austria
Goetheallee 35
01309 Dresden
Germany
dirk.proske@boku.ac.at

ISBN: 978-3-540-79554-4   e-ISBN: 978-3-540-79555-1

DOI: 10.1007/978-3-540-79555-1

Library of Congress Control Number: 2008930227

Typesetting: by the author and Integra, India

Cover design: eStudio Calamar, Spain

Printed on acid-free paper

9 8 7 6 5 4 3 2 1

springer.com

Preface

Since the German edition of this book, the topic of risk has experienced even greater attention, not only in the world of science but also in other fields, such as economics and politics. Therefore, many new publications have evolved. To keep with the idea of an encyclopedia for the topic of risk, this book has been completely reworked. Not only are many updated examples included in chapter "Risks and disasters" but also new chapters have been introduced, such as the chapter "Indetermination and risk". This new chapter was developed since the question "Is it possible for risks to be completely eliminated, and if not why?" has become a major point of concern. Therefore, especially in this chapter, the focus of the book has extended from a simple mathematical or engineering point of view to include much broader concepts. Here, not only aspects of system theory have to be considered, but also some general philosophical questions start to influence the considerations of the topic of risk.

The main goal of this edition, however, is not only the extension and revision of the book, but also the translation into the English language to allow more readers access to the ideas of the book. The author deeply hopes that the success the book made in the German edition continues and that readers experience a major gain from reading the book.

Expression of Thanks

Humans who dedicate their entire working life to one topic might reach a deep intellectual penetration of the topic. The obtained understanding of the topic can be reflected in an extremely clear exposition of the subject matter. If these people can be convinced to capture their considerations about that topic on paper, tremendous books would see the light of day. Intensive intellectual penetration is the natural basis for the preparation of good books. Therefore, like humans, good books require time to grow. It is not surprising that both subjects need such precious time since both are rooted in each other.

Obviously, due to the age of the author, this book was not created under such a favourable condition. Does the book then lose its value? No, because there may be other potential alternative routes. The diverse support of other people is one of these possible routes.

This support has to be considered even greater, since it was done for free. From a materialistic point of view, the creation of this book has been a complete waste of resources. However an author is to a certain extend a dreamer, a person who believes that the reader will start communication with the author and this is the major measure of success of a book.

What can this book tell, however, if it is not only considered as entertainment? Well, the goal of this book is quite impressive. It describes the value of our life for us. What are we going to do with our given time? This sounds funny considering the title of the book. But, from the author's point of view, this question is only an extension of the question of life risks. The question "How safe do we live?" yields to the question, "Why do we live like that?"

Of course, the decision as to what is important for an individual human is mainly connected to the persons themselves. But, this freedom is partly taken away by some circumstantial forces of living in a society. The primary values of life are, therefore, a mixture of individual and social goals. The importance of the individual elements can differ dramatically from human to human. This is not only valid for different humans but for individual humans as well, since their values change over their lifetime.

For example, people who know about their end rapidly change their life values. This phenomenon is called reframing and seems to be completely understandable. Since we know that all humans are mortal creatures, would it not be recommendable to check the individual values of life regularly?

During his youth, the author learned in school that humans are social creatures. Since that time, the naïve protest changed to the conviction that this discovery cannot be denied. The biggest problems that humans experience are social problems. The biggest risks to humans are social risks. But also the biggest successes are social achievements. If we consider this as a hypothesis, then we discover the biggest human task, which actually seems to be a human duty: the social responsibility for fellow human people. This does not only include living humans, but also ancestors as well as descendants.

But, what is meant by responsibility? If one looks closely at the term responsibility, one finds the term response, also to answer. The duty to answer has been, since the earliest civilisations, a part of religion, for example the religions of the Egyptians and Christians.

Only people, however, who are conscious about a question can answer. A good answer can only be given with a deep understanding and a great level of intellectual clarity. Then, it might be worthwhile to answer with a book. The reader can decide whether that worked out well.

As already mentioned, the author experienced the luck of strong support from many sides. He, therefore, would like to firstly thank his wife Ulrike and son Wilhelm, who accepted the many working hours without complaining. Additionally, his parents, Gerhard und Annelies Proske, supported him so strong. He further thanks Dr. med. Katrin Knothe, Helga Mai and Petra Drache for the revision of the German script and Dr.-Ing. Peter Lieberwirth, Dr.-Ing. Harald Michler, Dr.-Ing. Sebastian Ortlepp, Dr.-Ing. Knut Wolfram and Mr. Wolf-Michael Nitzsche for the photos. Prof. Jürgen Stritzke and the society of friends of civil engineering at the University of Technology Dresden are thanked for the financial support to the first edition. Furthermore, he strongly thanks Prof. Konrad Bergmeister, Prof. Udo Peil and Prof. Pieter van Gelder. The translation was supported by Kendra Leek and Charlotte Fletcher, who the author thanks as well. Finally, he thanks the publisher Springer for the support to the book.

After a look into the past, there should be a look into the future: the author wishes the reader a comprehensible and enjoyable book.

Preface German Edition

All outward forms are characterized by change. If we look out of a window, we see clouds, which pass in only a few minutes. If we return to a city, we see a changed face, the roads are different and sometimes the people living there have changed their culture. If we talk to geologists, they inform us about how mountains grow and disappear and oceans existed, where now forests or deserts are. Astrophysicists explain the beginning of the universe and the solar system, but they also point out, that one day the sun will stop shining.

The world is ruled by change. From the smallest elements to those the size of galaxies, beginning and end can be observed. This rule also applies to living matter. We can see the growth of trees or the grain ripen on the fields and we have to realize, that this law of change is also valid for humans. Whether we like or not: humans are time-limited creatures. Life consists of a beginning and an end. The question, when a human life begins and when it ends are one of the most difficult questions for human ethics. Some people, like the creatures of the current German Embryo protection law think, life begins with the fusion of a sperm and egg cell. For others the nesting is the beginning and for the majority of people probably birth is the beginning. Independent from such discussion is that the gain of life is a major event, not only for the individual human, but also for others, like the parents. Being alive is a major gift of nature and therefore our decisions are quite often based on the protection of this gift.

But if the birth is a great success, usually the death is the most horrible event in the life of a human. Also here the question of when a human is dead is sometimes extremely difficult to answer. The reader should be reminded of accident casualties, who were given artificial respiration for years. Independent from these examples we have to accept the limitation of lifetime.

Usually the limitation of lifetime is not present in our daily life. "We live always like we assume, that we have unlimited time, that the reached would be final and the virtual stability and the illusionary eternity are a major impulse for our actions." wrote Fritz Riemann in his famous book "The basic forms of fear". Faith into the future and faith in our existence frame our daily work and plans.

But can the awareness of mortality also be an impulse for action? Can the thought that our life can be finished virtually every moment – media in vita morte sumus – also push us forward? As already mentioned we share the nature of mortality with all living creatures. But Max Frisch wrote "Not the mortality alone, which we share with the newts, but the consciousness of it, that makes our existence only human, makes it an adventure." Here a strong impulse for human actions becomes visible: the wish for everlastingness, becoming immortal in our own way, surviving in stories and in things we have done: "Linger moment, you are too beautiful, the signs of my days on earth should never fade away." says Goethe's Faust. Here in the fight against transience one of the major roots for the uniqueness of humans can be found. It sounds incredible: denial and acceptance of the limitation of human lifetime can yield to motivation for humans.

The acceptance of the limitation of human lifetime is the basis for this book. The book actually goes even further. It describes direct and indirect causes for the passing away of humans. This description is mainly given in numbers, which include when, how many and why people died. It sounds terrible, but it is only a summary of data.

Is mortality, however, more than a single data point? Is the loss of a human not the biggest sorrow, which humans can experience? How can such a loss be put into numbers? And what is the gain developing such numbers? The author is aware about the fact, that numbers can not express sorrow. Numbers as a representative of objectivity (see Chap. 1) are unable to describe sorrow subjectively. Numbers themselves are neither good nor bad. They are a neutral tool. If they are a neutral tool then it is our responsibility to use them adequately. And numbers might indeed be useful for the prevention of sorrow. If we grasp how many people were killed by a certain cause, we might identity major causes of death and therefore spend more resources on preventative measures. If we detect the highest risks posed to humans in the past, we might be able to fight these risks in the future. This could yield not only to an increase in ones life span but also to a better life. The past is a huge early warning system, which should not just left unnoticed.

In the first chapter of this book, some comments about the limited objectivity of numbers and languages are given, which is important for the understanding of risk parameters. Additionally the terms risk, hazard, danger and disaster are introduced. In the second chapter risks are classified, identified and examples are given. Then risk parameters are introduced to describe risk quantitatively. These parameters follow a line of evolution. In the fourth chapter subjective influences on risk perception are discussed. This topic is of high importance, since people do not decide on risk parameters, but on subjective risk assessment. In the fifth chapter quality of life parameters are introduced. The author understands them as most sophisticated risk parameters, not only because the way we live decides how long we life, but also because humans do not basically decide on risk, they always compare risk and advantages, and quality of life can be expressed in the simplest way as access to advantages. Therefore such parameters can be used for decisions about the efficiency of risk mitigation measures. The question about how much resources should be spent against risks can be answered. But here again questions of ethic arise. How can one speak about limited resources if one has to save humans? Is it not the highest duty to spend all resources to save human lives? And what is understood as a safety measure? Is school a safety measure if we know that better educated people have a longer life? What does help mean? How much resources do we need to protect not the others, but ourselves? Don't we have to save some resources so that we have something to eat? And don't we need some sleep? How about a holiday? If alls these things are mitigation measures, do we have to completely optimize our life? Just one remark: we will later see, that optimization of safety measures is a drawback for safety. But in general we can observe that virtually all actions of humans are safety measures. Therefore we need some methods to evaluate all the actions. Are we then able to answer the following question: How can the available resources be utilized to save a maximum number of people?

After such an investigation, safety or mitigation measures can be divided into useful and non-useful actions. Useful actions yield to an increase in the quality of life. What that means can be read in this book. In general, the utilization of resources can yield to an improvement. But how is this proceeding today?

If one only observes the latest developments of laws and regulations in the German health sector. Here only regulations are considered, which intervene the measures for safety in a preventive or therapeutic way. The health sector itself can be understood as a safety measure (probably one of the biggest and most successful). But politicians, who have to follow some political circumstances since they act in that system, make the majority of laws. Therefore the regulations consider only very limited such effectiveness considerations. One of the examples is the treatment of diabetic's in Saxony, a federal state of Germany. Over the last few years, an intensive treatment program was established there to minimize late consequences of the disease. But caused by short-term lack of money, the program was stopped. This does a not mean that money is saved, since the late consequences will be much more expensive compared to the prevention program, but the political conditions at that time yielded to that result. The author does not know of an objective investigation. Therefore the choice of a right negotiation strategy, the choice of a good moment (window of opportunity), personal relations or lobbyism is more important then saving many people's live. Subjective assessment is actually more important then objective measures. Whether or not that is good will be discussed in this book. In general it can be stated, that there is a huge demand for techniques to compare the efficiency of mitigation measures. Many measures are just a waste of money, where others could save the lives of thousands of people with only a small increase in the budget.

Of course, it is impossible for this book to evaluate all known and all possible safety measures. That is not the goal of the book. But it can give a push for a more just procedure for the distribution of resources for safety measures over all sectors of a society. It does not matter, which industry for example medicine, nuclear power industry, structural industry, chemistry, waste disposal, fighting crimes, car producers or aviation and space industry. For all these areas universal procedures are now available to compare the efficiency of the safety measures. But if the industries are affected by the procedures, what about laws or regulations? Should they not follow the law of efficiency, or are there limits? And what does it mean for societies?

However no book can give an ultimate discussion of an issue. It belongs to humans "their is one drop of water enough to image the existence and diversity of an ocean" how Arkadi and Boris Strugazki have put it in the book "The waves suffocate the wind" to extend the work. Therefore the topic of this book remains open.

Contents

1 Indetermination and Risk

We have now not only traversed the region of the pure understanding and carefully surveyed every part of it, but we have also measured it, and assigned to everything therein its proper place. But this land is an island, and enclosed by nature herself within unchangeable limits. It is the land of truth (an attractive word), surrounded by a wide and stormy ocean, the region of illusion, where many a fog-bank, many an iceberg, seems to the mariner, on his voyage of discovery, a new country, and, while constantly deluding him with vain hopes, engages him in dangerous adventures, from which he never can desist, and which yet he never can bring to a termination.

Immanuel Kant: The Critique of Pure Reason, 1787 (Cramer 1989).

1.1 Introduction

The most common definition of the term risk comprises two components: (1) indetermination and (2) damage. The first component, indetermination, is discussed heavily in this chapter, whereas the term damage will be discussed in the chapter "Risks and disasters".

If the assessment of risks is considered a human technique to handle indetermination, a question would arise, "Does there have to be some indetermination in nature?". To discuss this question, this chapter delves into philosophy, linguistic theory, mathematical logic and system theory. Terms like causality and complexity will be discussed to some extent. While the first two sections deal with non-evaluated indetermination, the third section introduces terms of negative evaluated indetermination, such as (un)safety, risk, damages and hazards. Although disasters are not related to indetermination, data from them can be used for risk assessment, and they are a passage to the next chapter.

1.2 Indetermination

The quality of scientific models to describe certain processes in nature has made major progress over the last few hundred years. Based on these models, man has increasingly improved his understanding of nature and has fundamentally changed society and environment. The final prediction of the future behavior of nature, however, still remains impossible. The power of prediction shows a great diversity depending on the field of science. While in some fields a high level of prediction quality is reached, such as the Newton's law of motion in physics, no reliable models have yet been developed to predict the motion of societies (McCauley 2000). For example, economic developments over a period of several years are still unpredictable.

Unfortunately, the traditional division of sciences, including their special types of investigative systems, such as mechanical systems or social systems, does not improve the quality of scientific models at a certain level. For example, consider an engineer investigating the safety of a particular bridge using widely accepted codes of practice based on laws of physics and material sciences. If an accident happens on another bridge, society might impose more stringent safety requirements. A consequence of this can be new regulations, and thus the engineer must re-evaluate the safety of the initial bridge. Through this re-evaluation, an insufficiency in the safety of the bridge might be found, even if the bridge itself has not changed. This shows that the investigation of the safety of the bridge is influenced by social developments, which are currently not considered in the engineering model explicitly.

This section tries to move the general focus of engineers and scientists away from one specific point of view to embrace a more generalist view of system types, complexity and indetermination. First though, some examples and definitions of indetermination are provided.

1.2.1 Examples of Indetermination

Indetermination is apparent in everyday life. We might want to go shopping, but the shop is closed due to a holiday. We might book a holiday and there are concerns about the uncertainty of the weather. The issue of uncertainty or indetermination has been a point of discussion for centuries. Early civilizations tried to describe indetermination through the will of God. They introduced oracles, such as the one in Delphi, or they had a shaman.

One of the first historical examples that described indetermination in detail was Buridan's donkey (Schlichting 1999). Within this story, a donkey is located exactly between two stacks of hay (in Fig. 1-1, two dishes of oranges are shown for ease of drawing). The question that now arises is, in which direction will the donkey move and why? Of course, if there would be circumstances influencing the decision of the donkey been known they might be excluded. This yields to the query, "Can all causes be identified or are there some limits to identification?" If the latter is true, indetermination exists under all conditions.

Fig. 1-1. Buridan's donkey (Schlichting 1999)

Another example shown in Fig. 1-2 is a symmetrical water wheel. On the wheels are several bins, which are constantly filled with water coming from the top. The bins are partly open at the bottom, allowing water to flow out. In the beginning, the wheel is balanced in a stationary position. Therefore, the wheel will turn neither to the left nor to the right. As the bins fill with water and the forces increase, a certain level is reached where an extremely small disturbance in the balance of the wheel can cause the wheel to start moving in one direction. The situation is like a shift of dimensions, where suddenly the dimensions are zoomed in. It might also be called scale cascade. If the wheel might balance in the region of centimeters or millimeters, it might not in nanometers. The amount of information, energy and technology producing a balanced wheel over nearly all dimensions will grow exponentially (Schlichting 1999).

Of course, the behavior of the wheel is strongly connected to the volume of water flowing in the bins. After the wheel has started to rotate, it behaves quite regularly, but changes in rotation might occur now and then (Schlichting 1999).

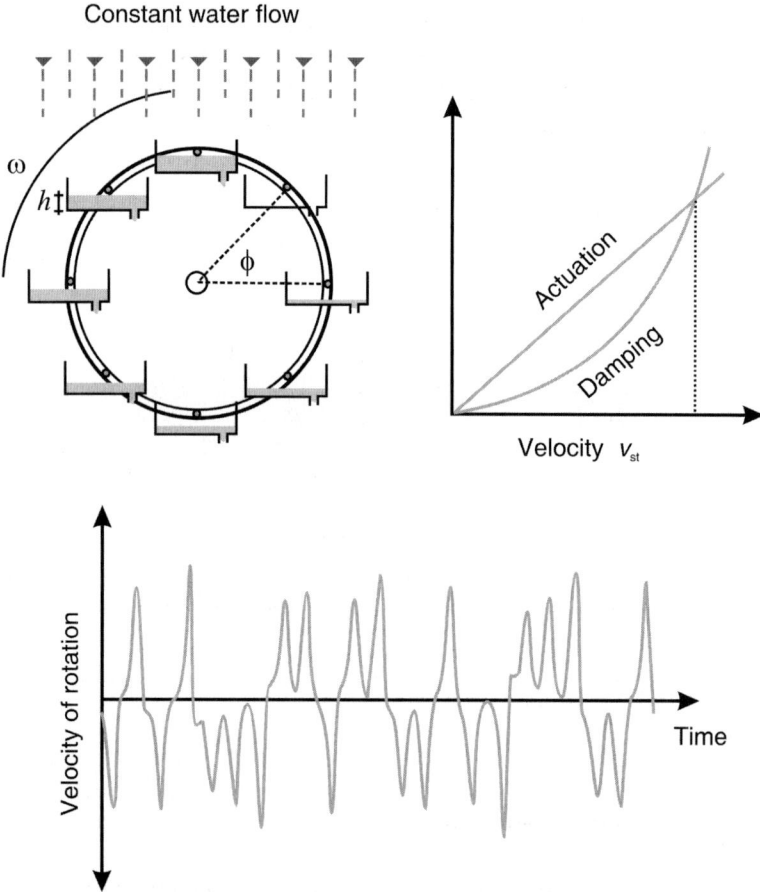

Fig. 1-2. Chaotic wheel: The figure at top left shows the system, and the figure at top right shows the function of the actuation and the damping. The bottom figure describes the velocity of rotation over time (Schlichting 1999)

   To illustrate this point with a more practical example, the solar system is considered. Usually the solar system or other astronomic systems are clearly considered causal systems with high amounts of predictability and low amounts of indetermination. However, the number of elements in a solar system is rather high, assuming about 100,000 bigger elements. For Newton it was impossible to find a solution for such a big system; therefore, he considered only the bigger bodies such as the sun (Weinberg 1975).

   Although Newton's description of gravitational forces was a milestone, it does not describe the entire solar system. The limits of his simplification is clearly shown by the astronomer Shoemaker: "…the solar system is not at all the eternal, unvarying mechanism envisioned by Isaac Newton, but a carnival – a dynamic, evolving cloud of debris, filigreed with bands and shells of shrapnel…" (Preston 2003). Considering that such instability exists, the solar system or other astronomical objects are not as predictable as one may assume (Fig. 1-3).

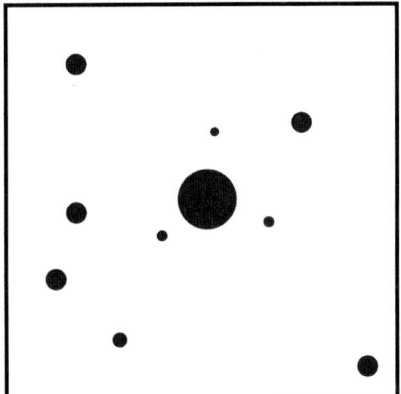 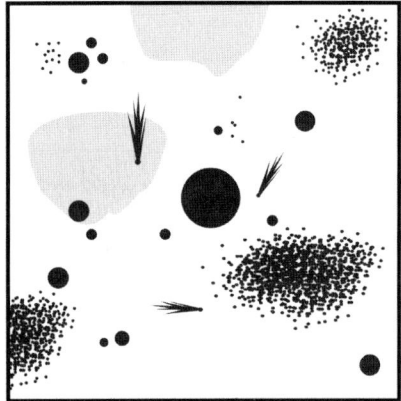

Fig. 1-3. Simplification of the solar system (*left*) by neglecting smaller elements (*right*) according to Weinberg (1975)

1.2.2 Types of Indetermination

There are many different words used to describe indetermination, such as "vagueness", "fuzziness", "randomness", "uncertainty" or "ambiguity". Most of these terms deal only with a certain aspect of indetermination. For example "vagueness" is mostly used for the indetermination of words or terms. "Uncertainty" or "ambiguity" is a type of indetermination, where people are aware of their limitations of knowledge (Kruse et al. 1994).

Klir & Yuan (1995) and Klir & Folger (1998) have broken down indetermination into four types:

• Non-specification (absence of information)
• Uncertainty (absence of accuracy)
• Dissonance (absence of arbitration)
• Confusion (absence of comprehension)

The first type of indetermination in the list includes for example an unclear definition of a term, whereas the second type considers the uncertainty of a subjective judgment. The third type includes the question whether something will happen or not. Originally statistics and probability were introduced only for this type of indetermination. Fortunately, later, the definition of probability has been extended to consider subjective elements, like Bayes theorem. The fourth type considers cases of lack of understanding.

When humans detect indetermination, they try to lower external indetermination to a certain extent. This can be seen through protection measures. Humans, however, do not only accept internal indetermination, they seem to need it as shown in Fig. 1-4. The internal indetermination of humans is the possibilities, chances and freedom of an individual. In the case of fear or emergency situations, this internal indetermination disappears. Humans are then dependent on a certain set of circumstances or do not have the chance to carry out decisions. Especially in the case of risk acceptance, this internal indetermination becomes visible. People tend to accept higher indetermination in the environment when paid with an additional internal indetermination. Internal indetermination does not mean that people live under uncertain conditions (external indetermination). Internal indetermination also becomes clearly visible in cases of power, where humans can decide based on their individual preference and current mood. It also fits very well to von Förster's ethical imperative: "Decide always, that the opportunities expand" (von Förster 1993).

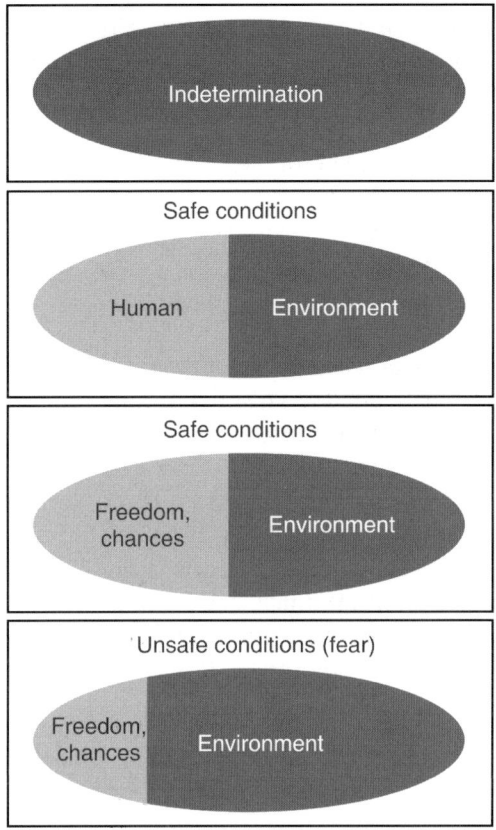

Fig. 1-4. Visualisation of internal (freedom, chances) and external indetermination (environment)

1.2.3 Indeterminate Verbal Language

As already mentioned, the vagueness of words or terms can be seen as indetermination. If the following properties are fulfilled, a term or an item is considered vague (Keefe & Smith 1997, Smith 2001, Edgington 1999):

- The item needs a borderline.
- There exists a defect in the borderline.
- The item is fragile to Sorites paradox.

Fig. 1-5. Sorites paradox shown in a rain cloud

Sorites paradox seems to be the summary of the first and second properties. It deals with the vague borderline of objects and definitions, and can be explained with a simple example as shown in Fig. 1-5. Let us assume the introduction of the term "rain cloud" as a certain number of rain drops in air with a certain geometric formation. If, in the definition of the rain cloud, the number of the required rain drops would change, the term rain cloud might still be applicable. However, if the change in the number of rain drops is repeated several times, one ends up with only a small number of rain drops, certainly not forming a rain cloud anymore. Here, the clear definition of the required number of rain drops remains difficult (Smith 2001).

Cantor has defined a set as: "Under a set I understand in general all many, which can be considered as one that means certain elements, which can be linked by a law to a whole." Also, Gottlieb Frege described in 1893: "A term has to be limited. A vague term would represent an area without clear borders and the limits would be at least partially blurred. This can not be an area and therefore such terms can not be called a term. Such terms and constructions can not be used for logic theories, since it is impossible to develop laws from such constructions" (Pawlak 2004).

However, such situations are common in languages, and should therefore be dealt with in this section. The major goal of language is the transmission of information. Using different elements of the language, one can freely choose the way the information is transmitted. Hegel once proclaimed in showing the diversity of possible expressions, "A building is not finished when its foundation is laid; and just as little, is the attainment of a general notion of a whole the whole itself. When we want to see an oak with all its vigor of trunk, its spreading branches, and mass of foliage, we are not satisfied to be shown an acorn instead" (Schulz 1975).

Figure 1-6 tries to show the quality of meaning of a word or term. At the peak of the curvature, the definition fits very well to the term, but the closer one comes to the edges, the lesser the word is appropriate in describing the information. Some of the edges are very sharp and there is a clear border, but some of the edges have a soft slope and it is difficult to decide whether the term is still appropriate or not (Riedl 2000, Laszlo 1998).

The basis of languages is words and terms, which are connected according to certain rules. The goal of the introduction of terms is to boost the velocity of communication, e.g. the flow of information. A definition is the assignment of content to a term. A term contains all properties that belong to a thinking unit (DIN 2330 1979). A definition of a term should be true, useful and fundamental. Every term includes objective parts (denotation) and subjective parts (Fig. 1-7). Since humans first started to learn languages and terms by subjective assessment, over the lifetime, the objective part increases. This can be easily observed by children (Kegel 2006, Wulf 2003). Wittgenstein (1998) has put it like that, "The limits of my language means the limits of my world." A possible relationship between terms and knowledge is shown in Fig. 1-8. In this figure, the extension of a symbol or sign towards knowledge is used, showing that initial assumptions are required.

If that is true, then terms, definitions and the language itself is correlated to the amount of individual (ontogenetical) and social (phylogenetical) knowledge. Since there is no absolute true meaning of terms, the adaptation is unlimited. One can observe this when young people change the meaning of words. An example of the vagueness of words can be seen in the following dialogue between a driver and his passenger:

- Driver: "Do I have to turn left?"
- Passenger: "Right!"

Most scientific books or papers start with the introduction of some terms. This is a general requirement to allow for a proper scientific discussion. If people have a different understanding of terms, a discussion will lead nowhere. In contrast to this, in the beginning of this book, the vagueness of terms is highlighted.

If the indetermination or vagueness of a language is known, is it possible to be removed?

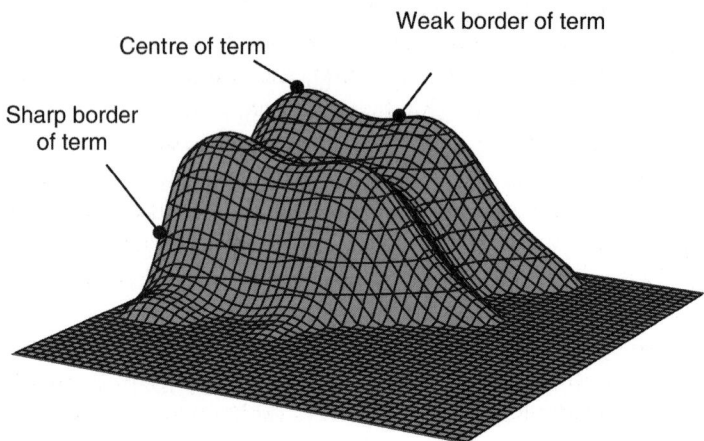

Fig. 1-6. Landscape of term fitting to some properties in the real world (Riedl 2000)

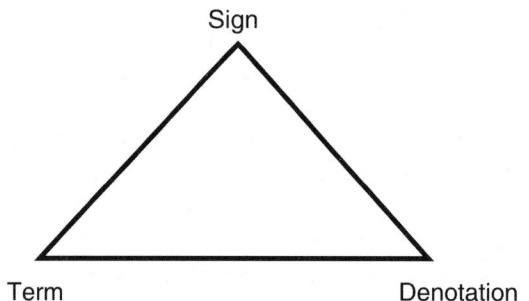

Fig. 1-7. Triangle of terms, sign and denotation (Weber 1999)

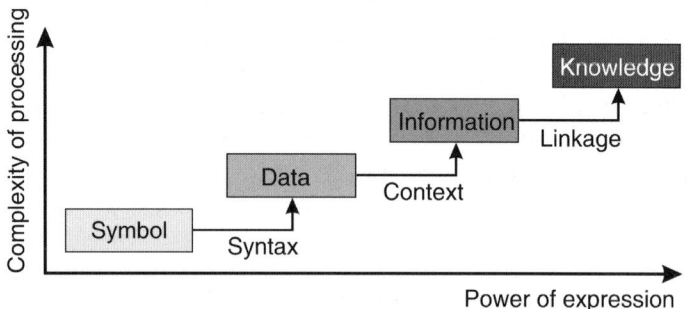

Fig. 1-8. Complexity of processing and the power of expression (Husemeyer 2001)

One of the best examples for the failure of absolute and vagueness-free languages is computer languages. In these, all terms are defined clearly, and the application of terms is only possible under clear grammar conditions. If the conditions are not fulfilled, the program is aborted and the statement has to be corrected by the programmer so that it fully complies. If computer languages are introduced, one would suggest that they could be used without limitations. Looking into the information technology sector, however, the number of such languages is widespread and keeps changing all the time. Why? Because the environment is continually changing. One might introduce a perfect language, but this language has to fully describe the known and unknown world. Otherwise, if something new is discovered, the language has to be expanded or, even worse, some of the terms are erroneous and have to be changed. If that is true, then languages are not absolute, because an absolute language has to fully model the nature (Kegel 2006).

This continual development and adaptation to changing environment can also be found in Pidgin and Creol languages (Hypercreolisation). Such languages were introduced under some special social conditions, e.g. colonization. Usually when colonized countries became independent, the languages disappeared and only some artifacts remain (Kegel 2006).

Therefore, all languages have to stay partly in vagueness. This is required, because languages, as humans, follow Muenchhausens plait. Muenchhausen told that he pulled himself out of a marsh by his own plait. In general, one would say this is ridiculous, but actually only from a mechanistic point of view. There exist no objective criteria to properly define how words and grammar should be used. Therefore, humans subjectively developed words and grammar. The best way by which terms can be identified is through what is most commonly understood – basically what an average person will understand the term to mean. That changes over time, as people distinguish themselves by how individuals use words. This can easily be seen among teenagers or in the different fields of science. In many fields of science, terms have a special meaning. As seen later, for mathematicians, quality of life parameters often are differentiable functions that can be computed in many ways, whereas for psychologists, quality of life parameters have to be identified through questionnaires.

1.2.4 Indeterminate Mathematical Language

Since mathematical language is often understood as an absolute language with minimum vagueness, a few general remarks should be given here. First of all, mathematical language is indeed a more objective type of communication as compared to a common language. However, that should not give us the impression that numbers and formulas are an absolute language at all. Let us consider Fig. 1-9. Here, the number of car accidents in connection with children is given. The crude number, however, is related in two different ways: (1) the number of children and (2) the number of vehicles. The difference in the interpretation of the data is impressive. While the drop in accidents over time is not as significant in the children, based ratio, the drop is extreme in the vehicle-based ratio. Obviously, the frame within which both numbers and mathematics are placed determines the results.

A second example is shown in Fig. 1-10. The picture shows four apples, but if, for example, a businessman applies the standard definition of trade according to the European Union, he might see only three apples. This sounds simple, but many law cases have discussed such issues in detail, and many people work and earn money discussing how many apples are there.

Since all mathematical work is embedded in a common language, the statement of vagueness in language is also true for mathematical language. Porzsolt (2007) put it this way: the cheating is already done when the mathematician starts his/her work. This limitation of initial assumptions is of utmost importance for mathematical proofs (Fig. 1-11). Models are, under all conditions, wrong; they might just describe some effects in a useful way (Box 1979).

Even though the following examples are not taken from pure mathematics, they show the importance of boundary conditions. Lewis Thomas responded about cloning: "Assuming one wants to clone an important and very successful diplomat…One has to take one cell. Then one has to wait 40 years…Additionally one has to reconstruct the entire environment, perhaps into the smallest detail. Environment then would also mean fellow men. Therefore one has not only to clone the diplomat itself. What the word environment really includes are other people…the dense crowd of loved ones, who listen, who talk, who smile or glower, who give or who take, open or hidden, who love or who beat. Independent from the information content of the genome these people are important for the forming of the human. If one only had the genome one would probably grow up a certain vertebrate, not more…One had to clone the parents as well, actually the entire family. In the end, one has to clone the entire world, nothing less" (Geissler 1991).

Eugen Wigner (1963) stated about physics: "Physics does not endeavour to explain nature. In fact, the great success of physics is due to a restriction of its objectives: it only endeavours to explain the regularities in the behaviour of objects…The regularities in the phenomena which physical science endeavours to uncover are called the laws of nature. Just as legal laws regulate actions and behaviour, the laws of physics also determine the behaviour of its objects of interest only under certain well defined conditions, but leave much freedom otherwise."

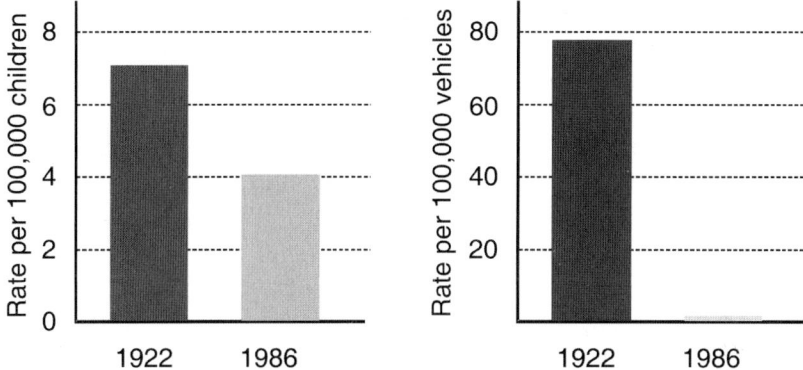

Fig. 1-9. Number of car accidents with children based on different ratios (Adams 1995)

Fig. 1-10. Three or four apples? That depends on the definition of an apple

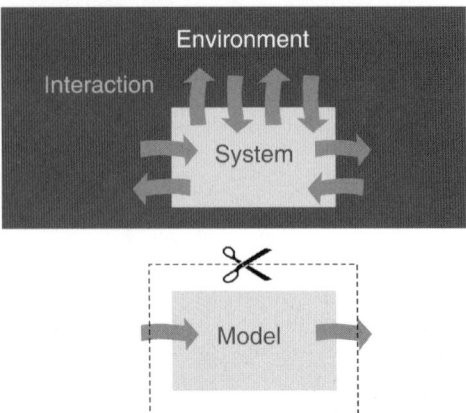

Fig. 1-11. Separation of a system from the environment to establish a model

This raises a question whether or not mathematics can solve all problems neglecting the uncertainty of the boundary conditions. Here, a new problem appears. Even with theoretically known mathematical procedures, not all such procedures might be applicable under real-world conditions. Therefore, the so-called practical computable truth has been introduced as a subset of the mathematical truth, the decidable truth and the computable truth (Fig. 1-12). The practical computable truth can be further separated according to different types of mathematical problems (Fig. 1-13) (Barrow 1998).

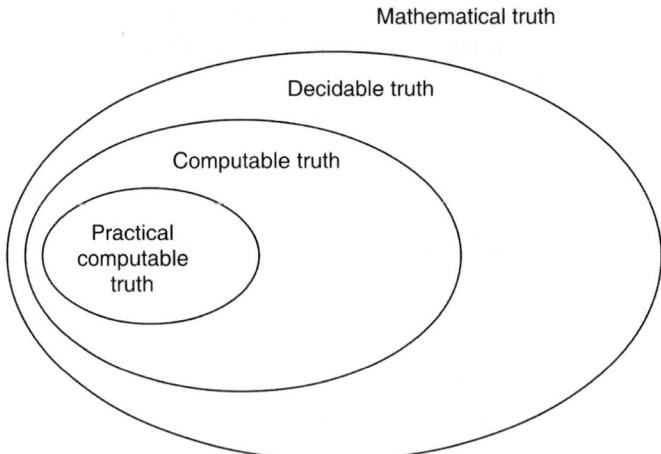

Fig. 1-12. Practical computable truth as a subset of the computable, decidable and mathematical truth (Barrow 1998)

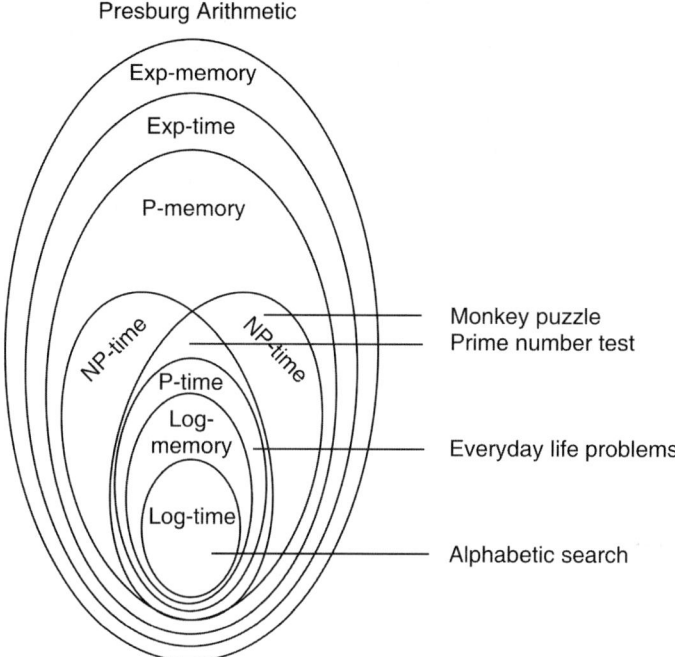

Fig. 1-13. Practical computable truth in detail (Barrow 1998)

So far, it has been shown that mathematical language is embedded in a verbal language, which introduces indetermination to mathematical language; that mathematical models include simplifications; and that even pure mathematical truth cannot often be directly determined, which yields to further indetermination.

Furthermore, Gödel's incompleteness has to be mentioned. Gödel's incompleteness theorem states that the validity of a class of formulas inside the system built by the formulas cannot be proven. Originally Gödel's works considered only natural numbers since he used them for his proof, but have been generalised in the form of "all consistent axiomatic formulations of the theory of numbers include non-decidable statements." The extended conclusion yields to the fact that the validity of mathematical statements cannot be proven with the tools of mathematics. Even in Greek times, this logical problem was contemplated. Probably the best known example is the statement by a Cretan saying, "All Cretans are liars." Here, the problem is easy to understand since the Cretan attacks his own statement, but in many other cases, the problem remains hidden. Coming back to Gödel's work and summarizing it in a different way: "Provability is not truth." This statement remains also very strong to the theory of Karl

Popper (1994) that science can only falsify, never prove validity (Cramer 1989).

Summarizing these thoughts, mathematical language is not an absolute language and includes indetermination. This is sometimes forgotten; however, many mathematicians are aware of this fact. There is a mathematical research project called "Robust mathematical modelling" (Société de Calcul Mathématique 2007) that also shows the limitations of mathematical modelling by giving the following statements:

"1. There is no such thing, in real life, as a precise problem. As we already saw, the objectives are usually uncertain, the laws are vague and data is missing. If you take a general problem and make it precise, you always make it precise in the wrong way. Or, if your description is correct now, it will not be tomorrow, because some things will have changed.

2. If you bring a precise answer, it seems to indicate that the problem was exactly this one, which is not the case. The precision of the answer is a wrong indication of the precision of the question. There is now a dishonest dissimulation of the true nature of the problem."

1.2.5 Indeterminate Causality

Languages and their elements, and here mathematics is understood as a language as well, show indetermination or vagueness. The relation of elements itself is also partly indeterminate. In the first instance, it might seem to be unrealistic, since the major experience of humans is based on strong relations between cause and reaction, which is nothing else but the introduction of a relation. A simple example can already show the limits of the relation of elements: someone causes something to happen, which takes 50 years for the result to become observable. Clearly here, humans have difficulties to find the causal relationship, because 50 years is not a marginal time span as compared to the lifetime of humans. Humans will have changed over that time period. Therefore, humans can observe only some sorts of relations (so far). This statement said differently: the possibility of the detection of causality depends on the type of system.

Different types of systems show different amounts of indetermination. The already mentioned example of Newton's law of motion for simple physical objects and the absence of such a law for societies is an example supporting that statement.

The disappearance of causality according to different types of systems can be seen in Fig. 1-14. Here, different domains of science and their objects of interest are shown. The more the objects increase their complexity, the more the causality disappears. The relation between complexity and indetermination will be discussed later. However, if one observes cultural developments, it becomes clear that the cognition of causal dependencies is virtually impossible. Only for some systems, e.g. complicated or simple systems, the formulation of causal relationships is possible. Such systems usually have the property that they can be cut out from their environment (Fig. 1-11). For example, systems can be investigated in a laboratory. It is then assumed that, for example, the mechanical systems in a structure behave comparably, and small changes in the initial conditions can be neglected. This is shown in Fig. 1-15. In the left figure, the behavior is alike a classical mechanical system: the actual system is not investigated in a laboratory; however, conclusions from comparable systems yield comparable results and can thus be used. Other systems might not behave so well-conditioned. First of all, there are many systems that cannot be cut out, for example social systems. Furthermore, such systems might give an unexpected response under comparable situations. This property is shown in Fig. 1-16. In the left figure, the small changes in initial conditions finally yield only small changes in response, whereas on the right figure, one function with slightly changed initial conditions gives a completely different response. That yields to the requirement of a total control of initial conditions, which is also known as Laplace demon (Mikulecky 2005). As already shown in the example of the Buridan's donkey and the waterwheel, more and more initial conditions have to be observed in smaller dimensions. However, there are some physical boundaries of quantum physics, such as Planck's "distance" and "time" reached.

The consequences of this theory were stated clearly by Heisenberg as follows: "In the sentence: If we know the present, we can predict the future, is not the apodosis wrong but the intention. We will never know all parts of the present. Hence the perception is a choice of a richness of possibilities and a limitation of the future possibilities. Because the statistical character of quantum theory is tied so strongly to the imprecision of the perception, one could assume that behind the perceived statistical world is a real world still including causal relationships. But such assumptions seem to be hopeless and senseless. The physic can only describe the perception in a formal way. Perhaps we can describe the true circumstances like this: due to the fact that all experiments have to accomplish the uncertainty principle, the invalidity of the causal relationships is proven for quantum mechanics" (TU Berlin 2006).

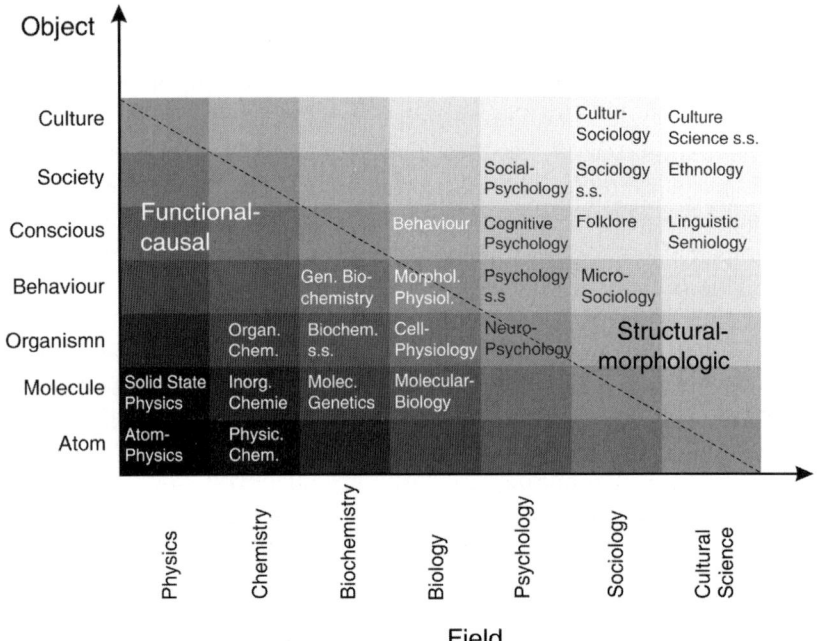

Fig. 1-14. Classification of objects and fields of science. There is a general loss of causality from *bottom left* to *top right* (Riedl 2000, please see also Anderson 1972); s.s. – stricto senso

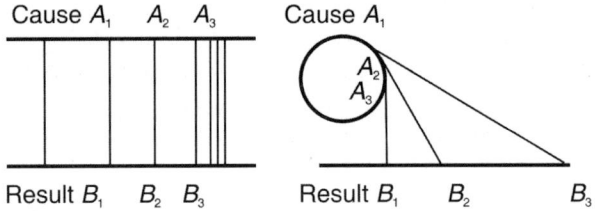

Fig. 1-15. Different types of causalities (Davies & Gribbin 1996)

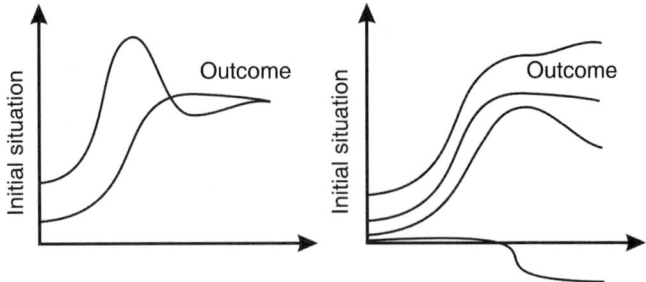

Fig. 1-16. Influence of changes in initial conditions based on the type of system

1.2.6 Goal of Prognosis

If indetermination is an irremediable property of the known world, does it make sense to attempt to predict the behaviour of complex systems? To show that, under some conditions, indeed good predictions can be developed, a few successful examples will be mentioned. In the book "The world in 100 years", several scientists (around the year 1910) have tried to predict the world in the year 2010 (Brehmer 1910). One statement says: "Everyone will have his/her own pocket telephone, with which he/she can contact, whoever and wherever they want…Kings, chancellors, diplomats, bankers, clerks or directors can carry out their business, giving their signatures…They will see each other, talk with each other, exchange their files and sign like they would be together at one location" (Brehmer 1910). In the year 1974, a Russian dissident predicted the fall of the Soviet Union in 1990. In 1999, in the journal "The futurist", superterrorism on the area of the US was forecasted. It said that "…the US will be victim of a high-tech terrorist attack on their own territory." Van Gelderen and de Wolff predicted technological waves. These waves were later called Kondratieff-waves. They were observed during the introduction of the steam machine, the introduction of the loom, the development of steel structures, the development of the chemical industry and lately the introduction of the Internet. In 1865, Jules Verne presumed a flight to the moon, and Roger Bacon was believed to have told in 1290: "Cars without horses will drive with unbelievable speed" (Proske 2006).

On the other hand, if one considers the predictions of economic growth in the US or in Germany, the indetermination becomes significant. Usually such predictions are permanently adapted to current developments, such as rise in oil prices, some disturbances in financial markets or some rumours. In the above-mentioned book "The World in 100 years", many prognoses failed completely. There is no prediction of gender equality or the end of colonialism. Are the mentioned successful examples real flukes? In general, the prediction of some technical developments seems to be easier than the prediction of social systems. However, they might also fail as shown in a statement by Sextus Julius Frontius, Clerk under emperor Vespasian (69–79 A.C.): "I will not note any new works and new war machines, since technology has reached its limits and I have no hope, that it can be further improved" (UAS Mittweida 2006).

On the other hand, some general developments might become visible in social systems. Over the centuries, there seems to be a constant improvement in the living conditions of humans. But, detailed predictions or predictions for special situations remain impossible. This fact should be kept in mind when politicians talk about, for example, economic growth based on their actions. If this would be true, then they are able to predict the behavior of social systems and they would be personally responsible.

In general, predictions are unable to delete the indetermination component of the world. However, projects of prediction might be able to visualise indetermination.

This yields us to the second goal of predictions. In general, one would assume that the goal of predictions is a true description of future situations or circumstances. This seems to be common sense, but it is wrong. What is gained by predicting a flood, an earthquake or an economic crisis? The goal of predictions is the preparation of possible future situations. One does not want to wait and see whether the prediction was right. But, predictions are used to actively design the future (Torgersen 2000, ESREF 2008).

The major goal of the human design of the future is the improvement of the world (for someone). Karl Popper has described this accurately by saying: "Every living thing quests a better world" (Popper 1994). Aristotle reached the same result two millenniums before: "In general one can state: humans don't look for the life of their ancestors, but they look for a better life" (Nussbaum 1993). The statement that every kind of living matter searches for improvements might even function as a definition of life. The ability to assess situations is of overwhelming importance for living creatures. Humans search for useful interaction, for useful electrons, for useful weather, for useful planets, useful feelings and useful worlds (Hildebrand 2006). The assessment then permits actions. Methodical actions follow the goal of improvement as well as the goal of preservation of life. Many actions of complex systems can be justified on the prevention of damage or injuries.

Assuming that actions are either based on the goal of improvement or on the prevention of disadvantages, recent developments, as shown later in this book, consider improvements of risk measures. Such measures for improvement might be quality of life parameters, which can be described in their simplest way as access to advantages. On the other hand, risks are considered a possibility of damage or disadvantage. It seems that both items cannot be considered independently, yet they seem to be the same. Independent from this consideration, risk measures need some mathematical description of indetermination.

1.2.7 Mathematical Models for Indetermination

As shown so far, indetermination and uncertainty seem to be the inherent properties of our world. Whether that statement will still be valid in the far-future is another question; however, for the near future, humans have to develop certain strategies to deal with indetermination. Starting from the earliest technique of the belief in God, within the last few hundred years, the number of scientific models, mainly mathematical, has increased substantially. While the first techniques, probabilistic and statistic models, were introduced a few hundred years ago, in particular, innumerable newer techniques were developed in the recent hundred years. Most of the newer techniques are only a few decades old, such as Fuzzy sets, Rough or Grey numbers (Fig. 1-17). Examples of different mathematical procedures are shown in Figs. 1-18 to 1-24.

The development of mathematical techniques follows a chain of mathematical proofs dealing with uncertainty. Such works include Heisenberg's indetermination or uncertainty principle, Gödel's incompleteness theorem and Arrow's impossibility theorem. The results of these works can be seen as a curtain, which one can not look behind (Planck distance, Planck time, light speed), reaching even the understanding of our own society (Arrow's impossibility theorem) and showing the disability of our tools (Gödel's incompleteness).

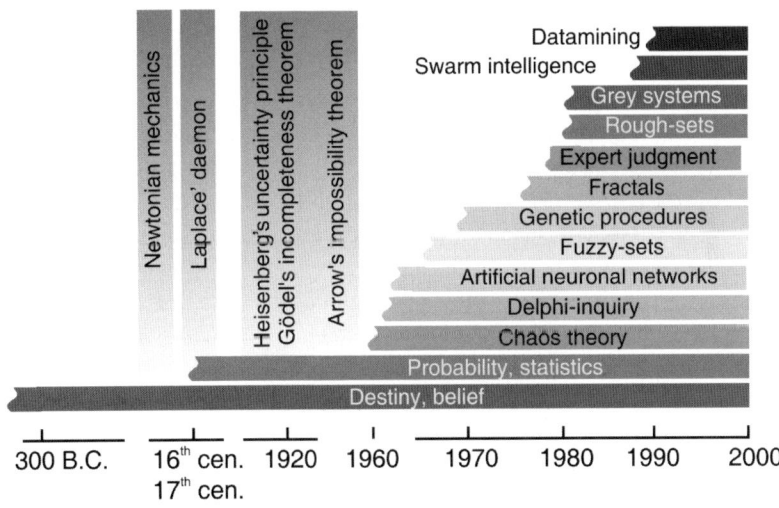

Fig. 1-17. Development of mathematical techniques concerning indetermination

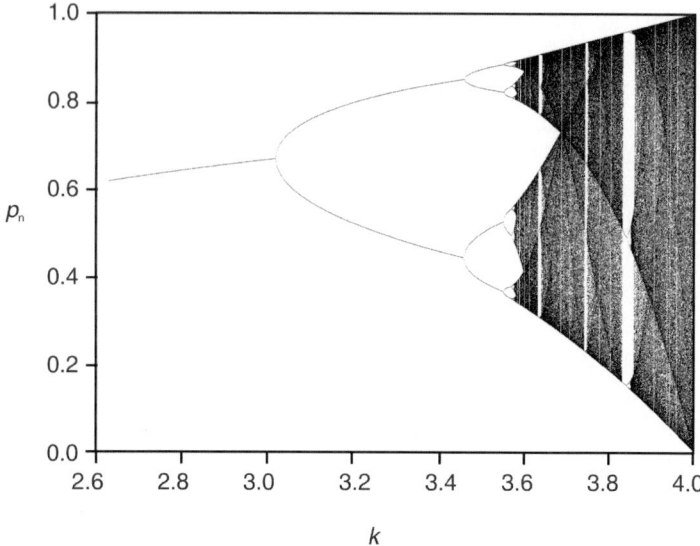

Fig. 1-18. Feigenbaum's diagram as an example of Chaos theory produced with the program ChaosExplorer (Riegel 2006)

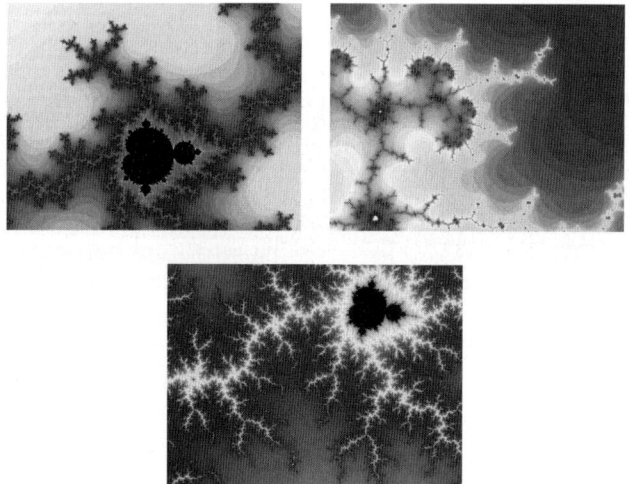

Fig. 1-19. Mandelbrot sets as an example of Chaos theory created with the program ChaosExplorer (Riegel 2006)

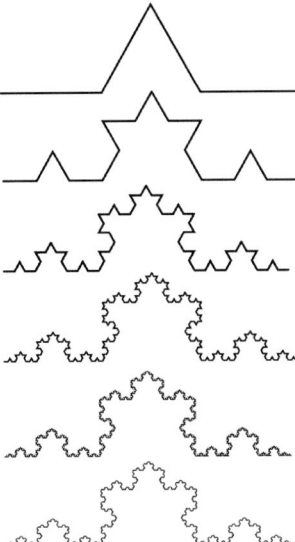

Fig. 1-20. Koch curvature in the Fractal theory

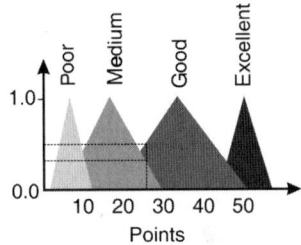

Fig. 1-21. Membership functions in the concept of Fuzzy sets (Zadeh 1965)

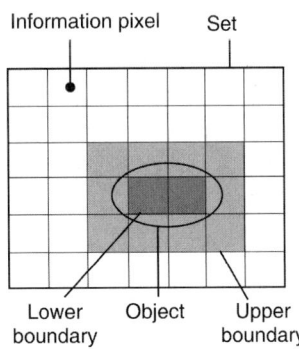

Fig. 1-22. Information pixel in the concept of Rough sets (Pawlak 2004)

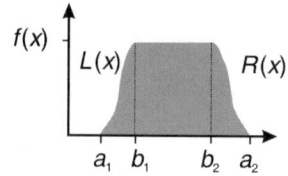

Fig. 1-23. Example of a Grey number in the concept of Grey systems (Deng 1988)

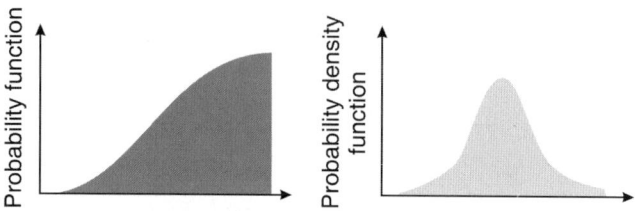

Fig. 1-24. Probability function and probability density function

So far, the term indetermination has been discussed. However, other terms are frequently used, such as "complexity" or "systems". Therefore, such terms will be discussed subsequently, considering that these terms are related to non-evaluated indetermination. Afterwards, terms related to evaluated indetermination (e.g. risk) are discussed. As this chapter deals with mathematical tools for indetermination, the first term discussed is stochasticity.

1.3 Terms Related to Non-Evaluated Indetermination

1.3.1 Stochasticity

Stochastic, as a generic term in the fields of statistics, probability and time series, is of major importance for dealing with uncertainty. It has leavened daily life. Most people know the meaning of mean value or standard deviation. The entire mathematical model is based on randomness. The term randomness shall be discussed in this section. Generally speaking, randomness exists in bifurcation situations (Fig. 1-25). Such bifurcations are events. During such events, the causality reaches a minimum and the indetermination reaches a maximum. Between the single points of bifurcations, there might be complete causal processes permitting predictions. Usually one tries to repress the event and transfer it into a process. This sometimes helps in reducing uncertainty, but usually fails to fully break uncertainty. Figures 1-26 and 1-27 visualize these thoughts.

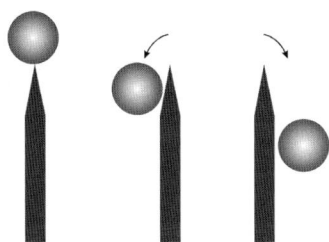

Fig. 1-25. Bifurcation situation

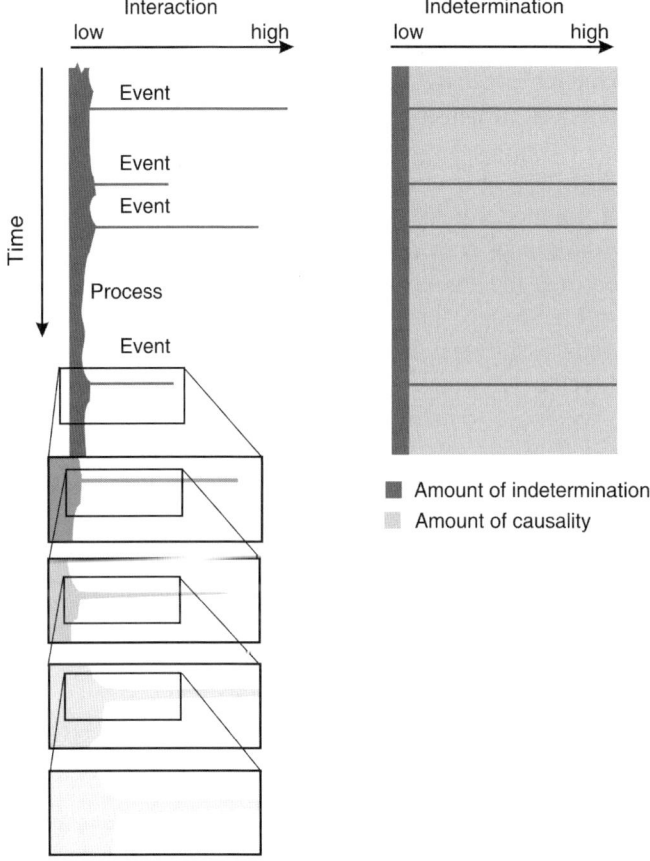

Fig. 1-26. Relation between events and processes, and indetermination and causality

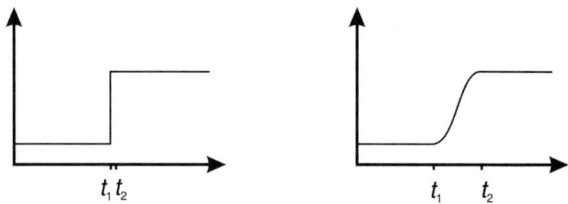

Fig. 1-27. Attempt to transfer an event into a process

As an example, let us assume a process of deterioration. This process might be described by a change rate. Sometimes, however, sudden changes of interaction occur, for example by an overload. Such sudden changes of actions are characterized by a high amount of indetermination, thus repressing the amount of causality in the system. One may then try to transfer the event into a process by modeling the behavior of the structure under overload (Fig. 1-27). This might be successful to a certain extent, but the complexity of the model increases dramatically, requiring further data. At a certain level, the model does not produce an improvement in the quality of predictions, because the amount of required input information transfers more uncertainty into the model than reducing it. The transfer of the event into a process then fails as shown in Fig. 1-26, where the curvature fades away.

In the 19th century, Poincare has defined randomness as: "A very small cause, which we do not notice, causes a considerable effect, which we can not ignore, and then, we say, the effect is coincidental." Nowadays, this is called chaos. Chaitin has defined randomness as lack of order (Longo 2002). Von Weizecker has called probability "the way of possible knowledge about the future" (Longo 2002). Eagle has called randomness an indetermination (Longo 2002). Some people have suggested that randomness is an indicator of the limitation of the human brain in understanding the world. Others have called order the complement to randomness. Von Mises has considered randomness as independence.

Kolmogoroff has defined randomness as complexity (van Lambalgen 1995). The Kolmogoroff complexity considers rules of numbers to describe a certain system. If one considers, for example, a simple number like 0.123456789101112131415, one can find a rule to construct the number again. The rules to reproduce the number might, at a certain point, require the same amount of information as the number itself. Then, there is no gain in developing rules since the rules have become so complex. Figure 1-28 gives two extreme examples of patterns and two examples from real world conditions. Of course, based on the introduced definition, randomness depends on the capability to identify patterns. Even patterns apparently complex might include simple rules as shown in

Fig. 1-29. Simplifying the rule of Kolmogoroff, one can state: the more complex something is, the more random it will behave. For simplification, the term randomness here might be extended to the term indetermination, saying that complexity will yield to indetermination. Further models introducing the concept of stochastic complexity will not be discussed here (Crutchfield 1994)

Fig. 1-28. Examples of different patterns. The top left square shows a complete randomly pattern of points; the top right square shows a clear order of the points. The left bottom square shows the lines of colour change in Belousov–Zhabotinsky reaction as an example of the mixture of order and disorder; the bottom right square shows strong order elements, like in a parking lot.

Fig. 1-29. Patterns produced with formulas $F(x) = a \cdot x + (1-a) \cdot 2 \cdot x^2 /(1+x^2)$, $x_{n+1} = b \cdot y_n + F(x)$ and $y_{n+1} = -x_n + F(x_{n+1})$ and for changing values ranging only from –0.7 to –0.8 for a and 1.0 to 0.95 for b (Aleksic 2000)

1.3.2 Complexity

As stated in the previous section, randomness and complexity are related, and in extending randomness to indetermination, then indetermination and complexity are also related. If this statement is true, one has only to look for complexity to identify indetermination in a system. For example, if properties of complexity are known, then these properties have to be identified instead of measuring data from the system. This means that the reaction of a system does not necessarily have to be investigated, but the system itself will give some indications about the indeterminate behaviour of the system.

Therefore, a definition of the term complexity is required. Such a definition is rather challenging. Some scientists refuse to define complexity, because it shows a great variety of properties, and is relative and polymorph (Riedl 2000). In addition, it depends on the viewpoint. For example, a butcher might not consider a brain of a cow as complex, whereas a neurobiologist might consider it as extremely complex (Flood & Carson 1993, Funke 2006). This dependency on the point of view can also be seen during the development of complexity over history. Several hundred years ago, many problems were considered complex, for example the law of falling bodies (Noack 2002, Schulz 2002). These laws are nowadays taught in schools and are not considered complex anymore. Most of the historical problems considered only a small number of elements and interactions. This started to change during the 19th century. For example, Bolzmann started to deal with systems with an incredibly large number of single elements (Noack 2002). He mainly looked into systems with unorganised complexity, such as molecules in the air (Weaver 1948). Earlier scientists, however, like Galileo dealt with more so-called trivial systems (Noack 2002). Between these two systems, a third arm exists (Fig. 1-30): systems with organised complexity (Riedl 2000, Weaver 1948). This type of systems is of major interest, and will be discussed in detail later. In general, the awareness of such systems in different fields of science started the development of complexity science. Approximately a hundred years ago in the fields of physics, mathematics, biology, chemistry or economics, the search for a general description of systems yielded to cybernetics, information and automation theory, general system theory and self-organisation theory (Fehling 2002). Even though such theories do not only deal with complexity, it becomes clear that, from the many different roots, there exists a great variety of different definitions of the term complexity (Patzak 1982, Klir 1985).

Nevertheless, most definitions are based on the listing of properties of complex systems, the explication is obviously simpler than the definition (Fehling 2002).

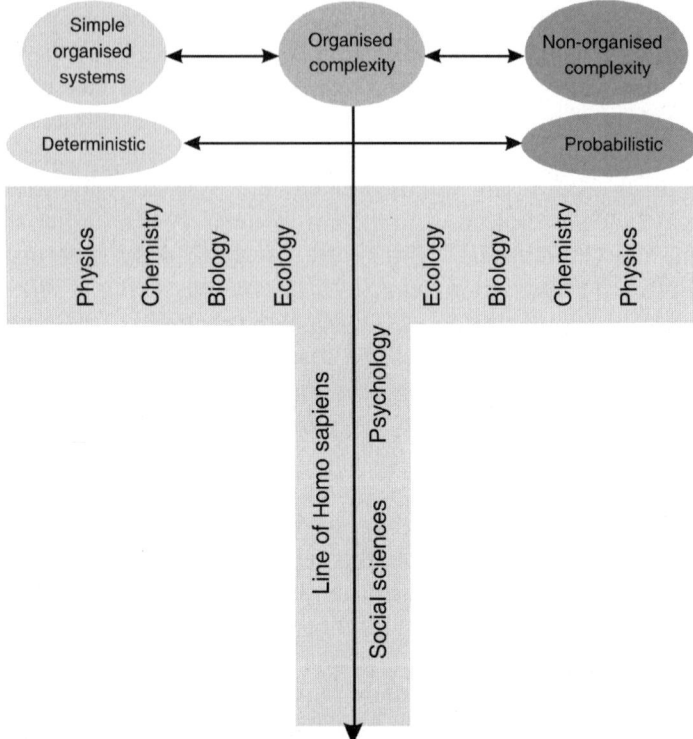

Fig. 1-30. Two-dimensional classification of systems according to Riedl (2000)

The term complexity comes from the Latin word "complectari", meaning embrace or contain. Some definitions of complexity are:

- Complexity is the product of the number of elements multiplied with their connections (Funke 2006, Fehling 2002, Patzak 1982).
- Complexity is an excess of possibilities produced by a system (Wasser 1994).
- Complexity is the logarithm of the number of possible stages of a system (Cramer 1989).
- Complexity is related to dynamics and high rates of changes (Fehling 2002).
- Complexity of a system is the shortest program for the description of the system (Schulz 2002, Weiss 2001).
- Complexity is intransparency. Complex systems cannot be fully described and therefore include, under all conditions, indetermination.

In addition, a complex system can be defined if one of the following properties is valid (Kastenberg 2005):

- The system is more than the sum of the parts. Therefore, the system is not divisible.
- Small changes might cause great effects on the system. This yields to chaotic behavior.
- The system consists of properties that can only be assessed using subjective judgment.

In contrast, Mikulecky (2005) and Perrow (1992) discuss complicated systems as:

- The system can be fully understood if all single elements are understood.
- There is a clear causal relationship between cause and action.
- The system behaves independent from the observer.
- The system is computationable.

According to Perrow (1992), in Fig. 1-31, there is a classification of different systems based on their interaction and coupling. This diagram can also be used for the identification of complexity. A relationship between the degree of randomness and the degree of complexity based on the type of systems is shown in Fig. 1-32 (Weinberg 1975). As mentioned in the beginning of this section, there is a relationship between complexity and randomness or indetermination. However, Weinberg (1975) considers the behaviour of unorganised systems as maximum degree of randomness. Considering again Fig. 1-14 from Riedl (2002) and Fig. 1-33 from Barrow (1998), there seems to be a different interpretation possible. Here, the highest degree of indetermination is reached in living and social systems. There are many further examples of complex systems, such as cells, biospheres, economic systems, climate or the human brain. The human brain is currently probably the most complex system known to human society.

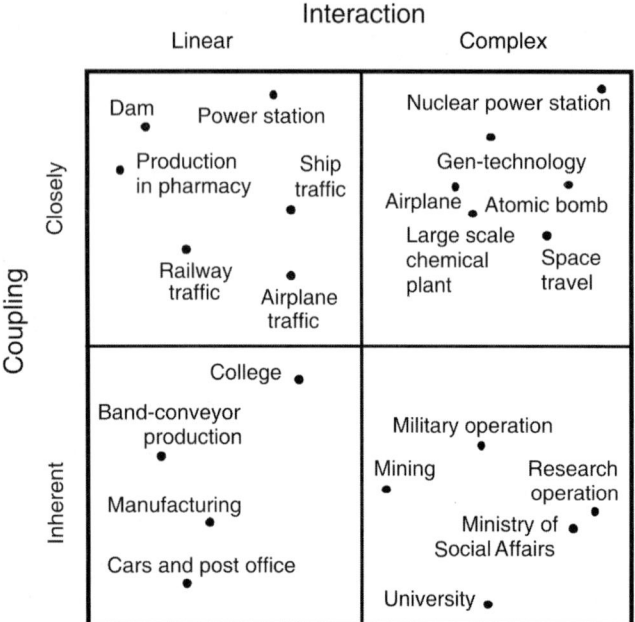

Fig. 1-31. Definition of complexity using the type of interaction and the type of coupling (Perrow 1992)

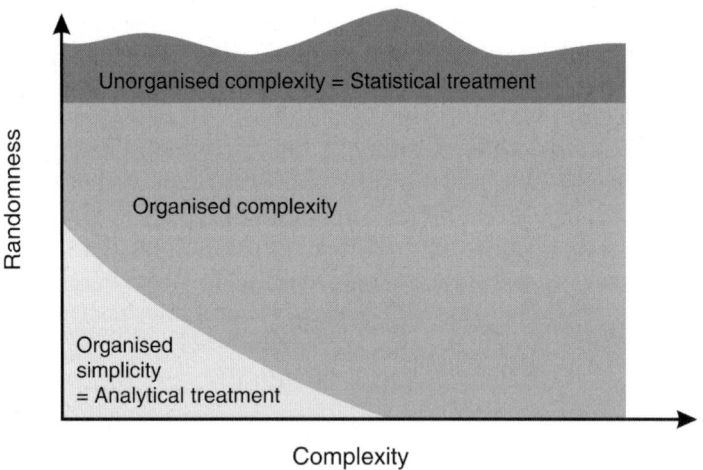

Fig. 1-32. Complexity versus randomness of systems (Weinberg 1975)

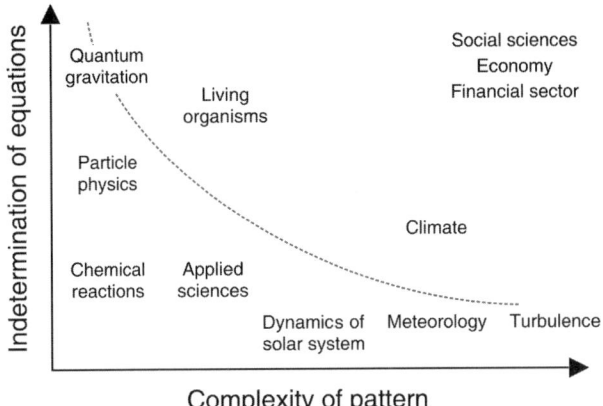

Fig. 1-33. Complexity of patterns of systems versus indetermination of equations (Barrow 1998). Please compare figure with Fig. 1-14

Table 1-1. Properties of complex systems in the human brain and some technical systems (Pulm 2004, Bossomaier 2000, Crossley 2006)

Object	Properties
Number of senso-neurons (external connection)	10^7 Number of elements
Number of inter-neurons (intern)	10^9–10^{11} Number of elements
Number of moto-neurons (external connection)	10^6 Number of elements
Number of connections per cell	100–10,000
Number of connections in the entire brain	10^{14}
Number of switches per neuron	10^3 per second
Number of switches per brain	10^{12}–10^{13} per second
Number of elements in Pentium IV	10^7
Number of elements in a modern plane	4–5×10^6

Table 1-1 includes information not only about the brain but also about some technical systems, such as planes and computer processors. However, before these systems are discussed, the brain data should be observed closer. In general, a major task of the brain is the reduction of complexity (Krüger-Brand 2006). This task is carried out in different steps: receive data, compute data and respond. These steps are similar to the ones found in different kinds of neurons in the brain: senso-neurons, which bring the input to the brain, for example, from the eyes. Then, there are inter-neurons, which process the incoming information, and then there are moto-neurons, which transport the results back to the world in terms of controlling muscles. If one looks at the number of elements, it becomes clear that inter-neurons are dominant in number by a huge factor. This effect is called separation, and is typical for complex systems. Sometimes the system starts to do something with the information that was never observed before. Such effects can be observed easily when people discuss the beauty of drawings or music. The increased complexity and separation becomes visible with the growth of the human brain during childhood. The older humans get, the more and more they separate. The effect of separation is indeterminated behaviour or, as Van Mises had defined randomness, independence. The fact of creating a complex system to deal with another complex system is nicely summarised by Lyall Watson: "If the brain would be so simple, we could understand it, we could not understand it."

Table 1-1 reveals further information. Currently, technical systems are far away from the complexity of the brain, and might still be considered as complicated systems. The computing performance of different items is given in Fig. 1-34 (Barrow 1998, Moravec 1998). The computer processors, especially, show a rapid growth in the number of elements (Fig. 1-35). Presumably in 20 or 30 years, the number of elements of the brain and that of a computing system would be same. Therefore, processors will then become complex systems. If it will become complex, it will also become partly indeterminate. This is already experienced nowadays when computers do something surprising, for example, when a computer crashes or the crash is extremely difficult to repeat. Also computation results might show unexpended differences, for example when distinct finite elements programs are used for the simulation of car crashes, on the same types of computers using the same software different results might be achieved.

The concept of complexity cannot only describe future behaviour of computers, it goes even further. Consider a state with about 30,000 laws concerning taxes and 20,000 or 50,000 people working with these laws. Even when the vagueness of human language is neglected, one will still end up with a complex system, and the results of that system are partially indeterminate. That is one reason why laws constantly have to be reworked by courts.

According to the examples, it would be useful to introduce a parameter for complexity, which can be used as a substitute indicator for indetermination. Such measures of complexity should (Randić et al. 2005):

- be independent from the type of system
- have a theoretical background
- be able to consider different complexity planes
- consider connections stronger than the number of elements
- monotonically grow with growing complexity properties
- follow the intuitive idea of complexity
- not be too complicated
- be useful under practical conditions

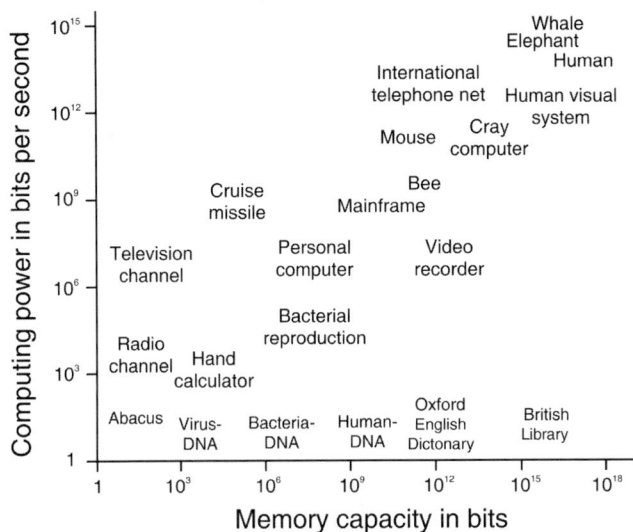

Fig. 1-34. Memory capacity versus computing power of systems (Barrow 1998, Moravec 1998)

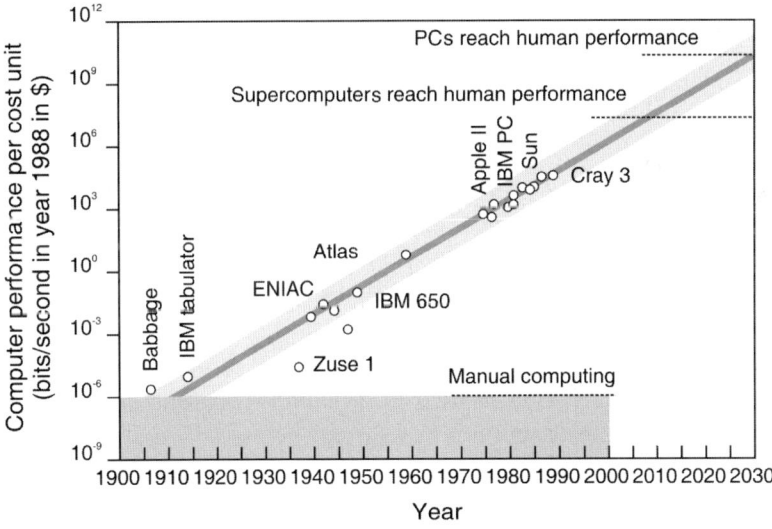

Fig. 1-35. Development of computing performance over time (Barrow 1998, Moravec 1998)

Some complexity measures have existed since the investigation of information masses of equidistant one-dimensional character chains. The computation of the complexity of information of the character chains is carried out in sequences of a certain length L. Such sequences will be moved over the entire character chain (Fig. 1-36). For the words with the length L the probability of occurrence can then be computed (Wolf 2000).

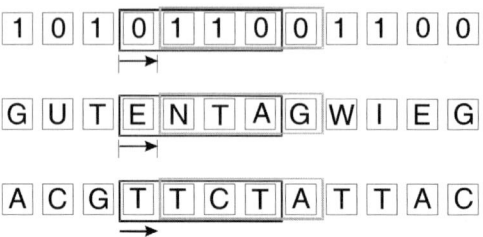

Fig. 1-36. Example of complexity computation (Wolf 2000)

Based on this idea, in 1948, Shannon developed a measure for the predictability of information in such chains. This measure is called Shannon's entropy. It is assumed that a certain number of questions have to be asked to identify a character: for example, the question would be of the type: is the character in the upper or lower part of the alphabet? Then, the number of questions can be put into a formula (Wolf 2000):

$$N = 2^i \qquad (1\text{-}1)$$

If one takes the logarithm of the equation, one obtains

$$I = \log_2 N = \log_2(1/p) \qquad (1\text{-}2)$$

where I is the information content, N is the number of characters and p is the probability of occurrence of a certain character. If different probabilities of the different characters in the alphabet are considered, then the formula changes slightly and the probability is capital-dependent:

$$I(z_i) = \log_2(1/p_i) = -\log_2 p_i \qquad (1\text{-}3)$$

The average information content of an alphabet is then just the sum of all the character information:

$$H = \sum_{i=1}^{N} p_i \cdot I(p_i) = -\sum_{i=1}^{N} p_i \cdot \log_2 p_i \qquad (1\text{-}4)$$

Table 1-2. Probabilities of occurrence of different characters in the alphabet (Föll 2006)

Character	p_i
Blank	0.151
E	0.147
N	0.088
R	0.068
I	0.063
...	
Y	0.000173
Q	0.000142
X	0.000129

The probabilities for different characters are given in Table 1-2. The measure H is called the entropy.

As an example, the drawing of balls from an urn should be considered. An urn contains an equal number of two types of balls. Therefore, the probability is 0.5 that one type of ball will be drawn verses another. The entropy per ball is given by $H = -0.5 \cdot \log_2 0.5 = -0.5 \cdot -1 = 0.5$. In the next example, there should be 99 white balls and one black ball. Then, the computation of the entropy yields $H = -0.99 \cdot \log_2 0.99 = -0.99 \cdot -0.1449957 = 0.01435$ for the white balls and $H = -0.01 \cdot \log_2 0.01 = -0.01 \cdot -6.6438 = 0.0664$ for the black ball. Obviously, the entropy reaches a maximum of 1 for both balls if the probability of occurrence is 0.5. In contrast, the entropy goes down to 0 if the probability is close to 0 or 1. This can be interpreted as the highest amount of uncertainty when the probability is 0.5 as shown in Fig. 1-37.

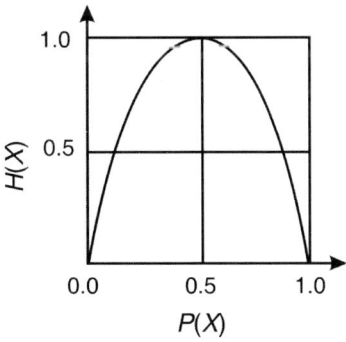

Fig. 1-37. Entropy versus probability of occurrence (Wolf 2000)

The question may arise why the name entropy from thermodynamics is used for information volume. Indeed, conversion is possible from information volume to thermodynamic entropy by saying that 0.957×10^{-23} JK^{-1} represents one bit of information (Föll 2006).

Based on Shannon's entropy, several complexity measures were introduced, such as the Rényi complexity measure (Wolf 2000). Other complexity measures were developed in the field of networks. Here, several such parameters should be mentioned (Bonchev & Buck 2005):

- Global Edge Complexity
- Average Edge Complexity
- Normalized Edge Complexity
- Sub-Graph Count (SC)
- Overall Connectivity (OC)
- Total Walk Count (TWC)
- A/D Index
- Complexity Index B

Examples of the application are shown in Fig. 1-38.

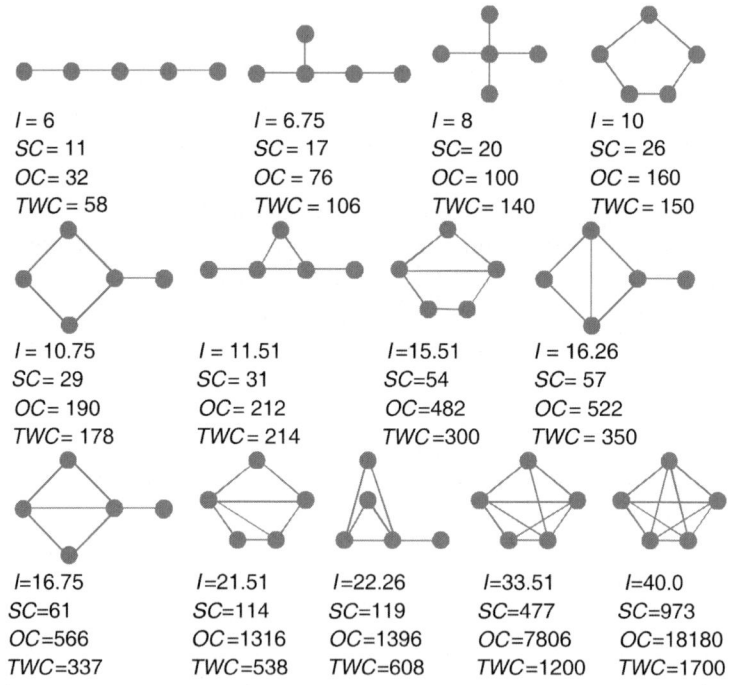

Fig. 1-38. Examples of complexity measures for the combination of five points (Bonchev & Buck 2005)

1.3.3 Systems

If one tries to compute complexity in an observed environment, it makes sense to classify some elements of the environment. Such a classification yields to the term systems: "A system is a certain way to observe the world" (Weinberg 1975). Even if, at first, this seems to be a rather useless definition, it describes quite clearly the background. Depending on the individual's history, humans will identify different systems. As already mentioned, the butcher considers the brain of a cow only as a simple system, whereas the neurobiologist might identify many more systems in the brain. Other definitions describe "a system as a set of elements which are in interaction" (Noack 2002), or "systems are objects which consist of different elements, but can be seen as one unit. Systems usually possess a boundary to the environment" (Schulz 2002). Further definitions are given by Bogdanow (1926), von Bertalanffy (1976), Patzak (1982) and Forrester (1972). It becomes clear that systems include a number of elements. These elements are interconnected in different ways. Based on these interactions and the system's behavior, systems can be classified.

The simplest forms of systems are "disordered systems". The molecules in the air form such a system. There are different types of disordered systems, but this will not be dealt with here. Usually the prediction of these systems is carried out over mean values, for example temperature (Weinberg 1975).

The next comes "trivial systems". A trivial system is distinguished by classical, clear causal relationships. These relationships are mainly linear, and prediction of the systems is possible. Such systems are sometimes called deterministic systems. All the classical analytical mathematical tools like differential equations are situated in the area of these systems. Most mechanical systems belong to this type (Weinberg 1975).

The systems much more difficult to describe are the so-called "non-trivial systems". The behavior of these kinds of systems is difficult to predict. They show non-linear behavior and self-reference. This means that output of the system also acts as input for the system. This is sometimes called causal nexus (Seemann 1997). In the section on "Complexity", the human brain was introduced as a complex system, and a special note was made regarding the great number of inter-neurons of the brain. There, it was called separation of the system, which becomes obvious here. When the system starts to act in a causal nexus, the system separates to a certain extent from the environment. Even if the systems are difficult to predict, there are signs of order.

"Autopoietic systems" reach an even higher degree of complexity. These systems renew permanently under the goal of preservation. Living creatures, economic systems and social systems are autopoietic systems. Such systems act autonomously to prevent actions jeopardizing their sheer existence. Therefore, such systems have to organize themselves. They have to be disturbance-resistant and need some damping procedures to cut-off disturbances. Such systems reach a certain dynamic balance with the environment (Fig. 1-39) (Haken & Wunderlin 1991).

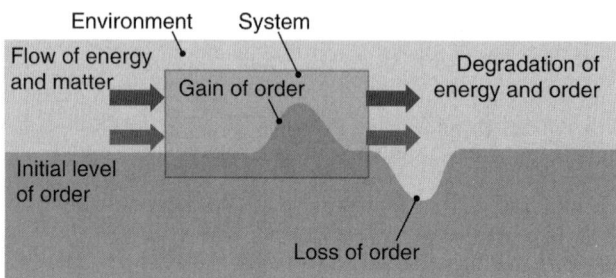

Fig. 1-39. Export of disorder and uncertainty to the environment to keep a complex system functioning (Riedl 2000)

One of the major properties of autopoietic systems is emergence. The term is very common in system theory. Emergence comes from the Latin word "emergere" meaning to appear or to present. This term tries to grasp the effect of new properties of the system, which are not predictable only from the knowledge of the single elements. The system develops some new reactions and techniques (Laughlin 2007). Actually, this term can be easily related to indetermination (Rayn 2007, Corning 2002, Crutchfield 1994, Cucker & Smale 2006, Scholz 2004).

The introduced classes of systems can be ordered according to their diversity of patterns and their momentum. The result is shown in Fig. 1-40. Obviously, the autopoietic systems show the greatest diversity of patterns since they develop new patterns. Moreover, their momentum is high since they are under continuous change. These properties can also be described as complex behavior. Extending the thought, systems can be classified according to their complexity. A general definition of a complex system says, "Complex systems consist of many particles, objects or elements, which are either equal or different. These single components interact in a more or less complicated, mostly non-linear, way" (Schulz 2002).

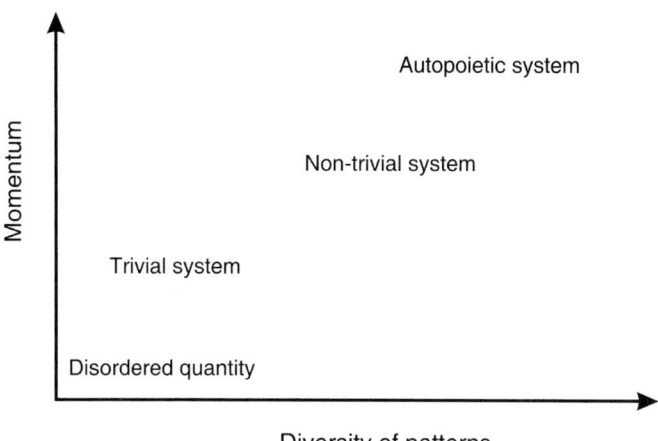

Fig. 1-40. Diversity of patterns versus momentum of different kinds of systems (Weinberg 1975)

Relating the different types of systems to different fields of science, one obtains something like Fig. 1-30 (page 30). This picture clearly shows that different fields of sciences have a strong correlation to different classes of systems. And, if it is true that the tools that can be applied in dealing with systems depend on the system, it becomes clear that different fields of science stick to different tools or techniques, as shown in Fig. 1-14 (page 18).

As already mentioned, the different systems, including the different types of causality, and therefore different tools, might yield to problems in understanding work in other fields of science. Since the degree of causality has a strong influence on the power of prediction, the design times for different systems show a great diversity.

1.3.4 Time

As seen in the sections "Complexity" and "Systems", the loss of causality becomes visible in a loss of time horizon for prediction. For astronomical systems, predictions of several million years are possible, whereas for the design of building structures, this value is in the range of 100 years. Economical systems can only be predicted for one or two years, and daily-life might include planning horizons of hours or days. Therefore, the dimension of time is of utmost importance in the understanding of indetermination. Table 1-3 shows some time scales for different processes.

Table 1-3. Characteristic time span for different processes (Frisch 1989)

	Characteristic time span in years
Astronomical process	$1-10^9$
Geological process	10^4-10^9
Biological process	$< 10^9$
Higher organized life	550×10^6
Development neocortex	$3 \times 10^4-10^5$
Ecological process	$10-10^4$
Economical process	$10-5 \times 10$
Political process	4–5 (legislature period) or up to 40 years (dictatorship)
Technical process	1–20
Human organism	Up to 100

Ernst Mach stated about time as "our concept of time phrases the most radical and general relation of things" (Neundorf 2006). This statement is an excellent description of the content of time. On the other hand, the understanding of the dimension of time usually eludes persistent. Most people actually consider time itself as timeless. Already earlier philosophers, however, assumed that time was created with the world. One stated, "The time was born with the world, the world was not born in the time." If this is true, then this would also have been an additional indicator for indetermination and limitation of the causal concept. Some physics simply avoid the question of the beginning of time by introducing some imaginary time (Seemann 1997).

If there is something as real time, it should be measurable. However, time, like space, can only be measured through references. Usually the dimension of time is introduced in some mathematical formula describing a certain process by considering an initial situation, some functional parameters and the special parameter of time. As already mentioned, the parameter of time can only be measured indirectly by some other processes. Therefore, a second process is required, which should be independent of the first process. Such a second process could be the ticking of a watch or the movement of the earth around the sun. Indeed we give the length of some processes in comparison to the number of revolutions around the sun. Therefore, if independence of the processes, as well as consistency of time in both systems is assumed, then one can include the processes in each other. This sounds surprising since, on one hand, independence and, on the other hand, consistency are required. That is probably causing one part of the difficulties about the time parameter. In general, it means that there exist both dependent and independent processes (Neundorf 2006). Considering some relativistic effects, it is known that even independent processes like time show dependencies under some circumstances.

The measurement of time has already been mentioned. Usually time is measured by the movement of masses, such as the earth, sun and so on. However, sometimes this might not be sufficient. Then, the term "changes" might be more efficient. Changes itself are strongly connected to interactions. In other terms, the existence of time is not independent, and depends on the existence of interactions, like the earth moving around the sun. This period or these irreversible processes are preferable for time measurements. If one imagines a time period of 100 years without any changes in the world, then, of course, there would be no way to measure it. Of course, it is impossible to stop the world, but what happens if the type of interactions changes. Usually gravitational forces are used for time measurement, such as the movement of the earth. Sometimes this type of interaction is not used, but nuclear interactions are used instead. Both types of interactions differ in their distances and in their magnitude by about 40 dimensions. If indeed such varieties exist in the processes for the time measurement, then time, as an independent process, might be affected as well. Actually these effects become visible: they are the so often discussed strength behaviour of quantum elements. The traditional coupling of processes as introduced in the beginning disappears, and therefore so does our traditional concept of the world. Fortunately, there still remain some relations independent from the great distance of orders.

Neundorf (2006) has stated: "Everything is related to everything and depends on each other. There are systems, which are related to other systems, but not every system is directly related to every system." Indeed time can only be measured if time is the only relation between the two processes. One can only measure time if the watch is not affected by another process. For example, if the time would be dependant on the number of people in a room, the time measurement would not be successful. But, considering the fact that all processes and changes are related, there is no absolute time. That means, indeed, the time measurement depends on the number of people in a room. But, if there is no absolute time, then there is no absolute causality and there is indetermination.

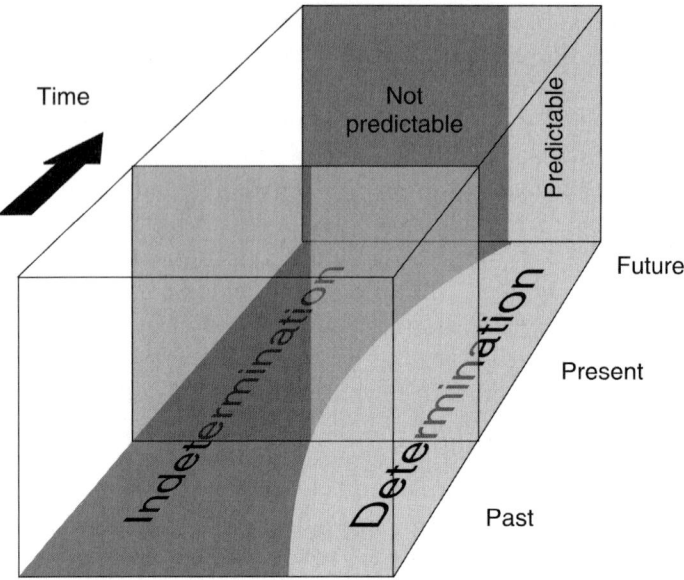

Fig. 1-41. Visualization of the amount of indetermination over time

The question whether indetermination or disorder increases or decreases in the world is of major philosophical interest. If order increases, the indetermination of the world would decrease and the quality of predictions would increase. On the other hand, if disorder or the lack of rules increases, then the quality of predictions will decrease.

Looking into several fields of science, one finds different answers to this question. The first law comes from the field of thermodynamics, which stated that entropy can only increase. Because entropy is a measure of disorder ($O = 1/E$) or uncertainty, order will decrease. This can actually be used for the identification of time directions.

On the other hand, in mathematical science, there exists Ramsey's mathematical law. This law describes an increase in order by declaring that every system with a sufficient amount of elements (complex system) will show new elements of order. Examples are randomly drawn lines on a piece of paper. Accidentally some of the lines form triangles, and therefore introduce new types of orders (Ganter 2004).

If indetermination depends on time, time could be used as an indicator of indetermination reaching a minimum in the present (Fig. 1-41). This would fit to the theory of Karl Popper calling history "open" (Popper 1994). In general, it seems to be that both determination and indetermination are needed to explain the basic physical items like energy, matter or time (Fig. 1-42). Determination and indetermination exist in our world, and are required to understand the world. Even indetermination alone is not a negative property since complex systems like humans demand them. On the other hand, indetermination might endanger living creations since future conditions may exceed the possible performance of the living creature. Therefore, living creatures develop protection to provide safety.

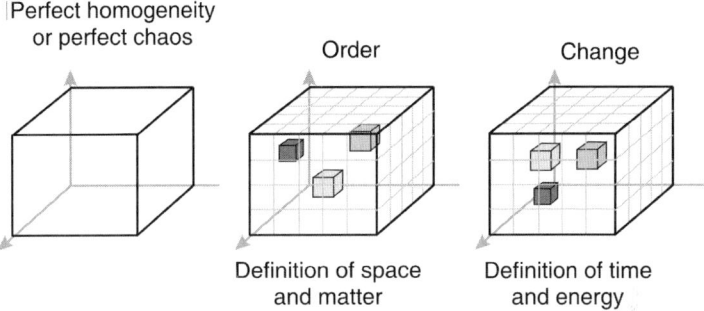

Fig. 1-42. Determination and indetermination are required to introduce space, matter, time and energy. Under perfect homogeneity or perfect chaos, such items are not required

1.4 Terms Related to Negative Evaluated Indetermination

1.4.1 Safety

Although "safety" is not a term related to negative evaluated indetermination, most terms related to negative evaluated indetermination are used to achieve safety. The goal of safety is the preservation of existence of an individual or a community. Since many general requirements of humans have found their way into laws, the preservation of psychological and physiological functioning of humans can be found in many constitutions and in the UN Charter of Human Rights.

Although the term safety can be found in many laws, this does not necessarily mean that the content of the term is clearly defined. Many humans have a different understanding of the term. Some common descriptions are (Hof 1991, Murzewski 1974):

- Safety is a state in which no disturbance of the mind exists, based on the assumption that no disasters or accidents are impending.
- Safety is a state without threat.

- Safety is a feeling based on experience that one is not exposed to certain hazards or dangers.
- Safety is the certainty of individuals or communities that preventive actions will function reliably.

1.4.1.1 Safety Concepts in Structural Engineering

Safety requirements and safety concepts have a long history in some technical fields. This can easily be seen in the code of Hammurabi, in which, as early as the 1700 B.C., pressure was put on builders through the introduction of strong penalties in case of collapsing structures. It has been estimated that the first application of a global safety factor in structural engineering dates back to 300 B.C. by Philo from Byzantium (Shigley & Mischke 2001). He introduced the global safety factor in terms of:

$$\gamma = \frac{\text{Resistance}}{\text{Load}} \tag{1-5}$$

Even empiric geometrical rules remained valid for nearly the next two millenniums. Only in the last few centuries, the applications of safety factors have become widespread. Over time, several different values were developed for different materials. In most cases, the values dropped significantly during the last century. For example, in 1880 in brick masonry, a factor of 10 was required, whereas 10 years later, a factor between 7 and 8 was required. In the 20th century, the values have changed from a factor of 5, then to 4, and now for the recalculation of historical structures with that material, a factor of 3 is chosen (Proske et al. 2006). This decline of safety factors can also be observed for other materials like steel. Especially with the development of new materials, an increase in the concern over the safe application of these materials has arisen. The development of safety factors for different materials led, in the beginning of the 20th century, to the initial efforts in developing material-independent factors, such as those shown in Tables 1-4 and 1-5.

Table 1-4. Global safety factor according to Visodic (1948)

Safety factor	Knowledge of load	Knowledge of material	Knowledge of environment
1.2–1.5	Excellent	Excellent	Controlled
1.5–2.0	Good	Good	Constant
2.0–2.5	Good	Good	Normal
2.5–3.0	Average	Average	Normal
3.0–4.0	Average	Average	Normal
3.0–4.0	Low	Low	Unknown

Table 1-5. Global safety factor according to Norton (1996)

Safety factor	Knowledge of load	Knowledge of material	Knowledge of environment
1.3	Extremely well-known	Extremely well-known	Likewise tests
2	Good approximation	Good approximation	Controllable environment
3	Normal approximation	Normal approximation	Moderate
5	Guessing	Guessing	Extreme

As the tables show, a further decline of global safety factors seems to be limited; otherwise, the major requirement – the safety of structures – might not be fulfilled anymore. Therefore, more advanced changes might be considered to meet the demanding requirements of economic and safe structures. Such developments would include special safety factors for the different column heads in the table, for example a safety factor for load and a safety factor for material. This indeed is the idea of the partial safety factor concept. It does not necessarily yield lower safety factors, but it does yield a more homogenous level of safety. The proof of safety is carried out by the simple comparison of the load event E_d with the resistance of the structure R_d.

$$E_d \leq R_d \tag{1-6}$$

Subsequently, the load event can be built upon several single elements, such as the characteristic dead load G_{kj} connected with a special safety factor $\gamma_{G,j}$ only for dead load, and the characteristic life load Q_{kj} connected with a special safety factor $\gamma_{Q,i}$ only for life load:

$$E_d = \sum_{j\geq 1} \gamma_{G,j} \cdot G_{k,j} + \gamma_{Q,1} \cdot Q_{k,1} + \sum_{i>1} \gamma_{Q,i} \cdot \psi_{0,i} \cdot Q_{k,i} \tag{1-7}$$

The same can be done with the resistance site:

$$R_d = R\left(\alpha \cdot \frac{f_{ck}}{\gamma_c}; \frac{f_{yk}}{\gamma_s}; \frac{f_{tk,cal}}{\gamma_s}; \frac{f_{p0,1k}}{\gamma_s}; \frac{f_{pk}}{\gamma_s}\right) \tag{1-8}$$

Unfortunately, the solid allocation of the safety factor to a single variable does not yield safe structures under all conditions. Therefore, if so-called non-linear calculations are carried out, there is something like a global safety factor for the resistance reintroduced:

$$R_d = \frac{1}{\gamma_R} R(f_{cR}; f_{yR}; f_{tR}; f_{p0,1R}; f_{pR}) \tag{1-9}$$

To give the reader an impression about such partial safety factors for resistance, Table 1-6 lists some of them for several materials.

Table 1-6. Material partial safety factors

Material	Limit state of ultimate load	Accidental load conditions	Limit state of serviceability
Concrete (up to C 50/60)	1.50	1.30	1.50
Non-reinforced concrete	1.80	1.55	–
Non-reinforced concrete	1.25		
Precast concrete	1.35		–
Collateral evasion	2.00	–	–
Reinforcement steel	1.15	1.00	1.15
Prestressing steel	1.15	1.00	1.15
Steel yield strength	1.10	1.00	
Steel tensile strength	1.25	1.00	
Steel tensile strength	1.00	1.00	
Wood	1.30	1.00	1.00
Masonry (Category A)	1.7 (I)/2.0 (II)	1.20	
Masonry (Category B)	2.2 (I)/2.5 (II)	1.50	
Masonry (Category C)	2.7 (I)/3.0 (II)	1.80	
Masonry – Steel	1.50/2.20		
Masonry	1.50–1.875	1.30–1.625	
Anchoring (C. A-C)	2.50	1.20	
Floatglass/Gussglass	1.80	1.40	1.40
ESV-Glass	1.50	1.30	1.30
Siliconglass	5.00	2.50	2.50
Carbon fibre	1.20		
Carbon fibre	1.30[1]		
Carbon fibre cable	1.20[1]		
Carbon fibre glue	1.50[1]		
Cladding	2.00		
Aluminum yield strength	1.10		
Aluminum tensile strength	1.25		
Bamboo as building material	1.50		
Textile reinforced concrete	1.80[2]		
Concrete under multiaxial loading	1.35–1.70		
Granodiorit tensile strength	1.50–1.70		
Historical masonry arch bridges	2.00[3]		
Historical non-reinforced arch bridges	1.80[3]		

[1] Consider construction conditions.
[2] Recent research indicates lower values.
[3] A partial safety factor for a system.

Although the partial safety factor concept was fi...
World War II, it took quite some time to become prac...
The first application can be found in steel design, where...
ple in the field of structural concrete is the ETV concrete,
in East-Germany (ETV is a German abbreviation for Un...
Codes). The ETV concrete was developed during the 197...
mented around the beginning of the 1980s.

The development of partial safety factors is strongly connected to the development of the probabilistic safety concept in structural engineering. The first proposals about probabilistic-based safety concepts were found by Mayer (1926) in Germany and Chocialov in 1929 in the Soviet Union. In the 1930s, the number of people working in that field had already increased, including Streleckij in 1935 in the Soviet Union, Wierzbicki in 1936 in Poland and Prot in 1936 in France (Murzewski 1974). Already in 1944 in the Soviet Union, the introduction of the probabilistic safety concept for structures was forced by politicians (Tichý 1976). The development of probabilistic safety concepts, in general, experienced a strong impulse during and after World War II, not only in the field of structures but also in the field of aeronautics. In 1947, Freudenthal published his famous work about the safety of structures. For further investigations into the development of safety concepts in structural engineering, refer Beeby (1994) and Pugsley (1966).

A model code for the probabilistic safety concept of structures has been published by the Joint Committee of Structural Safety (JCSS 2006). This safety concept is explained in detail. Safety of structures is defined as the capability of structures to resist loads (E DIN 1055-100, DIN ISO 8930). Due to the fact that no building can resist all theoretically possible loads, the resistance has to reach only a sufficient level (E DIN 1055-9). The decision whether a structure has qualitative property safety can only be answered by using a quantitative measure. For structures, "reliability" is one such measure used. The reliability is interpreted as a probability of failure (DIN ISO 8930, 1.1. & 1.2).

$$p_f = \int \cdots \int_{g(\mathbf{X}) \leq 0} f_{\mathbf{X}}(x)\,dx \tag{1-10}$$

A visualization of this formula is given in Fig. 1-43. This value can be related to a time period n:

$$p_f(n) = 1 - (1 - p_f)^n \tag{1-11}$$

Since the probability of failure is extremely low due to the high safety requirements, a substitute measure, the safety index, has been heavily applied instead of the probability. The safety index is defined as:

$$\beta = \Phi^{-1}(p_f) \tag{1-12}$$

and represents a change in the numerical integration towards an extreme value computation as shown in Fig. 1-44. The conversion is shown in Table 1-7.

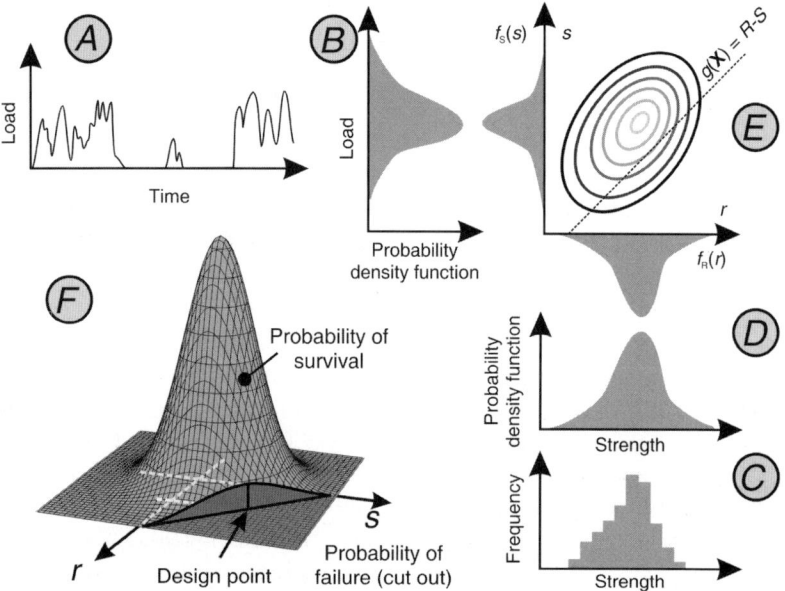

Fig. 1-43. Development of probability of failure based on the statistical uncertainty of load and resistance values

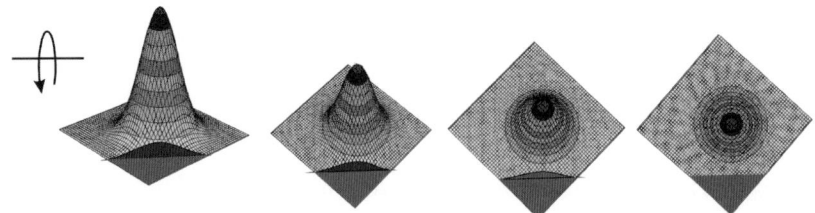

Fig. 1-44. Transformation of the integration in an extreme value computation

Table 1-7. Conversion of the probability of failure into the safety index

Probability of failure	10^{-11}	10^{-10}	10^{-9}	10^{-8}	10^{-7}	10^{-6}	10^{-5}	10^{-4}	10^{-3}	10^{-2}	10^{-1}	0.5
Safety index		6.71	6.36	5.99	5.61	5.19	4.75	4.26	3.72	3.09	2.33	1.28	0.0

Table 1-8. Required values for safety index according to the JCSS Modelcode

Costs of safety measure	Consequence of failure		
	Low	Medium	High
Low costs	3.1	3.3	3.7
Medium costs	3.7	4.2	4.4
High costs	4.2	4.4	4.7

Therefore, the proof of safety can be calculated by a comparison of the current probability of failure with the permitted probability of failure or the required safety index (Table 1-8). A summary of required safety indexes can be found in Proske (2004).

Yet another safety concept in structural engineering is Fuzzy-probabilistic procedures (Möller et al. 2000). Figure 1-45 shows a system of safety concepts in structural engineering. Incidentally, the parameters are describing unsafe conditions rather than safety. Additional to the reliability-based safety concepts are the codes for structures known as risk-based safety evaluations (Eurocode 1, 2.1-2.2, 3.1-3.2, special 3.2 (2), E DIN 1055-9, NBDIN (1981), 3.1, CEB). These evaluations, however, are mainly used for structures under accidental loads, for example earthquake impacts. The property safety is then fulfilled if the existing risk is less than an acceptable risk (E DIN 1055-9, Abs. 5.1 (3), Eurocode 1). This definition can be found in many other codes for technical products, such as DIN EN 61508-4, DIN EN 14971 or DIN VDE 31000 Part 2 (1984).

$$\text{vorh } R \leq \text{zul } R \Rightarrow S \qquad (1\text{-}13)$$

$$\text{vorh } R > \text{zul } R \Rightarrow \cancel{S}$$

Sometimes in literature, the definition of safety is described as an inverse of risk:

$$S = 1 - R \qquad (1\text{-}14)$$

This does not fit to the general understanding of safety as a qualitative measure. The situation is either safe or not. This will be discussed in the next section.

1.4.1.2 General Understanding of Safety

After summarizing the state of knowledge of safety concepts in one field, the author would like to introduce some additional considerations about the term safety. Obviously most definitions concerning safety include something such as "peace of mind" or "freedom from threats". These actually are terms for describing nothing else than a state without required actions. Furthermore, it considers the spending of resources including time, energy or money. This yields to the conclusion that safety is a state, where no further resources have to be spent to eliminate a hazard or danger. Actually here, safety is dependent not only from the hazard but also from the decision-making process about safety. Assume that the condition of structures has not changed; however, the awareness about safety has changed due to some media reports, thus it might happen that people start to feel the buildings are no longer safe. Moreover, people might have dramatically increased their resources and consider spending them. Here, it might happen that they spend some resources for mitigation action even though there is no additional information as before.

Summarizing these considerations, safety starts at the maximum point of the curvature of a "freedom of resources" versus "degree of distress" function (Fig. 1-46 bottom). Additionally, the horizon of planning can be used as a substitute parameter for safety. Under hazardous or dangerous conditions, people usually only have a limited horizon of planning, such as seconds or minutes; whereas under safe conditions, people might also plan in terms of years or even decades (i.e. pension fund).

If one considers, additionally, the Yerkes-Dodson curvature as a function of maximum performance versus degree of distress, it turns out that a state of maximum freedom of resources and horizon of planning does not necessarily yield maximum performance of people. Actually, it means that the state of safety might cause loss of safety, whereas an unsafe state might cause safety since people develop prevention actions. This statement fits very well to some recommendations given in the chapter "Subjective risk judgment", saying that slight hazards cause maximum safety behavior of humans.

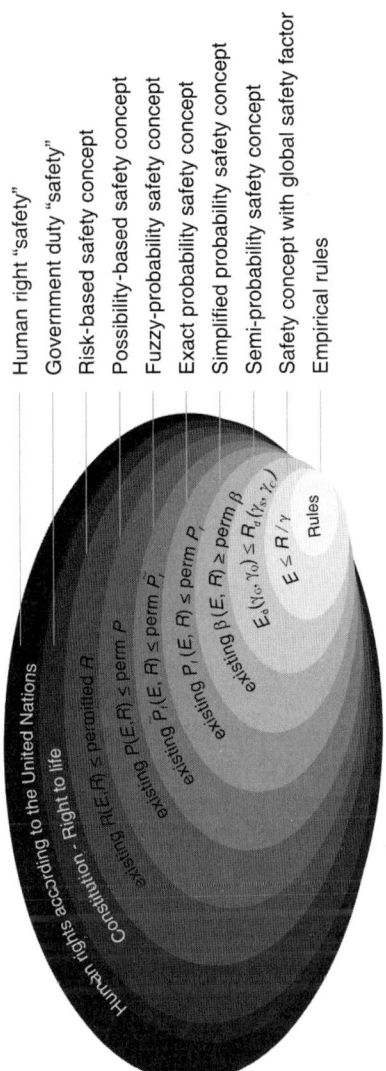

Fig. 1-45. The frame of safety concept in structural engineering

Unfortunately, all of these concepts are difficult to apply in numerical models. Therefore, the substitution of the qualitative statement "safety" to a quantitative statement "risk" seems to be preferable for numerical investigations of safety.

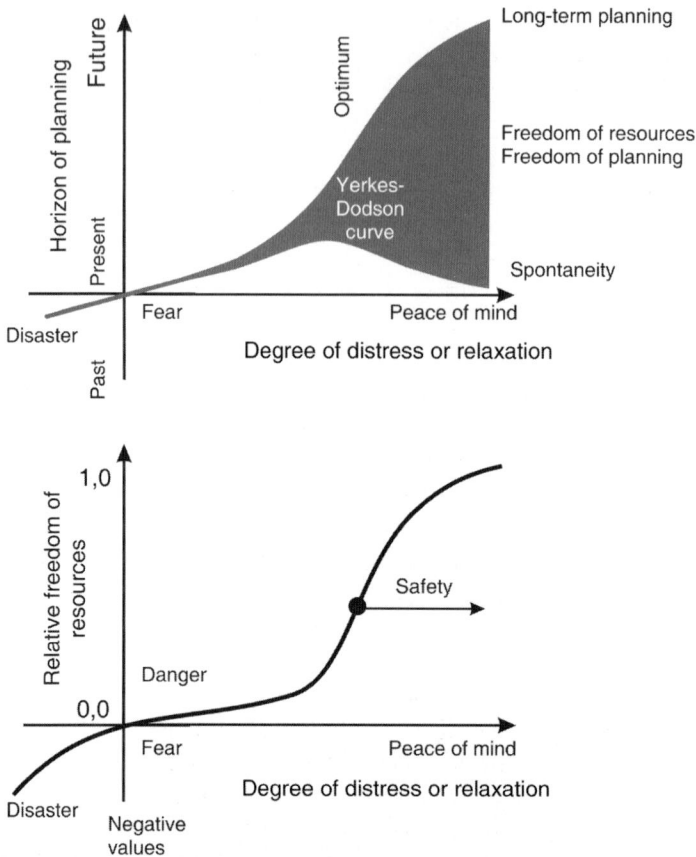

Fig. 1-46. Definition of safety through the freedom of resources versus degree of distress function. Here, safety starts at the maximum change in slope (bottom diagram). The top diagram shows the relationship between horizon of planning and degree of distress. Additionally, the Yerkes-Dodson curve is included

1.4.2 Risk

1.4.2.1 History of the Term

As mentioned earlier, not only humans but all living creations search for safety. To reach that goal, risk prevention or mitigation measures might be applied. Risk prevention measures are not only a part of the human society; animals also develop protection strategies. While these strategies mainly develop in animals through evolutionary processes, humans as Lamarck's creatures use additional techniques.

According to Covello & Mumpower (1985), the first simple systematic techniques to carry out risk analysis were made by the Asipu group and found at the Tigris-Euphrates valley around 3200 B.C. The Asipu group were people supporting others in decision processes under uncertain conditions. Of course, the inputs for the analysis were signs from Gods, which were interpreted in a certain way. Eventually, a final report was created on a clay tablet. Such types of prediction techniques were continuously developed over the next millenniums, for example the oracle of Delphi.

A term comparable to the word risk, however, was not introduced until the 14th century. The major increase in shipping and trading during that time created the need to develop new terms to deal with venture risks, hazardous enterprises and the possibility of loss. Shipping was, during the medieval ages, the most efficient means of transport. In contrast, traveling on land was ten times as expensive as sea traveling and five times as expensive as river traveling (Ganshof 1991). The growth of trade can be explained by dramatic changes at the beginning of the 11th and 12th century in Europe. It was at this time upheavals in agriculture and the rise of bourgeoisie started. The number of cities, for example, grew in Germany from only about a 100 in the 11th century to more than 3,000 in the 14th century (Heinrich 1983).

Earlier tracks of the term risk can be found in the Italian words "risco" and "rischio" (danger and risk) derived from the words "rischiare" and "risicare". Before the introduction of the word risk, only terms like *virtù* and *fortitudo* were used (Recchia 1999, Mathieu-Rosay 1985).

In the 16th century, the term risk was widely used in Roman languages, and written approval about the use of the word can be found. For example, Scipio Ammirato in Venice 1598 used the term "rischio" to describe the possibility to publish his information source. In the same year, Giovanni Botero also used the word (Luhmann 1993).

Further historical origins of the word risk are under discussion. Many publications mention that the word risk might be derived from the Greek term "rhizia" (root). Originally the term "rhizia" was used for the root of trunks. Later in Creta, the application was extended to cliffs at the root of mountains yielding to the term "rhizicon". Therefore, the meaning of the word changed from roots of trunks to cliffs that represent danger, a risk for shipping (Mathieu-Rosay 1985).

Other publications assume that the term can be traced back to the old-Persian term "rozi(k)". This term represents something like daily income, daily bread or destiny. This term was then altered to "rizq," which means living depending on God and destiny (Banse 1996).

Origin of the term from the Spanish and the African word "aresk" are also disputed.

1.4.2.2 Definition of Risk

There exist several different viewpoints with respect to the term risk. They are mainly based on the discipline in which the viewpoint was created and the requirements of that discipline. Renn (1992) has introduced different views of the problem of risk:

- Insurance-statistic based view
- Toxicological-epidemiological view
- Engineering-technical view
- Economical view
- Psychological view
- Social-theoretical view
- Cultural-theoretical view

Here, the shorter division with only four groups from Weichselgartner (2001) will be followed. The first group is represented by the statistical-mathematical formulation. It is the main content of chapter "Objective risk measures". The formulation is based on the general formula:

$$R = C \cdot P \tag{1-15}$$

where R is the risk, C is the negative consequence measure (damage or disadvantage) and P is an indetermination measure. Of course, all components can have different units. For example, the negative consequences can be given in monetary units, in time required, loss of space, loss of humans, loss of creatures, loss of energy, loss of environment and so on. The same is valid for the indetermination measure. Here, many different mathematical theories dealing with uncertainty and indetermination can be used as stated before. Of course, the most common technique is statistics and probability, but also Fuzzy numbers, Grey numbers and entropy measures are possible. Incidentally, here again the limitations of mathematics become visible, since the objectivity of such parameters depends very much on the assumptions used in the models. Corrupt data might heavily influence the results, such as outliers, censored data or multi-modale data. Additionally, humans have problems in understanding such numbers if they are directly affected by risks. In this case, other views on risk might be more appropriate to understand human behavior.

This yields to the second view of risks considering the properties of social systems and their reactions to risk, which are described in the social-cultural formulation of risk.

The third view of this kind is included in the psychologist-cognitive formulation discussed in chapter "Subjective risk judgment". Here, psychological based preferences of humans are considered when dealing with risks. This is of major importance since humans make all decisions. So even if the impression of risks is extremely biased based on some subjective preferences, the results and consequences of such an imaginary risk are real. Although the objective risk measures are called objective, the subjective risk judgment is actually more objective in the consequences. Here, the definition of Sandman (1987) fits, that Risk = Hazard + Outrage.

Last but not least, there is a geographic-spatial formulation of risks. This type is heavily used for natural hazards, which are connected to some natural processes. In turn, these processes are connected to some environmental conditions. For example, an avalanche requires some mountains, or a flood requires some water. Information on this is given in chapter "Risks and disasters".

Finally, it should be stated that risks are not constant values. First of all, the efficiency of mitigation measures and the value at risk change over time, as shown in Fig. 1-47 (Fuchs 2006). Furthermore, based on changed conditions advantages might turn to disadvantages and vice versa. This is shown in Fig. 1-48, where a benefit changes over time to a disadvantage.

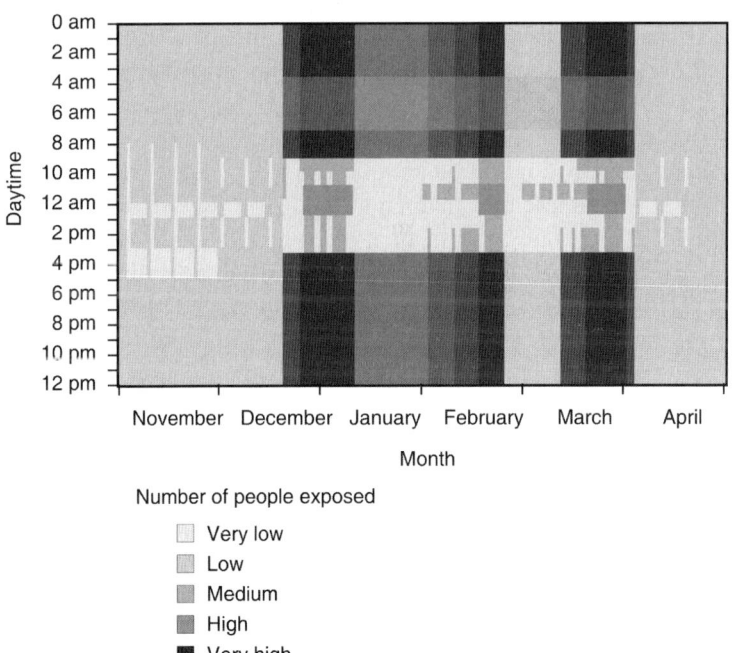

Fig. 1-47. Fluctuations of the number of exposed persons in a certain winter tourist area (Keiler et al. 2005)

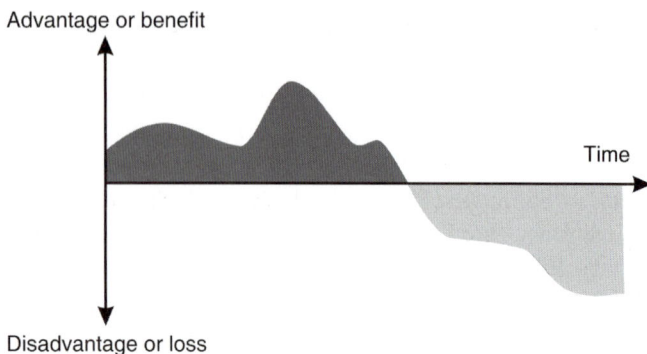

Fig. 1-48. Development of benefits or losses over time

1.4.3 Hazard and Danger

The terms safety and risk have strong relations to the terms hazard and danger. The term hazard implies the occurrence of a condition or phenomenon, which threatens disasters to anthropogenic spheres of interest in a defined space and time (Fuchs 2006). In general, a hazard is a natural, technical or social process or event that is a potential damage source (DIN EN 61508-4). There exist many further definitions of hazards; however, these will not be discussed here.

In general, a hazard might be completely independent from the activities of humans, for example an avalanche or a debris flow. However, if people are at the location of the process, then these people might be in danger.

A danger is a situation that yields without unhindered development to damage (HVerwG 1997). On one hand, danger can be seen as the opposite of safety, where no resources have to be spent. Luhmann (1997) stated that risks and hazards are opposites in terms of human contribution. Risks require, under all conditions, possible human actions, whereas hazards are independent from human actions. If that is true, however, then more technologies or more human resources simply means more risks due to the increased volume of human actions and decisions. This fits very well to the work of Lübbe (1989), who stated that awareness of uncertainty and risk increases with increasing resources and capabilities of societies.

1.4.4 Vulnerability and Robustness

Vulnerability is a term that permits an extension of the classical risk definition by only two terms. Instead further properties can be incorporated into the term vulnerability, which itself is then part of the risk definition (Fuchs 2006):

$$R = f(p, A_o, v_o, p_o) \qquad (1\text{-}16)$$

The equation includes R as risk, p as probability of this event, A_O as value of an object, v_O as vulnerability of the object during the event and p_O as probability of exposure of the object during the event. This equation can be further extended to several event scenarios and objects. However, the term vulnerability remains to be defined. Again, as with many other terms, the variety of definitions of vulnerability is virtually unmanageable. Examples of definitions are given in Table 1-9.

Table 1-9. Examples of the definitions of vulnerability taken from Weichselgartner (2001) and GTZ (2002)

Author	Definition
Gabor & Griffith in 1980	Vulnerability is the threats (to hazardous materials) to which people are exposed (including chemical agents and the ecological situation of the communities and their level of emergency preparedness). Vulnerability is the risk context.
Timmerman in 1981	Vulnerability is the degree to which a system acts adversely to the occurrence of a hazardous event. The degree and quality of the adverse reaction are conditioned by a system's resilience (a measure of the system's capacity to absorb and recover from the event).
UNDRO in 1982	Vulnerability is the degree of the loss to a given element or set of elements at risk resulting from the occurrence of a natural phenomenon of a given magnitude.
Petak & Atkisson in 1982	The vulnerability element of the risk analysis involves the development of a computer-based exposure model for each hazard and appropriate damage algorithms related to various types of buildings.
Susman et al. in 1983	Vulnerability is the degree to which different classes of society are differentially at risk.
Kates in 1985	Vulnerability is the "capacity to suffer harm and react adversely".
Pijawka & Radwan in 1985	Vulnerability is the threat or interaction between risk and preparedness. It is the degree to which hazardous materials threaten a particular population (risk) and the capacity of the community to reduce the risk or the adverse consequences of hazardous material releases.
Bogard in 1989	Vulnerability is operationally defined as the inability to take effective measures to insure against losses. When applied to individuals, vulnerability is a consequence of

(Continued)

Table 1-9. (Continued)

Author	Definition
	the impossibility or improbability of effective mitigation and is a function of our ability to detect hazards.
Mitchell in 1989	Vulnerability is the potential for loss.
Liverman in 1990	Distinguishes between vulnerability as a biophysical condition and vulnerability as defined by political, social and economic conditions of society. She argues for vulnerability in geographic space (where vulnerable people and places are located) and vulnerability in social space (who in that place is vulnerable).
UNDRO in 1991	Vulnerability is the degree of the loss to a given element or set of elements at risk resulting from the occurrence of a natural phenomenon of a given magnitude and expressed on a scale from 0 (no damage) to 1 (total loss). In lay terms, it means the degree to which an individual, family, community, class or region is at risk from suffering a sudden and serious misfortune following an extreme natural event.
Dow in 1992	Vulnerability is the differential capacity of groups and individuals to deal with hazards, based on their positions within physical and social worlds.
Smith in 1992	Human sensitivity to environmental hazards represents a combination of physical exposure and human vulnerability – the breadth of social and economic tolerance available at the same site.
Alexander in 1993	Human vulnerability is a function of the costs and benefits of inhabiting areas at risk from natural disaster.
Cutter in 1993	Vulnerability is the likelihood that an individual or group will be exposed to and adversely affected by a hazard. It is the interaction of the hazard of place (risk and mitigation) with the social profile of communities.
Watts & Bohle in 1993	Vulnerability is defined in terms of exposure, capacity and potentiality. Accordingly, the prescriptive and normative response to vulnerability is to reduce exposure, enhance coping capacity, strengthen recovery potential and bolster damage control (i.e. minimize destructive consequences) via private and public means.
Blaikie et al. in 1994	By vulnerability we mean the characteristics of a person or a group in terms of their capacity to anticipate, cope with, resist and recover from the impact of a natural hazard. It involves a combination of factors that determine the degree to which someone's life and livelihood are put at risk by a discrete and identifiable event in nature or in society.

Table 1-9. (Continued)

Author	Definition
Green et al. in 1994	Vulnerability to flood disruption is a product of dependency (the degree to which an activity requires a particular good as an input to function normally), transferability (the ability of an activity to respond to a disruptive threat by overcoming dependence either by deferring the activity in time, or by relocation, or by using substitutes), and susceptibility (the probability and extent to which the physical presence of flood waters will affect inputs or outputs of an activity).
Bohle et al. in 1994	Vulnerability is best defined as an aggregate measure of human welfare that integrates environmental, social, economic and political exposure to a range of potential harmful perturbations. Vulnerability is a multilayered and multidimensional social space defined by the determinate, political, economic and institutional capabilities of people in specific places at specific times.
Dow & Downing in 1995	Vulnerability is the differential susceptibility of circumstances contributing to vulnerability. Biophysical, demographic, economic, social and technological factors, such as population ages, economic dependency, racism and age of infrastructure, are some factors that have been examined in association with natural hazard.
Gilard & Givone in 1997	Vulnerability represents the sensitivity of land use to the hazard phenomenon (Amendola 1998). Vulnerability (to dangerous substances) is linked to the human sensitivity, the number of people exposed and the duration of their exposure, the sensitivity of the environmental factors, and the effectiveness of the emergency response, including public awareness and preparedness.
Comfort et al. in 1999	Vulnerability is those circumstances that place people at risk while reducing their means of response or denying them available protection.
Weichselgartner & Bertens in 2000	By vulnerability we mean the condition of a given area with respect to hazard, exposure, preparedness, prevention, and response characteristics to cope with specific natural hazards. It is a measure of capability of this set of elements to withstand events of a certain physical character.
GTZ 2002	Vulnerability denotes the inadequate means or ability to protect oneself against the adverse impact of external events, on the one hand, and to recover quickly from the effects of the natural event, on the other. Vulnerability is made up of many political-institutional, economic and sociocultural factors.

Harte et al. (2007) have introduced a quantification of vulnerability V for objects based on the damage parameter D:

$$V = \int_{\text{Lifetime}} D\, dt \qquad (1\text{-}17)$$

The damage can, for example, be described by the change of the eigen-frequencies f of a structure:

$$D = 1 - \frac{f_{\text{Damaged}}}{f_{\text{Non-damaged}}} \qquad (1\text{-}18)$$

To permit better comparisons, a normalized vulnerability can be introduced based on a loading factor λ:

$$V^* = \int_{\lambda=0}^{\lambda=1} D(\lambda) \cdot (1-\lambda) \cdot \frac{3}{D_{\text{Max}}} \qquad (1\text{-}19)$$

The parameters are visualized in Fig. 1-49. Here, the relationship between the load parameter and the damage parameter becomes visible. For normal conditions of structures, the limit state of serviceability with slight changes of the structure and the ultimate load bearing limit state could be numerically reached.

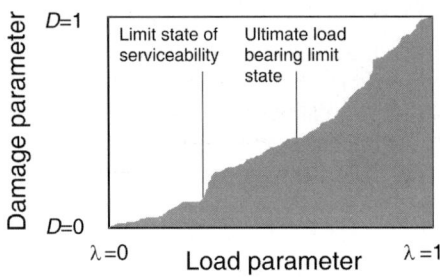

Fig. 1-49. Load parameter versus damage parameter. A load parameter of 1 represents collapse of the structures, whereas a load parameter of 0 means a virgin-like structure (Harte et al. 2007)

Assuming that robustness is the antonym of vulnerability, then the investigation of robustness might yield further information about vulnerability. Robustness is the ability of systems to survive unexpected or exceptional circumstances or situations. For structures, it means the structure would still function even when under improper use, short overloads or extreme conditions (Harte et al. 2007).

Robustness can be quantified by (Harte et al. 2007):

$$R = \frac{\lambda_{\text{Fail}}}{\lambda_{\text{Design}}} \cdot f(D)$$

(1-20)

Based on the assumption of contradiction between vulnerability and robustness, normalized robustness is defined based on normalized vulnerability (Harte et al. 2007).

$$R* = 1 - V* = 1 - \int_{\lambda=0}^{\lambda=1} D(\lambda) \cdot (1-\lambda) \cdot \frac{3}{D_{\text{Max}}}$$

(1-21)

With this concept, damage can be either directly or indirectly related to the term risk over the term vulnerability.

1.4.5 Damage and Disadvantage

In most definitions, damage is considered to be a part of the term risk. Damage can be understood as the occurrence of a change for the worse (difference theory), physical harm that is caused by something, a detraction of goods caused by an event or circumstance, and a depreciation of value, forfeit or loss caused by an event (DIN VDE 31000-2). Most definitions of damage can be found in the jurisprudence, since that is the major topic of this industry.

Nevertheless, the author does not consider "damage" as a part of risk, but "disadvantage". As already mentioned, damages are mainly described based on the difference theory. In this theory, a value of an object is calculated twice; if the value of the early time step is higher than the latter one, then the object has lost value, which can be interpreted as damage.

This theory, however, only compares values over time; it does not consider changes in the environment like people do. For example, conditions might change completely and the object is not of use anymore. Obviously that circumstance should be considered in the price, but then the loss of a value is not automatically due to damage. The definition of disadvantage includes the property of a less favorable position. That includes a direct comparison to other people.

The difference between damage and disadvantage shall be shown with a simple example. Imagine a teacher who arbitrarily gives everyone in the class a "B" grade after a written examination. Only one student receives a "C" grade. The marks are independent from individual performance and are only based on the teacher's personal preferences. The one student who received a "C" grade complains and requires to conduct a new examination for the entire class. This is carried out, and the marks are now different for most pupils. The average mark should stay constant. One student again receives a "C" grade based on proper marking of the written examination. Therefore, for the one student, the value has not changed, which means no damage at all, assuming that the average mark was constant. But, that one student, of course, will feel much more comfortable. Even though there was no damage in the first case, there was a disadvantage.

1.4.6 Disaster

Disasters, on the other hand, are events with not only disadvantages but also severe damages. In contrast to the term risk, they describe an event in the present or past, and do not deal with the future. Since, in the parameter risk, the uncertainty term disappears, only the damage parameter is able to describe different levels.

Here, different concepts have been developed, which are also connected to the term vulnerability (ISDR 2004, BZS 2003). However firstly, example definitions using damage numbers shall be mentioned. For example, one such parameter is the damage expressed in terms of percentage of the gross domestic product of a country or region. While in developing countries damages caused by disasters often reach more than 10% of the gross domestic product, in developed countries, the values are rarely higher than 3% (Mechler 2003).

Another example for the application of simple damage values is given below. A disaster is defined here by the minimum damage with (Mechler 2003):

- More than 100 fatalities
- Financial damage higher than 1% of the gross domestic product
- More than 1% of the population is affected by the event

or

- More than 10 fatalities
- More than 100 people are affected
- Emergency case announced
- The country asks for international help

The examples show that there exists no common definition for the term disaster. Some further examples are given here. The United Nations describe a disaster as "the disruption of the functionality of a society, the loss of life, real values and environmental values and exceedance of the capability of a society to cope with such an event" (DKKV 2002).

The Permanent Conference for Disaster Prevention suggested the following definition (DKKV 2002): "A disaster is an extremely serious and extensive, mostly surprising event that endangers the health, life and living conditions of humans and damages real values and can not be managed only with local or regional recourses."

The law for disaster protection in Schleswig-Holstein (State Disaster Protection Law) from the year 2000 defines a disaster in § 1, (1) as: "A disaster in the sense of the law is an event which endangers the life, health and vital supply of many humans, endangers important real values or environmental values or damages in such a magnitude, that sufficient protection and help can only be supplied if several forces and institutions work together under the leadership of a disaster protection agency."

Further definitions can be found in Quarantelli (1998); for example, "Disasters are events in societies or their larger subsystems (e.g. regions, communities) that involve social disruption and physical harm. Among the key defining properties of such events are (1) length of forewarning, (2) magnitude of impact, (3) scope of impact and (4) duration of impact", or "A disaster is an event concentrated in time and space, in which a society or one of their subdivision undergoes physical harm and social disruption, such that all or some essential functions of the society or subdivisions are impaired" (taken from Krebs 1998, see also Portfiriev 1998).

The common characteristic in the definitions is the requirement for external help. A disaster is an endangering situation, where the exposed person or social system needs further support (Fig. 1-46). For example, a breaking of a bone was a disaster for humans hundreds or thousands of years ago. However, efficient mitigation measures might simplify the access to such external support systems, like the health system. Nowadays, a broken bone might be dealt with in only a few hours. Even afterwards, still some support is required, which is easily accessible.

Finally it should be mentioned that the term disaster originates from the Greek tragedy meaning sudden turn. However a sudden turn does not have only to include doom, it often also offers new chances and may be the origin of new developments (Eder 1987).

References

Adams J (1995) Risk. University College London Press

Anderson PW (1972) More is Different: Broken Symmetry and the Nature of the Hierarchical Structure of Science, Science 177 (4047). pp 393–396

Aleksic Z (2000) Artificial life: Growing complex systems. In: TRJ Bossomaier & DG Green (Eds), Complex Systems. Cambridge University Press, Cambridge

Barrow JD (1998) The limits of science and the science of limits. Oxford University Press, New York (German translation 1999)

Banse G (1996) Herkunft und Anspruch der Risikoforschung, in: Banse G (Ed) Risikoforschung zwischen Disziplinarität und Interdisziplinarität. Von der Illusion der Sicherheit um Umgang mit Unsicherheit, Edition Sigma, Berlin, pp 15–72

Beeby AW (1994) 'γ-factors: A second look', The Structural Engineer, 72(2)

Bogdanow AA (1926) Allgemeine Organisationslehre (Tektologie), Organisation Verlagsgesellschaft mbH Berlin

Bonchev D & Buck GA (2005) Quantitative measures of network complexity. In: D Bonchev & DH Rouvray (Eds), Complexity in Chemistry, Biology and Ecology. Springer Science + Business Media, New York, pp 191–235

Bossomaier T (2000) Complexity and neural networks. In: TRJ Bossomaier & DG Green (Eds). Complex Systems. Cambridge University Press, Cambridge

Box GEP (1979) Robustness in the strategy of scientific model building. In: RL Launer & GN Wilkinson (Eds), Robustness in Statistics. Academic Press, New York, pp 201–235

Brehmer A (Hrsg) (1910) Die Welt in 100 Jahren. Verlagsanstalt Buntdruck GmbH, Berlin 1910, Nachdruck Hildesheim 1988

BZS (2003) Bundesamt für Zivilschutz. KATARISK: Katastrophen und Notlagen in der Schweiz. Bern

Corning PA (2002) The Re-Emergence of "Emergence": A Venerable Concept in Search of a Theory, Complexity 7 (6), pp 18–30

Covello VT & Mumpower J (1985) Risk analysis and risk management: An historical perspective. Risk Analysis, 5, pp 103–120

Cramer F (1989) Chaos und Ordnung – Die komplexe Struktur des Lebendigen. Deutsche Verlags-Anstalt, Stuttgart

Crossley WA (2006) System of Systems: An Introduction. Purdue University Schools of Engineering's Signature Area

Crutchfield JP (1994) The calculi of emergence: Computation, dynamics, and induction, Physica D, Vol 75, Issue 1-3, pp 11–54

Cucker F & Smale S (2006) The Mathematics of Emergence. The Japanese Journal of Mathematics, Vol. 2, Issue 1, pp 197–227

Davies P & Gribbin J (1996) Auf dem Weg zur Weltformel – Superstrings, Chaos, Komplexität: Über den neuesten Stand der Physik. Deutscher Taschenbuchverlag, München

Deng J (1988) Essential Topics on Grey Systems: Theory and Applications. China Ocean Press, Bejing

DIN 2330 (1979) Begriffe und Benennungen. Allgemeine Grundsätze

DIN VDE 31000–2 (1984) Allgemeine Leitsätze für das sicherheitsgerechte Gestalten technischer Erzeugnisse – Begriffe der Sicherheitstechnik – Grundbegriffe. Dezember 1984

DKKV (2002) Deutsches Komitee für Katastrophenvorsorge e.V. Journalisten-Handbuch zum Katastrophenmanagement -2002-. Erläuterungen und Auswahl fachlicher Ansprechpartner zu Ursachen, Vorsorge und Hilfe bei Naturkatastrophen. 7. überarbeitete und ergänzte Auflage, Bonn

E DIN 1055-100: Einwirkungen auf Tragwerke, Teil 100: Grundlagen der Tragwerksplanung, Sicherheitskonzept und Bemessungsregeln, Juli 1999

E DIN 1055-9: Einwirkungen auf Tragwerke Teil 9: Außergewöhnliche Einwirkungen. März 2000

Eder G (1987) Katastrophen in unserem Sonnensystem. Erdgeschichtliche Katastrophen – Öffentliche Vorträge 1986. Verlag der Österreichischen Akademie der Wissenschaften, Wien

Edgington, D (1999) Vagueness by degrees. In: Keefe & Smith (Eds), pp 294–316

ESRIF (2008) – European Security Research and Innovation Forum. http://www.esrif.eu/

Eurocode 1 – ENV 1991: Basis of Design and Action on Structures, Part 1: Basis of Design. CEN/CS, August 1994

Fehling G (2002) Aufgehobene Komplexität: Gestaltung und Nutzung von Benutzungsschnittstellen. Dissertation an der Wirtschaftswissenschaftlichen Fakultät der Eberhard-Karls-Universität zu Tübingen

Flood RL & Carson ER (1993) Dealing with Complexity: An introduction to the Theory and Application of System Science. 2nd ed. Plenum Press, New York

Föll H (2006) Entropie und Information. http://www.techfak.uni-kiel.de/matwis/amat/mw1_ge/kap_5/advanced/t5_3_2.html

Forrester JW (1972) Principles of Systems. Pegasus Communication

Freudenthal AM (1947) Safety of Structures, Transactions ASCE, 112, pp 125–180

Frisch B (1989) Die Umweltwissenschaften als Herausforderung an die Politik. In: G Hohlneicher & E Raschke (Eds), Leben ohne Risiko. Verlag TÜV Rheinland, Köln, pp 267–278

Fuchs S. (2006) Probabilities and uncertainties in natural hazard risk assessment. In: D Proske, M Mehdianpour & L Gucma (Eds), Proceedings of the 4th International Probabilistic Symposium, Berlin, pp 189–204

Funke J (2006) Wie bestimmt man den Grad von Komplexität? Und: Wie sind die Fähigkeiten des Menschen begrenzt, damit umzugehen? http://www.wissenschaft-im-dialog.de

Ganshof FL (1991) Das Hochmittelalter. Propyläen Weltgeschichte, Band 5, Propyläen Verlag, Berlin

Ganter B (2004) Die Strukturen aus Sicht der Mathematik – Strukturmathematik. Wissenschaftliche Zeitschrift der Technischen Universität Dresden, 53, 3–4, pp 39–43

Geissler E (1991) Der Mann aus Milchglas steht draußen vor der Tür – Die humanen Konsequenzen der Gentechnik. Vom richtigen Umgang mit Genen. In: E-P Fischer und W-D Schleuning (Eds), Serie Piper Band 1329, Piper München

GTZ (2002) – German Technical Cooperation. Disaster Risk Management. Working Concept. Eschborn

Haken H & Wunderlin A (1991) Die Selbststrukturierung der Materie – Synergetik in der unbelebten Welt. Vieweg, Braunschweig

Harte R, Krätzig WB & Petryna YS (2007) Robustheit von Tragwerken – ein vergessenes Entwurfsziel? Bautechnik 84, Heft 4, pp 225–234

Heinrich B (1983) Brücken – Vom Balken zum Bogen. Rowohlt Taschenbuch Verlag GmbH, Hamburg

Hildebrand L (2006) Grundlagen und Anwendungen der Computational Intelligence, Informatik I, Universität Dortmund, http://ls1-www.cs.uni-dortmund.de/~hildebra/Vorlesungen

Hof W (1991) Zum Begriff Sicherheit. Beton- und Stahlbetonbau 86, Heft 12, pp 286–289

Husemeyer U (2001) Heuristische Diagnose mit Assoziationsregeln. Fachbereich Mathematik/Informatik der Universität Paderborn, Dissertation

HVerwG (1997) – Hessischer Verwaltungsgerichtshof: Urteil vom 25.3.1997: Az. 14 A 3083/89

ISDR (2004) Living at Risk. A Global Review of Disaster Reduction Initiatives. Geneva

JCSS (2006) – Joint Committee of Structural Safety. www.jcss.ethz.ch

Kastenberg WE (2005) Assessing and Managing the Security of Complex Systems: Shifting the RAMS Paradigm. In: Proceedings of the 29th ESReDA Seminar, JRC-IPSC, Ispra, Italy, October 25–26 2005, pp 111–126

Keefe R & Smith P (Eds) (1997) Vagueness: A Reader. MIT Press, Cambridge, Massachusetts

Kegel G (2006) Der Turm zu Babel – oder vom Ursprung der Sprache(n) http://www.psycholinguistik.uni-muenchen.de/index.html?/publ/sprachursprung.html

Keiler M, Zischg A, Fuchs S, Hama AM & Stötter J (2005) Avalanche related damage potential – changes of persons and mobile values since the mid-twentieth century, case study Galtür, Natural Hazards and Earth System Sciences 5, pp 49–58

Klir GJ & Folger T (1998) Fuzzy Sets, Uncertainty, and Information. Prentice Hall

Klir GJ & Yuan B (1995) Fuzzy sets and fuzzy logic: theory and applications. Prentice Hall International, Upper Saddle River

Klir GJ (1985) Complexity: Some general observations. Systems Research 2(2), pp 131–140

Krebs GA (1998) Disaster as systemic event and social catalyst. In: EL Quarantelli (Ed), What Is a Disaster? Perspectives on the Question: A Dozen Perspectives on the Question. Routledge, London, pp 31–55

Krüger-Brand H (2006) Generation Research Program: Die unbekannte Generation plus. Deutsches Ärzteblatt, 1-2/2006, pp A26–A28

Kruse R, Gebhardt J & Klawonn F (1994) Foundations of Fuzzy Systems. Wiley

Laszlo E (1998) Systemtheorie als Weltanschauung – Eine ganzheitliche Vision für unsere Zukunft. Eugen Diederichs Verlag, München

Laughlin RB (2007) Abschied von der Weltformel, Piper, München

Longo G (2002) Laplace, Turing and the "imitation game" impossible geometry: Randomness, determinism and programs in Turing's test. In cognition, meaning and complexity, Roma, June 2002, CNRS et Dépt. d'Informatique. École Normale Supérieure, Paris et CREA, École Polytechnique, http://www.di.ens.fr/users/longo, 2006

Lübbe H (1989) Akzeptanzprobleme. Unsicherheitserfahrung in der modernen Gesellschaft. G Hohlneicher & E Raschke (Hrsg) Leben ohne Risiko. Verlag TÜV Rheinland, Köln, pp 211–226

Luhmann N (1993) Risiko und Gefahr. Riskante Technologien: Reflexion und Regulation – Einführung in die sozialwissenschaftliche Risikoforschung. Hrsg Wolfgang Krohn und Georg Krücken. 1. Auflage. Frankfurt am Main: Suhrkamp

Luhmann N (1997) Die Moral des Risikos und das Risiko der Moral. In: G Bechmann (Hrsg): Risiko und Gesellschaft. Westdeutscher Verlag, Opladen, pp 327–338

Mathieu-Rosay J (1985) Dictionnaire etymologique marabout (Reliure inconnue), Marabout

Mayer M (1926) Die Sicherheit der Bauwerke und ihre Berechnung nach Grenzkräften anstatt nach zulässigen Spannungen. Springer, Berlin, Verlag von Julius

McCauley JL (2000) Nonintegrability, chaos, and complexity. ArcXiv:cond-mat/0001198

Mechler R (2003) Natural Disaster Risk Management and Financing Disaster Losses in Development Countries. Dissertation. Universität Fridericiana Karlsruhe

Mikulecky DC (2005) The circle that never ends: can complexity made simpler. In: D Bonchev & DH Rouvray (Eds), Springer Science + Business Media, New York, pp 97–153

Möller B, Beer M, Graf W, Hoffmann A & Sickert J-U (2000) Modellierung von Unschärfe im Ingenieurbau. Bauinformatik Journal 3, Heft 11, pp 697–708

Moravec H (1998) Robot: Mere machine to transcendent mind. Oxford University Press

Murzewski J (1974) Sicherheit der Baukonstruktionen. VEB Verlag für Bauwesen, Berlin

Neundorf W (2006) How does the clock knows the time (in German)? http://www.neuendorf.de/zeit.htm

Noack A (2002) Systeme, emergente Eigenschaften, unorganisierte und organisierte Komplexität, 28.11.2002, http://www-sst.informatik.tu-cottbus.de/~db/Teaching/Seminar-Komplexitaet-WS2002/Thema1-Slides_Andreas.pdf

NBDIN (1981) – Normenausschuß Bauwesen im DIN: Grundlagen zur Festlegung von Sicherheitsanforderungen für bauliche Anlagen. Ausgabe 1981, Beuth Verlag

Norton RL (1996) Machine Design: An Integrated Approach. Prentice-Hall, New York

Nussbaum M (1993) Non-Relative Virtues. An Aristotelian Approach. In: MC Nussbaum & A Sen (Eds), The Quality of Life, Clarendon Press, Oxford, pp 242–269

Patzak G (1982) Systemtechnik – Planung komplexer innovativer Systeme: Grundlagen, Methoden, Techniken, Springer, Berlin, Heidelberg, New York

Pawlak Z (2004) Some issues on rough sets. In: JF Peters et al. (Eds), Transactions on Rough Sets I, Springer-Verlag, Berlin, pp 1–58

Perrow C (1992) Normale Katastrophen – Die unvermeidlichen Risiken der Großtechnik, New York

Popper KR (1994) Alles Leben ist Problemlösen. Piper, München

Porfiriev BN (1998) Issues in the definition and delineation of disasters. In: EL Quarantelli (Ed), What Is a Disaster? Perspectives on the Question: A Dozen Perspectives on the Question. Routledge, London, pp 56–72

Porzsolt F (2007) Personal Communication. University of Ulm

Preston R (2003) Das erste Licht – Auf der Suche nach der Unendlichkeit. Knaur, München

Proske D (2004) Katalog der Risiken. Dirk Proske Verlag, Dresden

Proske D (2006) Unbestimmte Welt. Dirk Proske Verlag, Dresden – Wien

Proske D, Lieberwirth P & van Gelder P (2006) Sicherheitsbeurteilung historischer Steinbogenbrücken. Dirk Proske Verlag, Wien Dresden

Pugsley AG (1966) The Safety of Structures. Arnold, New York

Pulm U (2004) Eine systemtheoretische Betrachtung der Produktentwicklung. Dissertation, Lehrstuhl für Produktentwicklung der Technischen Universität München

Quarantelli E. (1998) What Is a Disaster? Perspectives on the Question: A Dozen Perspectives on the Question. Routledge, London

Randić M., Guo X, Plavšić D & Balaban AT (2005) On the complexity of fullerenes and nanotubes. In: D Bonchev & DH Rouvray (Eds), Springer Science + Business Media, New York, pp 1–48

Rayn A (2007) Emergence is coupled to scope, not level. Complexity, Volume 13, Issue 2, pp 67–77

Recchia V (1999) Risk communication and public perception of technological hazards. November 1999, FEEM Working Paper No. 82.99

Renn O (1992) Concepts of risk: A classification. In: S Krimsky & D Golding (Hrsg) Social theories of risk. Praeger, London, pp 53–79

Riedl R (2000) Strukturen der Komplexität – Eine Morphologie des Erkennens und Erklärens. Springer Verlag, Heidelberg

Riegel R (2006) Programm ChaosExplorer 1.0.0., http://www.roland-riegel.de/chaosexplorer

Sandman PM (1987) Risk Communication: Facing Public Outrage. EPA Journal, November 1987, pp 21–22

Schlichting J (1999) Die Strukturen der Unordnung. Essener Unikate 11/1999, pp 8–21

Scholz J (2004) Emergence in cognitive Science: Clark's Four Proposals to the Emergence. PICS – Publication of the Institute of Cognitive Science, University of Osnabrück

Schulz M (2002) Statistische Physik und ökonomische Systeme – Theoretische Econophysics. 3. Juni 2002

Schulz U (Hrsg) (1975) Lebensqualität – Konkrete Vorschläge zu einem abstrakten Begriff. Aspekte Verlag, Frankfurt am Main

SCM (2007) Société de Calcul Mathématique. Robust mathematical modeling, http://perso.orange.fr/scmsa/robust.htm

Seemann FW (1997) Was ist Zeit? Einblicke in eine unverstandene Dimension. Wissenschaft & Technik Verlag, Berlin

Shigley JE & Mischke CR (2001) Mechanical Engineering Design. 6th ed. McGraw Hill. Inc., New York

Smith NJJ (2001) Vagueness. PhD. Thesis, Princeton University

Tichý M (1976) Probleme der Zuverlässigkeit in der Theorie von Tragwerken. Vorträge zum Problemseminar: Zuverlässigkeit tragender Konstruktionen. Weiterbildungszentrum Festkörpermechanik, Konstruktion und rationeller Werkstoffeinsatz. Technische Universität Dresden – Sektion Grundlagen des Maschinenwesens. Heft 3/76

Torgersen H (2000) Wozu Prognose. Beitrag zu den Salzburger Pfingstgesprächen 2000: Ist das Ende absehbar? – Zwischen Prognosen und Prophetie. 6.8.2000

TU Berlin (2006) http://www.physik.tu-berlin.de/~dschm/lect/heislek/html/unbestimmt.html

UAS (2006) University of Applied Sciences Mittweida. Department of Social Work: Economy and Social History, Teaching material, http://www.htwm.de/~sa/

van Lambalgen M (1995) Randomness and infinity. In HW Capel, JS Cramer, O Estevez-Uscanga, CAJ Klaassen & GJ Mellenbergh (eds), Chance and Uncertainty, Amsterdam, Amsterdam University Press, pp 9–27

Visodic JP (1948) Design Stress Factors. Vol. 55, May 1948, ASME International, New York

Von Bertalanffy L (1976) General Systems Theory, Foundations, Developments, Applications, George Braziller, New York

Von Förster H (1993) Wissen und Gewissen. Suhrkamp Verlag, Frankfurt am Main

Wasser H (1994) Sinn Erfahrung Subjektivität – Eine Untersuchung zur Evolution von Semantiken in der Systemtheorie, der Psychoanalyse und dem Szientismus. Königshausen & Neumann, Manuskript 1994

Weaver W (1948) Science and complexity. American Scientist. 36, pp 536–544

Weber WK (1999) Die gewölbte Eisenbahnbrücke mit einer Öffnung. Begriffserklärungen, analytische Fassung der Umrisslinien und ein erweitertes Hybridverfahren zur Berechnung der oberen Schranke ihrer Grenztragfähigkeit, validiert durch einen Großversuch. Dissertation, Lehrstuhl für Massivbau der Technischen Universität München

Weichselgartner J (2001) Naturgefahren als soziale Konstruktion – Eine geographische Beobachtung der gesellschaftlichen Auseinandersetzungen mit Naturgefahren. Dissertation, Rheinischen Friedrich-Wilhelms-Universität Bonn

Weinberg GM (1975) An Introduction to General Systems Thinking, Wiley-Interscience

Weiss M (2001) Data Structures & Problem Solving Using Java. Addison Wesley

Wigner EP (1963) Events, laws of nature, and the invariance principles. Nobel Lecture, December 12, 1963, http://nobelprize.org

Wittgenstein L (1998) Logisch-philosophische Abhandlung. Tractatus logico-philosophicus. Suhrkamp, Original 1922

Wolf F (2000) Berechnung von Information und Komplexität in Zeitreihen –
    Analyse des Wasserhaushaltes von bewaldeten Einzugsgebieten. Dissertation.
    Universität Bayreuth. Fakultät für Biologie, Chemie und Geowissenschaften
Wulf I (2003) Wissen schaffen mit Quantencomputern,
    http://www.heise.de/tp/r4/artikel/15/15398/1.html, 21.08.2003
Zadeh LA (1965) Fuzzy sets. Information and Control 8, pp 338–353

2 Risks and Disasters

2.1 Introduction

Generally data about historical events strongly influence the awareness of risks. It is especially through disasters that the basis for risk estimation has been developed. This chapter classifies disasters based on their primary cause and gives examples of natural, technical, health and social disasters. The terms risk and disaster are interchangeably used here since historical disasters can indicate future risks.

2.2 Classes of Risks and Disasters

The classification of disasters based on their primary cause is an often-used technique, but presents major drawbacks as well. As disasters are considered to cause harm and damage to humans and their properties, there arc always some human or technical preventive actions possible. There fore, even disasters that are commonly accepted as natural disasters include, to a certain extent, some social or technical failures. Figure 2-1 shows that people consider there to be a human component for virtually all types of disasters. What is interesting, however, is the fact that people also consider social disasters, for example insurrection, to be not entirely human-made, but including some natural components as well.

Therefore, the classification of risks into distinct categories, such as ones shown below, is never really possible:

- Natural
- Natech (natural hazards triggering technical disasters)
- Man-made and

- Deliberate acts
  or
- Natural (volcano, earthquake, flood, storm)
- Technical (dam failure, airplane crash, car accident)
- Health (AIDS, heart attack, black death…) and
- Social (suicide, poverty, war, manslaughter)

Already the Natech risks – a technical disaster caused by a natural disaster, show the interaction and overlap between different categories of risk. Furthermore, risks to humans can be classified according to the exposure paths: inhalation risks, ingestion risks, contamination risks, violence risks and temperature risks, including fire. Another way of classifying risks is shown in Fig. 2-2, which classifies disasters and risks not only according to natural and man-made influence but also according to the voluntariness and intensity of the disaster.

Although other classifications of risks are possible, the system distinguishing natural, technical, health and social risks is used for the presentation of disasters within this chapter.

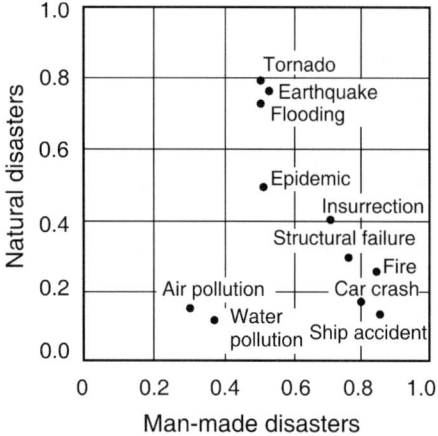

Fig. 2-1. Perception of causes of disasters (Karger 1996)

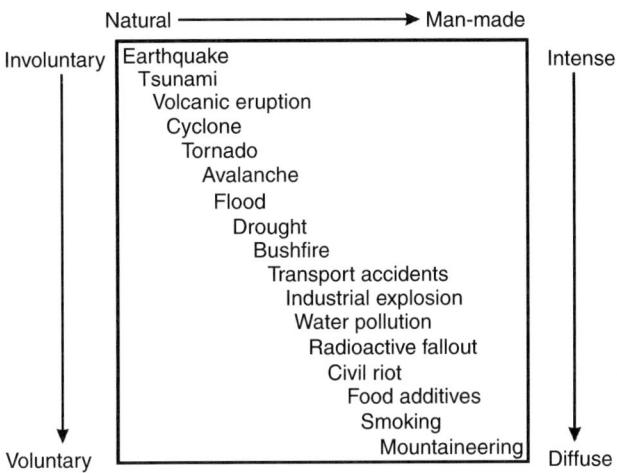

Fig. 2-2. Classification of causes of disasters (Smith 1996)

2.3 Natural Risks

2.3.1 Introduction

Natural hazards are caused by natural processes independent from the existence of the humans. Natural risks cause natural hazards when humans are exposed to them. Such hazards can be snow storms, hail, drought, flooding, hurricanes, cyclones, typhoons, volcanic eruptions, earthquakes, climate change, bites from animals or meteorites (ICLR 2008, IRR 2008). Natural hazards contribute substantially to the threat to humans. Table 2-1 lists the 40 most deadly disasters between 1970 and 2001 and Table 2-2 lists the most costly disasters for the same time period. Obviously natural disasters dominate this list. The development of the frequency of natural risks is shown in Fig. 2-3. The overall cost in the past decade attributed to these natural disasters exceeds 50 billion dollars. Of course, one has to keep in mind that over this time period the population and the number of goods on earth has increased substantially, and finally the quality of recording data has improved as well.

In the 20th century, an estimated 60 million fatalities were caused by natural disasters. At the end of the 20th century, on average about 80,000 fatalities per year were caused by natural hazards (Fig. 2-4), although in 2004 the major Tsunami event killed probably up to 250,000 people.

Table 2-1. The 40 most deadly disasters from 1970 to 2001 (Swiss Re 2002)

Date	Country	Event	Fatalities
14.11.1970	Bangladesh	Storm surge	300,000
28.07.1976	China, Tangshan	Earthquake 8.2 [1]	250,000
29.04.1991	Bangladesh	Tropical cyclone in Gorky	138,000
31.05.1970	Peru	Earthquake 7.7	60,000
21.06.1990	Iran	Earthquake in Gilan	50,000
07.12.1988	Armenia, UDSSR	Earthquake	25,000
16.09.1978	Iran	Earthquake in Tabas	25,000
13.11.1985	Columbia	Volcano eruption Nevado del Ruiz	23,000
04.02.1976	Guatemala	Earthquake 7.4	22,000
17.08.1999	Turkey	Earthquake in Izmit	19,118
26.01.2001	India, Pakistan	Earthquake 7.7 in Gujarat	15,000
29.10.1999	India, Bangladesh	Cyclone 05B hits federal state Orissa	15,000
01.09.1978	India	Flooding after monsoon	15,000
19.09.1985	Mexico	Earthquake 8.1	15,000
11.08.1979	India	Dam failure in Morvi	15,000
31.10.1971	India	Flooding in the Bay of Bengal	10,800
15.12.1999	Venezuela, Columbia	Flooding, landslides	10,000
25.05.1985	Bangladesh	Cyclone in Bay of Bengal	10,000
20.11.1977	India	Cyclone in Andra Pradesh/Bay of Bengal	10,000
30.09.1993	India	Earthquake 6.4 in Maharashtra	9,500
22.10.1998	Honduras, Nicaragua	Hurricane Mitch in Central America	9,000
16.08.1976	Philippines	Earthquake in Mindanao	8,000
17.01.1995	Japan	Great-Hanshin- Earthquake in Kobe	6,425
05.11.1991	Philippines	Typhoon Linda and Uring	6,304
28.12.1974	Pakistan	Earthquake 6.3	5,300
05.03.1987	Ecuador	Earthquake	5,000
23.12.1972	Nicaragua	Earthquake in Managua	5,000
30.06.1976	Indonesia	Earthquake in West-Irian	5,000
10.04.1972	Iran	Earthquake in Fars	5,000
10.10.1980	Algeria	Earthquake in El Asnam	4,500
21.12.1987	Philippines	Ferry Dona Paz collides with tanker	4,375
30.05.1998	Afghanistan	Earthquake in Takhar	4,000
15.02.1972	Iran	Storm and snow in Ardekan	4,000
24.11.1976	Turkey	Earthquake in Van	4,000
02.12.1984	India	Accident in a chemical plant in Bhopal	4,000
01.11.1997	Vietnam and others	Typhoon Linda	3,840
08.09.1992	India, Pakistan	Flooding in Punjab	3,800
01.07.1998	China	Flooding at Jangtse	3,656
21.09.1999	Taiwan	Earthquake in Nantou	3,400
16.04.1978	Réunion	Hurricane	3,200

[1] Richter scale.

Table 2-2. The 40 most costly disasters 1970–2001 (Swiss Re 2002)

Insured damage[1]	Date	Event	Country
20.185	23.08.1992	Hurricane Andrew	USA, Bahamas
19.000	11.09.2001	Assault on WTC and Pentagon	USA
16.720	17.01.1994	Northridge-Earthquake	USA
7.338	27.09.1991	Typhoon Mireille	Japan
6.221	25.01.1990	Winter storm Daria	France, UK
6.164	25.12.1999	Winter storm Lothar	France, CH
5.990	15.09.1989	Hurricane Hugo	Puerto Rico, USA
4.674	15.10.1987	Storm and flooding in Europe	France, UK
4.323	25.02.1990	Winter storm Vivian	West/Central Europe
4.293	22.09.1999	Typhoon Bart	South Japan
3.833	20.09.1998	Hurricane Georges	USA, Caribbean
3.150	05.06.2001	Tropical cyclone Allison	USA
2.994	06.07.1988	Explosion of Piper Alpha	Great Britain
2.872	17.01.1995	Great-Hanshin Earthquake	Japan, Kobe
2.551	27.12.1999	Winter storm Martin	France, Spain
2.508	10.09.1999	Hurricane Floyd, rains	USA, Bahamas
2.440	01.10.1995	Hurricane Opal	USA and others
2.144	10.03.1993	Blizzard, Tornado	USA, Mexico, Canada
2.019	11.09.1992	Hurricane Iniki	USA, North Pacific
1.900	06.04.2001	Hail, flooding and tornados	USA
1.892	23.10.1989	Explosion in a petrochemical plant	USA
1.834	12.09.1979	Hurricane Frederic	USA
1.806	05.09.1996	Hurricane Fran	USA
1.795	18.09.1974	Tropical cyclone Fifi	Honduras
1.743	03.09.1995	Hurricane Luis	Caribbean
1.665	10.09.1988	Hurricane Gilbert	Jamaica and others
1.594	03.12.1999	Winter storm Anatol	West/Northern Europe
1.578	03.05.1999	> 70 tornados in the middle west	USA
1.564	17.12.1983	Blizzard, cold wave	USA, Canada, Mexico
1.560	20.10.1991	Forest fire, city fire, drought	USA
1.546	02.04.1974	Tornados in 14 states	USA
1.475	25.04.1973	Flooding of Mississippi	USA
1.461	15.05.1998	Wind, hail and tornados	USA
1.428	17.10.1989	Loma-Prieta Earthquake	USA
1.413	04.08.1970	Hurricane Celia	USA, Cuba
1.386	19.09.1998	Typhoon Vicki	Japan, Philippines
1.357	21.09.2001	Explosion at a fertilizer plant	France
1.337	05.01.1998	Cold wave and ice disaster	Canada, USA
1.319	05.05.1995	Wind, hail and flooding	USA
1.300	29 10 1991	Hurricane Grace	USA

[1] In million US dollars, 2001.

In addition to their development over time, natural risks also show some geographical correlation. In Fig. 2-5, the distribution of natural hazards for the territory of the US is shown. Dilley et al. (2005) presented the geographical distribution of natural hazards on a worldwide scale.

Since the earliest appearance of civilization, the humans have described natural disasters, such as the big flooding in The Bible. Probably one of the earliest natural risks were astronomical risks.

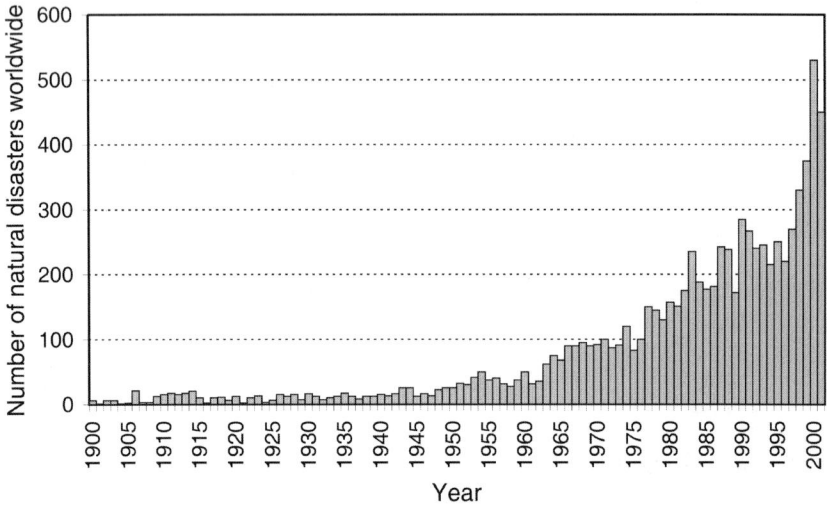

Fig. 2-3. Number of recorded natural disasters since 1900 (EM-DAT 2004)

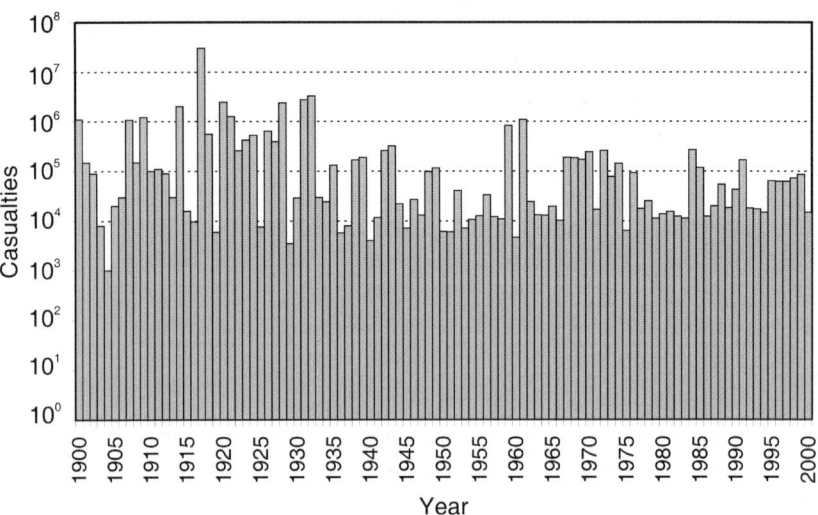

Fig. 2-4. Number of fatalities by disasters worldwide 1900–2000 (Mechler 2003)

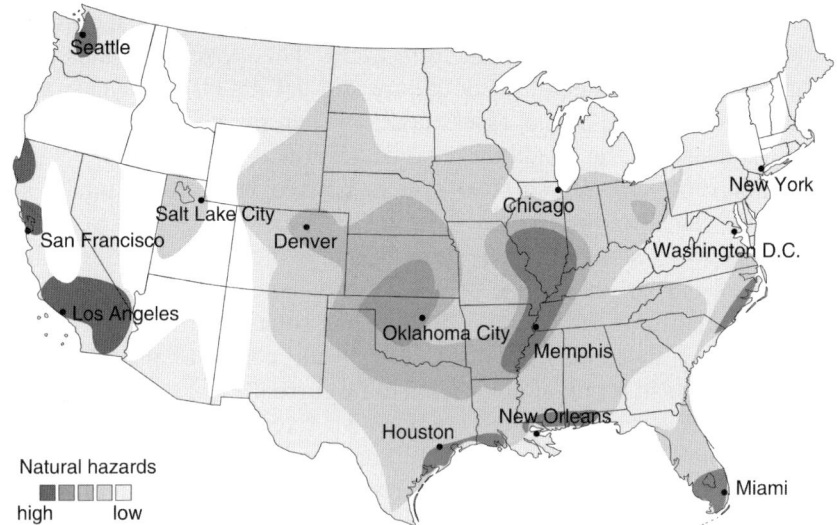

Fig. 2-5. Distribution of natural hazards in the US (Parfit 1998)

2.3.2 Astronomical Risks

2.3.2.1 Solar System

If one looks into the sky on a clear night, the number of stars seems to be innumerable. The stars visible to the naked eye mainly belong to the Milky Way galaxy, which contains up to 200 billion stars (Lanius 1988, NGS 1999). With a little bit of luck, one can see the Andromeda galaxy that is two million light years away. In general, if one observes the astronomical systems, worlds with other time scales become visible. By extending the power of the human eye through sophisticated telescopes, objects at greatest distance from earth, which are essentially the physical borders of our space, may become visible. Such objects reach distances of up to 13 billion light years from the earth. Actually, one has seen only some historical objects, since when dealing with such distances, the speed of light becomes important. Within this distance of 13 billion light years, mankind, so far, has found approximately 100 billion galaxies (NGS 1999). Considering such a huge number of objects, there might, of course, be objects that pose a hazard to humans. The question is, are we sufficiently far enough to be not at risk. If events like supernovas happen in the neighbourhood of the solar system, this might result in the extinction of mankind. Evidence has also been found of huge collisions between galaxies (NASA 2003). It seems, however, that such events have happened at a sufficiently large distance from the earth. As already mentioned, Andromeda is at a distance of about two million light years from earth. The closest galaxies to the earth, the irregular Magellanic cloud, are still a 150,000 light years away. Therefore, in the case of risk assessment, one can mainly focus on our galaxy and the stars in the neighbourhood of our sun.

The adjacent stars of our solar system behave comparably to our sun. The closest star, Proxima centaury, is at a distance of 4.25 light years. The relative speed of stars is rather slow although they move rather quickly around the centre of the Milky Way. The stars have rotated about 20 times around the centre of the galaxy. The average rotation time is 225 million years. Whether there exists a black hole in the centre of the galaxy and what that means in terms of risk seems to be currently unknown.

Mankind, however, can focus on our own fixed star. According to different computations, the sun is 4.5 billion years old (Lanius 1988). This value corresponds very well to the age of stones from the moon or from meteorite stones. It has been estimated that the fuel of the sun, hydrogen, will exhaust in about five billion years (Fig. 2-6). Then, the core of the sun will consist of helium, and a fusion of helium, to carbon will occur. The sun will evolve through fusions from hydrogen to helium, then to carbon, neon, oxygen, silicon and at the end iron. The probable death of the sun will be a supernova, leaving a neutron star. Examples of supernovas can be found in historical documents. Japanese and Chinese sources around the year 1054 describe the sudden flash of a star from Taurus (Lanius 1988). Pictures taken in that region today show a cloud. This fact supports the supernova theory. Such a huge energy release in the neighbourhood of our solar system would probably destroy all types of life in the system.

Coming back to the sun, although the sun will exist five billion more years, in about three billion years, the sun will already be a red giant causing unfavourable conditions. It will be 100 times as big as its current size. The drop in surface temperature from 5,800°C to 4,000°C will probably not help much. Considering such conditions as a disaster, one has to investigate how precise are our models about the development of stars. Fortunately, for such models, not only theoretical considerations but also observations can be used. Very old structures in the universe, especially, can give hints about the possible development of stars. In some globular clusters, different age clusters of stars were observable. In general, however, the rule of thumb is valid – that the stars of a region have the same age. Therefore, the chance of risk of a sudden explosion of stars in our neighbourhood or the explosion of our sun in a few decades or centuries should be extremely low.

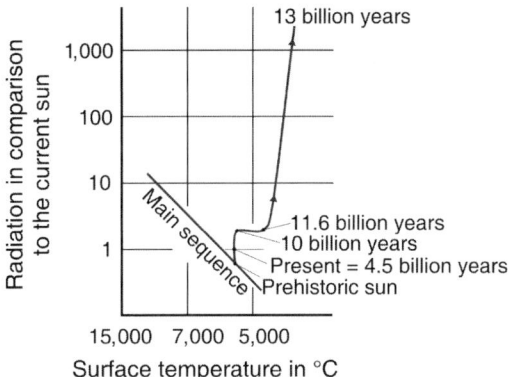

Fig. 2-6. Development of the sun in a Hertzsprung–Russel diagram (Lanius 1988)

Probably the earth will change extensively over the lifetime of the sun, because the earth is not a static, solidified mass. This becomes clear during events such as volcanic eruptions or earthquakes. The earth itself belongs to a group of planets. A simple tabalation of the distance of the planets from the sun is the Titus Bode rule (Table 2-3). Here, the distance is given as a factor of the Astronomical unit. The Astronomical unit represents eight light minutes or about 150 million kilometers (Gellert et al. 1983). The observed distances show rather high values, and therefore collisions or other negative impacts from these planets seem currently neglectable, although Pluto crosses the orbit of Neptune. Here, the latest developments, for example excluding Pluto from the group of planets, are not considered.

Table 2-3. Distance of planets from the sun according to the Titius–Bode rule (Gellert et al. 1983)

Planet	N	a_n	a
Mercury	$-\infty$	0.4	0.39
Venus	0	0.7	0.72
Earth	1	1.0	1.00
Mars	2	1.6	1.52
Small planets	3	2.8	2.78
Jupiter	4	5.2	5.20
Saturn	5	10.0	9.55
Uranus	6	19.6	19.20
Neptune	–	–	30.09
Pluto	7	38.8	39.5

Titius–Bode rule: $a_n = 0.4 + 0.3 \times 2^n$; where a_n is planet-sun distance in Astronomical units and a is observed planet-sun distance.

However, the great distances in our solar system might not be sufficient to provide safety. According to the Milankovich theory, astronomic conditions influence the climate on earth. The main engine for the climate on earth is solar radiation. This depends on at least some rotation properties of the system earth-sun (Williams et al. 1998, Allen 1997):

- Eccentricity is the difference between the earth's orbit around the sun and a circle. It changes in a cycle with a period of 100,000 years.
- The earth is lopsided. It changes with a period of 41,000 years.
- Gyration has a period of 19,000–23,000 years to complete a cycle.

These might have an effect on humans and might impose risks. Space weather, also called space hurricane season, has only recently been recognied as a risk. Especially since the occurrence of a power failure in 1989 in Canada and one historical event earlier in 1859, it was thought that there might be an interaction between the solar wind and the earth's magnetic field. Such interactions might endanger astronauts during missions as well as affect electric and electronic equipment on earth (Plate 2003, Romeike 2005). Currently, the NASA installs satellites to improve prediction of such events, and there are some Aurora forecast reports available (Geophysical Institute 2007). Additionally, man's activities on the planet might lead to consequences of astronomical risks, like the damage of the ozone layer.

2.3.2.2 Meteorites

The great distances between the planets might create an illusion that direct hits between astronomical bodies virtually do not exist in the solar system.

Observing, however, the common topography of most moons, planets or asteroids with solid surfaces proves this theory wrong: short-time and high-energy impacts are the main and, in many cases, the only geological surface modelling processes known so fast (Koeberl 2007).

An example of a large impact structure is the Mare Orientale on the moon with a diameter of 930 km (Langenhorst 2002). Figure 2-7 shows the number of craters on the moon related to the age of moon. In the year 1999, an impact on the moon was observed. On Jupiter, the impacts of pieces from the comet Shoemaker-Levy-9 attracted worldwide attention. The planet Mercury must also have experienced a major impact in its history, as not only the impact side of the planet changed but also the reverse side of it changed shape.

The earth is also exposed to such bombardment from space. Currently, there are about 170 known impact structures worldwide (Fig. 2-8). The size of such craters varies from 300 km to less than 100 m (Koeberl 2007). Historically, there have been much bigger impact events, creating, for example, the moon (Langenhorst 2002). On average, the earth is hit by

20,000 meteors per year. Some sources estimate even higher values, but the estimation remains difficult. The overall mass of meteors has been estimated as 4×10^5 per year (Langenhorst 2002). The speed of meteors ranges from 10 to 70 km/s (Gellert et al. 1983). Twenty percent of meteors are particles from comets. Some meteor showers occur regularly at certain times of the year; examples are listed in Table 2-4.

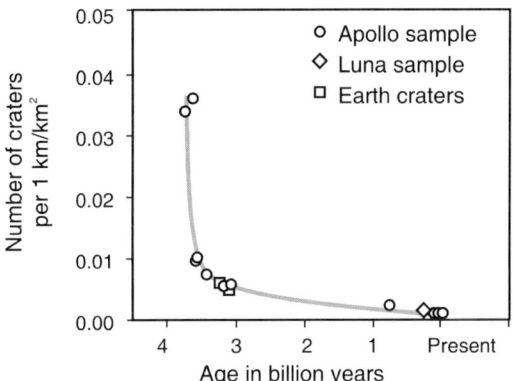

Fig. 2-7. Number of craters on the moon related to the age of the moon (Langenhorst 2002)

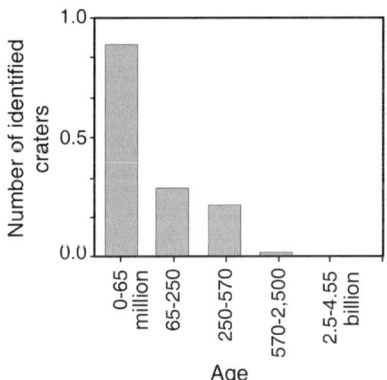

Fig. 2-8. Number of identified craters per million years (Langenhorst 2002)

The October-Draconids is a good example of a regular meteor shower. The shower dates back to the Giacobini–Zinner comet, which appeared for the first time on 9 October 1926. The following years of 1933 and 1946 were extremely rich in meteor showers. On 9 October 1933, an average of 14,000 falling stars were detected. Falling stars are meteorids that break up at nearly 70–120 km before reaching earth. The contribution of falling stars to the meteorite volume that reaches earth is only about 0.05%.

Ninty-nine percent of the meteorite material reaching the earth comes from micrometeorites. That explains why the estimation of the overall volume of meteorites hitting the earth is so difficult. On the other hand, however, the 170 known impact structures prove that bigger impacts are possible. Tables 2-5 and 2-6 give an overview about historical meteorite impacts. The listing of only major impacts from pre-historical times is owed to the fact that other surface-forming processes are active on earth, such as wind, water, ice and so on. Additionally, small craters are much more difficult to identify than bigger ones. For example, the Barringer crater in the US with a diameter of 1 km is much easier to identify than a crater of over a 100 m diameter in a forest.

Although Table 2-6 finishes with the year 1969, there were many meteorite events afterwards. For example, in 2000, a meteorite impact occurred in Canada/Alaska (Deiters 2001). In April 2002, a meteorite was found in Bavaria, Germany. The so-called Neuschwanstein meteorite had a mass of 1.75 kg, but it probably belonged to another body of mass more than 600 kg (Deiters 2002). One recent example is the Pennsylvania Bolide. This meteor was observed on 31 July 2001 at around 6 p.m. The meteor was visible in daylight from Canada to the state of Virginia; houses were shaken by the explosion.

One speaks of meteorites if the body hits the earth surface; otherwise, one speaks of meteors. Both can yield major destruction since the explosion of meteors might cause damages as severe like a direct hit.

In general, the human experience with heavy meteorite impacts is rather low since the beginning of human civilization. It seems to be a rare event with a low probability of occurrence, but the consequences could be disastrous. Table 2-7 shows the relationship between the return periods of impacts versus the damage potential in terms of explosion power. For example, the explosion power during an impact of a body with a diameter of 10–20 km, such as the Chicxulub Asteroid in Mexico, is about 10^8–10^9 megatons TNT (NASA 1999). For comparison purposes: the current explosion power of all atom bombs is about $10^{5.5}$ megatons TNT. It has been estimated that the so-called nuclear winter needs approximately 10^4 megatons TNT, and the Hiroshima bomb had about 10^2 megatons TNT (NASA 1999).

An explosion power of 10^8 megatons TNT would probably leave the entire civilization and biosphere heavily shattered. A hit by a body with a diameter of several kilometers would presumably kill all life on that side of the planet. The explosion wave of temperature over 500°C and a speed of 2,000–2,500 km/h would move around the world (NN 2003). If the ocean would be hit, there would be Tsunamis with a height of several kilometers. Additionally, the impact would yield to earthquakes and volcanic eruptions. Since the impact would bring huge masses of dust into the atmosphere, the temperature would drop rapidly. This is observed to a lesser degree for some volcanic eruptions. In general, under such situations, there is a possibility that mankind would become extinct. Quite some studies have investigated such conditions mainly based on the assumption that such conditions could exist as an aftermath of a World War (Chapmann et al. 2001).

Table 2-4. Yearly returning meteore showers (Gellert et al. 1983)

Season	Name	Origin	Classification
3 January	Bootiden (Quadrattiden)	Unknown	Rich flow
12 March–5 April	Hydraiden	Part Virginiden	Weak flow
1 March–10 May	Virginiden	Ecliptically	Strong flow
12–24 April	Lyriden	Comet 1861 I	Moderate flow
29 April–21 May	Mai-Aquariden	Comet Halley	Rich flow
20 April–30 July	Scorpius-Sagittariiden	Ecliptically	Weak flow
25 July–10 August	Juli-Aquariden	Ecliptically	Intensive flow
20 July–19 August	Perseiden	Comet 1862 III	Strongest flow
11-30 October	Orioniden	Comet Halley	Intensive flow
24 September–10 December	Tauriden	Ecliptically	Moderate flow
10–20 November	Leoniden	Comet 1866 I	Moderate flow
5–19 December	Geminiden	Elliptical	Intensive flow

Table 2-5. Examples of meteorite impacts according to Gellert et al. (1983), Munich Re (2000), Koeberl & Virgil (2007), Pilkington & Grieve (1992), Langenhorst (2002) and Woronzow-Weljaminow (1978)

Name and country	Age (years)	Diameter (km)
Vredefort structure, South Africa	2,023 million	300
Sudbury structure, Canada	1,850 million	250
Clearwater lakes, Canada	290 million	32/22
Manicouagan, Canada	212 million	100
Aorounga, Chad	200 million	17
Gosses Bluff, Australia	142 million	22
Deep Bay, Canada	100 million	13
Chicxulub, Mexico	65 million	170
Mistastin Lake, Canada	38 million	28
Nördlinger Ries crater, Germany	15 million	
Steinheimer Becken, Germany	15 million	
Kara-Kul, Tajikistan	10 million	45
Roter Kamm, Namibia	5 million	2.5
Bosumtwi, Ghana	1.3 million	10.5
Wolfe Creek, Australia	300,000	0.87
Barringer Crater, Arizona, USA	50,000	1.2
Chubkrater, Canada		3.2

Table 2-6. Examples of meteorite impacts according to Gellert et al. (1983), Munich Re (2000), Koeberl & Virgil (2007), Pilkington & Grieve (1992), Langenhorst (2002) and Woronzow-Weljaminow (1978)

Location and country	Date	Comments
Naples, Roman Empire	79 A.D.	
Ensisheim, Germany	1492	
Cape York, Greenland	1895	33 tonnes iron meteorite
Kanyahiny, CSSR	09.06.1866	
Pultusk, Poland	30.01.1868	Rain of about 100,000 stones
Long Island, Kansas, USA	1891	564 kg
Tunguska Meteorite, Russia	30.06.1908	About 1,000 km hearable, 7 million tonnes heavy, 1,200–1,600 km^2 forest area destroyed
Hoba, South-West Africa	1920	60 tonnes iron meteorite
Sikhote-Alin Meteorite, USSR	12.02.1947	200 craters, biggest 27 m in diameter, 70 tonnes overall material
Furnas Co, Nebraska, USA	18.02.1948	About 1,000 stone meteorites, one of them 1,074 kg
Allende Meteorite, Mexico	08.02.1969	Overall 4 tonne of particles

Table 2-7. Return period of meteorite impacts (NASA 1999)

Size	Return period	Explosion power	Example
10 km	50–100,000,000 years	10^8 megaton TNT	Chicxulub, Mexico
1 km	1,000,000 years	10^5 megaton TNT	Mistastin Lake, Canada
100 m	10,000 years	10^2 megaton TNT	Barringer Krater, USA
10 m	1,000 years	10^{-1} megaton TNT	Tunguska Meteorite, Russia
1 m	1 year	10^{-2} megaton TNT	
1 mm	30 s	10^{-10} megaton TNT	

However, it is not that only natural elements are causing such hazards. About 11,000 large man-made objects in space are currently observed. It has been estimated that 100,000 smaller objects exist and the number of objects of size 0.1–1 cm is over one million (McDougall & Riedl 2005, NRC–Committee on Space Debris 1995). Of course, these objects are rather a risk to space ships but might also re-entry earth's atmosphere.

2.3.2.3 Mass Extinctions

Meteorite impacts might be one possible cause for the so-called mass extinctions during prehistoric times. An example of this was the Cretaceous–Tertiary boundary extinction event, probably caused by the impact of the Chicxulub Asteroid in the area of Mexico. As a consequence of this impact, about 17% of all biological families, including the dinosaurs, became extinct. Whether the asteroid was the one and only cause of extinction is still debated.

The idea of a major asteroid impact is found in some work by Alvarez in the 1970s. Alvarez investigated the quantity of Iridium in the maritime mud. In general, the results of the investigation, which used some material from Italy (Gubbio), showed the expected value of 0.3 ppb (parts per billion) Iridium. Some soil layers, however, showed dramatically higher values of Iridium in the order of 10 ppb. Afterwards, a material from Denmark (Stevn's Klint) was tested. Here, the Iridium values were even higher, reaching up to 65 ppb. This Iridium peak was found worldwide in layers representing the time of the Cretaceous–Tertiary boundary, which is now sometimes called the Cretaceous–Paleogene boundary. Iridium itself is a noble earth element. It mainly comes from meteorites or partly from volcanic eruptions. This yields to the hypothesis that, probably during the Cretaceous–Tertiary boundary, a huge meteorite impacted the earth. Based on this consideration, the impact body should have had a size of about 10 km and hit the earth approximately 65 million years ago (Alvarez et al. 1980).

The next step in the validation of this theory would be the identification of a crater. Unfortunately, terrestrial processes like sedimentation, erosion or plate tectonics might cover or erase the typical surface expression of an impact event. An important witness of an impact, however, which is not so easily altered over time, is the affected rocks. Craters are filled with melted, shocked or brecciated rocks. Some parts of the rocks have been transported during the impact from the crater to the surroundings. This transported material is called ejecta. Most parts are found in the surroundings within a distance of less than five times the diameter of the crater (Montanari & Koeberl 2000, Melosh 1989). Therefore, it is thought that the increase in Iridium was due to abundant impact debris. The identification of the local crater was done using microtektites. These are small tektites, not more than 1 mm in diameter, generated during high pressures. Such high pressures might be shock waves caused by a meteorite impact.

Summarizing the results, it shows that there was a possibility of a major meteorite impact 65 million years ago. Other explanations are also possible. Some explain the increase in Iridium by heavy volcanic eruptions. About 65 million years ago, there was an extremely active volcanic region at the Deccan Traps. The amount of lava released could have covered half the area of modern India. Due to the possible huge releases of gases, there was an average temperature increase of up to 8°C around half million years before the meteorite impact. Iridium can also be found abundant, as mentioned before, in volcanic material (NGS 1998a,b).

Paleontologists also refute the fact that the extinction of the entire classes of animals was based solely on one disastrous event. There seems to be some current discussion, however, on whether some species survived, for example some dinosaurs. There are indications that, for example, some hadrosaurids might have survived the event for some time. But, under such conditions, it might be that the animal class cannot recover (Jablonski 2002). Additionally, it is not clear if the fossils might have been moved to later sediment. Although paleontologists do not like the idea of sudden extinction events, such phenomena have occurred several times in the history of the earth as shown in Table 2-8.

Table 2-8. Mass extinctions in the history of the earth (Morell 1999)

Age	Before (million years)	Percentage of extinct biologic families
Ordovician	440	25%
Devon	370	19%
Perm	250	54% [1]
Trias	210	23%
Cretaceous	65	17% (C/T boundary extinction event)
Average	88	Probability of 1.14×10^{-8} per year

[1] During Perm, about 90% of all marine families became extinct (Hoffmann 2000).

Figure 2-9 is a graphical visualisation of the diversity of marine families that have existed throughout time. The first interesting fact is the diversity of marine life at present. There have never been so many different families at one time. The data, however, have not yet considered the influence of humans, which is, of course, decreasing the diversity of marine life. The second interesting fact is the repeated occurrence of mass extinction events. Probably the most significant event was the Perm extinction. Some references state that, around 100,000 years back, 50% of all biological families and 90% of all marine families became extinct (Morell 1999, Newson 2001). Other references give lower percentages (NGS 1998a,b). The oldest event was the Ordovician extinction. During this event, about 25% of all biologic families disappeared (Morell 1999). Other references, however, mention that up to 75% of all animal families became extinct (NGS 1998a,b). During the Devonian event, the number of extinctions was about the same.

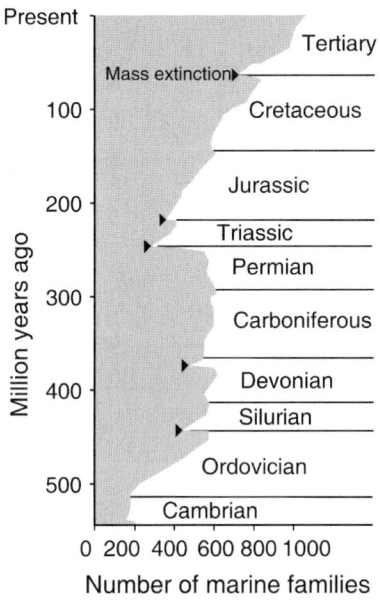

Fig. 2-9. Number of marine families over time (NGS 1998a,b)

The current number of species has been estimated at about 1.5 million. However, since the human population has shown a rapid growth over time, many species have became endangered. In 1980, the overall biomass was estimated to be about $2,000 \times 10^6$ tonnes. Assuming that the human biomass was about 2.5% of the overall biomass and farm stocks was approximately 325×10^6 tonnes, then about $1,625 \times 10^6$ tonnes were left for the 1.5 million species. Furthermore, it should be assumed that the amount of biomass remains constant, and mankind and farm stock constantly grow between 1.7 and 2.1%; then in the year 2020, only about $1,370 \times 10^6$ tonnes will be left for all other species. One can easily imagine that this will yield to a decrease in the number of species. Furthermore, since the decrease is at very fast pace, it could be considered a mass extinction (Füller 1980).

Many examples can be given for the extinction of species. For example, the wild pigeon, Extopistes Migratorius, was very common at the end of the 19th century in the eastern states of the US. According to different estimations, the overall number of this bird species must have been several billions. However, due to the good taste and the good preservability of the meat, the birds were hunted since about 1870. In 1879, in Michigan alone, more than one billion birds were captured. The last brooding nest was seen in 1894, the last wild pigeon was seen in 1907, and in 1914 the last pigeon captured died. Nowadays, only a few primed specimens remain, for example in Jena, Germany (Füller 1980).

A further example was the Quagga. This was a type of zebra living in southern Africa. Because the Boers thought it was a competitor for their own farm stocks, Quaggas were killed. In 1870, only three Quaggas were left in zoos, and in 1883, the last one died (Bürger et al. 1980). Further examples of exctinct species are listed in NGS (1998a,b) and Sedlag (1978).

2.3.3 Geological Risks

While the causes of disasters considered so far were either extra-terrestrial or unknown, the following disasters are connected to the composition of the earth. These disasters show that the earth is not a static, unchangeable planet but rather something that develops over time. Such developments might release huge amounts of energy or materials and, therefore, heavily influence life on the planet (Coch 1995, Zebrowski 1997).

2.3.3.1 Volcano

The earth is built up on different chemical layers. The layers include the upper silicate solid crust, a viscous mantle, a liquid outer core and a solid inner core. Even though most rocks are only around several hundred million years old, some rocks have been found to be more than 4.4 billion years old, indicating the age of the crust.

The thickness of the crust lies between 5 and 70 km deep, depending on whether it is an oceanic or continental crust. The crust permanently moves, mainly driven by the circulation of viscous and liquid materials in the layers beneath. If the crust opens or ruptures, the fluid, hot rock and also gases and ash might escape, creating volcanism. The maximum depth of the volcanic material is the border of the outer core and the lower mantle, which is between 2,500 and 3,000 km deep. Volcanism is a geologic, earth surface-forming process. The number of volcanos is estimated at several thousands; however, only a few hundreds are active (Fig. 2-10 and Table 2-9). In general, the number of small eruptions is about 50 per year and the number of major events is about one in 10 years. (DKKV 2002)

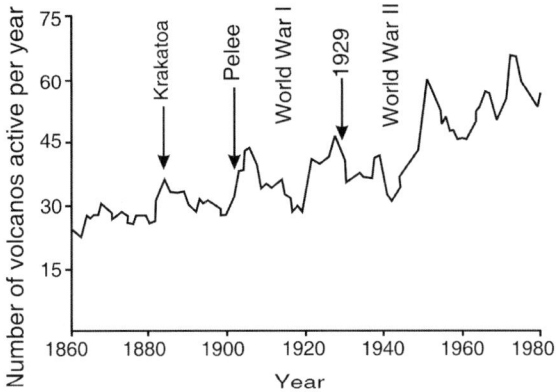

Fig. 2-10. Development of the number of active volcanos per year. Notice the drop in active volcanos during the World Wars (Smith 1996)

Table 2-9. Strongest volcanic eruptions in the last 250 years according to Graf (2001)

Volcano	Year	Explosivity	Opacity	SO_2 (Mt)
Laki, Island	1783	4	2,300	100[1]
Tambora, Indonesia	1815	7	3,000	130[1]
Cosiguina, Nicaragua	1835	5	4,000	
Askja, Island	1875	5	1,000	
Krakatoa, Indonesia	1883	6	1,000	32[1]
Tarawera, New Zealand	1886	5	800	
Santa Maria, Guatemala	1902	6	600	13[1]
Ksudach, Kamtschatka	1907	5	500	
Katmai, Alaska	1912	6	500	12[1]
Agung, Indonesia	1963	4	800	5 ± 13[1]
Mount St. Helens, USA	1980	5	500	1
El Chichón, Mexico	1982	5	800	7
Pinatubo, Philippines	1991	6	1,000	16 ± 20

[1] Estimated.
Mt = mega tonnes

   The intensity of volcanic features can be described by different measures. The volcanic explosivity index (VEI) has been introduced by Newhall & Self (1982). It is a measure comparable to the Richter scale, which is used to measure the intensity of earthquakes. The VEI is based on the volcanic volume released and the height of the smoke pillar (Table 2-10). Critic against the VEI is, for example, mentioned in Smolka (2007). Another volcanic intensity measure is the Tsuya-index. Additionally, for many volcanic eruptions, the opacity is given, which is a measure that describes the influence of the volcanic eruption on global climate.

Table 2-10. Volcanic explosivity index (Newhall & Self 1982, USGS 2008)

Category	Volume erected in m^3	Column height in km	Examples
0	$<10^4$	<0.1	Nyiragongo, Tanzania (1977)
1	$10^4–10^6$	$0.1–1$	Unzen, Japan (1991)
2	$10^6–10^7$	$1–5$	Nevado del Ruiz, Colombia (1985)
3	$10^7–10^6$	$3–15$	El Chicon, Mexico (1982)
4	$10^8–10^9$	$10–25$	Mount St. Helens, US (1980)
5	$10^9–10^{10}$	>25	Krakatoa, Indonesia (1883)
6	$10^{10}–10^{11}$	>25	Tambora, Indonesia (1815)
7	$10^{11}–10^{12}$	>25	
8	$>10^{12}$	>25	Taupo, New Zealand

In the last 100,000 years, there were probably two volcanic eruptions with a VEI of 8: the Taupo eruption in New Zealand approximately 25,000 years ago with 1,200 km^3 of released material, and the Toba eruption in Sumatra, Indonesia, about 75,000 years ago with about 2,800 km^3 of released material. Additionally, in Yellowstone, there were probably volcanic eruptions with a VEI of 8 (Newhall & Self 1982). For comparative purposes, the released material of Mount St. Helens explosion was of the order of 0.6 km^3. Statistics about extreme volcanic eruptions are given in Mason et al. (2004).

However, even volcanic eruptions of lower intensity can have severe impacts on humans (Sparks & Self 2005). For example, the eruption of Mont Pelé in 1902 on the French Caribbean island Martinique claimed about 30,000 lives (Newson 2001). Only two citizens of the town St. Pierre survived the eruption, which included one prisoner. Lacroix investigated the Mont Pelè eruption and published a famous work about volcanic explosions. However, already in the 18th century, researchers like Hutton started to develop theories about the changing and evolving earth (Daniels 1982). In 1815, the Tambora eruption in Sumbawa, Indonesia, claimed about 10,000 lives. Afterwards, an additional 82,000 fatalities occurred in the region due to famine and disease. Furthermore, the weather worldwide was affected, and in the summer of 1816 the coldest temperatures ever were recorded. In combination with the Napoleon wars, the harvest loss was a heavy burden on the European population. In the New England states in North America, freezing temperatures were observed in summer this year. The Krakatoa eruption yielded to an average worldwide temperature drop of about 0.3°C. On average, volcanic eruptions can affect climate by the same magnitude as anthropogenic effects.

For some volcanoes, there exists quite a strong historical record with eruptions occurring sporadically over the last 2,000 years. For example, the "Vesuvius" volcano in Italy erupted in the years 79 (perish of Pompeii), 203, 472, 512, 685, 787, five times between 968 and 1037, 1631 (about 4,000 fatalities), nine times between 1766 and 1794, 1872, 1906 and 1944. Malladra said that "the Vesuvious is sleeping, but his heart beats" (Daniels 1982).

There are different types of volcanic eruptions: effusive eruption; steam explosion, submarine eruption, Stromboly-like eruption, Plinian eruption, glow eruption, paraoxysmale explosion, flooding of basalt and ash stream (Daniels 1982). Additionally, models of the eruption potential of magma are related to the concentration of silicic acid and water within the magma. Depending on the composition of these substances, the magma is either more fluid or more viscous. In combination with high steam pressure, viscous magma can enable heavy eruptions.

2.3.3.2 Earthquakes

Based on the theory of plate tectonics (Wegener 1915), the small-width asthenosphere layer moves on the approximately 100-km-thick lithosphere layer. The lithosphere – a layer of solid earth is broken into several plates (Fig. 2-11). The internal thermal engine of the earth drives the motion of the plates (Sornette 2006). The borders of the plates are not sharp boundaries, but instead they are complex systems, sometimes reaching several hundred kilometers long. Therefore, the estimated number of plates has changed over the last few decades, since microplates have been found in boundary areas. While previously only 14 plates were known, newer works suggest that 42–52 plates exist. The plate size follows a "power law". The limitation of the earth surfaces limits the size of the plates. Furthermore the size of the convection cells in the upper mantle also influences the size of the plates. (Sornette 2006).

Not only geometrical reasons, e.g. the size of the plates, but also energy consumption limit the seismic moments. Therefore, the well-known Gutenberg-Richter law (Gutenberg & Richter 1954) for earthquake seismic moment releases was modified for large earthquakes (Hosser et al. 1991). It is thus concluded that the maximum earthquake energy realised can be found, if external events like meteorite, impact are not considered. The two strongest earthquakes in the 20th century (Chile 22 May 1960, 1.1×10^{26} erg energy released, and Alaska, 28 March 1964, 4×10^{25} erg energy released) nearly reached that maximum value. It is quite interesting to note that the average yearly seismic energy of all other earthquakes is about 3×10^{24} erg. The recurrence time of a typical maximum earthquake is less than 100 years (Sornette 2006). Table 2-11 lists the magnitudes of some earthquakes in the previous century. The magnitude is related to the energy released, and is described on the Richter scale as the logarithm of the maximum amplitude of 0.05 Hz measured with a standard Wood-Anderson seismograph on a solid ground 100 km away from the epicenter. Scales for the effects of earthquakes on buildings are the MSK-Scale (Medvedev-Sponheuer-Karnik), the Mercalli-Scale and the EMS-Scale.

Table 2-11. The largest earthquakes from 1900 to 2004 (Munich Re 2004a)

Date	Magnitude	Region
22.5.1960	9.5	Chile
28.3.1964	9.2	USA, Alaska
9.3.1957	9.1	USA, Aleutian
4.11.1952	9.0	Russia, Kamchatka
26.12.2004	9.0	Indonesia
31.1.1906	8.8	Ecuador
4.2.1965	8.7	USA, Aleutian
15.8.1950	8.6	India, Assam
3.2.1923	8.5	Russia, Kamchatka
1.2.1938	8.5	Indonesia, Banda Sea
13.10.1963	8.5	Russia, Kuril Islands

Table 2-12. The 15 deadliest earthquakes from 1900 to 2004 (Munich Re 2004a)

Date	Event	Mag.	Region	Fatalities	Economic losses in million US-dollars
27/28.7.1976	Earthquake	7.8	China, Tangshan	242,800	5,600
16.12.1920	Earthquake, landslide	8.5	China, Gansu	235,000	25
1.9.1923	Earthquake	7.8	Japan, Tokyo	142,800	2,800
8.10.2005	Earthquake		Pakistan, India	88,000	5,200
28.12.1908	Earthquake	7.2	Italy, Messina	85,900	116
25.12.1932	Earthquake	7.6	China, Kansu	77,000	
31.5.1970	Earthquake, landslide	7.9	Peru, Chimbote	67,000	550
30.5.1935	Earthquake	7.5	Pakistan, Quetta	50,000	25
20/21.6.1990	Earthquake	7.4	Iran, Gilan	40,000	7,100
23.5.1927	Earthquake	8.0	China, Gansu	40,000	25
26.12.1939	Earthquake	7.9	Turkey, Erzincan	32,900	20
13.1.1915	Earthquake	7.5	Italy, Avezzano	32,600	25
25.1.1939	Earthquake	8.3	Chile, Concepción	28,000	100
26.12.2003	Earthquake	6.6	Iran, Bam	26,200	500
7.12.1988	Earthquake	6.7	Armenia, Spitak	25,000	14,000

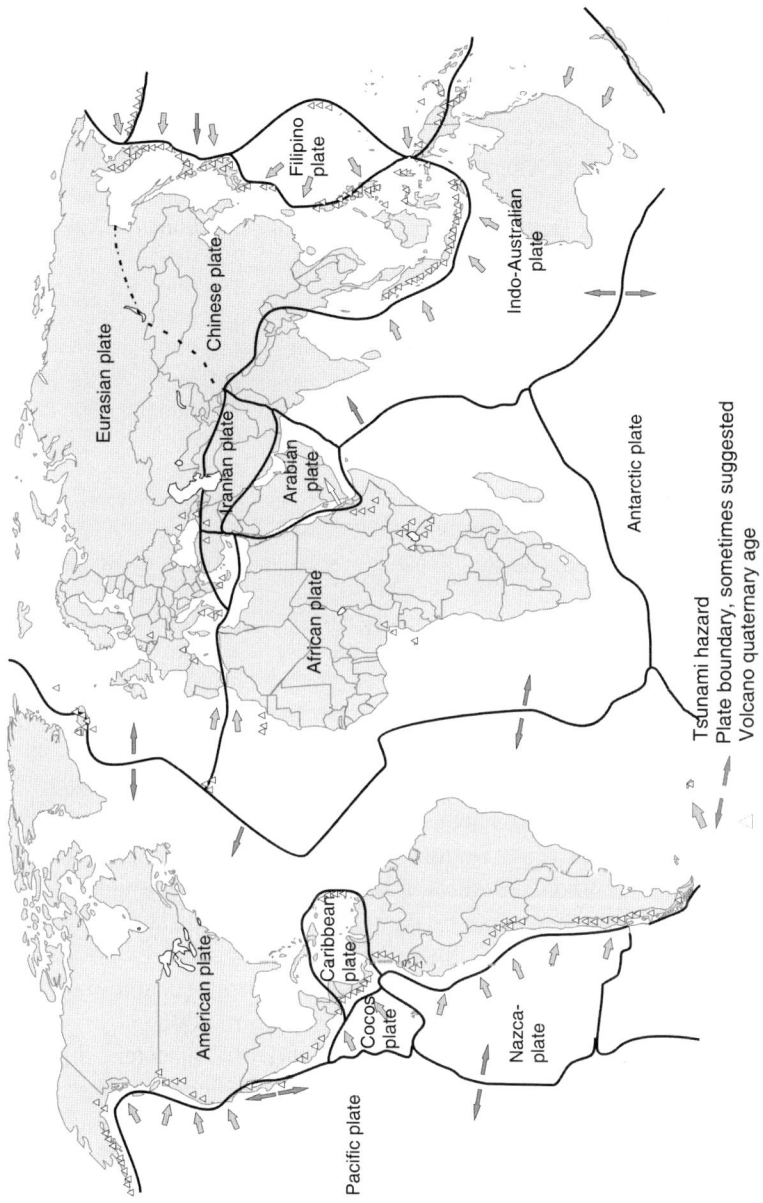

Fig. 2-11. World map of tectonic plates and volcano and earthquake prone areas according to Diercke (2002)

Table 2-13. The 20 deadliest earthquakes of all times (USGS 2001)

Year	Country, region	Fatalities
23.01.1556	China, Shansi	830,000
27.07.1976	China, Tangshan	255,000
09.08.1138	Syrien, Aleppo	230,000
22.05.1927	China, Xining	200,000
22.12.856	Iran, Damghan	200,000
16.12.1920	China, Gansu	200,000
23.03.893	Iran, Ardabil	150,000
01.09.1923	Japan, Kwanto	143,000
05.10.1948	USSR, Turkmenistan, Ashgabat	110,000
September 1290	China, Chihli	100,000
28.12.1908	Italia, Messina	70,000–100,000
8.10.2005	Pakistan, India	88,000
Nov. 1667	Kaukasus, Shemakha	80,000
18.11.1727	Iran, Tabriz	77,000
01.11.1755	Portugal, Lisabon	70,000
25.12.1932	China, Gansu	70,000
31.05.1970	Peru	66,000
1268	Italia, Asia Minor, Sicily	60,000
11.01.1693	Italia, Sicily	60,000
30.05.1935	Pakistan, Quetta	30,000–60,000

Based on our understanding of the mechanism of earthquakes, it becomes clear that earthquakes are connected to some regions: the borders of the tectonic plates. In these regions, about 3 billion humans live (DKKV 2002). Tables 2-12 and 2-13 show a list of deadliest earthquakes. In some areas, earthquake activities can be dated back over thousands of years. The city of Antioch in Turkey was hit by severe earthquakes in the years 115, 458, 526, 588, 1097, 1169 and 1872. It was estimated assumed that the earthquake of 458 resulted in 300,000 victims (Gore 2000).

However, not all earthquakes yield such dreadful disasters. In fact, the current developments in engineering have significantly decreased the number of fatalities. Research has been focused on strengthening the structures for the load case earthquake in California for at least 50 years. Many risk analysis studies on earthquakes have been conducted, for example Lomnitz & Rosenblueth (1976), PELEM (1989), Grünthal et al. (1998), FEMA-NIBS (1999) and Tyagunov et al. (2006). In the previous decades in the US, severe earthquakes have caused relatively as low a number of 130 fatalities. To achieve this result, about 25 billion dollars were spent for protection actions (Parfit 1998). The progress becomes even more visible when the damage consequences of two earthquakes are compared as shown in Table 2-14. The topography of the hit regions in Armenia as well as California is comparable: the ratio of lowlands to mountains, the distribution of the population, or the ratio of new civil structures to the overall number of structures (Bachmann 1997). However, the consequences of earthquakes show great differences.

Since the damage to property for Armenia was unknown, two further examples of damage should be given for developed countries. In 1994, the Northridge earthquake in California caused an estimated damage of 44 billion US dollars. Even more damaging was the Kobe earthquake in 1995, with about 6,300 fatalities and an economic loss of 100 billion US dollars (Mechler 2003).

Table 2-14. Casualty and damages from two earthquakes (Bachmann 1997, Newson 2001)

	Spitak earthquake	Loma-Prieta earthquake
Date	7 December 1988	17 October 1989
Magnitude	6.9	7.1
Region	Armenia	Northern California
Fatalities	>25,000	67
Injured	31,000	2,435
Homeless	514,000	7,362
Damage of property	Unknown	~7,8 billion US dollars

2.3.4 Climatic Risks

2.3.4.1 Climate Change

In general, geological changes do not only threaten on a short-term scale, like earthquakes, but also over the long-term as they may affect climate and create changes to the weather. It is now very well known that the climate on earth has changed permanently. For example, 65 million years ago, the average climate on earth was much warmer than today, and within 55 million years, a maximum average temperature value was reached. Figure 2-12 shows the development of average earth surface temperature with a logarithmic time scale over 65 million years. At the peak value, the polar ice was completely molten and tropical forests reached up to the Arctic region (NGS 1998a,b). The ancestors of crocodiles lived up to Greenland, and the ground temperature of the oceans, which is now about 4°C, reached 17°C. The temperature gradient between equator and polar regions was nearly neglectable. Of course, geographic land distribution has heavily influenced such climatic conditions. Approximately 60 million years ago, Australia and Antarctica got separated. This created a new ocean, which cooled Antarctica. At the same time, Africa bounded into Europe and India bounded into Asia. In these regions, new mountains were created, which changed the climate conditions, for example lowering the average sea-level and increasing the size of land. Since land has lower heat-storage capabilities, this could therefore have contributed to a drop in temperature. Additionally, it is assumed that the formation of coal and oil might have deprived the atmosphere of global warming gases. Therefore, from about 55 to 35 million years ago, earth climate experienced a cooling phase yielding polar ice caps. Following this, a warming phase occurred in the next 14 million years, and then a cooling phase started again a few thousand years ago. The polar ice caps reached three times a size than nowadays.

Based on the measurement of tree rings, coral reefs, ice drillings or other sources of historical information, one can well estimate the climate changes on earth over the last 20,000 years (Wilson et al. 1999). Compared to such climatic changes, the fluctuations are rather low. However, fluctuations do occur, as one may make out from the size of glaciers, for example. Figures 2-13 and 2-14 show the Aletsch glacier in Switzerland. This glacier has lost about 100 m of its in height at the Konkordia place over the last 100 years. Moreover, the overall ice volume on earth has decreased by about 10% since 1960.

The development of the average temperature in the northern hemisphere over the last 2,000 years is shown in Fig. 2-15. However, this curve has been heavily debated by the scientific community due to uncertain data. Therefore, the figure includes not only the average data but also the confidence interval (grey area). The figure shows the well-known warm temperatures over the medieval times as well as the small ice age. During the warm time, settlements from the Vikings existed in Greenland, which had to be abandoned during the ice age later (NGS 1998a,b, Baladin 1988).

However, Fig. 2-15 also shows the increase of average temperatures over the last 50 years. Based on these data, such a rapid change has never been observed before. The 1990s were probably the warmest decade over the last 1,000 years. The average surface temperature of the earth has increased over the last century by approximately 0.6°C ± 0,2°C (IPCC 2001).

Considering the temperatures based on seasonal data for Europe, values might be even more alarming. The summer of the year 2003 was about 2° higher than average. This was probably the hottest summer in the last 500 years, followed by the summer of 1747. However, since 1750, there were rather warm summers, whereas the 20th century started rather cool. From 1923 to 1947, a first warming period occurred, which was followed by a cooling period until 1977. Since then, an increase of the average temperature has been observed. (NZZ 2004a,b).

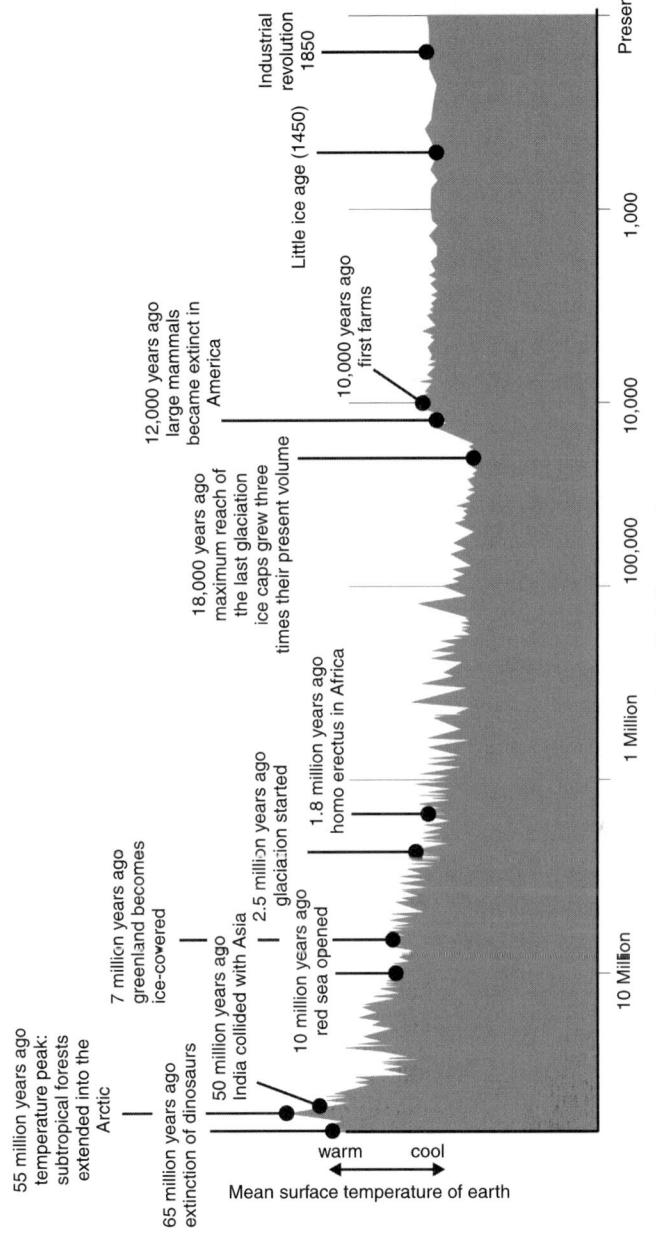

Fig. 2-12. Development of mean earth surface temperatures according to NGS (1998a,b)

Fig. 2-13. The Aletsch glacier from the Konkordia place. The Aletsch glacier is the longest in the Alps

Fig. 2-14. The Konkordia lodge at Grünegg firn. Here, the Altesch glacier lost about 100 m thickness from 1877

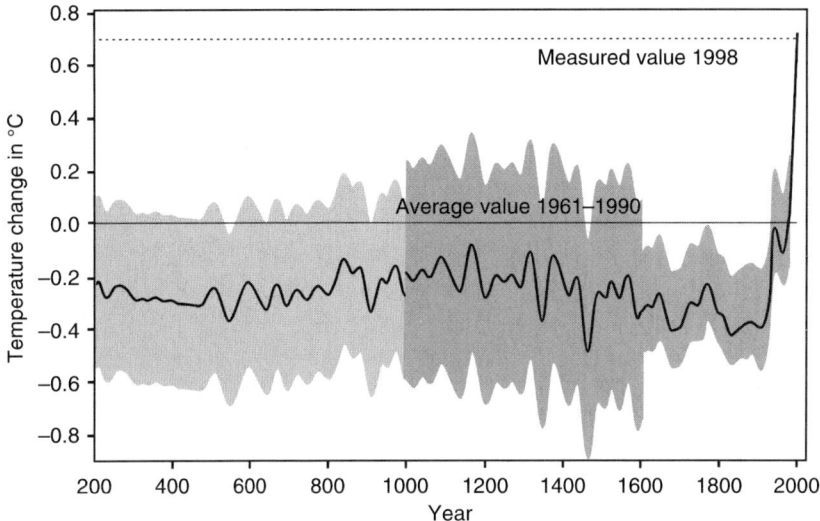

Fig. 2-15. Mean temperature of the northern hemisphere over 2,000 years (IPCC 2001)

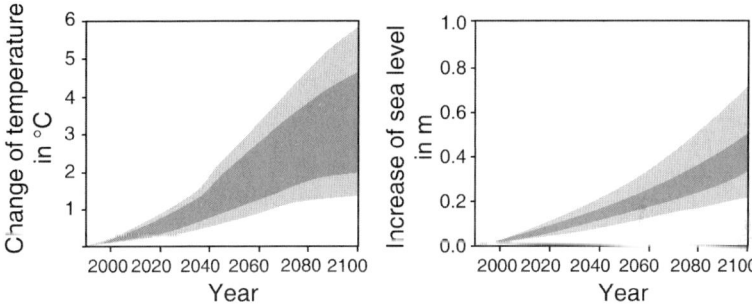

Fig. 2-16. Estimation of change of temperature and increase of sea level over this century (IPCC 2001)

For the winter, an increase of the average temperatures in Europe has been identified. The period from 1973 to 2002 has been considered the warmest over the last 500 years. The increase of the winter temperatures is especially interesting for risk considerations, since the loss of temperature yields to less snow and less long-term high-pressure areas in Central Europe. These high-pressure areas originally anticipated the movement of stormy low-pressure regions from the Atlantic to Central Europe. Besides, the coldest winter in Europe, based on some measurement occurred in 1708/1709.

However, not only the measurement of temperature but also the observations of other climatic effects, such as the appearance of storm or the appearance of El Nino, seem to indicate a temperature increase. Nevertheless, the proof that a systematic climate change is currently under way is difficult to obtain. The climate system of the earth is extremely complex and related to many other systems, such as social systems, astronomical systems, maritime systems and so on. Mankind is currently influencing the climate system by the massive burning of fossil fuels.

If the trend continues as shown in Fig. 2-16 (left), then temperature increase will not only directly influence living conditions and agriculture, but also that the sea level will rise. The consequences of a sea level rise of about 0.5 m is shown in Table 2-15. About 1/5th of the world's population lives within a distance of 30 km from the sea, and the majority of Mega cities are coastal cities or are located closest to the sea (Geipel 2001).

Table 2-15. Estimated loss of area caused by an increase of sea level of 0.5 m for North America and Japan, 0.6 m for Indonesia and 1 m for all other countries according to Abramovitz (2001)

	Loss of area		Affected population	
	km^2	%	Million	%
Egypt	2,000	< 1	8	11.7
Senegal	6,000	3.1	0.2	2.3
Nigeria	600	< 1	> 3.7	3
Tanzania	2,117	< 1	–	
Belize	1,900	8.4	0.07	35
Guyana	–		0.6	80
Venezuela	5,700	0.6	0.06	< 1
North America	19,000	< 1	–	–
Bangladesh	29,846	20.7	14.8	13.5
India	5,763	0.4	7.1	0.8
Indonesia	34,000	1.9	2.0	1.1
Japan	1,412	0.4	2.9	2.3
Malaysia	7,000	2.1	> 0.05	> 0.3
Vietnam	40,000	12.1	17.1	23.1
Netherlands	2,165	6.7	10	67
Germany	–	–	3.1	4

Climate change might not only cause flood risks or storm risks; it might also yield to additional gravitational risks. For example, when the temperature of permafrost ground increases, it might lose strength yielding to failure. Historical data of mass movement events in the Alps show a peak at the end of the 19th century, which is related to the end of the small ice age (Totschnig & Hübl 2008). The failures of permafrost might also cause glacier lake dam failure yielding to floods (Kääb et al. 2006).

2.3.4.2 Temperature Extremes – Heat and Cold

Of course, direct temperature changes also affect people. In August 2003, not only France but also many European countries were hit by a heat wave. From the 4th to 12th of August 2003 in Paris, all measured temperature values (minimum, highest and average values) were higher than ever recorded since 1873. The heat wave in connection with high air pollution resulted in an increase in mortality. Already on the 4th of August, about 300 people died, which was more than statistically expected. On the 12th of August, 2,200 more people died. Over the entire period, about 14,802 people died – a number much greater than the statistical average. Mainly the elderly, 75 years of age and greater, contributed to the mortality rate (70%). The highest incidence in the rate of mortalities was found in nursing homes. Also, people between the age of 45 and 75 were hit with higher mortality rates (30%) (VMKUG 2004).

Additionally, the mortality increase was regionally dependent. While in some rural areas only a slight increase was found, in Paris and in the suburbs, a dramatic increase (more than 130%) was observed. Other countries were also affected, such as Italy and Portugal (VMKUG 2004). Figure 2-17 shows a relation between mortality and maximum daily air temperatures. It is clear that high temperatures increase the mortality rate. However, these consequences do not consider a shortage of water supply.

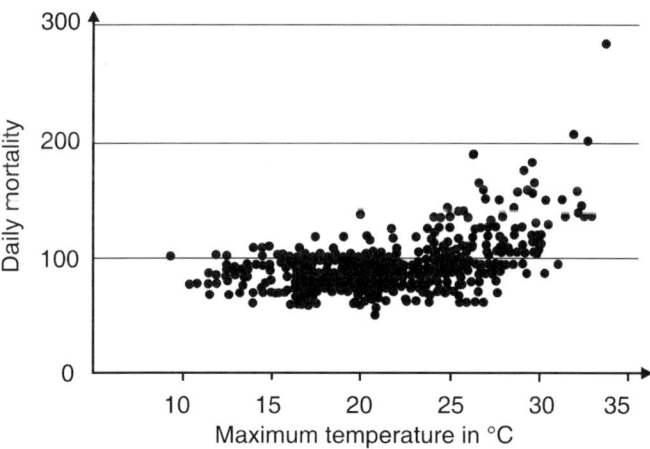

Fig. 2-17. Relation between daily mortality and maximum daily temperatures in a city (Munich Re 2003)

2.3.4.3 Drought and Forest Fire

High temperatures can cause shortage of water with dramatic consequences, including droughts. Since droughts are chronic conditions in many countries or regions, they are often not considered as natural disasters or risks. However, within this book, they are assumed as a special climatic condition and, therefore, classified as a natural risk. For example, in some parts of Africa, droughts happen regularly within a cycle of several years. Drought contributes 98% to the fatality distribution by disaster origin in Ethopia (Tadele 2005). Table 2-16 lists heavy droughts in combination with famines.

There is a general discrepancy between the amount of water flowing into the region and the amount of water flowing out. If more water flows out, then the region is arid. Approximately 40% of the human population lives in areas with water shortages. It has been estimated that, by the year 2025, only half the amount of drinking water will be available per capita worldwide (Fig. 2-18). This will create major problems for millions of people living in arid regions. Nevertheless, if a region is arid, there are still some techniques to provide a sufficient amount of water to the public. For example, in the Middle East, water desalination devices are used, and closed water circuits is another technique. Quite often, however, safety measures against droughts are either not carried out or are carried out insufficiently.

Famines might be caused not only by droughts but also by combinations of droughts and heavy fires destroying agricultural lands or forests. The heavy forest fires in October/November 2003 and 2007 in California are a good example of this. In 2003, about 20 people were killed by the fire and 2,200 houses destroyed, although 13,000 fire fighters tried to extinguish the fire. The financial damages were enormous. Damage information is also available for the Oakland Hill fire storm in October 1991. It resulted in insurance claims totaling approximately 2.3 billion US dollars. Two fires in 1993 caused a loss of about 1 billion dollars. Further examples are a fire in Arizona (Rodea-Chediski) with a loss of 125 million dollars and in New Mexico (Cerro Grande) for about 150 million dollars (Munich Re 2004b).

Forest fires are no more a local problem but a worldwide problem. For example, the forest fires in Indonesia in the first decade of the 21st century became known worldwide due to its resulting pollution in the air. In December 2001 in Sydney, the so-called black Christmas fire occurred with more than 15,000 fire fighters in action. The name black Christmas came into use since Sydney was covered by thick smoke clouds from that fire. Other big historical fires connected to droughts in Australia were the Tasmanian fire in 1967 and the Victorian fire in 1939 (Fig. 2-19). Data about forest and wildland fires in Europe can be found in Plate (2003).

Even though droughts provide the required conditions for massive fires, quite often arson can be the cause (McAneney 2005). Drought is, in many cases, related to anthropogenic hazards like civil conflicts, which then trigger the famines (Tadele 2005).

Table 2-16. Heavy droughts with famine (NN 2004)

Year	Location	Consequences
1064–1072	Cairo, Egypt	Non-appearance of Nile floods causing famine where probably 4,000 people starved to death
1069	Durham, England	Probably 50,000 people starved to death
1199–1202	Cairo, Egypt	Non-appearance of Nile floods causing famine where probably 100,000 people starved to death
1669–70	Surat, India	Probably three million people starved to death
1769–70	Delhi, India	18 months of drought, an estimated three million people starved to death
1790–91	Bombay, India	Famine in India, probably several thousand people starved to death, cannibalism occurred
1833	Guntur, India	Drought and famine with about 20,000 victims
1866	Raipur, India	Drought in Bengal, Orissa and Bihar, probably 1.5 million people starved to death or have died by diseases
1868	Bhopal, India	
1876–77	Madras, India	Supposedly the most severe famine with about three million people starving to death and another three million dying from cholera
1877–78	Tschangtschun, Manchuria, China	Drought over several years caused famine in northern and middle China with about 1.3 million victims
1898	Punjab, India	Probably one million people starved to death
1921–22	Nischni Nowgorod, Wolga region, Russia	Prolonged drought causes famine, several million people affected
1932–33	Kiev, Russia-Ukraine	Economic rearrangement and drought yielded to famine, several million people were affected
1932–1940	Dodge City, Kansas, USA	Drought in middle west of the USA, 350,000 people left the region
1962	Parana, Brazil	Drought for several months yielded to severe fires in the coffee production regions
1967–1970	Biafra	Drought and war yielded to famine, about eight million people affected
1969–1974	Gao Mali, Sahel zone, Africa	Drought and political difficulties yielded to famine and disease
1972	Nagpur, India	Heat wave of more than 40°C for several months, heavy damage to agriculture
1984–1985	Mekele, Ethiopia	Prolonged drought and war yielded to famine in several African countries. Ethiopia was the most severely hit

(Continued)

Table 2-16. (Continued)

Year	Location	Consequences
1992	Bulawayo, Zimbabwe	Drought and famine hit 30 million people
1994	Grafton, New South Wales, Australia	90% of the wheat harvest was lost due to drought

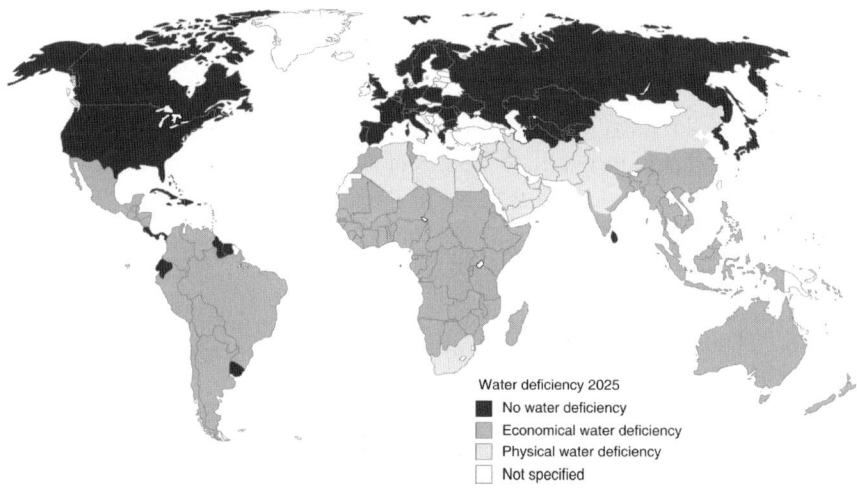

Fig. 2-18. Prognosed regions of water deficiency in 2025 (Geipel 2001)

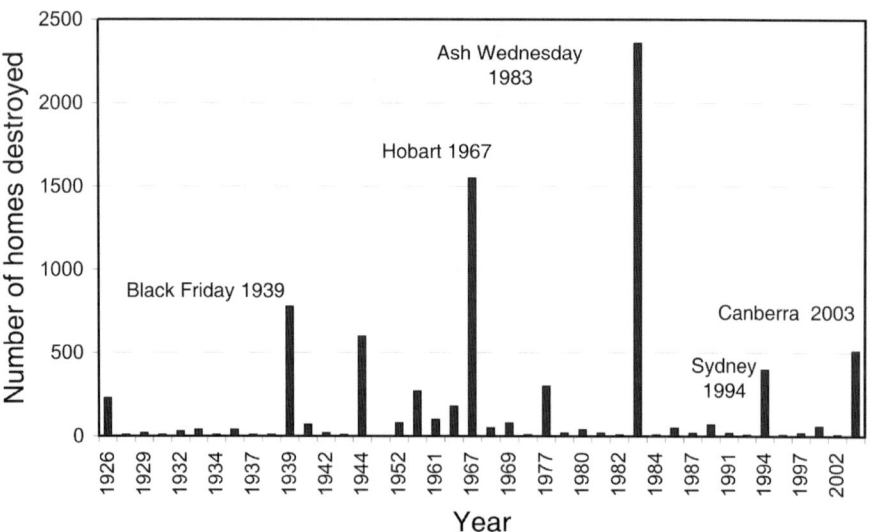

Fig. 2-19. Major forest fires in Australia based on the number of houses destroyed according to McAneney (2005)

2.3.4.4 Famine

Famines are temporally limited situations where there is perceptible lack of food for major parts of the population. Famines have occurred recently in Zimbabwe and North Korea. The chronic lack of food for single social layers is not understood as famine. Nevertheless, the latter can be found worldwide in many regions (Figs. 2-20 and Fig. 2-21). For example in Brazil, it has been estimated that about 1/4th of the population is unable to obtain a sufficient amount of food (NZZ 2004). In Ghana, for example, 1/3rd of the population is considered malnourished (Jelenik 2004). The Food and Agriculture Organization of the United Nations estimated the number of malnourished people to be approximately 840 million. Many of them are children. Undernourishment in the year 2000 caused about 3.7 million deaths worldwide. In Africa, 50% of the deaths of children of age less than 5 are related to malnutrition. It has been estimated that undernourishment is responsible for about 10% of the global burden of disease (FAO 2003).

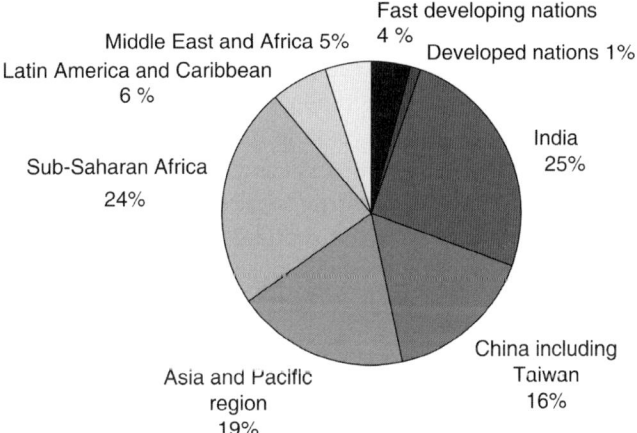

Fig. 2-20. Geographical distribution of undernourished people worldwide according to FAO (2003)

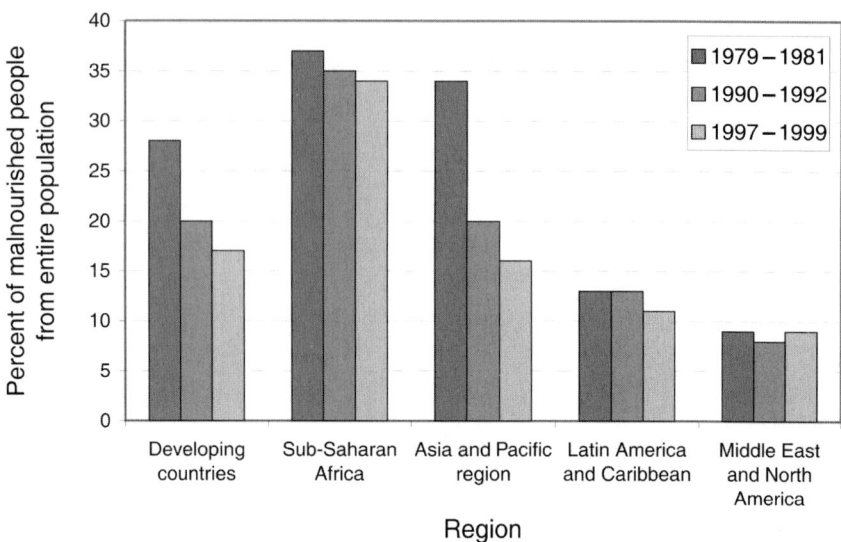

Fig. 2-21. Time development of percent of undernourished people in different regions according to FAO (2003)

The nutritional conditions are identified based on different anthropometrical parameters adjusted to the type of undernourishment. Acute undernourishment, also called wasting, is mainly identified by a ratio of weight to height measurements. Here, values less than 70% or an age-dependent diameter of the upper arm indicate undernourishment. To determine chronic undernourishment, a ratio of height to age is mainly used. Again, diameter of the arm can be used for identification. Last but not least, pathogenic signs such as edemas or starvation dystrophy can be used to identify undernourishment. Such edemas occur on the arms, legs and in the face. The skin becomes fragile and hairs can be pulled out without pain. With a heavy loss of fatty tissue and muscular tissue, the ribs, spine and scapula become visible. Generally, severely malnourished children get the expression of old people. Another epidemiological measure for undernourishment is the amount of energy provided. In general, an energy amount between 2,000 and 2,600 kcal is required per capita per day.

There are records of famines in Switzerland in 1438, 1530, 1571–1574, 1635–1636, 1690–1694, 1770–1771 and 1816–1817 (Mattmüller 1987, Kurmann 2004). Famines probably occurred more frequently than the records can show (Abel 1974, Labrousse 1944). This is because records for many regions became only available since medieval times, and many famines were probably limited in spatial terms. Famines were considered one of the major threats to societies before industrialisation. Nevertheless, famines occurred even at the end of the 19th and at the beginning of the 20th century. One very well-known famine was the Irish potato famine. Probably more than one million people died and several hundred thousand people emigrated.

The question of whether famines are a natural or a social risk remains. Historically, usually a crop failure or harvest failure triggered a famine. But, this does not have to necessarily yield to a famine. Mostly after such harvest failures, the heavy prices for food can cause an economic crisis. Therefore, economic reactions to harvest failures heavily influence the development of famines. This fact becomes visible now with better means of transportation. With the increase in shipping especially around the end of the medieval times, the number of famines dropped since ships were used excessively to bring food, which positively acted in the market, at least in coastal regions.

Examples of famine as a social risk were the famines in Switzerland in 1770–1771 (Mattmüller 1982) and 1816–1817. Here, an economic crisis fint occurred in the textile industry. The unemployed people were unable to pay for the rising prices of food, and therefore a famine resulted for most of the unemployed. This fact is also valid for some African countries, where the per capita food production would be just sufficient to avoid undernourishment, but the portion of malnourished children remains constant. A good indicator here is the percentage of income spend on food (Ziervogel 2005).

Generally, not only reducing undernourishment but also increasing the safety of food is a major goal. This means that people have access to sufficient, harmless and nutritious food, which fulfills the psychological requirements and permits an active and healthy life. Undernourishment does not necessarily mean a lack of food – diseases might also be a cause. The general term for undernourishment is negative energy balance in the human body. The minimum energy required by a human is composed of energy necessary to maintain the body temperature, permit active work and fight diseases.

2.3.5 Aeolian Risks

After discussing some climatic risks and their consequences to humans in terms of famines, further climatic risks like aeolian and hyraulic risks are introduced here.

Winds are moving air masses in relation to the earth surface (Bachmann 2003, Höffner et al. 2005). They are powered by the energy supply from the sun. Storms can be considered a type of severe wind condition found in all continents (Munich Re 2005a). Storms are classified into barocline, tropical and convective storms. Additionally, storms of the same type

might have different names in different parts of the world: hurricanes, typhoons and cyclones are all tropical cyclones (DKKV 2002). Figure 2-22
shows the structure of a tropical cyclone.

The frequency of storms is geographically not equally distributed. Some
regions have a high storm frequency as compared to others. Figure 2-23
shows the distribution of tornados worldwide. In the US, about 1,000
storms are recorded per year. In general, the number of fatalities is rather
low. In 2002, storms caused 55 deaths in the US.

In general, intensive research has been carried out on the investigation
of wind and its forces, as it is important for sailing and structures. Figure
2-24 shows the spectral density of wind speed indicating some microme-
tereological and macrometereological conditions.

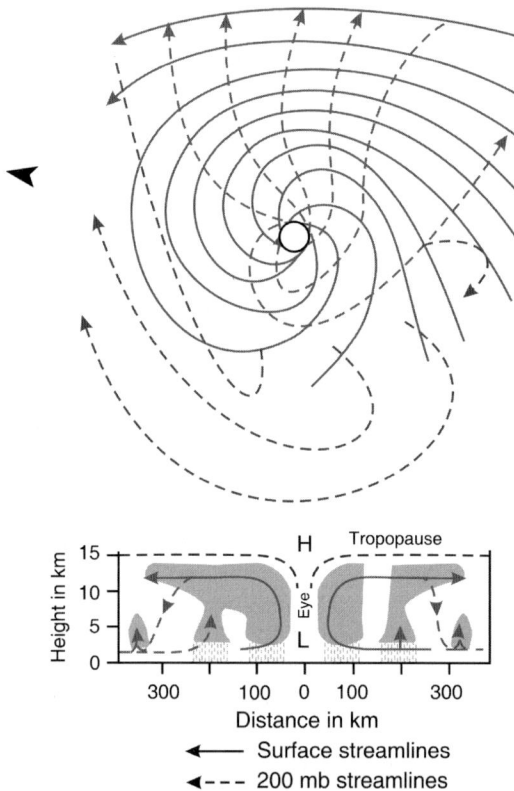

Fig. 2-22. A model of the arial and vertical structure of a tropical cyclone taken
from Smith (1996)

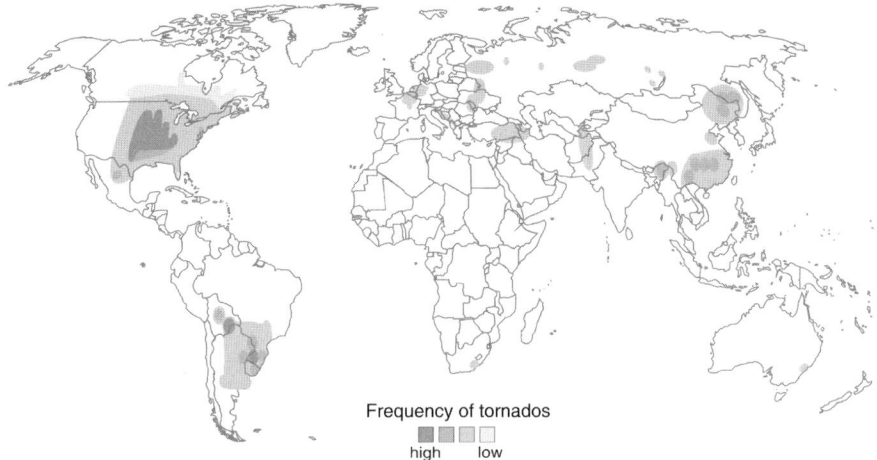

Fig. 2-23. Frequencies of tornados worldwide (Vesilind 2004)

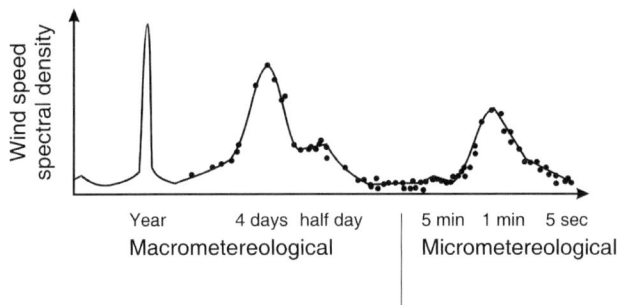

Fig. 2-24. Spectral density of wind according to van der Hoven (1957)

Depending on the type of storm, different storm intensity measures have been introduced. The Fujita scale describes tornado intensity by wind speeds and damages (Table 2-17). Another intensity measure is the Saffir–Simpson scale, which was developed in the 1970s (Table 2-18). This scale is used to measure hurricane intensity based on wind speeds and atmospheric pressure. Further measures are also known or under development, for example the Torro scale, the Hurrican Intensity Index or the Australian-Tropical-Cyclone-Severity Scala (Munich Re 2006b). The Beaufort scale is widely used for winds that are less intense than hurricanes (Table 2-19), and is the oldest intensity measure originating around the beginning of the 19th century.

Table 2-17. Fujita-scale (Vesilind 2004)

Fujita-scale	Description	Wind speed in km/h
F 6	Inconceivable	>512
F 5	Incredible	420–512
F 4	Devastating	333–419
F 3	Severe	254–332
F 2	Significant	182–253
F 1	Moderate	117–181
F 0	Gale tornado	64–116

Table 2-18. Saffir–Simpson scale (Kantha 2006)

Category	Maximum wind speed (m/s)	Surface pressure (mb)	Storm surge (m)
1	33–42	> 980	1.0–1.7
2	43–49	979–965	1.8–2.6
3	50–58	964–945	2.7–3.8
4	59–69	944–920	3.9–5.6
5	70+	< 920	5.7+

Table 2-19. Beaufort scale (Lieberwirth 2003)

#	Description	m/s	km/h	Land conditions
0	Calm	0–0.2	0	Smoke rises vertically
1	Light air	0.3–1.5	1–5	Wind motion visible in smoke
2	Light breeze	1.6–3.3	6–11	Wind felt on exposed skin
3	Gentle breeze	3.4–5.4	12–19	Leaves in constant motion
4	Moderate breeze	5.5–7.9	20–28	Small branches begin to move
5	Fresh breeze	8.0–10.7	29–38	Smaller trees sway
6	Strong breeze	10.8–13.8	39–49	Umbrella use becomes difficult
7	Moderate gale	13.9–17.1	50–61	Effort needed to walk against the wind
8	Fresh gale	17.2–20.7	62–74	Twigs break from trees
9	Strong gale	20.8–24.4	75–88	Light structure damage
10	Storm	24.5–28.4	89–102	Considerable structural damage
11	Violent storm	28.5–32.6	103–117	Widespread structural damage
12	Hurricane	> 32.7	> 118	Considerable and widespread damage

Webster et al. (2005) have described the possible changes in storm intensity in a warmer climate. Already the effects of winterstorms in Europe have been discussed in Sect. 2.3.4.1 (Lieberwirth 2003, Munich Re 2006a). Even under the current climatic conditions, storms represent a major hazard. In 1970, a storm in Bangladesh together with a water storm surge killed about 300,000 people. This was probably one of the biggest natural disasters in the 20th century (O'Neill 1998). In October 1998, Hurricane Mitch reached Central America and killed 5,700 people in Honduras alone. More than 600,000 people were affected by the hurricane. The damages reached up to 80% of the gross domestic product. In Peru, about 300 bridges were heavily damaged (Williams 1999). Another severe storm reached southern Africa in the year 2000. In February and March, the cyclone Eline crossed Mozambique, South Africa, Botswana, Swaziland, Zimbabwe, Malawi and Zambia, causing more than 1,000 fatalities and more than 660 million US dollars lon of property (Munich Re 2001). In 1992, Hurricane Andrew in the US caused a financial damage of 15 billion US dollars. Also, the winterstorms in Europe in 1990 costed about 15 billion US dollars (Mechler 2003).

Major hurricanes in the US in terms of fatalities occurred in Galveston with 8,000 fatalities in 1900, on the Lake Okeechobee in 1928 with 2,500 fatalities, Savannah, Georgia in 1893 with 1,500 fatalities and in Cheniere Caminada with 1,250 fatalities in 1893 (Munich Re 2006b). This numbers show the enormous damage capacity of storms. Therefore, many risk assessment studies for storms or strong winds were carried out in the last decades. Examples are works by Leicester et al. (1976), Petak & Atkisson (1992), Khanduri & Morrow (2003) and Heneka & Ruck (2004).

2.3.6 Hydraulic Risks

Besides earthquakes, floods are the biggest natural hazard in terms of fatalities. Probably the biggest natural disaster ever was the Henan Flood in China in 1887 with between 900,000 and 1.5 million victims. The biggest natural disaster in Europe was probably the "Gross Manndränke" in the year 1362. Here, between 30,000 and 100,000 victims were counted. And, the flood in summer 2002 in Central Europe was probably the costliest European natural disaster over (Mechler 2003). First estimations ranged from 20 to 100 billion Euros (Kunz 2002). Figure 2-25 shows the Dresden railway station, Germany, during a flood in 2002. The station is usually more than 1 km from the river. About 300,000 people were affected in Germany alone. However, a flood in China in 1991 affected about 220 million people (Newson 2001).

An incomplete list of severe floods is given in Table 2-20. A cyclone in Bangladesh in 1970, which caused about 300,000 victims, was already mentioned (Mechler 2003). However, the flood in the 1950s in the Netherlands with about 2,000 fatalities should be mentioned. The flood resulted in an intensive program to avoid floods in this country where about 40% of the population live behind dams and below sea-level. An intensive study about the loss of life in floods has been carried out by Jonkman (2007).

Historical data of floods in Europe have shown that floods are corre-lated. In fact, so called "time clusters" could be found. There exist several theories to describe such effects, for example climate cycles or the fatigue of the water storage capability of a landscape after a flood. Therefore, the next flood can occur even with a lower water supply. An example of a cluster could be shown for the French rivers Ter, Segre and Llobregat. There are data available since the year 1300. Peaks of floods were ob-served between 1582 and 1632, with maximum values in 1592 and 1606. Moreover, in the period from 1768 to 1800 and between 1833 and 1868, clusters of floods could be found (Huet & Baumont 2004, Baladin 1988). Historical data for the river Elbe are shown in Table 2-21. Figure 2-26 shows flood marks on a house in Bad Schandau, Germany. More data can be found in NaDiNe (2007). Interestingly, floods show not only a temporal correlation but also a geographic correlation, as shown in Jorigny et al. (2002)

Fig. 2-25. Dresden central railway station in summer during the 2002 flood

Table 2-20. List of severe floods at the German North Sea coast and worldwide (NN 2004, Schröder 2004).

Date	Location	Number of victims
2200 B.C.	Hyderabad, India	
26.12.838	East Frisian coast	2,437
1099	Boston, England	Thousands
17.02.1164	East Frisian coast	20,000
16.01.1219	Jütland, Denmark	Thousands – 36,000
14.12.1287	East Frisian coast	Thousands
1287	Dunwich, East Anglia, England	< 500
1332–1333	Peking, China	Several millions
16.01.1362	Schleswig, Germany	30,000–100,000
09.10.1374	East Frisian coast	–
1421	Dort, The Netherlands	100,000
26.12.1509	East Frisian coast	–
31.10.1532	East Frisian coast	
1.11.1570	East Frisian coast	< 4,000
1606	Gloucester, England	> 2,000
1634	Cuxhaven, Germany	> 6,000–8,000
1717	Den Haag, Netherlands	11,000
1824	St. Petersburg, Russia	10,000
3.-4.2.1825	East Frisian coast	200
1851–1866	Shanghai, China	Several millions
1887	Henan, China	900,000–1.5 millions
1890	New Orleans, Louisiana, USA	
1911	Shanghai, China	20,000
1927	Cairo, Illinois, USA	300
1931	Nanking, China	130,000–several millions
1935	Jérémie, Haiti	2,000
31.1-1.2.1953	Hollandflut	2,000
1954	Wuhan, China	40,000
1955	Cuttack, India	1,700
16.2.1962	East Frisian coast	330
1999	Ovesso Monsun Flood, India	10,000
1999	Venezuela	25,000–50,000

Table 2-21. Historical maximum water levels of the river Elbe at Dresden (Fischer 2003)

#	Date[1]	Water level (m)	Volume m (^{3}/s)
1	17 August 2002	9.40	4,700
2	31 March 1845	8.77	5,700
3	1 March 1784	8.57	5,200
4	16 August 1501	8.57	5,000
5	7 February 1655	8.38	4,800
6	6 – 7 September 1890	8.37	4,350
7	3 February 1862	8.24	4,493
8	24 February 1799	8.24	4,400
9	2 March 1830	7.96	3,950
10	17 March 1940	7.78	3,360
11	20 February 1876	7.76	3,286

[1]Further major floods occurred in 1015, 1318 and 1432, however detailed information about the water level and volume is missing.

The floods mentioned so far were storm surges, sea floods caused by storm and tides, or river floods caused by the heavy release of water either through snow or ice melting or heavy rain. Both types cause the so-called "surface waves". Another type of sea surface waves is "freak waves". They are explained by spatial and temporal interference, and are defined as being at least three times the size of an average storm wave (Rosenthal 2004).

In contrast to such surface waves, tsunamis are deep-sea waves caused by under-sea earthquakes. They have a great wave length and wave speed, and usually are not noticeable at sea. However, at coastal regions, they can cause great damages as the event in December 2004. In fact, it was not the first tsunami wave to cause severe damages (Table 2-22). Many tsunami events were known in the Pacific region, for example in Hawaii (1960 and 1974) or in Papua New Guinea (PTM 2004, Synolakis et al. 2002, Synolakis 2007, Yeh et al. 1995). Tsunamis are also common in Norway. For example, events that occurred in 1905, 1934 and 1936 killed about 174 people. It was assumed that in Holocene times, slides at the Norwegian Continental Slope caused huge tsunamis in the North Sea regions (NGI 2007).

Fig. 2-26. Flooding marks on a house in Bad Schandau, Germany

Table 2-22. Major tsunamis (Munich Re 2004a, Smolka & Spranger 2005, Schenk et al. 2005)

Date	Magnitude	Region	Fatalities
26.12.2004	9.0	Indonesia, Sri Lanka, India, Thailand	>223,000
1883		Krakatau, Indonesia	36,400
1.11.1755		Portugal, Marocco	>30,000
15.6.1896		Japan, Sanriku	27,000
1815		Indonesia	>10,000
17.8.1976	8.0	Philippines	4,000
2.3.1933	8.3	Japan, Sanriku	3,060
21.5.1960	9.5	Chile, Hawaii, Japan	3,000
28.3.1964		USA, Alaska, Hawaii, Japan, Chile	3,000
12.12.1992	7.5	Indonesia, Flores	2,500
17.7.1998	7.1	Papua New Guinea	2,400
20.12.1946	8.1	Japan, Nankaido	2,000
5.11.1952		Russia, Paramushir Island	1,300
7.12.1944	8.0	Japan, Honshu	1,000
31.1.1906	8.2	Ecuador, Colombia	500

2.3.7 Gravitational Risks

2.3.7.1 Introduction

The Alpine regions are heavily exposed not only to climatic hazards but also to gravitational risks. Such risks consider mass movements driven by gravitation. Examples are landslides, debris flow, torrents and avalanches. Figure 2-27 shows a system of such different processes depending on the amount of water and the distribution of solid materials. Table 2-23 gives time estimation of the processes and the warning time.

Table 2-23. Types of processes and their properties (Holub 2007)

Process	Speed	Warning time
Flood	20 km/h	Minutes to hours
Debris flow	40 km/h	Minutes
Spontaneous slope failure	4 km/h	Seconds to minutes
Permanent slope failure	0.0001–1 m/year	Months to years
Rock fall	110–140 km/h	Seconds
Flowing avalanche	40–140 km/h	Seconds to minutes
Powder avalanche	110–250 km/h	Seconds

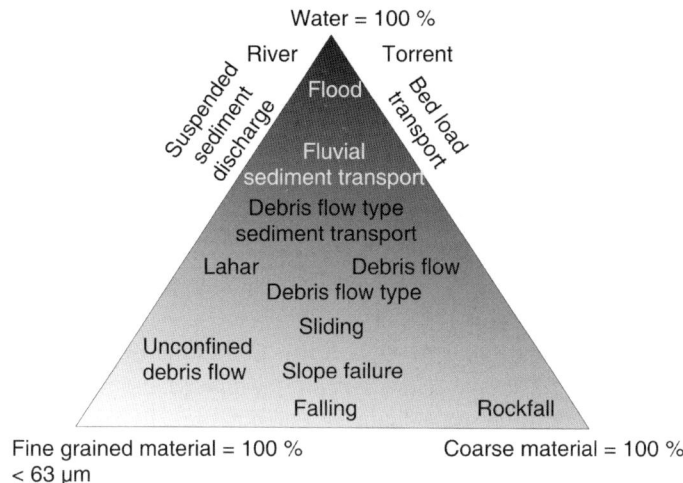

Fig. 2-27. Displacement processes in torrents (Hübl 2007)

2.3.7.2 Rock Fall

A rock fall is a fast downward movement of rocks or an unconstrained coarse material reaching a certain total volume and where partial loss of ground occurs. When the contact between the falling rocks remains, it is called a rock slump or a rock slide (Fig. 2-28). Additionally, it is important to note that the water content is negligible. The negligible water content yields to a lower travelling distance as compared to debris flow (Abele 1974, Selby 1993).

In the German language, in particular, a variety of terms are used in dealing with the processes of rock falls. The different terms are separated according to the volume, the block size, the area covered or the released energy. The classification of rock falls is shown in Table 2-24 (Poisel 1997, Abele 1974).

Table 2-24. Classification of rock falls according to Poisel (1997)

	Volume (m^3)	Block size	Area covered (ha)
Pebble and block fall (Steinschlag)	0.01	20 cm	< 10
Rock fall	0.1	50 cm	< 10
Block fall	2	150 cm	< 10
Cliff fall (Felssturz)	10,000	25 m	< 10
Rock slide (Bersturz)	> 10,000		> 10

Rock fall   Rock slump

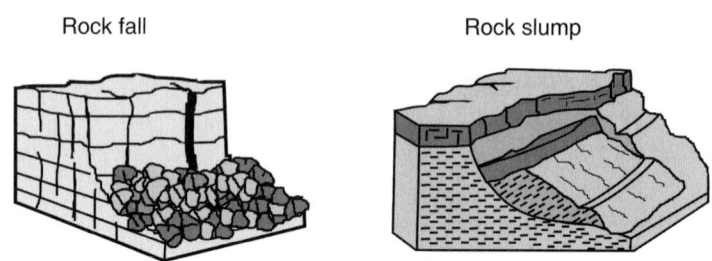

Fig. 2-28. Difference between a rock fall and a rock slump (Varnes 1978)

There are many causes for rock falls, including postglacial load removal, mechanical stress changes caused by disturbances, stress changes caused by sliding, temperature changes, earthquakes, hydrostatic or hydrodynamic pressures, load by trees, ice loads or anthropogenic changes. If the rock elements have started to move, there are different kinds of movements. Usually the rock fall starts with sliding, free-fall, bouncing and rolling. The difference between bouncing and rolling is that bouncing constitutes one full rotation of a stone block without ground contact. The first ground contact after free fall is very important, because during that contact up to 85 % of the kinetic energy can be assimilated by the ground. Of course, this depends very much on the type of ground, for example rock or scree.

Some computer programs were developed in the last decades to model rock falls, such as "Rockfall", "CRSP" or "Stone". Using these programs, the kinetic energy of a rock at a certain location can be determined and mitigation measures can be designed. Active mitigation measures might be fences, nets, walls, dams, galleries and also forests. Such active measures can be used for rock falls, but not for heavy rock slides. For severe rock slides, passive mitigation measures, for example regulating the use of areas, have to be implemented.

Some major historical rock avalanches occurred in 6000 B.C. at Köfelsberg, in 1248 at Mont Granier causing 2,000 fatalities, in 1348 in Dobratsch, caused by an earthquake, killing 7,000 people and in 1618 at Piuro, caused by illegal quarrying, killing 2,500 people (Erismann & Abele 2001).

2.3.7.3 Debris Flow

Debris flows are extremely mobile, highly concentrated mixtures of poorly sorted sediments in water (Pierson 1986). The materials incorporated in debris flows vary from clay-sized solids to boulders of several meters in diameter (Fig. 2-29). Debris flows can exceed the density of water by more than a factor of two. The front of a debris flow can reach velocities up to 30 m/s (e.g. Costa 1984, Rickenmann 1999), and peak discharges can be tens of times greater than for floods occurring in the same catchment (e.g. Pierson 1986, Hungr et al. 2001). It is difficult to quantify annual economic losses due to such phenomenon; however, in the year 2005, more than 80 million Euros were spent in Austria for protection measures against torrential hazards (including floods, bed load transport and debris flow) (WLV 2006).

In debris flow research, the flowing mixture is mostly divided into the liquid "matrix", composed of water and fine sediment in suspension, and the solid phase, consisting of coarse particles dispersed in the fluid. Depending on the relative concentration of fine and coarse sediments, the prefix "viscous" or "granular" is often used. Since the early 1970s, research increasingly focussed on the topic of debris flow behaviour (Johnson 1970, Costa 1984). Mud flows and debris flows consisting of a considerable amount of fine sediments are often regarded as homogeneous fluids, where the bulk flow behaviour is controlled by the "rheologic" properties of the material mixture (e.g. Julien & O'Brien 1997, Coussot et al. 1998, Cui et al. 2005). For debris flows mainly consisting of coarse particles and water, this simple rheological approach has limitations. In the last decades, geotechnical models have been employed to describe the motion of (granular) debris flows (e.g. Savage & Hutter 1989, Iverson 1997, Iversion & Denlinger 2001).

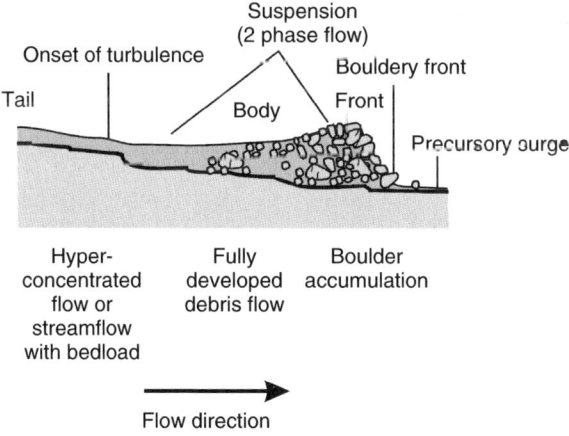

Fig. 2-29. Debris flow surge (Pierson 1986)

The behaviour of debris flows can be very variable, strongly depending on sediment composition and water content. Moreover, debris flow volume and bulk flow behaviour may change during travelling through a channel, for example by the entrainment of loose sediment and/or incorporation of water from a tributary. For this reason until now, no general applicable model used in praxis was able to cover the range of all possible material mixtures and event scenarios.

There were some rules of thumb developed to estimate return periods as well as intensities, including volumes and duration (Figs. 2-30 and 2-31). Table 2-25 shows some examples of extreme volumes. Due to their high density and fast movement, debris flows can jeopardize humans and residential areas. Table 2-26 mentions some calamities caused by debris flows.

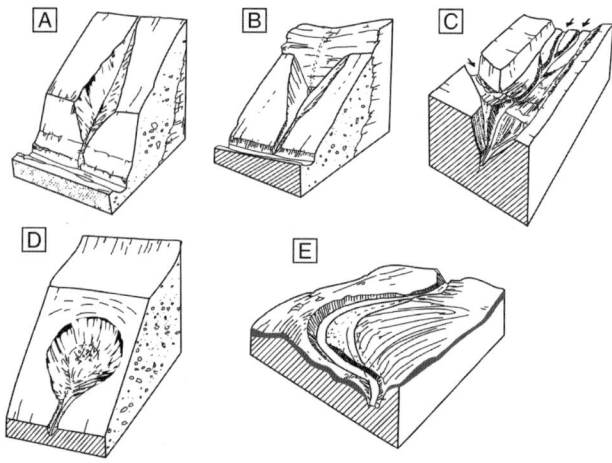

Fig. 2-30. Different types of erosion scars (Weber 1964)

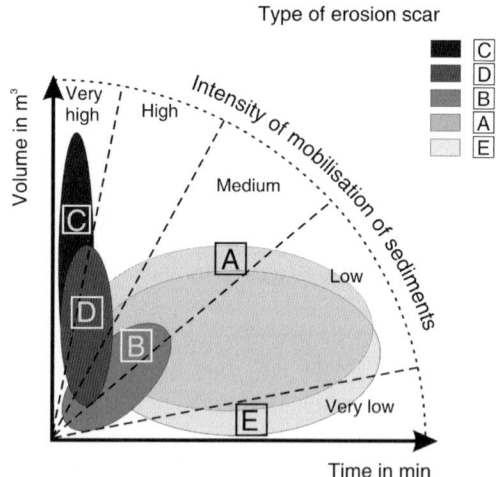

Fig. 2-31. Types of debris flow depending on the erosion scar (Hübl 2006)

Table 2-25. List of major debris flows, rock avalanches or landslides (Bolt et al. 1975)

Location	Time	Volume in billion cubic metres	Comments
Blackhawk Landslide, Mojave Desert	Several thousand years ago	320	8 km length
Vajont Dam, Italy	1963	250	
Axmouth, Devon, England	1839	40	No fatalities
Madison Canyon, Montana	1959	27	26 fatalities
Sherman Glacier Slide, Alaska	1964	23	Air-crushing mechanism
Iran, Saidmarreh	Prehistoric times	20	
Flims, Switzerland	Prehistoric times	15	
Rossberg or Goldau, Switzerland	1806	14	457 fatalities, 4 villages
Tsiolkovsky crater, Moon		3	100 km length (based on the lower mass, the volume of the moon was decreased by a factor of 216)
Usoy, Pamir, USSR	1911	2.5	54 fatalities

Table 2-26. Historical debris flow events with fatalities (Kolymbas 2001, Stone 2007, Pudasaini & Hutter 2007)

Year	Place	Fatalities
1596	Schwaz, Austria	140
1596	Hofgastein, Austria	147
1669	Salzburg, Austria	250
1808	Rossberg/Goldau, Switzerland	450
1881	Elm, Switzerland	115
1893	Verdalen, Norway	112
1916	Italy/Austria	10,000
1920	Kansu Province, China	100,000–200,000
1949	USSR	20,000
1962	Huascaran, Peru	5,000
1963	Vajont-Reservoir, Italy	2,043
1996	Aberfan, South Wales	144
1970	Huascaran, Peru	18,000
1974	Mayunmarca, Peru	451
1985	Stava, Italy	269
1985	Nevado del Ruiz, Peru	31,000
1987	Val Pola, Italy	30
2006	Leyte, Philippines	1,400

In view of the severe damage capabilities of debris flows, there have been some mitigation measures introduced. For example, torrential barriers have been employed in the former Austrian empire since the middle of the 19th century. Figure 2-32 shows a possible classification of such torrential barriers, and Figs. 2-33 and 2-34 show two examples.

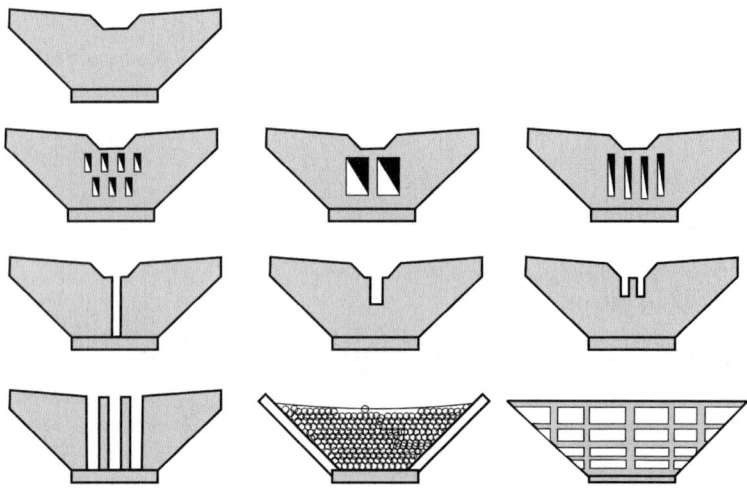

Fig. 2-32. Types of torrential barriers (Bergmeister et al. 2007)

Fig. 2-33. Example of the Christmas tree barrier

Fig. 2-34. Example of a barrier

2.3.7.4 Landslides

Landslides belong, like rock falls or avalanches, to the group of gravitation-initiated hazards. In contrast to debris flow, landslides are non-channel processes. Landslides of slopes and hillsides are mass movements by sliding or flowing under wet conditions. The soil is usually fine-grained and is able to absorb water. The mass moves or flows on a slide face. Very often the process changes over time. For example, a mass movement might start as a landslide, but then trigger a debris flow. A torrent might initiate a debris flow. There were many different attempts to define landslides or debris flows (Table 2-27). However, as explained in the first chapter, it seems to be questionable to reach an ultimate state of such definitions (DIN 19663, DKKV 2002, Crozier 1998, Cruden & Varnes 1996).

Table 2-27. Types of landslides (Cruden & Varnes 1996, Bell 2007)

Velocity class	Description	Velocity (mm/sec)	Typical velocity	Damage
7	Extremly fast	5×10^3	5 m per sec	Disaster, many fatalities
6	Very fast	5×10^1	3 m per min	Some fatalities
5	Fast	5×10^{-1}	1.8 m per hour	Evacuation possible
4	Moderate fast	5×10^{-3}	13 m per month	Impassible structures resist
3	Slow	5×10^{-5}	1.6 m per year	Mitigation measures
2	Very slow	5×10^{-7}	16 mm per year	Some structures can resist
1	Extremely slow			Movement not perceptible

In contrast to avalanches or rock falls, landslides, also sometimes in combination with other processes like rock avalanches, have caused some major accidents with enormous numbers of fatalities. For example, in December 1999, a terrific landslide killed about 30,000 people in Venezuela. The landslide occurred after two weeks of permanent rain. The water volume corresponded with the two years, average rain volume of this region (Abramovitz 2001).

Another landslide in combination with debris flow and volcanic eruption occurred at the Columbian volcano Nevado del Ruiz Armero on 13th November 1985. The event killed about 30,000 people (Newson 2001). However, several smaller tragedies have been caused by landslides as well. For example in Sweden, many smaller landslides have happened in the last centuries (Table 2-28). In Table 2-28, an event on 1st October 1918 is mentioned, which shows only a small volume of moved area (0.2 ha), but the number of victims is rather high. In this case, a train crashed into the moving slope and started to burn (Alén et al. 1999, Fréden 1994, SGU 2004).

Table 2-28. Examples of historical landslides in Sweden (Alén et al. 1999, Fréden 1994, SGU 2004)

Year	Location	Size (ha)	Number of fatalities
1150	Bohus, River Göta älv	37	
1648	Intagan, River Göta älv	27	> 85
1730	Gunnilse, River Lärjeån	30	
1759	Bondeström, River Göta älv	11	
1918	Getå, Bråviken Bay	0.2	41
1950	Surte, River Göta älv	22	1
1957	Göta, River Göta älv	15	3
1977	Tuve, Hisingen Island, Göteborg	27	9

Landslides occur in many countries. For example, landslides killed 73 people in Australia from 1842 to 1997 (QG 2004). In the USA, landslides cause an average yearly damage between 1.2 to 1.5 billion US dollars (1985) (Fernández-Steeger 2002). Damages and numbers of fatalities for Hong Kong are given by Shiu & Cheung (2003). In recent years, severe accidents have happened, for example, in Philippines due to landslides, killing up to 200 people.

Risk evaluation of landslides has found wide application, and might be further applied to avoid severe events in the future (Fell & Hartford 1997, Shiu & Cheung 2003). This is especially of importance since many victims could easily have been saved by sufficient land-use planning. More sophisticated are the methods where the indirect consequences of landslides, for example tsunamis, are considered (Minoura et al. 2005).

2.3.7.5 Avalanches

Avalanches are moving masses of snow. The movement depends on the type of the avalanche (Fig. 2-35). Flowing avalanches are sliding or flowing, whereas in powder avalanches, the material is suspended (flying). Usually a flowing avalanche starts by breaking and releasing snow slabs. These snow slabs accelerate by sliding. When the slabs break into pieces through tumbling and bounding, the movement of snow becomes a flow. The density of flowing avalanches is comparable with the density of normal snow (Table 2-29). The flowing avalanches might slide either on the ground itself or on another layer of snow. These are called ground avalanches or upper (top) avalanches. Additionally, ground avalanches might consist of non-snow materials. In contrast, powder avalanches have a density much lower than snow. However, they either develop from a flowing avalanche or, in most cases, are combined with flowing avalanches. Pure powder avalanches exist only if, through certain conditions, the powder avalanche part separates from the flowing avalanche part. The speed of a powder avalanche reaches up to 90 m/s, whereas flowing avalanches might reach a velocity of up to 40 m/s. Worldwide about 10×10^5 avalanches fall per year (Egli 2005, Mears 2006, Pudasaini & Hutter 2007). The impact

Table 2-29. Some characteristic properties of different avalanche types (William 2006)

Avalanche type	Flow density (kg/m^3)	Concentration of solid material (%)	Typical deposit density (kg/m^3)
Powder avalanche	1–10	0–1	100–200
Wet avalanche	150–200	30–50	500–1,000
Dry snow avalanche	100–150	30–50	200–500

of the avalanches depends not only on the type of the avalanche, but also of the size. The Canadian Avalanche Size Classification is one example to describe the size (McClung & Schaerer 1993). Currently intensive research work is carried out to develop models for avalanche impact forces (Gauer et al. 2008, Sovilla et al. 2008). For such investigations real size avalanche test sites are used, where impact forces are measured (Issler et al. 1998).

The origin of the word avalanche is still under discussion. Some researchers suggest that it came into use after a Spanish bishop, Isidorus, in the 6th century who mentioned of avalanches. He used the terms "lavina" and "labina". Therefore, the origin seems to be the Latin word "labi", which means sliding, slither or drift. There exist some dialects with "lavina" or "lavigna". Other theories consider the German word "lawen" as the origin. This word describes thaw, and might have been applied to moving snow masses caused by thaw (Schild 1982).

Roman writers were the first to mention avalanches. Livius noted the human loss caused by avalanches when Hannibal crossed the Alps in 218 B.C. After the fall of the Roman Empire, there were rare documentations about avalanches, for example the mentioned work by Isidorus. From the beginning of the 12th century, the number of avalanche documentations increased (Schild 1982).

The first known reports on avalanches were from Iceland in 1118 (Pudasaini & Hutter 2007) and from the European Alpine region in 1128 and earlier (Ammann et al. 1997). Totschnig & Hübl (2008) have collected data about more than 13,000 events of natural hazards for Austria, many of them avalanches. The data start with the year 325 and runs until 2005. Table 2-30 lists reports of damages by avalanches for Switzerland for a period of 500 years (Fig. 2-36).

The worst years in Norwegian avalanche history were 1679 when up to 600 people were killed, and 1755 when approximately 200 people were killed. Even in countries like Turkey avalanches can cause high number of fatalities like in 1992 with more than 200 victims (Newson 2001). However, the highest number of fatalities caused by avalanches happened in the wars. Fraser (1966, 1978) has estimated that avalanches killed about 40,000–80,000 soldiers during the times of the World Wars as avalanches were used as weapons. Avalanches on 13th December 1916 alone killed around 10,000 soldiers.

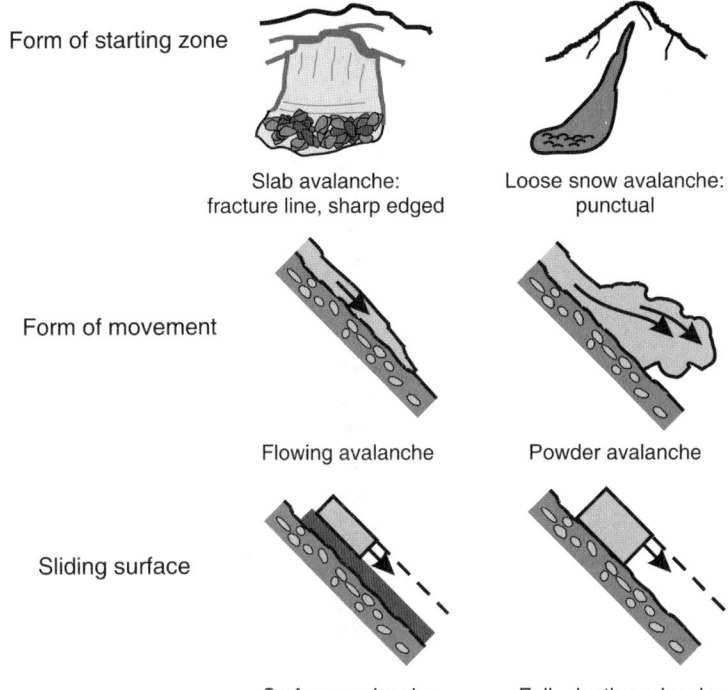

Form of starting zone

Slab avalanche:
fracture line, sharp edged

Loose snow avalanche:
punctual

Form of movement

Flowing avalanche

Powder avalanche

Sliding surface

Surface avalanche

Full‒depth avalanche

Fig. 2-35. Classification of avalanches (Munter 1999)

Table 2-30. Examples of avalanches and accidents in Switzerland (Ammann et al. 1997)

Date	Location	Number of fatalities and damages
January 1459	Trun, Disentis (Surselva/GR)	25 fatalities, church, some houses and stables destroyed
1518	Leukenbad (VS)	61 fatalities, many buildings destroyed
February 1598	Graubünden (Engadin) Livigno	50 fatalities, damage on buildings and livestock
	Campodolcino (Italy)	68 fatalities
January 1667	Anzonico (II) Fusio-Mogno (II)	88 fatalities, village mainly destroyed (event not certain)
January 1687	Meiental, Gurtnellen (UR)	23 fatalities, 9 houses and 22 stables destroyed, cattle killed,
	Glarnerland	many avalanches
February 1689	St. Antönien, Saas im Prättigau, Davos (CR)	80 fatalities, 37 houses and many other structures destroyed, damage on forest and cattle killed

(Continued)

Table 2-30. (Continued)

Date	Location	Number of fatalities and damages
	Vorarlberg, Tirol (Austria)	149 fatalities, about 1,000 houses and many other structures destroyed, great loss of livestock and forest
February 1695	Bosco Gurin (TI) Villa/Bedretto (TI)	34 fatalities, 11 houses and many stables destroyed, 1 fatality, church and some houses destroyed
January 1719	Leukerbad (VS)	55 fatalities, 50 houses and many other buildings destroyed
February 1720	Ftan, St. Antönien, Davos (GR)	40 fatalities, many buildings destroyed
	Ennenda, Engi (GL) Obergesteln (Goms/VS)	7 fatalities, 4 buildings destroyed many fatalities (between 48 and 88 depending on the source), about 120 buildings destroyed
	Brig, Randa, St. Bernhard (VS)	75 fatalities
March 1741	Saastal (VS)	18 fatalities and about 25 buildings destroyed
February 1749	Rueras, Zarcuns, Disentis (Surselva GR) BoscoGurin (II) Goms, Vispertäler (VS), Grindelwald (BE)	75 fatalities and about 120 buildings destroyed 54 fatalities, many avalanches
December 1808	Obermad/Gadmental (BE)	entire village destroyed, 23 fatalities, major damages on buildings, further 19 avalanches in Berner Oberland
	Zentralschweiz (mainly Uri)	20 fatalities and major damages
	Selva (Surselva/GR)	parts of a village completely destroyed
March 1817	Andereggl Gadmental (BE) Elm (CL), Saastal (VS), Tessin und Engadin (GR)	village destroyed, 15 fatalities, many avalanches with fatalities and damages
1827	Biel, Selkingen (Goms, NS)	51 fatalities and 46 houses destroyed
January 1844	Göschenertal (UR), Guttannen, Grindelwald and Saxeten (BE)	13 fatalities and damages on buildings
April 1849	Saas Grund (VS)	19 fatalities, 6 houses and 30 other buildings destroyed
March 1851	Ghirone-Gozzera (II)	23 fatalities and 9 buildings destroyed

Table 2-30. (Continued)

Date	Location	Number of fatalities and damages
January 1863	Bedretto (II)	29 fatalities and 5 houses and stables destroyed
Winter 1887/88	Three avalanche periods, main areas: North and Mittelbünden, Tessin, Goms	1,094 recorded avalanches claim 49 fatalities and 850 buildings destroyed
December 1923	Northern Alps, Gotthard, Wallis, Nord- and Mittelbünden	major avalanche damages in many parts of the Swiss Alps
Winter 1950/51	Two avalanche periods, main areas: Graubünden, Uri, Oberwallis, Berner Oberland, Southern Alps (Tessin, Simplon)	1,421 recorded avalanches claim 98 victims and 1,527 buildings destroyed
January 1954	Northern Alps, NordbündenVorarlberg (Austria)	258 recorded avalanches claim 20 fatalities and 608 houses destroyed, 125 fatalities and 55 houses destroyed or damaged
January 1968	Northern Alps and Graubünden, Davos	211 recorded avalanches claim, 24 fatalities and 296 houses destroyed
April 1975	Southern Alps	510 recorded avalanches claim, 14 fatalities and 405 houses destroyed
February 1984	Northern Alps, mainly Gotthard area, Samnaun	322 recorded avalanches claim, 12 fatalities and 424 houses destroyed

Fig. 2-36. After a heavy snow fall with up to 350 cm of fresh snow within one week in February 1999, the bottom of the Lötschen valley in Switzerland was filled by avalanches up to 10 m in height

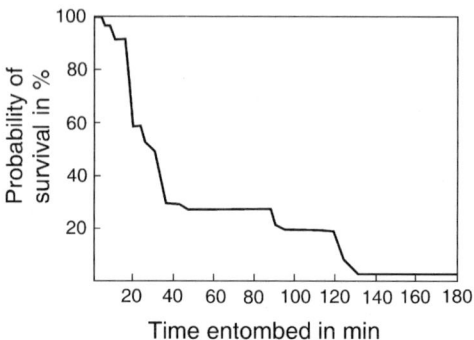

Fig. 2-37. Probability of survival in an avalanche according to Falk et al. (1994)

As with other gravitational hazards, there are possiblities to employ mitigation measures against avalanches. Such mitigation measures might range from nets to dams. In addition, avalanche warning using infrasound or laserscanning is possible (Prokop 2007). Currently, in most Alpine regions, avalanche reports are available to inform people entering mountains during snow times (EAS 2008). However, if people are hit by an avalanche, the chance of survival depends very strongly on the time entombed, as shown in Fig. 2-37.

2.3.8 Biological Risks

Humans are exposed not only to hazards caused by the non-living environment, but also to hazards caused by the living environment. Such hazards include poisonous plants or animals, predators and viruses. The last one will be dealt with more in Sect. 2.5. In this section, predators such as bears are of interest.

Great predators, such as bears, lions, tigers and wolves, are able to kill humans. However, in general, the number of fatalities is rather low. It is different though for polar bears. For example, in Spitzbergen where we find many polar bears, people have to carry weapons if they move out of human settlements (Fig. 2-38). In the last 10 years, polar bears have killed two people in Spitzbergen having an overall population of less than 2,000 people. Those killed, however, were mainly tourists. In general, polar bears are extremely aggressive. On the other hand, the number for killings by brown bears in Europe shows rather lower values (Table 2-31). The high number of fatal accidents caused by bears in Romania is mainly due to the artificial overpopulation for hunting, and is therefore not representative.

The great predators in Europe became extinct during the middle of the 19th century. Actually, in earlier times, lions lived in Europe; however, they were very early on hunted and wiped out by the Greeks and the Persians. Currently, there are talks about re-introducing the great predators into densely populated areas, such as the Alps. Whether this would result in an increase in the number of accidents is heavily discussed.

Fig. 2-38. Polar bear warning sign at Svalbard

Not only wild animals but , domestic animals too attack people. In Germany, an average of nearly two persons are killed per year by dogs. For the German population, this results in an individual risk of 2×10^{-8} per year. Table 2-32 gives examples of such killings in Germany for about 40 years.

Furthermore, people might be attacked by snakes, scorpions or other creatures. Even small insects can kill people. On average in Austria, about five people per year are killed by such accidents. This value is higher than the number of persons being killed by dogs.

More over, many people world over fall a victim to poisonous plants, mushrooms and bacteria.

Table 2-31. Statistics of killings by bears in Europe (Riegler 2007)

Country	Population in million	Bear population	Fatal accidents
Norway	4	230	1
Sweden	9	700	1
Finland	5.2	400	0
Russia European part	106	2,300	6
Albania	3.1	130	0
Poland	38.6	300	0
Slovakia	5.3	400	0
Romania	22.3	6,800	24
Yugoslavia	23.5	2,300	4
Italy	57.7	110	0
France	59.5	8	0
Austria	8	25	0
Greece	10.6	200	0
Spain	40.1	300	0
Total	492.2	14,203	36

Table 2-32. Statistics of killings by dogs in Germany since 1968 according to Bieseke (1986, 1988) and Breitsamer (1987)

Date	Location	Accident
18 November 1968	Landau, Pfalz	Dog kills a baby (14 days old)
18 March 1971	Wunsiedel	Dog kills a child (4 years old)
2 January 1972	Frankfurt/Main	Dog kills a man (73 years old)
23 August 1972		Dog kills a child (6 years old)
25 April 1973	Waiblingen	Two dogs kills a pupil (12 years old)
23 March 1974	Saarland	Dog kills a child (6 years old)
30 September 1974	Herne	Dogs kill a boy (12 years old)
6 October 1974	Dinslaken	Dog kills a child (8 years old)
1976	Schwarzwald	Dogs kill a demented woman
1976		Dog kills a drunken man
December 1977	Rödental bei Coburg	Dogs hurt and kill a child (6 years old)
January 1977	Karlsruhe	Dogs kill a child (5 years old)
5 April 1977	Berlin-Frohnau	Dog kills a child (3 years old)
August 1977	Delmenhorst	Dog or wolf kills a child (7 years old)
10 September 1979	Rothenburg o.d.T	Dog kills a woman (82 years old)
1982	Berlin	Two dogs kill a child (6 years old)
March 1983	Düsseldorf	Two dogs kill a 34-year-old woman
1983	Munich	Dog kills a baby (10 days old)
August 1984	Straubing	Two dogs kill a 79-year-old woman
16 January 1985	Hannover	Dogs kill an old woman

Table 2-32. (Continued)

Date	Location	Accident
January 1985	Nuremberg	Dog kills a young woman
28 January 1985	Giessen	Two dogs kill a young girl (10 years old)
8 February 1985	Straubing	Dogs kill a pensioner
18 March 1985	Flensburg	Two dogs kill a girl (11 years old)
2 August 1985	Bamberg	Dog kills girl a (3 years old)
6 August 1985	Berlin	Dog kills man a (48 years old)
January 1986	Gosier	Dog kills a pensioner
6 February 1986	Frankfurt/Main	Two dogs kill an man (61 years old)
February 1988	Bavaria	Dog kills a old woman
5 November 1988	Odenwald	Dogs kill a man
November 1989	Buchholz	Dog kills a baby
20 March 1989	Karlsruhe	Three dogs kill a child (4 years old)
19 May 1989	Ofterdingen	Dog kills a child (7 years old)
September 1990	Berlin	Dog kills a boy (11 years old)
October 1990	Rottal-Inn	Three dogs kill an old woman
12 July 1993	Hannover	Dog kills a girl
27 June 1994	Bad Dürkheim	Dog kills a taxi driver
3 November 1994	Halberstadt	Dogs kills a drunken man
June 1995	Frankfurt/Main	Dog kills a woman (86 years old)
9 April 1996	Arnsberg	Dog kills a child (5 years old)
10 June 1996	Berlin	Dog kills a woman (86 years old)
10 June 1996	Mörfeld-Waidorf	Dog kills a woman (63 years old)
26 June 1996	Frankfurt/Main	Dog kills a woman (86 years old)
23 July 1996	Bamberg	Dog kills a child (3 years old)
15 February 1997	Zwickau	Dog kills a baby (7 months old)
28 April 1998	Bützow	Dogs kill a child (6 years old)
11 May 1998	Uckermarkt	Dog kills a woman
14 February 1999	Stralsund	Dogs kill a two children
2 February 2000	Frankfurt/Main	Dogs kill a woman (51 years old)
4 March 2000	Gladbeck	Dog kills a woman (86 years old)
March 2000	Untergruppenbach	Dog kills a man (24 years old)
June 2000	Hamburg	Dog kills a child (6 years old)
8 August 2001	Pinneberg	Dog kills a girl (11 years old)
28 March 2002	Zweibrücken	Dogs kill a child (6 years old)
3 April 2002	Neuental	Dogs kill a man (54 years old)
16 November 2002	Pforzheim	Dog kills a baby
1 September 2004	Bremen	Dog kills a man (drug addict)

2.4 Technical Risks

2.4.1 Introduction

The term technology here describes procedures and methods of the practical application of the laws of nature. Therefore, by definition, technical and natural risks cannot be completely separated since technology is based on natural laws. Nevertheless, there is a common understanding of technology that permits a separation from natural processes. Such understanding of technology includes industries, such as the construction industry, transport industry, chemical industry, food industry, energy sector, information technology and so on. The risks connected with such sectors will be looked at in detail.

2.4.2 Means of Transportation

2.4.2.1 Introduction

Motion is a general need of humans. The first things that humans learn are actions that permit movement. While economic considerations assume that an increased capacity of transport and motion means more wealth, other works such as Knoflacher (2001) suggest that more transport capabilities does not necessarily improve wealth. At a certain level of transport capacity, transport decreases wealth. In general, however, if one looks at the development of different means of transportation, such as road traffic, railway traffic, ship traffic, air traffic or space traffic, one finds a constant increase in transport volume. Figure 2-39 illustrates this statement for Germany in terms of the domestic transport volume in billion-tonne-kilometer within the last few decades. As seen in the figure, there is a strong competition between the different means of transportation. Very often the choice of transportation depends not only on the competitiveness but also on some general properties, such as distance.

Figure 2-40 shows the choices of mode of transport taken from a survey in 1999 in the German city of Heidelberg. Clearly, walking and biking are mainly chosen for short distances, whereas the individual car traffic is chosen for activities within a radius of about 10 km. Public transport reaches two peaks, one for very short distances – which is probably bus transport, and the second – most likely railway transport, for longer distances, say 50 km. It is interesting to note that, during the medieval ages, most humans and goods did not have a larger transport radius than about 100 km. Nowadays, however, more or less the entire surface of the earth is reachable by transport means.

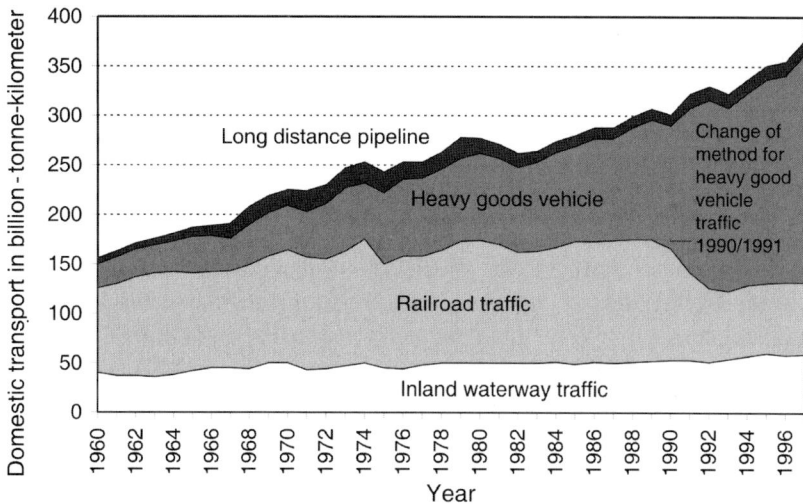

Fig. 2-39. Contribution of different means of transport in the development of internal transport in Germany from 1960 to 1997 (Proske 2003)

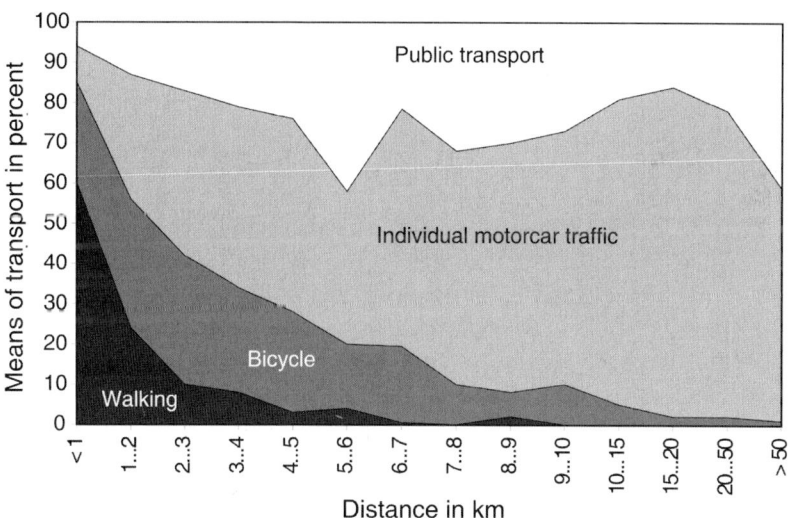

Fig. 2-40. Means of traffic versus distance after a survey in the year 1999 for the German city of Heidelberg (Heidelberg 1999)

2.4.2.2 Road traffic

Road transport is obviously older than motorcars. It has been assumed that since the invention of the wheel about 5,000–6,000 years ago, there was some form of road transport, even if those roads might not be comparable to what we see nowadays. However, the risk of travelling in old coaches will not be discussed here.

Modern road transport can be clarified in several ways. One possibility is the division according to the means of transport, such as bicycle traffic, motorcar traffic and so on. A second division considers public transport, individual transport, scheduled services and occasional traffic. Additionally, one can distinguish among the goals of transport, such as holiday transport, job transport, transport during spare time, transport for shopping and business transport. Figure 2-41 gives the portions and the development over one year for Germany based on the latter classification.

Table 2-33 shows that, over the last decades, the number of trips per person as well as the distance and time travelled have increased. Only within the last few years in Germany, there has been a small drop in individual travel due to the high unemployment rate and the increase in fuel prices. Additionally, the cold winter in 2005/2006 decreased the traffic volume.

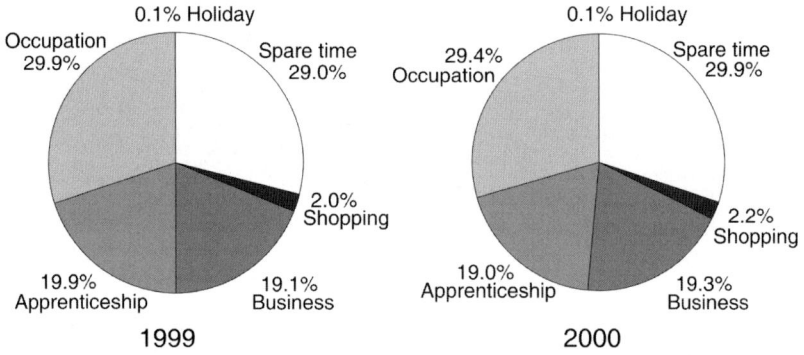

Fig. 2-41. Oriented type of transport for Germany in 1999 and 2000 according to Kloas & Kuhfeld (2002)

By neglecting these short time variations, the development of car traffic over the last century gives an impressive picture about the success and growth of that means of transport. Figure 2-42 shows the development of passenger cars in millions in Germany from 1914 up to 2003.

Table 2-33. Mobility indicators for Germany according to Chlond et al. (1998)

Indicator	1976	1982	1989	1992	1994	1995	1996	1997
Participation of population in traffic in %	90.0	82.2	85.0		91.9	93.9	92.9	92.0
Number of trips per person per day	3.09	3.04	2.75	3.13	3.32	3.39	3.46	3.52
Number of trips per mobile person per day	3.43	3.70	3.24		3.61	3.61	3.73	3.82
Number of motorcars per inhabitant				0.508	0.502	0.467	0.511	0.518
Travel time per day in hours:minutes	1:08	1:12	1:01		1:19	1:20	1:21	1:22
Kilometer per person per day	26.9	30.5	26.9	33.8	39.3	39.2	39.6	40.4
Average way length in km	8.7	10.0	9.80	10.8	11.8	11.5	11.5	11.5

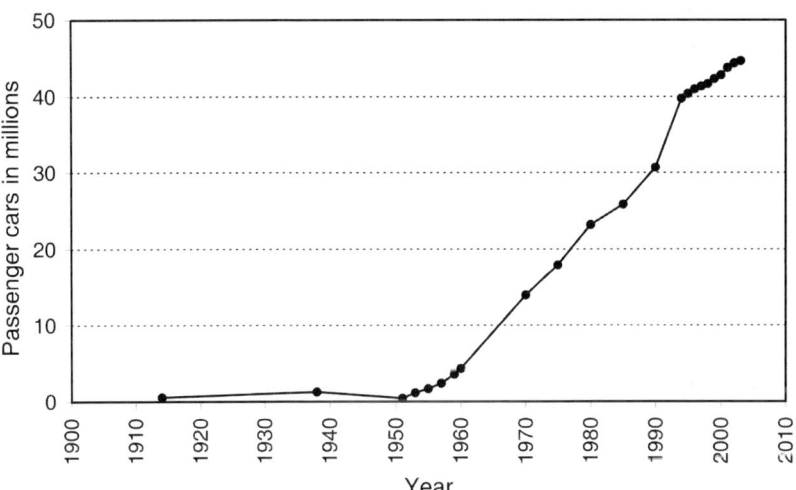

Fig. 2-42. Development of the number of passenger cars in Germany from 1914 to 2003 (KBA 2003)

The temporal development of road traffic fatalities for Europe, Poland, US, Japan and Germany is shown in Figs. 2-43, 2-44, 2-45 and 2-46. While Figs. 2-43 and 2-44 use a timescale of two decades, Fig. 2-45 shows the monthly development of fatality numbers in Poland. Highest values are usually reached in fall and lowest values are reached in spring. Figure 2-46 shows the fatality numbers for young drivers on a daily basis. As seen from the figure, the maximum fatality numbers occur on Saturday nights.

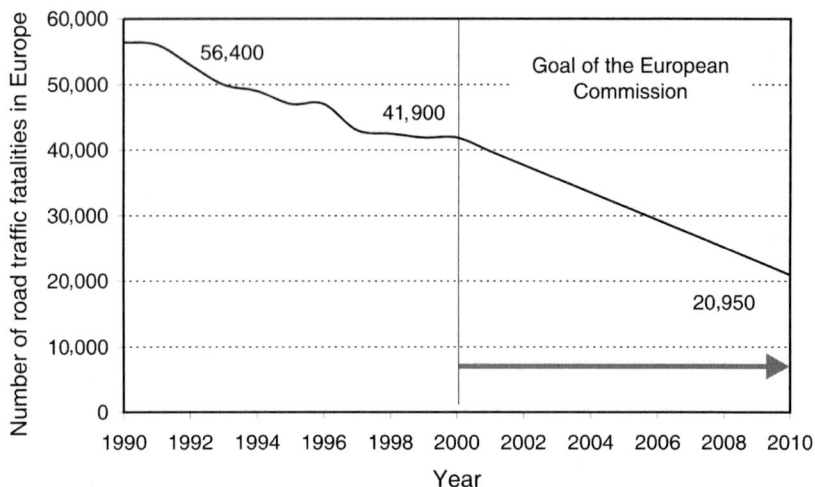

Fig. 2-43. Development of road traffic fatalities in Europe including goal values

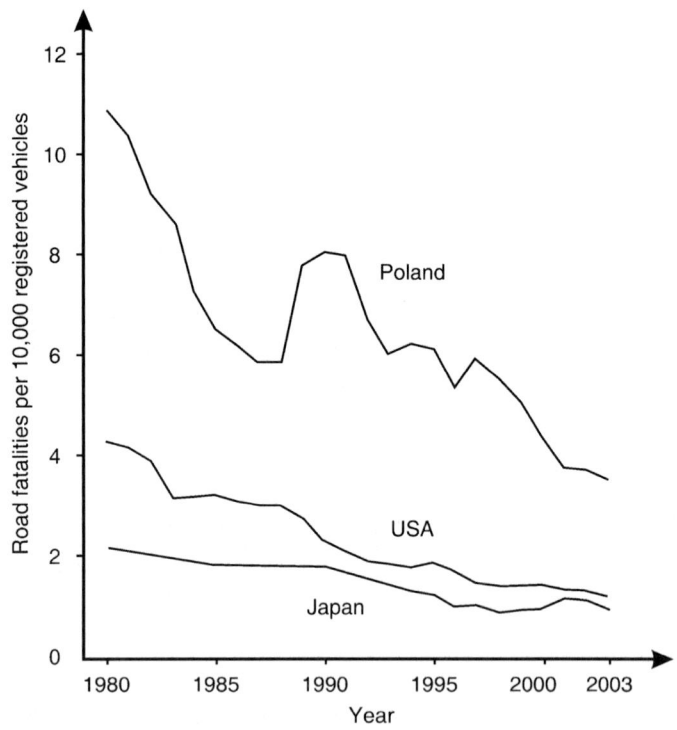

Fig. 2-44. Development of road traffic fatalities in Poland, USA and Japan according to Krystek & Zukowska (2005)

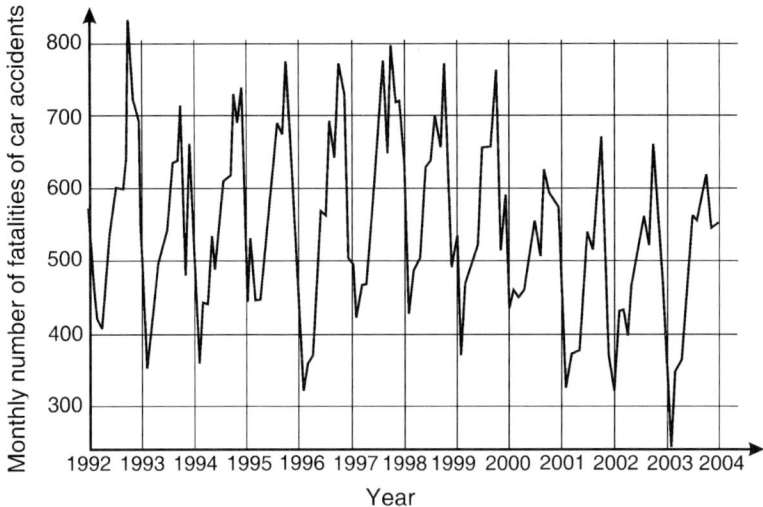

Fig. 2-45. Development of monthly numbers of road fatalities in Poland according to Krystek & Zukowska (2005)

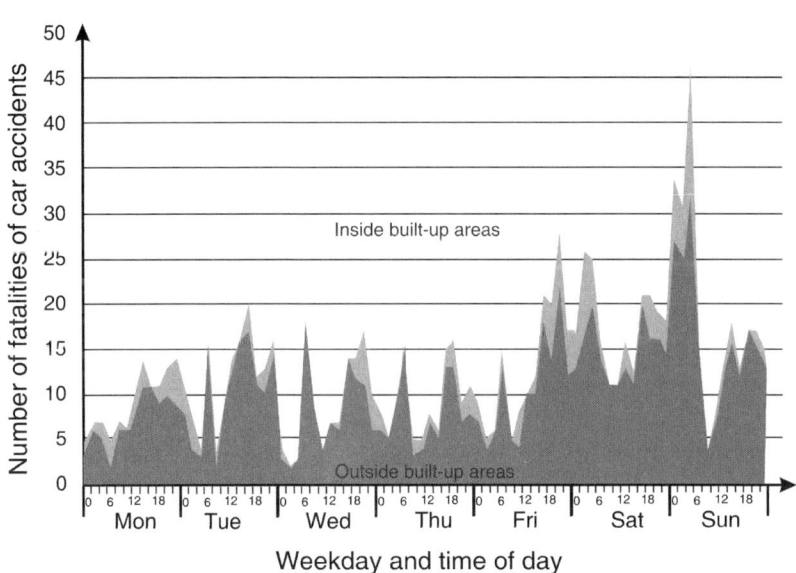

Fig. 2-46. Number of road fatalities involving people of age between 18 and 24 years over a week in Germany according to the SBA (2006)

About 1.2 million people worldwide are killed per year (over recent years) by road traffic – particularly in low-and middle-income countries (90%) and involving a significant number of young people (Peden et al. 2004). About 1,000 people under the age of 25 are killed per day world-wide in road accidents. Road accidents is the main cause of death on a global scale of children of age between 15 and 19. For the ages 10–14 years and 20–24 years, it is the second leading cause of death. Most young fatalities are vulnerable participants of the traffic. They are pedestrians, motor cyclists or passengers of public transport (Toroyan & Peden 2007).

Weather conditions also play an important role in traffic risks. Figure 2-47 shows the distribution of road surface temperatures related to some environmental conditions. The relationships between weather conditions and accidents has been investigated by Andrey & Mills (2003) and Andrey et al. (2001).

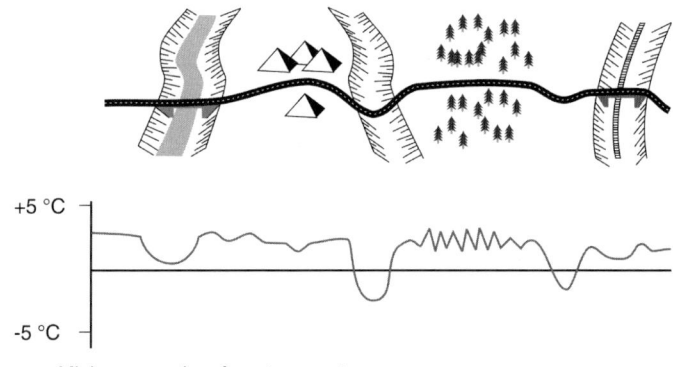

Minimum road surface temperature

Fig. 2-47. Pattern of minimum road surface temperature according to Smith (1996)

However, traffic itself can cause a risk structure as shown by two bridge collapses by lorry impact against bridges in Germany (Scheer 2000). Figure 2-48 shows force–time functions for different means of transport and debris flow impacts. On 28th June 1997, an overpass bridge collapsed over a motor highway in France, caused by a truck impact against a pier, resulting in three fatalities. In general, in France, the number of impacts of cars against piers or other parts of bridges is about 30 per year and about 20 for trucks. But, only in 1973, 1977, 1990, 1997 and 1998, collapses caused by impacts have occured (as far as the data reach). The probability of impact per bridge per year has been estimated at 0.0085 for cars and 0.006 for trucks. Considering the amount of traffic, the probability of an impact of a truck per passage is about 4×10^{-9} per year. The probability that a bridge collapses during an event is about 10^{-4}, not considering bridges that are designed for an impact. First design rules for impacts were introduced in 1966 and updated in 1977 and 1993 (Trouillet 2001, Holicky & Markova 2003).

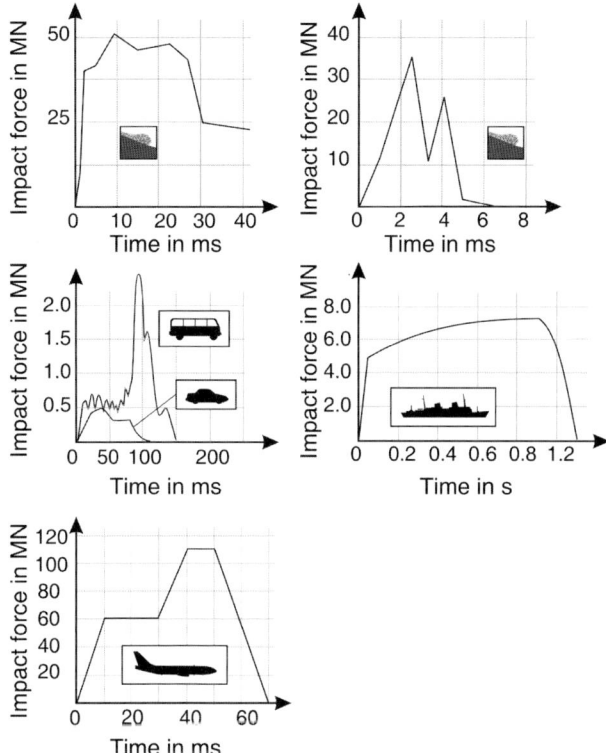

Fig. 2-48. Possible time impact–force functions by means of transport and natural processes (Zhang 1993, Rackwitz 1997, Proske 2003)

2.4.2.3 Railway Traffic

Trains are considered the safest way to travel as shown in Table 2-35, giving mortalities and accident rates for different means of transport. Following World War II, there were several years where no train passenger was killed. For example in Great Britain, for a period of 12 years after the World War II, no passenger was killed, excluding those who were killed jumping onto a moving train or those killed at crossings. However, the train operator was not responsible for these accidents. Since railway traffic started over 150 years ago in Great Britain, about 3,000 passengers were killed. Figure 2-49 shows the development of railway accidents for U.K. in recent years.

Table 2-34. Mortalities and accident frequencies for different means of transport according to Kafka (1999) and Kröger & Høj (2000)

Means of transport	Mortality or accident frequency
Railway (Japan)	7.69×10^{-13} passenger kilometer
Railway (goods traffic Germany)	5.00×10^{-7} goods kilometer
Road traffic (Japan)	6.67×10^{-11} passenger kilometer
Road traffic	4.67×10^{-6} car kilometer
Railway	2.13×10^{-5} train kilometer
Airplanes (scheduled flights)	4.19×10^{-9} flight kilometer

Fig. 2-49. Development of train accidents per million train-km in Great Britain over the last years (Evans 2004)

The soundness of railway traffic is due to the use of track chains, which mainly control the direction of the train. Along with this, the signal technology used nowadays is mainly automatic, providing a high level of safety.

However, in Europe, about 100 train passengers and about 800–900 people are killed per year (European Commission 2002). The latter number corresponds to fatalities at railroad crossings and suicides. Unfortunately in Germany, there have been some major accidents, strongly influencing the perception of the safety of railways. One such event was the the accident in Eschede, Germany, in 1998 with about 100 fatalities. This accident was the combination of two disasters: firstly the train derailed, and secondly a bridge was hit and collapsed. Another impact of a train against a bridge happened in 1977 in Australia, which claimed about 90 fatalities (Schlatter et al. 2001). Table 2-35 lists several major train accidents.

Table 2-35. List of several major accidents involving the railway according to Kichenside (1998), Preuß (1997) and DNN (2005)

Country, Location	Date	Number of fatalities	Remarks
Sri Lanka, Seenigama	26.12.2004	1,300	Tsunami hit a train.
France, Saint Michel	12.12.1917	660	A train with approximately 1,100 soldiers derailed due to overload and defect brakes.
Soviet Union, Tscheljabinsk	3.6.1989	645	A gas pipeline close to a train line explodes and hits two trains.
Italy, Balvana	2.3.1944	521	Due to overload of the steam engine, the train came to a standstill in a tunnel; people suffocated by the steam of the locomotive.
Ethiopia, Schibuti-Addis-Abeba	13.1.1985	428 fatalities, 370 injured	Train derailed due to high speed on a bridge, four railway carriages fell from the bridge.
Indonesia, Sumatra	8.3.1947	400	
Spain, Leon	16.1.1944	400	Train stoped in a tunnel.

2.4.2.4 Ship Traffic

The construction of ships is known since at least 10,000 years (Mann 1991). Such a long time of existence, however, does not mean that ships have been a safe form of transport during the last centuries. As a rough measure, it has been estimated that about 250,000 ships were lost on the coast of Great Britain (Wilson 1998). In the year 1852 alone, 1,115 ships were lost on this coast, claiming about 900 lives. One January storm lasting over five days took 257 ships, and 486 seamen lost their lives. The yearly maximum of lost ships was reached 1864 with 1,741 ships and 516 fatalities. But, the history of lost ships goes back to the beginning of this technology. In 255 B.C., a Roman fleet was returning from a battle in Cartago. The fleet came into a storm and about 280 ships were lost with 100,000 men (Eastlake 1998). In addition, the fall of the Spanish Armada in 1588 with the sinking of 90 ships and the loss of 20,000 men could be mentioned; however, this was not only due to natural hazards (Wilson

1998). A detailed description about the battle of the Spanish Armada can be found in Hintermeyer (1998).

In addition to these two famous events, there are many others, which are usually not considered. To create awareness about the risks of this technology, a few accidents are mentioned, which have been chosen based on the number of fatalities taken from Wilson (1998) and Eastlake (1998):

- In 1545, the "Mary Rose" was lost with 665 men.
- On 22nd October 1707, four ships run aground after they lost orientation due to a storm. About 1,650 men were lost.
- In 1852, the "Birkenhead" run aground claiming 445 fatalities.
- In 1853, the "Annie Jane" was lost with 348 fatalities.
- In 1854, the steam ship "City of Glasgow" was lost with 480 passengers.
- In 1857, the"Central America" was lost with 426 passengers.
- In 1858, the "Austria" was lost with 471 passengers.
- In 1859, the "Royal Charter" was lost with 459 fatalities.
- In 1865, on the "Sultana" the steam boiler exploded, killing approximately 1,600 people.
- In 1870, the "Captain" sunk due to a gust with a loss of 483 seamen.
- In 1873, the "Atlantic" was lost with 560 men.
- In 1874, the "Cosapatrick" was lost with 472 fatalities.
- On 3rd September 1878, the passenger ship "Princess Alice" was hit by a carbon steam ship, killing about 645 people.
- In 1898, the "La Bourgogne" and the "Cromartyshire" collided and 546 people were drowned.
- In 1904, one of the biggest ship disasters in New York was the fire of the "General Slocum", killing about 955 people, mainly woman and children.
- In 1912, the "Titanic" was lost with 1,503 fatalities.
- In 1914, the "Empress of Ireland" and the "Storstad" were lost with 1,078 men.
- 1917, the "Vanguard" was lost due to an explosion on the Orkney Islands, claiming nearly 670 fatalities.
- On 30th January 1945, probably, the biggest ship disaster happened. The "Wilhelm Gustloff" carrying German refugees was attacked by a submarine, claiming about 9,000 fatalities.
- In 1957, one of the biggest sailing ships built, the "Pamir", was lost with 80 men.

- In 1987, the biggest ship disaster in peace times was the tragedy with the "Dona Paz" on the Philippines. The "Dona Paz" was a passenger ferry designed for about 1,500 people, but was occupied by about 4,400 people. The ship was involved in a crash with the oil tanker "Vector". Unfortunately, the oil exploded, claiming nearly 4,400 fatalities.
- In 1994, the "Estonia" was lost claiming 757 fatalities.

Due to the high frequency of ship losses, especially during the middle of the 19th century, some safety measures were implemented. For example, on 29th May 1865, an organisation for the rescue of shipwreck survivors was established in Germany. Since then, the organisation has saved approximately 62,000 people (Hintermeyer 1998). Even nowadays, heavy accidents involving ships still occur. In 2006, a ferry sunk in Egypt with about 1,600 passengers. Figure 2-50 shows the loss of ship tonnage in the last years.

Much work has been carried out to develop models indicating the possible accident risk of ships (Gucma 2005, 2006 a, b, Gucma & Przywarty 2007). This becomes especially important if ships are used for the transportion of hazardous materials. A list of ship accidents during the transportion of LNG (Liquefied Natural Gas) is given in Table 2-36.

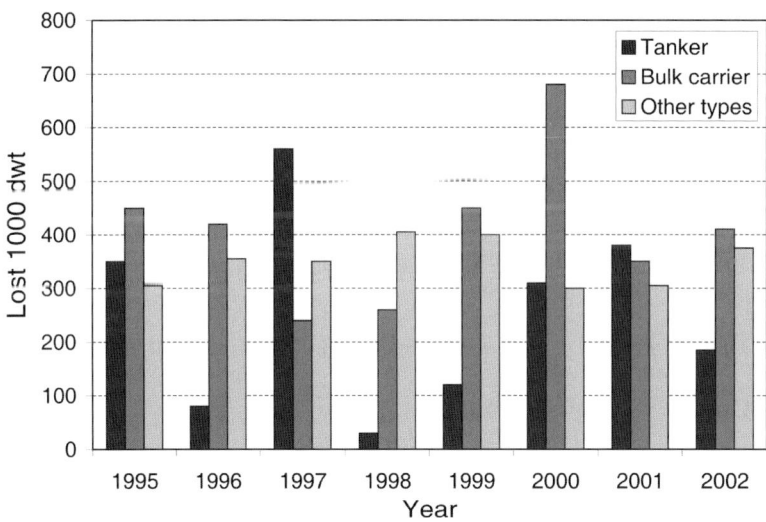

Fig. 2-50. Loss of death weight tonnage (dwt) of ships for several years (ISL 2007)

Table 2-36. Accidents or damages involving LNG carriers (Foss 2003)

Year	Ship	Description of event	Personal Injuries	Damage to ship	LNG release
1965	Jules Verne (now Cinderella)	Overfilling	None	Fractures in tank cover and deck	Yes
1965	Methane Princess	Valve leakage	None	Fractures in deck	Yes
1971	Esso Brega (now LNG Palmaria)	Pressure increase	None	Damage to the top of cargo tank	Yes
1974	Massachusetts (barge)	Valve leakage	None	Fractures in deck	Yes
1974	Methane Progress	Touched bottom	None	None	No
1977	LNG Delta	Valve failure	None	None	Yes
1977	LNG Aquarius	Overfilling	None	None	Yes
1979	Mostefa Ben Boulaid	Valve leakage	None	Fractures in deck	Yes
1979	Pollenger (now Hoegh Galleon)	Valve leakage	None	Fractures in tank cover and deck	Yes
1979	El Paso Paul Keyser	Stranded	None	Severe damage to hull and cargo tanks	No
1980	LNG Libra	Shaft moved against rudder	None	Fracture to tailshaft	No
1980	LNG Taurus	Stranded	None	Hull damage	No
1985	Gadinia (now Bebatik)	Steering gear failed	None	None	No
1985	Isabella	Valve failed	None	Fractures in deck	Yes
1989	Tellier	Broken moorings	None	Hull damage	Yes
1990	Bachir Chihani	Hull fatigue	None	Structural cracks	No
1996	LNG Portovenere	Firefighting system malfunction	Six dead	None	No
2002	Norman Lady	Collision with submarine	None	Minor hull damage	No
2003	Century	Engine breakdown	None	None	No
2003	Hoegh Galleon	Engine breakdown	None	None	No
2004	Tenaga Lima	Damage to stern seal	None	Minor repairs	No
2004	British Trader	Fire in transformer	None	Minor repairs	No
2005	Laieta	Engine breakdown	None	Minor repairs	No

Table 2-36. (Continued)

Year	Ship	Description of event	Personal Injuries	Damage to ship	LNG release
2005	LNG Edo	Gearbox vibration	None	Replacement gearbox	No
2005	Methane Kari Elin	Leaks in cargo tanks	None	Extensive repairs	No
2006	Catalunya Spirit	Damaged insulation	None	Extensive repairs	No

2.4.2.5 Air Traffic

Firstly, the air traffic risk for the uninvolved public shall be considered. This risk seems to be rather low. Table 2-37 mentions some incidents where airplanes hit uninvolved people on the ground. Figure 2-48 also includes the design impact–force function for an airplane impact. Between 1954 and 1983, about 5,000 airplanes crashed worldwide (not involving data from China and the Soviet Union). If one estimates that 1% of the entire world is habituated, then the probability of being hit by a crashing airplane is about 10^{-8} per year (van Breugel 2001).

From 1953 to 1986 in the Western world, about 8,300 jet aircrafts were built. In 1986, about 6,200 were in operation. The worst accident happened in 1977, when two Boeing 747 collided at the airport of the Canary Islands with nearly 600 fatalities (Gero 1996). However, in general, the number of airplane accidents has dropped since the beginning of the 1990s, as shown in Fig. 2-51. This statement has to be seen in the general context of air traffic development as shown in Fig. 2-52.

Statistical data about airplane crashes are given in Tables 2-38, 2-39, 2-40, 2-41, 2-42, 2-43. While Table 2-38 gives some general frequencies of airplane loss, Tables 2-39 and 2-40 give data based on flight stages. Table 2-41 gives accident data subject to the airplane type, not including passenger airplanes of the 4th generation. Tables 2-42 and 2-43 give data for some German airports. Figure 2-53 gives further information for Table 2-42. It should be mentioned here, however, that in Germany before the unification in 1990, helicopter crashes on the German border heavily contributed to the number of flight accidents (Weidl & Klein 2004).

The comparison of the risks of different means of transport depends very strongly on the parameter chosen. While in Table 2-34, airplanes showed a low risk by distance-depending fatalities, in Fig. 2-54, air transport includes a higher risk by trip-based risk parameters.

Table 2-37. Examples of airplanes hiting ground structures (van Breugel 2001)

Year	Description and location	Fatalities
1987	Small airplane crashed into a restaurant in Munich, Germany	6
1987	A-7 Corsair crashed into a hotel in Indianapolis, USA	14
1987	Harrier Jump-Jet crashed into a farm near Detmold, Germany	1
1988	A-10 Thunderbolt II hit 12 houses near Remscheid, Germany	6
1988	Boeing 747 exploded over Lockerby, Scotland; airplane parts hit petrol station and houses	280
1989	Boeing 707 crashed into a shanty town near Sao Paulo, Brasil	17
1990	Military airplane hit a school in, Italy	12
1992	Hercules Transporter crashed into a motel in Evansville, USA	16
1992	Boeing 747 crashed into 10-floor house in Netherlands	43
1992	C-130 crashed into a house in West Virginia, USA	6
1996	Fokker-100 crashed into houses in Sao Paulo, Brasil	98
2000	Military airplane crashed into a house in Greece	4
2000	Boeing 737-200 crashed into house near Patna, India	57
2000	Concorde crashed into a hotel near Paris, France	113

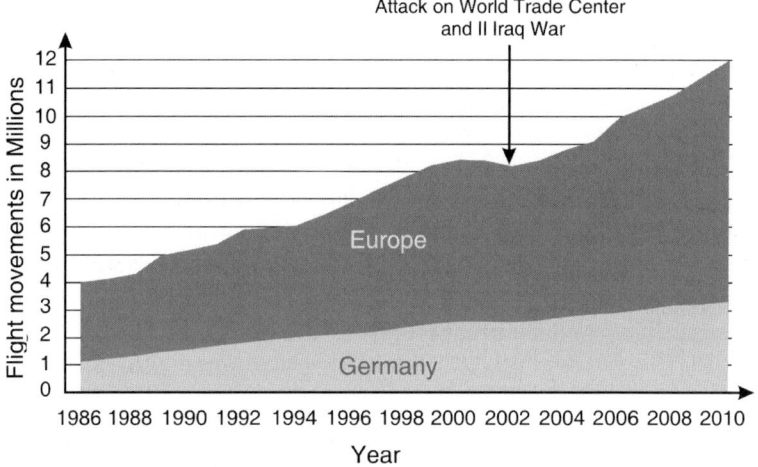

Fig. 2-51. Development of flight movements in Europe and Germany over the last decades (Konersmann 2006)

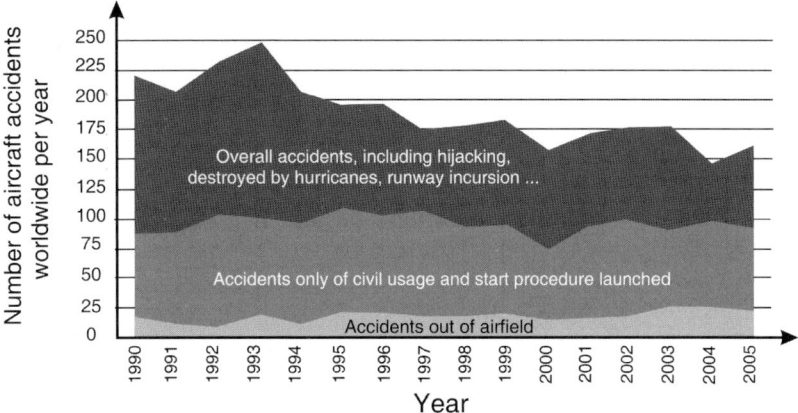

Fig. 2-52. Development of aircraft accidents worldwide considering only aircrafts with a minimum weight of 5.7 tonnes (Konersmann 2006)

Table 2-38. Frequency of airplane crashes according to Weidl & Klein (2004)

Source	Frequency
German risk study nuclear power plants (part 4)	2.5×10^{-12} per flight and km
German risk study nuclear power plants (phase B)	6.0×10^{-7} per flight
Bureau of Transportation Statistics – >20 tonnes	3.84×10^{-10} per flight and km
Bureau of Transportation Statistics – <20 tonnes	1.11×10^{-8} per flight and km
German Federal Bureau of Aircraft Accidents Investigation, Report 2001	6.5×10^{-7} per flight
International Air Transport Association – Ranking	4.24×10^{-10} per flight and km

Table 2-39. Distribution of flight accidents to flight stages based on data for the years 1959–1985 (Moser 1987)

Stage of flight	Share on flight time in %	Share on accidents in %	Trend
Take off	1	21.8	Falling
Climb	19	7.2	Growing
En route	37	5.5	Falling
Initial approach	14	6.1	Constant
Final approach	10	32.4	Constant
Landing	1	24.5	Falling
Rolling	18	2.5	Growing

Table 2-40. Number of air traffic accidents related to different traffic types using data from the year 1985

Accidents involving different types of flights	Number
Fatal accidents in passenger schedule air traffic	16
Fatal accidents in passenger non-schedule air traffic	5
Fatal accidents in passenger regional and shuttle air traffic	10
Fatal accidents in cargo traffic	9
Non-fatal accidents in passenger schedule air traffic	115
Non-fatal accidents in passenger non-schedule air traffic	5
Non-fatal accidents in cargo traffic	16

Table 2-41. Failure rates of different airplanes. However, be aware that some of the airplanes were operating under difficult conditions, such as the Fokker F.28 (Moser 1987)

Model	Percentage of airplanes involved in accidents from all produced airplanes	Number of fatal accidents per million flights
1. Generation I		
Aérospatiale Caravelle	11.8	
DeHavilland Comet	9.8	
Convair 880/990	8.8	
McDonnell Douglas DC-8	8.1	
Boeing 707	7.6	
Boeing 720	3.3	
2. Generation II		
Fokker F.28	7.4	3.38
British Aerospace One-Eleven	5.7	0.54
Vickers VC10	3.7	
McDonnell Douglas DC-9	3.5	0.49
Hawker Siddeley Trident	2.6	
Boeing 737	2.2	0.74
Boeing 727	2.1	0.51
3. Generation III		
McDonnell Douglas DC-10	3.0	2.87
Boeing 747	1.5	1.51
Lockheed 1011 TriStar	1.2	1.21
Airbus A300	0.8	

Table 2-42. Examples of accident rates for some airports (Konersmann 2006) – See also Fig. 2-52.

Take off weight class	1	2	3	4	5	6
Number based on stocktaking 2005	4,847	1,792	5,025	8,806	3,307	2,603
Average number of daily possible flights (take off and landing = one flight)	5	8	6	5	3	2
Average number of flights per year in million	8.845	5.232	11.004	16.070	3.621	1.900
Average number of risk relevant accidents	9	1	5	4	2	1
Reference accident rate $\times 10^{-7}$	10	1,9	4,5	2,5	5,5	5,3

Table 2-43. Accident numbers for the Frankfurt airport (Germany) based on Fricke (2006)

	2000	2015
Probability of accident per flight	5.75×10^{-8}	5.15×10^{-8}
Probability of accident per year	0.026	0.034

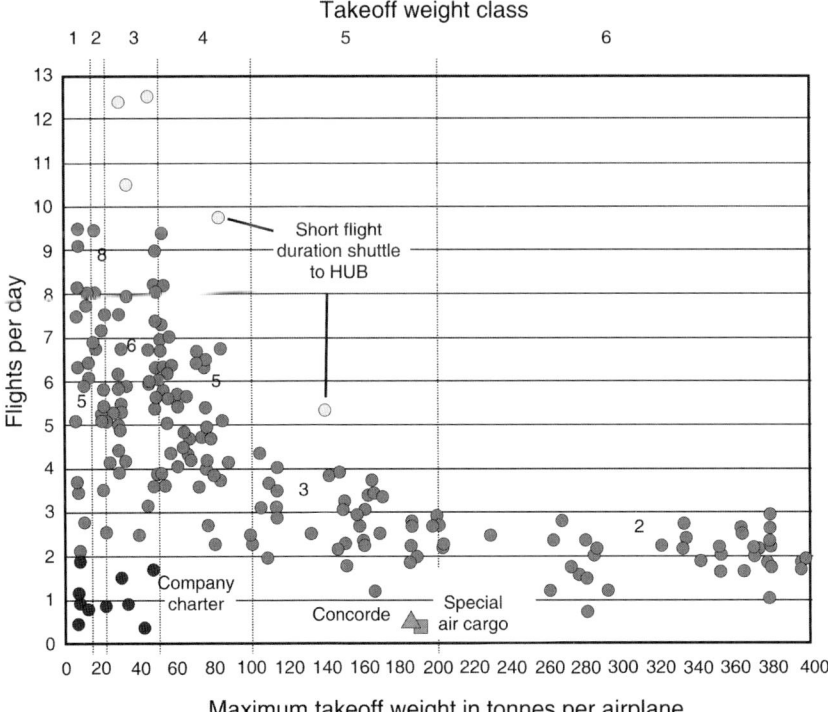

Fig. 2-53. Number of flight movements per day compared to takeoff weight (Konersmann 2006)

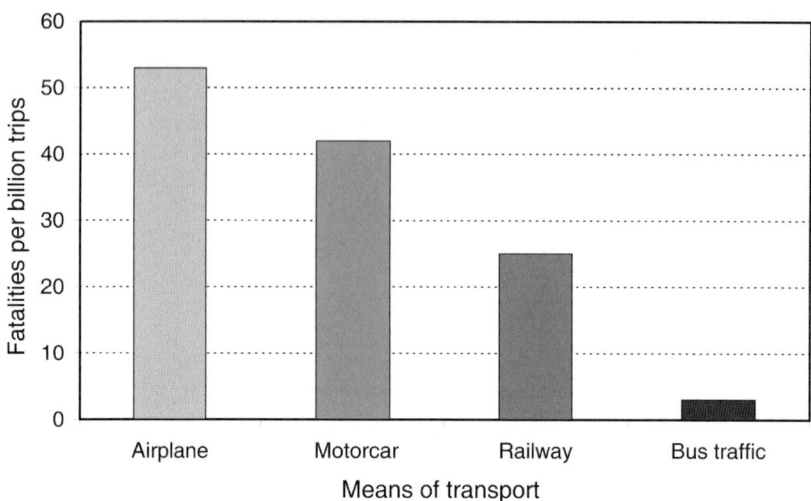

Fig. 2-54. Comparison of fatalities per billion trips for different means of transport (LIN 2003)

2.4.2.6 Space Traffic

In comparison to the other means of transport, space travel features some special properties. Firstly, the number of people moved by this means of transport is rather less. Additionally, the costs of transport are extremely high. However, in terms of risk, space travel contains a rather high risk. This has not only been shown by the losses of Space Shuttles, but was also valid for the first space flights. By 1979 about 92 astronauts had been to space. Four astronauts were killed by accidents during travel, while some astronauts were killed on earth during tests in 1967. Here, probably the best known accident was the fire of an Apollo capsule on ground. However, severe rocket explosions have also caused high numbers of fatalities. In April 1967, the Russian astronaut Komarow was killed when the parachute did not open during return. In 1970, the astronauts of Apollo 13 luckily survived the mission. In June 1971, three Russian astronauts were killed by a pressure loss during re-entry. (Mielke 1980)

Although the Space Shuttle concept was considered to improve the safety and to drop the probability of complete loss from 10^{-1} to 10^{-2} (Paté-Cornell & Fischbeck 1994), the explosion of the Challenger in 1986 and the loss of the Columbia in 2003 showed drawbacks in this Space Shuttle concept. Considering these two losses and approximaetly 130 missions, one obtains 2/130=0.015 fatality per mission. This might also be considered fatality if the number of passengers per mission would be equal and astronauts would not have a second mission. Considering the 450 people who have been to space so far and the nearly 20 fatalities, this yields to 20/450= 0.04 fatality per mission. Future space missiles should feature a further improvement of safety in terms of probability of loss as shown in Table 2-44.

Table 2-44. Probability of loss or damage for planned new missiles (Altavilla et al. 2000)

Projects	Probability of damage or loss
Hermes Project	Failure of missile per mission 10^{-4}
	Loss of parts per mission 10^{-3}
	Overall per mission 10^{-2}
Assured Crew Return Vehicle (ACRV)	Failure during mission and four years in space 3×10^{-3}
Crew Rescue Vehicle (CRV)	Failure during mission and five years in space 5×10^{-3}
Crew Transport Vehicle (CTV)	Loss of missile per mission 3×10^{-3}
	Heavy damage per mission 2×10^{-3}
	Risk to population on ground per mission 10^{-7}

2.4.3 Dumps

While the former sections discussed means of transport, this section discusses dumps, which are used for the spatial fixing of materials. Such materials are mainly solid wastes. The amount of waste materials generated has reached alarming values. For example, in the US, the amount of wastes has grown from 0.5 million tonnes per year during the 1940s to about 300 million tonnes in 1993 (Garrick 2000). In Germany, the overall amount of wastes produced has grown during the last decades too. While in 1983 about 30 million tonnes of debris was produced, in 1993 it reached 43 million tonnes (Merz 2001). Fortunately, over the last decades, the amount of wastes per capita has remained constant and the amount of special debris has fallen, even though the definition of hazardous waste has become wider. Hazardous waste is of particular interest, since it includes toxic materials.

The development of the construction of high-safety dumps or high-safety barriers designed for hazardous waste was intensively discussed during the 1980s and 1990s in Germany (HMUE 1992). This was due to the limited capacity for waste burning and disposal in Germany at the end of the 1980s. Since then, the regulations for opening new dumps have become even harder. Dump sites are obliged to obtain permission, which is now restricted for densely populated areas. Currently, it is assumed that about 1/3 of wastes can be burned or recycled. This is probably a fitting value for most developed countries. In addition, waste is also exported from many countries. These limitations of dump sites have yielded to increased efforts since the 1980s to reduce the overall amount of wastes. Waste at present, and into the future, is treated by separation, burning, gasification, hydration or biological treatment prior to dumping (Merz 2001).

Nevertheless, in Germany, 3,000 dumps still remain in use. Most of them are used for rubble and earth excavations, while only a small number (300) being used for domestic wastes. The standard quality of these remaining dumps, however, is still rather low: only 2/3 of the dumps, treat the waste gases and only 1/5 clean the waste water. In about 10% of all domestic waste dumps, the water is not treated in any way. Approximately 70% of the domestic waste dumps pump the water into a municipal clarification plant. In the former German Democratic Republic, about 10,000 waste dumps were assumed to exist. In 1990, about 6,000 were still in use, with only 120 of them complying with dump standards (Paffrath 2004).

One has to mention, however, that even in developed countries in the 1970s, terms like "waste separation", "dump foundation sealing", "seepage water capture", "degasing and control" and "aftercare" were unknown to dump designers. Nowadays, many different systems for dumps are developed and offered by companies. Such high-safety dumps include, for example, multi-barrier systems, and also the aftercare has been put into regulations. Nowadays, dumps are not seen anymore as passive storage but as reactors, in which chemical and biological processes occur. The lifetime of dumps is usually between 30 and 100 years. However, even nowadays absolute safety of dumps cannot be reached. Acceptable values about the release of materials can be found in different regulations. Here, only the requirements for the application of concrete as a second barrier against water-endangering fluids should be mentioned. Water-endangering fluids are those that which are able to contaminate bodies of water and change their properties. If such fluids are released, damage and harm to the environment and humans is assumed. Usually the so-called second barrier experiences only very rare contact with the fluid, e.g. about 10^{-4} per year. Based on some safety classes, acceptable failure risks in terms of volume released material per time are presented in Table 2-45 (Wörner 1997, Kiefer 1997).

Table 2-45. Acceptable released volume of water-endangering fluids per time (Kiefer 1997)

Safety class	Safety requirement for ground and water	Acceptable volume
1	Low safety requirement (for example, already contaminated ground)	1 m³/100 years
2	Usual safety requirements	1 m³/10,000 years
3	Special safety requirements (for example, drinking water reservoir)	1 m³/100,000 years

To calculate the final acceptable risk, the hazard of the fluid also has to be considered. This hazard rate is based on the toxicity and the migration capability of the fluid. Toxicity can be described using different measures for different animal groups. Here, the Median Lethal Dose (LD_{50}) has been used. The exposure of that dose yields to a mortality of 50%. As mentioned, such toxicity measures have been adapted to mammals, fishes and bacteria. In Germany, in 1998, about 2,700 cases of uncontrolled releases of such water-endangering fluids were registered. About 900 cases occurred in storage devices.

However, not all releases are documented or yield to further measures. Since 1912, in Segnitz (Germany), color pigments (Schweinefurter green) were produced. That yielded to an average arsenic concentration of 2,500 mg/kg in the soil and 9,300 mg/l in the groundwater. The concentration levels of copper reached 2,300 mg/kg in the soil. In Fürth, (Germany), a mirror production site was closed. The mercury concentration reached 20 µg/m³ in the air, 3,000 mg/m³ in the rubble and 4,000 mg/m³ in the soil. The tar lake in Rochlitz, Thuringia, is well known. Here, nearly 20 million tonnes of brown coal smolder tar as well as nearly 10 million tonnes of oil were processed. Additionally, during World War II, the plant was bombed and about 100,000 tonnes of materials was released. The ground and the groundwater were heavily contaminated, yielding to so-called swimming oil lenses in the groundwater. Many technologies have been developed to clean such locations, for example HOT-PACK, MEC-TOOL, XTRAX, BEST, mobile oven, biological techniques and stabilization techniques. As a last example, the Wismut Company for the mining of uranium in East Germany is mentioned. The cleaning-up of radioactive waste at dump sites had cost several billion Euros (Paffrath 2002).

Further examples of the uncontrolled release of wastes can be found in the mining industry. Here, great amounts of dump sites occur, which often include toxic materials and, therefore, represent a hazard to the environment. Often the waste is stored in lakes braced by artificial dams. If such dams fail, and the water of the lake is released into the environment, then an environmental, disaster could happen. Examples of dam failures are discussed in the following section.

2.4.4 Dam Failure

On 7th August 1975, in the central China province of Henan, after 26 h of heavy rain, the Banqiao dam, the Shimantan dam and others failed. The failure released approximately 600 million cubic-meters of water, which traveled with a velocity of 50 km/h over the valleys and plains. According to Chinese sources, about 85,000 people died within 24 h of the failure. Another 145,000 people died in the days after due to famine and disease. If these data are correct, then the failure of these dams was one of the biggest disasters in the world – only exceeded by the storm surge in Bangladesh in 1970, the Tangshan earthquake in 1976 and the Tsunami disaster in 2004. It can be seen as the biggest technical disaster of all times (Lind & Hartford 1999)

This disaster shows clearly the connection between technical and natural risks. The dam would probably not have failed under regular weather conditions. In addition, the relationship between primary and secondary disasters is visible. More people died as a result of the secondary disaster of insufficient food and disease. This is a clear sign of the collapse of the social system after the primary disaster. It is quite often described by the term vulnerability. If, during an earthquake, some bridges collapse, then the region might not be accessible by external help and is, therefore, extremely vulnerable. This has been intensively discussed. For example, in the region of Cologne in Germany, a vulnerability study has been carried out investigating the possible impacts of the collapsing of the Rhine bridges during either a flood or earthquake. In Japan, studies have considered failure of bridges after earthquakes. It does not, however, have to be only structural failures that increase the vulnerability of a region. For example, there might be a collapse of the telephone and electrical systems, and therefore no information about a disaster can be communicated. In this case, satellite observation could be used on identify sudden changes in regions.

Returning to the topic on dams, not all dam failures have resulted in such terrible disasters. From 1960 to 1996, of approximately 23,700 dams in the USA, about 23 dams failed (Lind & Hartford 1999). These failures claimed 318 lives. One of the latest dam failures occurred in Syria in June 2002 with about 22 fatalities and making 3,800 people homeless. Some dam failures are listed in Table 2-46. Figure 2-55 shows the Katse dam in Lesotho, Africa.

Table 2-46. Failure of dams (Pohl 2004)

Dam	Country	Year of accident	Number of fatalities
Puentes	Spain	1802	680
Dale Dyke Dam	U.K.	1864	250
Qued-Fergoug	Algeria	1881	200
South Fork/Johnstown	USA	1889	2,209
Bouzey	France	1895	100
Austin	USA	1910	100
Saint Francis	USA	1928	500
Möhne	Germany	1943	1,200
Vega de Tera	Spain	1959	400
Malpasset	France	1959	400
Vajont	Italy	1963	2,000
Vratsa	Bulgaria	1966	600
Rapid City	USA	1972	250
Macchu-Earth dam	India	1979	2,500

Fig. 2-55. Katse dam in Lesotho

A special dam failure case happened in 1963 in Vajont, Italy. This case was already mentioned in the section on debris flows and landslides. In this case, as a result of heavy rainfalls, a huge slope failure occurred. A volume of about 2 km × 1 km × 150 m moved into the lake behind the dam and caused a 70-m-high flood wave. The wave killed about 2,500 people in areas behind the dam. This disaster has been intensively discussed not only in the field of debris flow but also in the Courts. In this particular case, the dam itself did not fail, but the design of the overall dam system failed, because the possibility of the slide was not considered (Pohl 2004).

Even if dams completely meet their structurally intended goals, other types of risks are generated. In 1950, approximately 5,000 dams with a crown higher than 30 m existed. This number has grown to 45,000 by the year 2000. This has had major economical, ecological and social impacts. In order to build the dam on the Yangtze River in China, nearly two million people were evacuated. Currently, there is a discussion whether new massive dams can cause earthquakes (Klesius 2002).

2.4.5 Structural Failure

The failure of dams is a special case of the failure of structures. Usually the failure of structures is related to two causes: either extremely high loads, which are called accidental loads, such as impacts; or insufficient strength mainly related to human failures. Both can be found in developed and developing countries. For example, on 3 February 2004 in Turkey, a ten-story building collapsed; on 27 January 2004, a building collapsed in Cairo, Egypt, after a fire; and on 16 February 2004, a swimming pool building collapsed in Moscow killing 25 people. Also, in the spring of 2004, a part of the new terminal at an airport in France collapsed.

Examples from Germany are the collapse shown in Fig. 2-56 or the collapses of the Halstenbeker sports hall close to Hamburg in 1997 and 1998. The structure received the nickname the buckling egg. Since two collapses happened during the construction phase, nobody was injured, but the structure was never completed. The failure was due to a rather complicated structure type, which was very vulnerable to small changes to the geometry. A second example of structural failure in Germany occured on 2 January 2006 in Bad Reichenall. Again a sports hall collapsed in the afternoon killing 15 people, most of them children. The structural failure led to an intensive debate about the safety of public buildings in Germany. It was assumed that snow overload caused the failure, since also in Poland a hall collapsed for that reason in the same winter. This discussion echoed in the beginning of the 1970s, when many light-weight structures in East and West Germany failed under snow load. Initially, sabotage was assumed to be the cause of failure in East Germany; however, very optimistic assumptions about snow load and the change of the construction material from wood to steel were later identified as the causes. It was determined that the cause of failure in Bad Reichenhall was a combination of lack of maintenance, structural changes over time and the application of new bonding materials. As a consequence of that disaster, the Ministry for Buildings and Structures decided in December 2006 to change the law regarding the safety standards of public buildings by requiring additional inspections.

Still, even including that event, structures in Germany remain an extremely safe technical product. In Germany in 2004, about 17,293,678 residential buildings (Knobloch 2005) and about 5,247,000 non-residential buildings (SBA 2005) were in use. The entire population of 80 million people is virtually exposed more than 20 h per day. Thus, an average yearly fatality number, of 10 is extremely low.

This low risk is mainly reached by high control efforts – not only during the design process but also at the construction site itself. A few examples should be mentioned. In 2000–2001, an overpass over a freeway interchange was erected. During the erection of the superstructure, a rather low concrete strength was reached. Many experts were asked about the cause, and heavy discussions took place about the possible reasons and who was to be blamed. At last, it turned out additives for the concrete production were produced in a chemical plant, which had accidentally labeled textile softener as concrete additive. The textile softener caused an insufficient low strength of the concrete, and the superstructure blasted right after construction in the summer of 2001. Another example is a highway bridge in Saxony close to Dresden, Germany, where during the pouring of concrete for the superstructure, too few concrete workers were at the site. Therefore, the concrete was not properly compacted and the concrete structure through test drillings was shown to have hollow cavitations. This, of course, decreased the effective cross-section of the bridge and, therefore, strongly influenced the load-bearing capacity. Parts of the bridge thus had to be rebuilt again.

In comparison to events of accidental loads, such construction cases remain as rather individual problems. Earthquakes, in particular, cause accidental loads on structures that often claim many victims. For example, many buildings collapsed due to earthquakes in Morocco on 25 February 2004, and on 6 February 2004 in Bam in Iran. About 70% of the buildings of the city were destroyed, killing between 30,000 and 40,000 people. Already in the section on earthquakes, the difference between the consequences of earthquakes in developed and developing countries was discussed. Many of the fatalities in developing countries could have been avoided by the application of modern structural codes. However, comparable problems might be observed in developed countries as well, as the next example describes.

First of all, modern codes require that the design and construction of structures be carried out by specialist staff. However, if one actually looks at many construction sites, reality quite often gives a different impression. For example, there were many cases in Germany where nobody on-site could speak German, but all the plans were given in German. In one case, the gradient on a new bridge was wrong, and some workers were asked to grind off the concrete to meet the required gradient. Since the workers did not understand German, they grinded off the concrete until sparking. By then the reinforcement and, even worse, the prestressing elements of the bridge had already been reached.

The majority of building failures are caused by human error. The results of an investigation of 800 damaged buildings are summarized in Table 2-47 (Matoussek & Schneider 1976). The so-called damage report of the German Ministry of Buildings and Structures also supports the findings of this table. Ignorance is by far the most prominent cause of faults. A good example of ignorance was seen by the author during a visit to a hall construction site, when steel anchors were found in the dump. The site engineer then discovered that some workers had simply detached the anchors from a steel column because the concrete was poured too high. The anchoring elements were essential for the load-bearing behaviour of the column, but the workers just wanted to continue with their work.

The next major cause of damages is insufficient knowledge. This might actually be connected to ignorance either through limited communication capabilities or through the employment of cheap but badly trained workers. The negative impacts of badly trained workers cannot only be found in the field of structures but also in hospitals, nuclear power plants and other fields.

The distribution of the causes of structural damages related to the different stages of structural development can be seen in Table 2-48. About 50% of errors are made during the construction phase. This is only acceptable due to the major effort in quality control and inspection and the low vulnerability of structures under normal conditions. In general, structural materials such as steel, reinforced concrete or masonry are able to redistribute loads in case of local failures and, therefore, increase the safety of structures – especially if so called "static indeterminate structures" are designed with a high degree of robustness. The quality control itself includes a further 10% of error (Table 2-48), but in combination with the other values, it results in rather low probabilities of failure:

$$P(I|A)=P(I) \times P(A)=0.1 \times 0.1 = 0.01$$
$$P(I|E)=P(I) \times P(E)=0.1 \times 0.4 = 0.04$$
$$P(I|C)=P(I) \times P(C)=0.1 \times 0.5 = 0.05$$

Of interest to note is the distribution of errors between the architects and the civil engineers. The common explanation is the possibility of causal relations in the different professions. While civil engineers produce easily controllable static computations, architects create uncontrollable results – for example, the beauty or usability of a building. In the last few years, however, at least in Germany, this is changing. Architects have become more and more responsible for their products.

The above figures show that at least every tenth structure produced has faults. The majority of faults are caused by human errors. This is a latent phenomenon and does not have to result in disasters. As a result of this, control mechanisms have been developed and employed in many fields.

Fig. 2-56. The skewed house of Weimar (Germany)

Table 2-47. Causes of building damages (Matoussek & Schneider 1976)

	Percentage on overall damages
Ignorance, carelessness	37
Insufficient knowledge	27
Underestimate of influences	14
Forgetfulness and mistakes	10
Unwarranted abandonment on others	6
Objectively unknown situations and influences	6

Table 2-48. Probability of errors during design and construction of structures

Stage of project		P
Error of architect	$P(A)$	0.1
Error of design engineer	$P(E)$	0.4
Error during construction	$P(C)$	0.5
Error during control	$P(I)$	0.1

The different stages of development of a structure can be investigated for the occurrence of errors as well as for the usage time. Thus, damages and failures can be divided between occurring during the construction phase and occurring during the usage phase. Initially, this may seem rather strange since one would assume that heavier loads are experienced during the usage time of the structure. However, this is not always the case. For example, on a bridge a situation might arise where, during the time of construction, a heavy mass transport of earth from one side of the valley to the other side is carried out by big trucks over the unfinished bridge. Here, maximal loading is experienced during the construction phase. Also the structure might not be finished and, therefore, the proper statical system is not yet activated. Some part of the structure may, therefore, experience a load which it will never experience again in such a magnitude. This change of system and special load types during construction can be seen for a bridge built in incremental launching method. Here, the bridge is built on one side of the valley, and then moved section by section over the columns to the other side of the valley. The advantage of building with this technique is extremely good fabrication conditions. The bridge can then be produced at one place at the prefabrication area. However, the movement of pushing the bridge over the valley includes some special loads. In addition, since the bridge is pushed every week or every second week, the concrete material has not reached the final strength.

It is, therefore, not surprising that many bridges or houses collapse during the construction phase. Table 2-49 gives a list of known bridge failures. Here, the major cause of failure, approximately 25%, is during construction. If the collapse of the formwork is also considered a failure during construction time, then the figure reaches about 50%. In some cases, bridges also collapsed during load tests, as seen in Table 2-50. Another major cause of failure is connected to accidental loads, such as impacts. Only 1/4 of the failures happened under regular conditions.

Bridge failure cannot only be classified temporally but also according to the statical way the bridge failed. In the last decades, especially, stabilizing problems have caused bridge failures. For example, at the end of the 1960s and the beginning of the 1970s, some collapses of steel box girder bridges occurred. Examples of the failures of these bridges are the 4 Vienna Danube bridge in 1969, the Rhine bridge in Koblenz in 1971, the Cleddau bridge in Milford Haven in 1970, the Westgate bridge in Melbourne in 1970 and the reservoir bridge in Zeulenroda in 1973. Here, neglected initial deformations yielded to buckling during bridge feed.

Table 2-49. Causes for the failure of bridges (Scheer 2000)

S. No.	Cause during/by	Number of known cases
1	Construction	93
2	Usage under normal conditions	86
3	Ship impact	48
4	Impact of under- passing traffic	16
5	Impact of over- passing traffic	18
6	Flood and ice	32
7	Fire and explosion	15
8	Failure of formwork	48
Total		356

Table 2-50. Historical examples of bridge failures (Scheer 2000)

Year	Location	Remark
1209	Old London Bridge, England	Constriction of river cross- section yielded to destruction
1817	Dryburgh Abbey, Scotland	Chain suspension bridge destroyed by storm
1820	Union Bridge, Berwick, Scotland	Chain suspension bridge destroyed by storm
1830	Durham, England	Chain suspension bridge showed great deformation under live loads
1830	Yorkshire, England	Chain suspension bridge destroyed by a cow herd only a few months after construction
1876	Ashtabula, USA	Collapse during snow storm under train (80 fatalities)
1887	Bussey bridge close to Boston, USA	Steel framework bridge collapsed under train (26 fatalities)
1891	Mönchstein, Basel, Switzerland	Bridge collapsed under train (73 fatalities).
1893	Chester, USA	Framework bridge collapsed under train during rebuilding (40 fatalities)
1893	Louisville, USA	165- m- long framework bridge collapsed due to hurricane/gale (22 fatalities)
1907	Quebec, Canada	Framework bridge collapsed during construction (74 fatalities)
1931	Bordeaux, France	Suspension bridge failed during inauguration party and load tests (15 fatalities)

(Continued)

Table 2-50. (Continued)

Year	Location	Remark
1940	Frankenthal close to Mannheim, Germany	Failure of lift technique (42 fatalities)
1948	Pier at Stresa, Italy	Overload by about 1,000 people (12 fatalities)
1962	Bridge at Morace, Yugoslavia	Failure of bridge claimed 21 fatalities
1972	Naga City, Philippines	Bridge overloaded during procession (145 fatalities)
1977	Bundesstaat, Assam, India	Bridge failed under train (45 fatalities)
1982	Bridge over Brajmanbari, Bangladesh	Bridge failed under overloaded bus (45 fatalities)
1970	Westgate, Melbourne, Australia	Failure of bridge by buckling (34 fatalities)
1977	Pushkin bridge close to Moscow, Soviet Union	Insufficient maintenance caused failure (20 fatalities)
1981	Bridge over Totora-Oropeska river, Peru	During maintenance bridge was overloaded and load-bearing ropes failed (50 fatalities)
1994	Sungsusteel framework bridge close to Seoul, South Korea	Mistakes during design and overload caused failure (32 fatalities)

Although this list gives an impression that bridges collapse quite regularly, this is not true. Most of the 600,000 bridges in the US, about 120,000 bridges in Germany and about 150,000 bridges in U.K. are reliable technical products. Some of them still function well after being in use for more than a few hundred years. No other technical products are in use for such long periods that compare to buildings and structures (Proske et al. 2005).

2.4.6 Energy Production and Nuclear Power Plants

People are not only used to risks in structures, they are also exposed to risks in professions. However, there seems to be some professions more exposed to risks than others. For example, fishing is more dangerous than teaching. The topic of safety of the nuclear power industry seems to be of major public concern. However, this type of energy production might impose higher risks. All other types of energy productions should be compared in terms of the risks to humans (European Commission 2007).

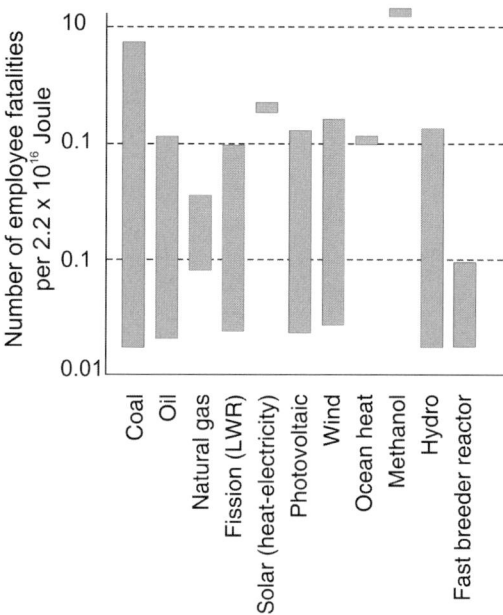

Fig. 2-57. Risks of different energy supply technologies in terms of employee fatalities per energy production (Inhaber 2004)

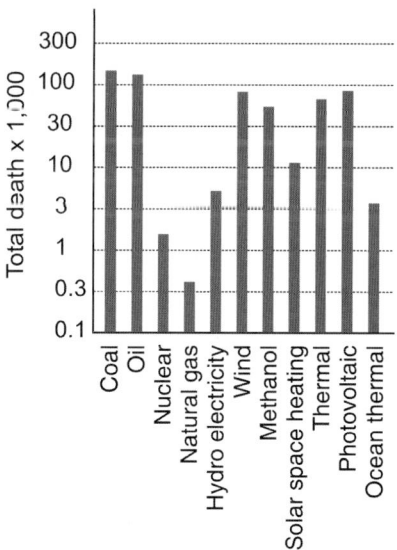

Fig. 2-58. Total number of deaths related to energy production technologies (Inhaber 2004)

Such risks include the failure of hydroelectric dams, the explosion of gas storages, the explosion in mines or accidents in nuclear power plants. Examples can be found in Inhaber (2004). Figure 2-57 and 2-58, taken from Inhaber (2004), give fatality numbers related to energy production technology.

However, nuclear power plant seems to feature some special properties. For example, the absolute damage potential for this technology is much higher compared to other energy production technologies known so far. A study of the 19 nuclear power plant locations in Germany has assumed damages in an area with a diameter of 2,500 km and an affected population of 670 million people. Early fatalities and damages were assumed only in an area with a diameter of 20 km. After an initial 20,000 fatalities, another 100,000 fatalities, were estimated with a probability of 5.9×10^{-5} per year (Hauptmanns et al. 1991). Many studies have been carried out to estimate the risks of this technology (GRS 1999). In the chapter "Objective risk measures", it will be shown that, due to these considerations, the nuclear power industry is a major precursor for risk assessment.

However, all of the risk studies could not have avoided some accidents in nuclear power plants, although they might have prevented many more other accidents. The accident of Chernobyl is probably known worldwide. Many different references describe the accident in detail (Jensen 1994, IAEA 1991). Depending on the source, the number of fatalities caused by this accident differs between 42 (Haury 2001), 9,000 (IAEA), over 264,000 fatalities (IPPNW) and up to 500,000 fatalities (Haury 2001). The second biggest accident was Three Mile Island (TMI) – fortunately with no external consequences (US-NRC 2004). The magnitude of such accidents can be classified according to the INES (Tables 2-51 and 2-52). One of the recent accidents occurred in a nuclear power plant in August 2006 in Sweden. In 2007, 442 nuclear power plants in 31 countries were in operation (Busse & Sattar 2007). Currently, there are more than 10,000 reactor years of operation (Carlier 2004).

Incidentally, it should be mentioned that in the Republic of Gabun, something like a natural nuclear reactor was found. Due to the coincidence of circumstances, a nuclear reaction occurred in six different zones for probably 150,000 years. Moderation was carried out by water for decreasing U 235 concentration. The natural reactor was found when the mining of uranium had started there (Rennert et al. 1988).

Table 2-51. International Nuclear Event Scale (INES) (IAEA 2007)

Level	Off-site impact	On-site impact	Defense degradation
7 Major accident	Major release: widespread health and environmental effects		
6 Serious accident	Significant release: likely to require full implementation of planned counter-measures		
5 Accident with off-site risk	Limited release: likely to require partial implementa-tion of planned counter-measures	Severe damage to reactor core and/or radiological barriers	
4 Accident without significant off-site risk	Minor release: public exposure of the order of prescribed limits	Significant damage to reactor core and/or radiological barriers/fatal expo-sure of a worker	
3 Serious incident	Very small release: public exposure at a fraction of pre-scribed limits	Severe spread of contamination/ acute health effects to a worker	Near accident
No safety layers remaining			
2 Incident		Significant spread of contamination/ overexposure of a worker	Incidents with sig-nificant failure in safety provisions
1 Anomaly			Anomalous beyond the authorized oper-ating regime
0 Deviation Out- of- scale event	No	Safety No safety relevance	Significant

Table 2-52. Examples of events in the nuclear event scale (IAEA 2007)

Level	Nature of event	Examples
7 Major accident	External release of a large fraction of the radioactive material in a large facility (e.g. the core of a power reactor) (in quantities radiologically equivalent to more than tens of thousands of terabecquerels of iodine-131).	Chernobyl NPP, USSR (now in Ukraine), 1986
6 Serious accident	External release of radioactive material (in quantities radiologically equivalent to the order of thousands to tens of thousands of terabecquerels of iodine-131).	Kyshtym Reprocessing Plant, USSR (now in Russia), 1957
5 Accident with off-site risk	External release of radioactive material (in quantities radiologically equivalent to the order of hundreds to thousands of terabecquerels of iodine-131). Severe damage to the installation.	Windscale Pile, UK, 1957 Three Mile Island, NPP, USA, 1979
4 Accident without significant off-site risk	External release of radioactivity resulting in a dose to the critical group of the order of a few millisieverts. Significant damage to the installation. Such an accident might include damage leading to major on-site recovery problems such as partial core melt in a power reactor and comparable events at non-reactor installations. Irradiation of one or more workers resulting in an overexposure where a high probability of early death occurs.	 Windscale Reprocessing Plant, UK, 1973 Saint-Laurent NPP, France, 1980 Buenos Aires Critical Assembly, Argentina, 1983
3 Serious accident	External release of radioactivity resulting in a dose to the critical group of the order of tenths of millisievert. On-site events resulting in doses to workers sufficient to cause acute health effects and/or an event resulting in a severe spread of contamination; for example, a few thousand terabecquerels of activity released in a secondary containment where the material can be returned to a satisfactory storage area.	

Level	Nature of event	Examples
	Incidents in which a further failure of safety systems could lead to accident conditions.	Vandellos NPP, Spain, 1989
2 Incident	Incidents with significant failure in safety provisions but with sufficient defense in depth remaining to cope with additional failures.	
	An event resulting in a dose to a worker exceeding a statutory annual dose limit and/or an event that leads to the presence of significant quantities of radioactivity in the installation areas not expected by design and that require corrective action.	
1 Anomaly	Anomaly beyond the authorised regime but with significant defence in depth re-maining. This may be due to equipment failure, human error or procedural inade-quacies and may occur in any area covered by the scale, e.g. plant operation, transport of radioactive material, fuel handling, waste storage.	
0 Deviation	Deviations where operational limits and conditions are not exceeded and which are properly managed in accordance with adequate procedures.	

2.4.7 Radiation

The release of radioactive material, and therefore radioactivity, is one of the major concerns about the safety of nuclear power plants. About 2,000 measurement stations observe radioactivity nearly equally distributed over Germany. Additionally, there are further measurement points around nuclear power plants. While such measurement nets have not shown significant release of radioactivity, current investigations show a higher leukemia rate for children in the vicinity of nuclear power plants (Kaatsch et al. 2008).

The reactor four in Chernobyl might have included radioactive material in the scale of 4×10^{19} B before explosion. About $1–2 \times 10^{18}$ Bq were released during and after the explosion. It has been assumed that all inert gases volatilised. Additionally, about 10–20% of the nuclides iodine, cesium and tellurium discharged. Other nuclides were released at lower portions. Twenty-five `percent of the released material exited the reactor on the first day. The rest of the material was released over the next nine days into the environment. About 115,000 people were evacuated after the accident. So far, only the unit Becquerel has been introduced as a measure of radioactivity. However, other measures with relation to radioactivity should be mentioned as well. "Gray" is the measure for the absorbed dose. While Becquerel is given per second, Gray (Gy) is given in Joule per kilogram. The dose can be used to quantify damages to creatures. Unfortunately, exact lethal dose cannot be given, and therefore the so-called median lethal dose (LD_{50}) given. This is the dose killing 50% of the exposed population. For humans with a good medical treatment this value reaches about 5 Gy, whereas excellent medical treatment, might shift the LD_{50} up to 9 Gy. Figure 2-59 shows the distribution of the percentage of death from a population for some dose. A further measure related to the dose is the Sievert. The Sievert is also given in Joule per kilogram, but it considers the different effects of different types of ionizing radiation. α-, β-, γ-radiation have different consequences, and therefore the Sievert gives the opportunity to compute an equivalent dose.

Of course, ionizing radiation has impacts on animals and plants. However, no effects for acute radiation exposure were found with less than 0.1 Gy (in terrestrial environment). A chronical radiation exposure of less than 1 mGy per day did not show effects. In aquatic environments, these values are 10 times higher (Paretzke et al. 2007, Wenz et al. 1980).

Returning to the accident of Chernobyl: on 26 and 27 April, the radiation reached about 10 mSv per hour. At that amount of radiation, it was decided to evacuate the population. People working in the rescue team might have been exposed up to 15 Gy. Extreme values of the radiation might have been in the order of 40 Gy. The average dose in the area lying 30 km around the nuclear power plant was 2 Gy. At least 237 people experienced severe radiation disease, and approximately 30 people died soon after the event by this disease.

Most people cleaning the area were exposed to an equivalent dose of 100–250 mSv (IPSN 1996). Most parts of the population were exposed to 15 mSv. From this amount of radiation, most parts come from the consumption of radioactive meat and milk products. About 700,000 people were exposed to 100 mSv. The related probability of cancer is about 0.005. Most countries of the northern hemisphere were affected by this release of radioactive material. However, the radiation values were rather low. For example, in Denmark, a dose of 0.02 mSv was reached as compared to the natural radiation dose of 2.4 mSv. Further information can be found in Jacob (2006). Kellerer (2006) mentions that the chance of cancer for the overall population in Russia changed from 20% to 22%. In Germany, about 2,000 additional cancer cases are assumed. However, the number of new cancer cases reaches about 330,000 per year.

Another example of a major accident, however not related to nuclear power plants, was the release of radioactive material in the Brasilian city of Goiania. Here, an X-ray apparate was stolen and dissembled in a scrap yard. About 249 people were contaminated, and about 120,000 people were controlled for contamination.

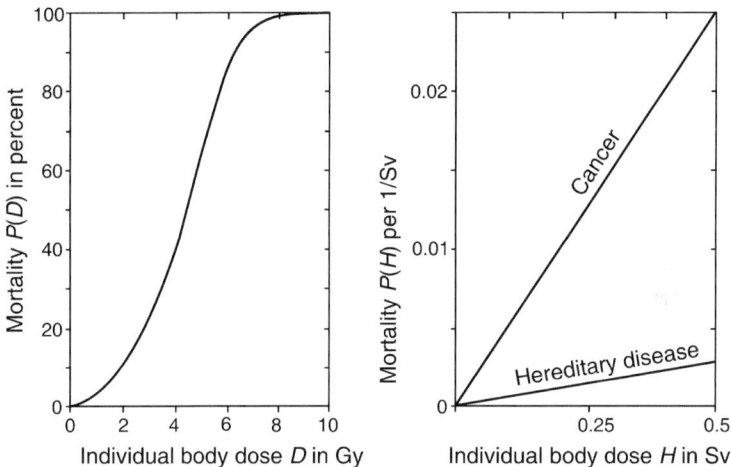

Fig. 2-59. Stochastic and deterministic relation between radiation dose and mortality (Jensen 1994, IAEA 1991)

However, radionuclides are common in the environment. The amount of radioactive material in a typical human body is shown in Table 2-53. Table 2-54 gives some radiaton values for Radon 226 in German beer and drinking water (Überkinger Quelle).

This natural radiation is not surprising since primordiale radionuclides can be found in natural decomposition. The origin of nuclides can be divided into two parts: the galactic cosmic processes, namely the background radiation and supernovas, are the first part. This part is the source of all heavy nuclides in man and the environment. The solar cosmic processes, namely solar activity and solar flares, are the second part. The sun is actually the source of all light nuclides in man and the environment (Van der Heuvel 2006).

Natural primordial radionuclides with their half-life period are given in the Tables 2-55 and 2-56.

Table 2-53. Natural radioactive material in a typical human (50 kg) in Bq (Paretzke et al. 2007)

Nuclide	Presence in Bq
Tritium	20
Carbon-14	3,500
Pottasium-40	4,00
Rubidium-87	600
Lead-210	18
Polonium-210	15
Radium-226	1.2
Uranium-238	0.5

Table 2-54. Content of Radon 226 in German beer and drinking water (Paretzke et al. 2007)

Nuclide	mBq/l
Schneider Weiße	147
Erdinger Weissbier	13
Pkantus Weizenbock	9
Paulaner Beer	33
Überkinger Quelle	296

Table 2-55. Half-life period of primordial radionuclides (Paretzke et al. 2007)

Nuclide	Half-life period in years
K-40	1.3×10^9
Rb-87	4.8×10^{10}
In-115	4.0×10^{14}
Te-123	1.2×10^{13}
Te-128	1.5×10^{24}
Te-130	1.0×10^{21}
La-138	1.4×10^{11}
Nd-144	2.1×10^{16}
Sm-147	1.1×10^{11}
Sm-148	7.0×10^{16}
Gd-152	1.1×10^{14}
Lu-176	3.6×10^{10}
Hf-174	2.0×10^{15}
Ta-180	1.0×10^{13}
Re-187	5.0×10^{10}
Os-186	2.0×10^{16}
Pb-190	6.1×10^{11}
Pb-204	1.4×10^{17}

Table 2-56. Half-life period of primordial radionuclides (Paretzke et al. 2007)

Nuclide	Half-life period
Tritium	12.3 years
Beryllium 7	53.3 days
Carbon 14	5,730 years
Sodium 22	2.6 years

Radon concentrations in the environment are widely discussed. (Brüske-Hohlfeld et al 2006). Umhausen (2,600 inhabitants) in Tyrol, Austria, is an example of high natural ionosing. Here are some radioactivity values measured in houses :

- Yearly average dose: $2,000 \ Bq/m^2$
- Extreme values: $40,000 \ Bq/m^2$
- Highest value ever measured: $274,000 \ Bq/m^2$

In the last 20 years, there were about 41 deaths due to lung cancer in Umhausen. Statistically, only six to seven cases were expected. The major cause for the high radioactivity was probably the very high permeability of the ground, which might have been caused by a huge rock or debris flow about 10,000 years ago. Radon is not only a major contributor to natural radiation in some regions, like in Umhausen or some parts of India. In Germany, Radon contributes to about 25% to the natural radiation (Brüske-Hohlfeld et al. 2006). Moreover, radiation during flying has been considered as risk (Schraube 2006). Although accidents have happened in nuclear power plants or nuclear weapon production and storage as shown in Tables 2-57 and 2-58, the overall contribution is rather low as shown in Table 2-59. Of greater concern is the increasing radiological radiation exposure by medical treatment, for example computer tomography (DA 2007) as shown in Fig. 2-60. Here, recommendations can be found at EANM (2007).

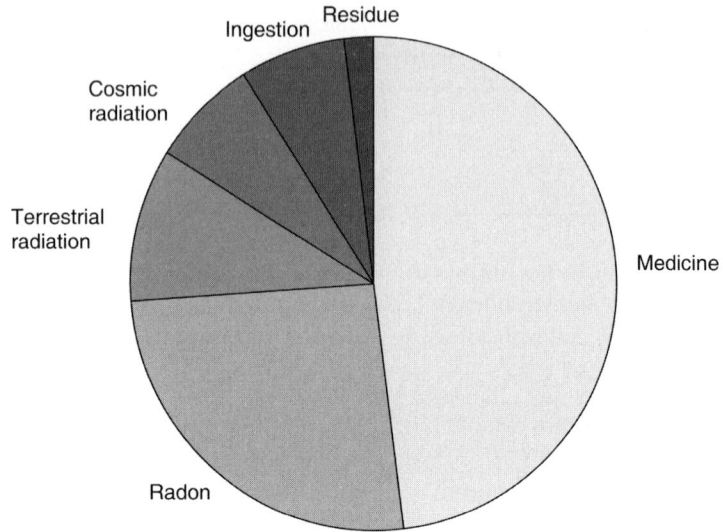

Fig. 2-60. Radiation exposition in Germany (BfS 2007)

Table 2-57. Major accidents and routine releases of radioactivity into the environment (Paretzke et al. 2007)

Activity	Mode of release	
	Routine action	Accidental event
Military purposes		
Weapon production	Hanford, USA (1944–1945) Chelyabinsk, USSR (1948–1956)	Techa River, USSR (1949–1951) Kysim, USSR (1957) Windscale, UK (1957) Rocky-Flats, USA (1969) Tomsk-7, USSR (1993)
Atmospheric tests	Nevada, USA (1951–1962) Semipalatinsk, USSR (1949–1962) Novaya Zemlya, USSR (1955–1962)	Altay, USSR (1949) Marshall Islands, USA (1954)
Nuclear fleet	Kola Peninsula, USSR	
Weapon transport		Palomares, Spain (1966) Thule, Greenland (1968)
Power production		
Reactor operation	Worldwide	Three Mile Island, USA (1979) Chernobyl, USSR (1986)
Fuel processing	Sellafield, UK La Hague, France	

Table 2-50. (Continued)

Activity	Mode of release	
	Routine action	Accidental event
Radioisotope use		
Loss of sources		Cuidad Juarez, Mexico (1982)
		Goiania, Brazil (1987)
Satellite re-entry		SNAP-9A, Global (1964)
		Cosmos-954, Canada (1978)

Table 2-58. Further information about some accidents in terms of the amounts of nuclide-specific radioactivity released in pBq (Paretzke et al. 2007)

Event	^{131}I	^{137}Cs	^{90}Sr	^{106}Ru	^{144}Ce	^{239,240}Pu
Techa River	6.5×10^5	12	12	10…20	≈10	11
Nuclear tests 1952–1962	0.7	910	600	12 000	30 000	0.07
Kystim	1 200	0.03	2	1.4	24	6.0×10^{-6}
Windscale		0.02	8	0.003	140	
Chernobyl		85		30	0.0002	
Goiania		0.05		0.01		

Table 2-59. Residue in Germany (Paretzke et al. 2007)

	mSv per year
Fallout Chernobyl	0.015
Fallout A-Bombs	0.01
Technology and Research	< 0.02
Nuclear Power Plants	< 0.001
Profession	< 0.01

Here, so far only ionising radiation has been considered a risk. Unfortunately, it is unclear whether non-ionising radiation might also include risks (Junkert & Dymke 2004). Such non-ionising radiation is heavily used in mobile phones. This is an example where humans are not concerned about the risk, as the advantage is so manifest and the disadvantage is so uncertain. This constellation has also been true for many chemicals.

2.4.8 Chemicals

Chemicals might impose hazards of explosion, poisoning, suffocation, fire, fire support, cauterisation, frostbite, infection or contamination of the environment.

Currently, there are about 10 million chemicals known. Between 50,000 and 70,000 of them are produced, stored, transported and used in great quantities. The European Inventory of Existing Commercial Chemicals includes about 100,000 materials or material groups that have been available in the European market since 18 September 1981. All so- called high production volume chemicals, with a volume of more than 1,000 tonnes have been investigated. These include approximately 2,600 materials. These materials were put into 140 groups, which were investigated in terms of risk (Ahrens 2001).

In the workplaces, there are about 500 scientifically based limits that set the allowable concentration values of certain chemicals; however, 5,000 substances are used in a significant amount at certain workplaces (Brandhofer & Heitmann 2007).

Hazardous chemicals have always been produced, transported, stored and used. As a toxicity measure in Fig. 2-61, the total dose for some chemicals is given. About 8,000 chemical plants exist in Germany. Usually chemical plants are quite often considered a hazard. In Germany per year, between 10 and 20 severe accidents happen in chemical plants. Therefore, the frequency of such an accident is $1–2 \times 10^{-3}$ per year per plant (Ruppert 2000). However, it seems that the number of accidents is decreasing (Ruge 2004).

One of the worst accidents in chemical plants was a major explosion in the BASF plant in Oppau in 1921. On 21st September, about 4,500 tonnes of ammonium nitrate exploded. The explosion created a crater of a diameter of 100 m and a depth of 20 m. About 500 people were killed. After this accident, the trade and production of ammonium nitrate was forbidden in Germany.

Incidentally on 21st September 2001, a major explosion of ammonium nitrate happened, but now in Toulouse, France. The number of victims reached about 30; however, the number of injured people came to thousands. About 30,000 flats, 700 public buildings and 112 schools were damaged. The explosion caused a panic, and the drinking water supply in Toulouse was interrupted for three days (Hubert et al. 2004, Munich Re 2004a,b).

Explosions in chemical plants occur regularly worldwide. Other examples are an explosion of fuel in Cleveland in 1944 with about 130 fatalities (Considine 2000) or an explosion in a pharmaceutical plant in Kingston, North Caroline, in 2003 with about 30 injured (Munich Re 2004a,b). One of the recent accidents with fires in plants occurred on 11th December 2005 in Bruncefield, Great Britain. The complete damage was about 500 million British pounds. There were 43 fatalities. However, the number could have been much higher, became fortunately the fire broke out on Sunday in an industrial area.

However, explosions and fire are not only the result of uncontrolled release of chemical substances. Such substances might also be toxic. For example, in 1976 in Seveso in Italy, about 2kg of dioxin was released into the environment. Cows and small animals died in the vicinity of the source. About 70,000 poisoned farm stock animals had to be killed, 200,000 people received medical treatment, and several houses heavily exposed to the poison had to be demolished.

Probably the worst chemical accident happened on 3rd December 1984 in Bhopal in India. About 40 tonnes of methyl isocyanate were released into the environment killing almost 4,000 people. According to different source, between 20,000 and 250,000 people experienced long-term permanent or partial disabilities (Broughton 2005, Union Carbide 2007).

As shown in this example, the effects of released chemicals can be categorised into long-term and short-term hazards. Long-term hazards are generated by exposure to a low concentration of a chemical over the long term, whereas short-term hazards are generated by exposure to a high concentration of a spontaneously released chemical presenting an immediate danger.

Another example of the consequences of unintended release of chemicals was the fire in November 1986 at the Swiss company Sandoz. The water for fire fighting washed toxic insecticides into the river Rhine. Over a river length of 400 km, heavy damages occurred to the river flora and fauna (Rütz 2004).

The final example in this section, which was related to mining, is an accident in 2003 in China in a gas field. Although this was not essentially a chemical plant, it was related to some chemical material. The release of the gas killed about 250 people, injured more than 9,000 people and caused the evacuation of about 60,000 people (Spiegel 2003).

Many more examples could be given from production plants. However, accidents can also occur during transport. In Austria, there are over 1.7 million transportations of hazardous chemicals on roads and about 130,000 railway wagons transporting hazardous chemicals per year. There are approximately 20 accidents with trucks carrying hazardous chemicals per year in Austria. About 20% of the container ship transport worldwide includes hazardous materials. This value is surprisingly high, however why so many transportations are carried out, however many chemicals are used in daily life. Figure 2-62 some shows some common warning signs seen on many household chemical bins. Several chemicals are hidden in technical equipment. For example, ammonia is heavily used as a refrigerant in skating sport centers or in cooling houses. Chloride is used in swimming pools for disinfection purposes. Currently, alternatives to chlorine are being developed to avoid the future application of chloride in public rooms by terrorists. Moreover, in agriculture, many hazardous chemicals are applied, such as fertilizers and pesticides.

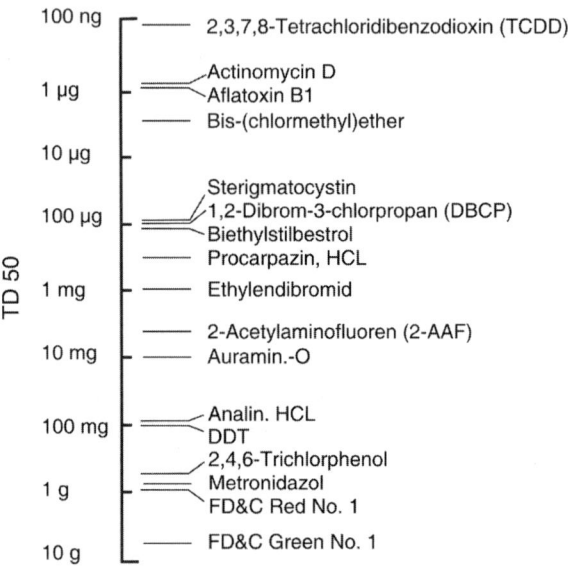

Fig. 2-61. Total dose 50 for different chemicals

Fig. 2-62. Examples of warning signs

2.4.9 Fire

As already mentioned, fire can be a risk related to some chemicals. In general, fire is defined as "uncertain type of burning which spreads out uncontrolled" (DIN 14011 and ÖNORM F 1000 Part 2). Fire requires heat, oxygen and burning material. A list of major fires is given in Table 2-60.

Besides, serval biggest conflagrations were intentionally caused during World War II. Some German cities that were heavily bombed with incendiaries experienced up to 100,000 fatalities, such as Hamburg in 1943 or Dresden in 1945.

As seen from Table 2-60, fire disasters already happened in times of early civilization. The first water pump for fire-fighting was invented in 250 B.C. by Ktesibios in Alexandria. Hero from Alexandria introduced a portable pump (NN 1997). Already, however, around 2400 to 2000 B.C., the old Egyptian language had signs for conflagration and burning city (Flemmer et al. 1999).

In 1518, the Augsburgian Goldsmith invented a drivable syringe (NN 1997). Fireplugs were introduced in Great Britain after the Great Fire of London in 1666. About the same time, the Dutch man van der Heijden invented the water hose and the hand pump for fire-fighting.

In 1676, the first fire insurance company, the Hamburgian General-Feur-Cassa, was founded. Most of the regional fire insurance companies in Germany were founded in the 18th century, for example the Lippian State Fire Insurance. Prior to insurance, people who lost their property to fire were regularly given permissions to beg. This however, did not help the people much (NN 1997).

However, technical inventions improved fire safety. Lightning rods were already known in antiquity, but were forgotten. They were reinvented by Benjamin Franklin in 1750. The first lightning rod was installed in 1769 at the Hamburgian church St. Jacobi (NN 1997). In 1829, hand pumps were substituted by machine pumps, and in 1888 in Chicago, the first turntable ladder was introduced (Flemmer et al. 1999). Especially, the introduction of sprinkler fire-extinguishing installation decreased the number of fire victims as shown in Fig. 2-63 for Canada.

Table 2-60. List of severe fires (Flemmer et al. 1999)

Time	Location	Victims	Remark
19 July 64 A.D.	Fire of Rome	Not known	Probably intentional
August 70 A.D.	Destruction of Jerusalem	1/4 of the Jewish population perished	Probably/partially intentional
September 1666	Great Fire of London	Only 8 fatalities, but 100,000 left	London had just experienced a pest

(Continued)

Table 2-60. (Continued)

Time	Location	Victims	Remark
		homeless	epidemic the previous year.
September 1812	Fire of Moscow		Intentional
17 January 1863	Fire in a church of Santiago de Chile	2,500 fatalities	
8 October 1871	Great Fire of Chicago	About 300 fatalities, but 90,000 left homeless	The damage had been estimated at about 200 million dollars yielding to the liquidation of 54 American fire insurances.
8 October 1871	Forest fire of Peshtigo, Wisconsin	2,682 fatalities	About 1,000 km^2 area of forest burnt.
8 December 1881	Theatre fire of Vienna	896 fatalities	
3 February 1901	Oil field fire of Baku	More than 300 casualties	
15 June 1904	Fire of the steamer "General Slocum" on Hudson river	More than 1,000 fatalities involving children and mothers	
10 March 1906	Mine fire of "Courrières"	1,205 fatalities	
22 September 1928	Fire in Novedades theater in Madrid	About 110 fatalities and 350 injured	
6 June 1931	Glass palace in Munich		
2 March 1934	Hakodate, Japan	More than 900 fatalities, 2,000 injured and about 150,000 homeless	The fire occured together with a snowstorm. Most of the fireplugs were frozen. An area of about 15 km^2 was completely destroyed.
May 1937	Fire of "Hindenburg" in Lakehurst	35 fatalities	With about 590 flights 16,000 passengers were transported over the years.
1937	Fire of a high school in London, Texas	294 children	Gas explosion
28 July 1945	Fire in Empire State		

Table 2-60. (Continued)

Time	Location	Victims	Remark
	Building, New York		
16 April 1947	Explosion of French Tanker "Grandcamp" in Texas City	More than 2,000 fatalities	90% of all houses in Texas city destroyed, 100 million US dollars loss.
28 July 1948	Fire in BASF in Ludwigshafen	178 fatalities and 2,500 injured	
9 June 1995	Collission of "Johannishus" in the English Channel		
1 December 1960	School fire in Chicago		
19 July 1960	Mine fire in Salzgitter		
17 October 1960	Ship fire on Rhine		
17 December 1961	Circus fire in Niteroi, Brasilia	323 fatalities and over 500 injured	2,500 people visited a circus performance, when the circus tent started to burn and fall down.
22 December 1963	Fire on "Lakonia"		
11 July 1978	Explosion of a tank truck at the camping site "Los Alfaques" in Spain	180 fatalities and 600 injured	
6 July 1988	Explosion of "Piper Alpha" in the North Sea	170 fatalities	
15 April 1997	Fire in Mina (close to Mekka)	343 fatalities and 2,000 injured	
1998	Fire of a toy production plant in Bangkok	210 fatalities and 500 injured	
17 October 1998	Pipeline fire in Nigeria	Probably more than 1,000 fatalities	An oil pipeline leaked and people attempted to take from a sea of oil. The oil lake exploded.
30 October 1998	Fire of discothèque in Göteborg	61 fatalities	
3 December 1998	Fire of orphanage in Manila	About 30 fatalities	

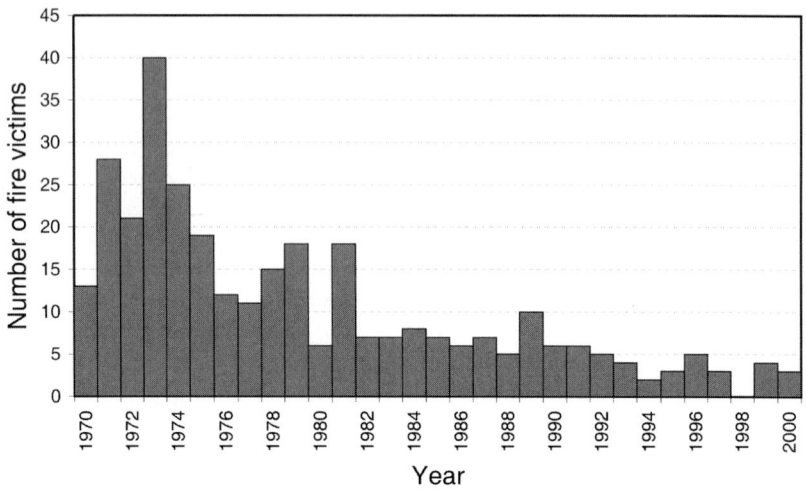

Fig. 2-63. Development of the number of fire victims in Canada according to Holdgate (2001) (taken from Maier 2006)

The main causes of death by fire are: suffocation by carbon monoxide (more than 60%), combustion (approximately 26%) and physical violation (more than 10%) (Schneider und Lebeda 2000). In Germany, fires cause an average yearly damage in the double-digit billion Euro range (Friedl 1998). Every year, about 220,000 fires are registered, but only full fires are collected in a database. That means there is an average of 2.75 fires per 1,000 capita (Mehlhorn 1997). Table 2-61 gives the direct damage of fires for different countries in relation to the gross national product. The distribution of fires subject to different building categories is shown in Table 2-62 for Finland. This distribution is partially reflected in the fire loads for different building categories shown in Table 2-63 and by the probabilities of fire for different building categories shown in Table 2-64. Causes of fires are given in Table 2-65, and Table 2-66 gives probabilities of fire accretion under different firefighting conditions.

Table 2-61. Fire damages in developed countries (Schneider und Lebeda 2000, Mehlhorn 1997, Tri Data Corporation 1998, World Fire Statistics Center London)

	Direct fire damages as percentage of gross national product	Number of fire fatalities per 1,000,000 inhabitants (Mehlhorn 1997)
USA		21–27
Finland	0.19	22–24
England	0.24	19–21
Sweden	0.21	16–20
Denmark	0.39	15–20
Belgium	0.45	20
France	0.26	13–19
Norway	0.34	17–18
Netherlands	0.20	6–13
Germany	0.19	9–13
Italy	0.15	7–9
Switzerland	0.25	5–6
Austria	0.16	7
Canada		18–22
Spain		7–12
Japan		14–16

Table 2-62. Number of fires from 1996 to 1999 and areas of different building categories in Finland according to Rahikainen & Keski-Rahkonen (2004)

Occupancy	Fires	Total area (m^2)
Residential buildings	4,361	231,565,978
Commercial buildings	356	18,990,450
Office buildings	140	16,354,516
Transport and firefighting and rescue service buildings	123	10,627,751
Buildings for institutional care	197	8,780,942
Assembly buildings	112	7,379,199
Educational buildings	122	15,801,759
Industrial buildings	1,038	40,321,357
Warehouses	405	7,434,710
Other buildings	2,650	2,437,960

Table 2-63. Excerpt from BSI (2003) and NFSC fire load densities taken from Weilert & Hosser (2007)

Occupancy	Fire load density (mJ/m^2)			
	Average		Fractile	
		80%	90%	95%
Dwellings	780	870	920	970
Hospital (room)	230	350	440	520
Hospital storage	2,000	3,000	3,700	4,400
Hotel (room)	310	400	460	510
Office	420	570	670	760
Shops	600	900	1,100	1,300
Manufacturing	300	470	590	720
Manufacturing and storage	1,180	1,800	2,240	2,690
Libraries	1,500	2,250	2,550	–
Schools	285	360	410	450

Table 2-64. Probability of fire for different building types (Schneider & Lebeda 2000)

Type of building	Country	Probability of fire per million m^2 floor space per year
Industry building	Great Britain	2
Industry building	Germany	2
Office building	Great Britain	1
Office building	USA	1
Residential building	Great Britain	2
Residential building	Canada	5
Residential building	Germany	1

Table 2-65. Causes of fires (Tri Data Corporation 1998)

Cause of fire	Percent of fires
Unknown cause	47.2
Arson	15.5
Open fires	6.6
Arson by children	2.9
Heating	3.8
Cooking	5.4
Electric distributor (cables)	4.5
Heat radiation from other sources	2.4
Smoking	3.2
Natural fires (lightning)	1.3
Electrical appliances in household	2.1
Explosions, fireworks	3.0
Other electrical appliances	2.0

Table 2-66. Probability of fire accretion according to firefighting equipment based on Bub et al. (1983)

Firefighting by	Probability that a fire will develop to a full size
Public fire brigade	0.1
Sprinkler	0.01
Well-equipped, factory-owned fire brigade with fire detector system	0.001–0.01
Sprinkler system and well-equipped, factory-owned fire brigade	0.0001

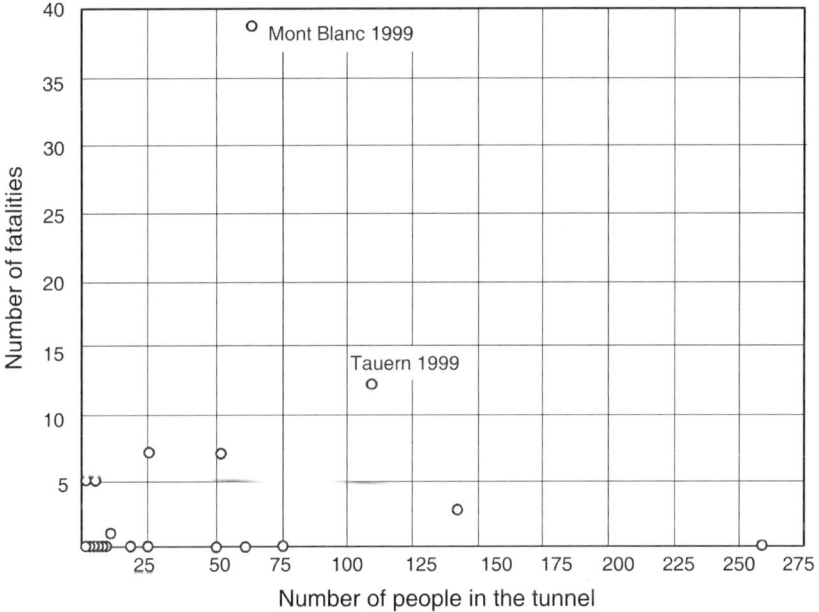

Fig. 2-64. Number of fatalities of fire accidents in tunnels (Rackwitz & Streicher 2002)

A special type of fire is fires in tunnels. Such events receive great public attention when they happen. A list of several tunnel fires can be found in Promat Tunnel (2008), and in combination with the transport of hazardous materials in Cassini & Pons-Ineris (2000) (Fig. 2-64). Much work was carried out after the accident in the Mont Blanc road tunnel in 1999 (Bergmeister 2006, 2007, Krom & de Wit 2006). Although high temperatures can be reached, the major problems with tunnel fires are the smoke and restricted orientation capability. It has been estimated that under an optical smoke density of 0.066 m^{-1}, the visibility is less than 4m, and therefore people unfamiliar with the tunnel are unable to find emergency exits. People who know the tunnel might find emergency exits with a visibility of 1m. However, in addition to limited visibility, the smoke also usually contains toxic materials like carbon monoxide or acids. With a carbon monoxide density between 600 and 800 ppm in air, humans become unconscious (Fileppo et al. 2004).

In some cases, the fire is directly related to car accidents and transportations of hazardous materials. Moreover, human behaviour in tunnels might increase risks. It is interesting to note that the type of lights in tunnels have to fulfill some requirements in order to not cause epileptic seizures.

2.4.10 Explosions

Explosions and fires are often connected. An explosion is an extremely fast force based on the expansion of gases or steam (VGB 2001). The velocity of an explosion is usually so high that humans are unable to identify the sequences of it. The speed of combustion can be used for the classification of fire and explosion, as shown in Table 2-67. Table 2-68 gives examples of distances travelled by explosion particles, and Table 2-69 lists the effects of overprenure caused by explosion.

Table 2-67. Classification of fire and explosion based on the combustion speed

	Combustion speed
Combustion	Millimeter per minute
Deflagration	Centimeter per minute
Explosion	Meter per second
Detonation	Kilometer per second

Table 2-68. Distance travelled by explosion particles at some explosions (van Breugel 2001)

	Year	Distance from explosion centre	Mass (kg)	Speed (m/s)
Romeo Village	1984	Fragments of tank 500 m high and 3,000 away	–	100–170
Crescent City		Fragments of tanker 100 m away	>10,000	85
Mexico City	1984	Fragments of tank 400 m away	13,000	60–150
Feyzin	1966	Slab elements 300 m away	70,000	–
Texas City	1978	Parts 230 m away	–	–

Table 2-69. Effects of overpressure caused by explosion

Overpressure in bar	Effect
0.60	Immediate fatalities
0.30	Failure of structures
0.20	Heavy damages to structures
0.14	Deadly injuries
0.07	Heavy injuries
0.03	Glass failure

Some examples of major explosions will now be provided. In 1645, a gun powder tower in Delft exploded. The explosion was heard within a 80-km radius. Nowadays, there is a market place at the origin of the explosion. In 1807, a boat loaded with gun powder exploded in the Dutch city of Leiden. About 150 people were killed, including 50 children. It was said that Napoleon visited the site of explosion and wrote an Imperial decree dealing with the location of factories inside cities based on their hazardous level. It is surprising to note that this still has not been successfully applied in practice even after 200 years. Therefore, since 1807, the number of explosions inside cities has continued to rise. For example, there was a large explosion of gun powder in 1867 in Athens (Ale 2003).

Two of the biggest explosions in history were connected to ships that were loaded with gun powder. The first of these two disasters was the explosion on 6 December 1917 in Halifax, Canada. The French transport ship Mont Blanc was scheduled to transport gun powder to the European battlefields. The ship was loaded in November in New York with gun powder that had an equivalent explosive force of 3kt TNT (Trinitrotoluene). The ship was supposed to enter the harbour of Halifax, Canada, prior to travelling to Europe. To sail to Halifax, the Narrow Canal had to be navigated before entering the city harbour. The Norwegian freighter Imo tried at the same time as the Mont Blanc to enter the Narrow Canal. Additionally, the Norwegian freighter was travelling at potentially double the allowed speed in the canal and rammed into the Mont Blanc. The collision parts of the Imo bored into the starboard side of the Mont Blanc. Benzene barrels on the deck of the Mont Blanc were damaged and leaked. Unfortunately, the captain of the Imo decided to free his ship again. As a consequence, benzene was ignited by the friction between the metal parts of the ships scrapping against each other, and further benzene barrels exploded on the deck of the Mont Blanc. Therefore, the entire front part of the Mont Blanc stood in flames. The firefighting capacity of the ship was overburdened and unable to extinguish the fire. The captain of the ship, however, did not decide to sink the Mont Blanc but to abandon it. The ship then did not just explode but floated towards Richmond, a suburb of Halifax. It reached the piers and enflamed them. Some attempts were made to tow the Mont Blanc away from the piers but they failed. In the end, the entire ship exploded half an hour after the collision. The explosion killed about 2,000 people, many of them watching the fire in the neighbourhood of the pier. About 9,000 people were injured. The ship itself was torn apart and parts of the ship were found several miles away. Several ships in the harbour were destroyed or heavily damaged. A railway bridge was also heavily damaged by the explosion and collapsed under a train. Wagons of the train fell. The explosion not only caused many casualties, but also left about 25,000 homeless (Korotikin 1988).

A comparable disaster happened during World War II in 1944 in Bombay. Here, a ship, the Fort Stikene, was loaded with 1,400 tonnes of gun powder. The ship was also transporting cotton and dry fish at the same time. It was scheduled to be unloaded on 13 April 1944. In the morning, the unloading began. In the afternoon, a fire in the cotton was found. For two and half hours, the crew tried to stop the fire, but they failed. Even worse, on the outside of the ship, a glowing spot had become visible. The fire spread out, reaching the sectors of the ship with the gun powder. Again, the ship was not sunk or pulled out of the harbour. When the ship exploded, around 13 ships in the neighborhood were lost. In a radius of 400 m, all buildings experienced heavy damages, including 50 storage facilities. The glowing parts of the ship ignited hundreds of fires in the city and began to spread. Two days later, on 15 April, the fires reached the centre of the city. The fire reached to such an intensity that it was visible from approximately 120 km away. To avoid the complete loss of Bombay, a fire protection stripe of a width of 400 m was formed. The explosion caused 1,500 fatalities, 3,000 injuries and left countless people homeless (Korotikin 1988).

Even though the Bombay explosion occurred over half a century ago, events in the last few years show that the hazards of explosions remain. While some explosions connected to the chemical industry are presented in Sect. 2.4.8 section, here the explosion of a fireworks plant in the Dutch city of Enschede in May 2000 is mentioned. This explosion completely destroyed close to 300 houses and killed 22 people.

If a plant using explosive materials is proposed, its location has to be considered. Tables 2-70, 2-71 and 2-72 can help in determining the proper distance of each plant from residential areas. There are three different distance classifications: explosive zone, damage zone and attention zone. The explosive zone is in close proximity to the explosion itself. It is defined by an overpressure higher than 0.6 bars. The damage zone is located between an overpressure of 0.6 and 0.07 bars. Here, deadly injures can occur. In the

Table 2-70. Criteria for land use classification according to the Italian Ministry of Public Works Decree 2001, taken from Uguccioni (2004)

Criteria for definition	Territorial	Class				
	A	B	C	D	E	F
Residential area (criteria: building index in m^3/m^2)	>4.5	1.5–4.5	1–1.5	0.5–1	<0.5	–
Places where there is a concentration of people with limited mobility, e.g. hospitals, retirement homes, nurseries, elementary schools	>25 beds >100 people	<25 beds <100 people				
Places where a significant concentration of people outdoors can occur, e.g. marketplaces or other commercial places	>500 people	<500 people				
Places where a significant concentration of people indoors can occur, e.g. commercial centres, office buildings, hotels, high schools, universities		>500 people	<500 people			
Places where a significant concentration of people can occur, with limited period of presence, e.g. theatres, churches, stadiums, etc.		>100 people outdoors, >1,000 people indoors	<100 people outdoors, <1,000 people indoors	Any number with max. weekly attendance		
Railway stations and transportation network nodes		>1,000 people per day	<1,000 people per day			
Industry, farming				<1,000 people per day		
Within plant fences, areas nearby within which there are no structures present and where the presence of people is normally foreseeable					Any dimension	x

thrid zone, the attention zone, no fatalities should occur, but injuries can happen. The pressure reaches up to 0.03 bars. Obviously, residential areas are only permitted in the attention zone.

As an example, an industry storage facility with 20 tonnes of explosive materials is considered. The explosive zone then reaches approximately 70 m, the damage zone reaches about 420 m and the attention zone reaches about 1,200 m. Therefore, the storage has to have a minimum distance of 420 m from the nearest residential area.

Table 2-71. Definition of damage threshold levels according to the Italian Ministry of Public Works Decree 2001, taken from Uguccioni (2004)

Accident scenario	High lethality	Beginning of lethality	Irreversible damages	Reversible damages	Structural damages/Domino effects
Stationary heat radiation	12.5 kW/m^2	$7\ kW/m^2$	$5\ kW/m^2$	$3\ kW/m^2$	$12.5\ kW/m^2$
BLEVE/Fireball (variable heat radiation)	Fireball radius	$350\ kJ/m^2$	$200\ kJ/m^2$	$125\ kJ/m^2$	200–800 m depending on storage type
Flash fire (instantaneous heat radiation)	LFL	1/2 LFL	–	–	–
VCE (peak overpressure)	0.5 bar (0.6 open spaces)	0.14 bar	0.07 bar	0.03 bar	0.5 bar
Toxic release (absorbed dose)	LC_{50} (30 min, hmn)	–	IDLH	–	–

Table 2-72. Land use categories compatible with industrial plants according to the Italian Ministry of Public Works Decree (2001), taken from Uguccioni (2004)

Probability class of events	Effect category			
	High lethality	Beginning of lethality	Irreversible damages	Reversible damages
$<10^{-6}$	DEF	CDEF	BCDEF	ABCDEF
10^{-4}–10^{-6}	EF	DEF	CDEF	BCDEF
10^{-3}–10^{-4}	F	EF	DEF	CDEF
$>10^{-3}$	F	F	EF	DEF

Much research has been carried out in the field of ammunition storages as special storage case of hazardous materials. Computer programs have been developed to compute risk values, like the program ESQRA-GE from the Ernst Mach Institute in Germany (Doerr et al. 2002). Further works can be found in Gürke (2002), Swisdak (2001), Kingery & Bulmash (1984), NATO (1997) and Prasse (1983). Ammunition is a major cause of explosions, both unintentional, as shown with examples, and intentional. The biggest non-nuclear explosion of all times occured on 18 April 1947, when 6,700 tonnes of ammunition were used to destroy the German, island Helgoland. The explosion was, however, not successful.

2.4.11 Professional Risks

People who deal with using, producing and storing ammunition are under certain professional risk. These risks can be classified as professional risks. Figures 2-65 and 2-66 show such risks in terms of accident rates for different professions. For example, mining remains a risky profession even if it is not included in Fig. 2-66. It has been estimated that more than 8,000 miners have a fatal accident per year in China. For example, in July 2001, about 200 miners died due to water breaking into a stannary. However, mining disasters have a long history. For example, in the Ruhr area, mining had already started in the 13th century. The exploitation of galleries started there in the 16th century. In 1881, an authority required that every mine has to have two vertical tunnels. This indicates that mitigation measures had been applied probably based on some fatal accidents. However, even so, the severity of such accidents has not decreased over the years, as shown in Table 2-73.

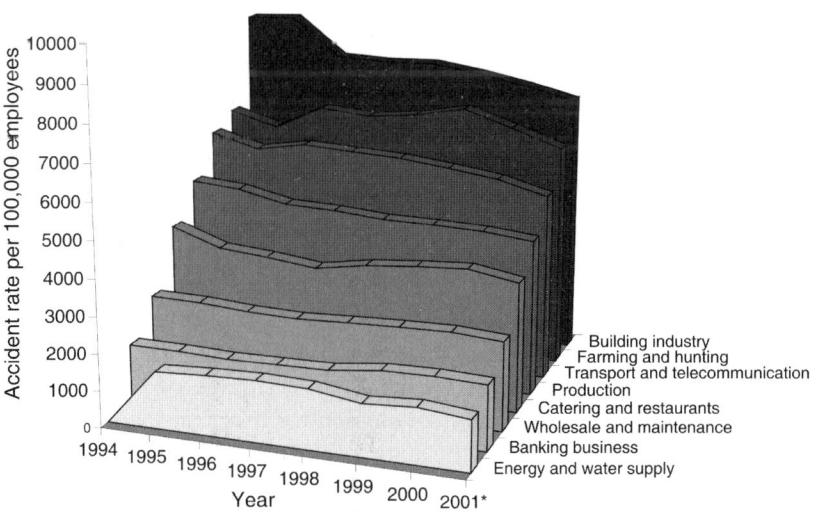

Fig. 2-65. Accident rates for different professions (Bergmeister et al. 2005)

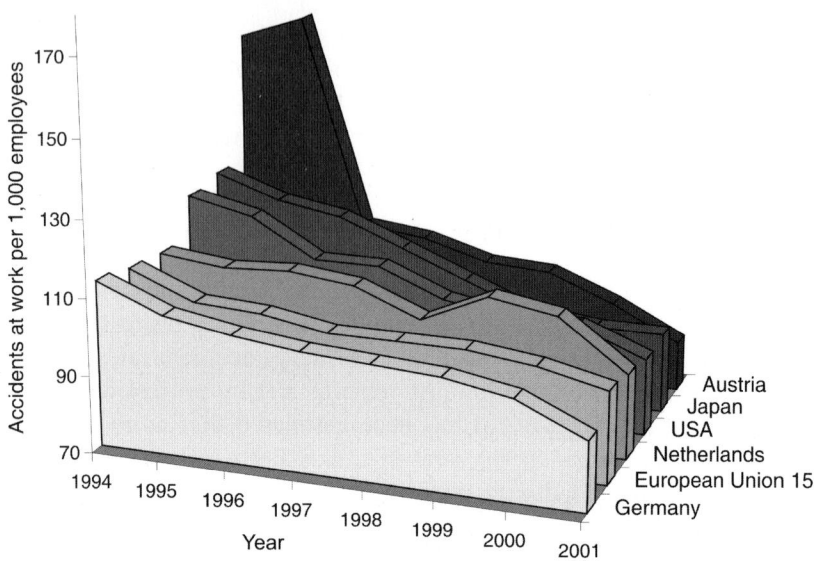

Fig. 2-66. Accident rates for different countries (Bergmeister et al. 2005)

Table 2-73. Major mining accidents (Kroker & Farrenkopf 1999)

Year	Location	Number of fatalities
1942	Benxihu, China	1,500
1906	Courrières, France	1,100
1960	Shanxi, China	680
1960	Coalbrook, South Africa	440
1913	Colliery, Wales	440
1946	Grimberg, Germany	405
1866	Yorkshire, England	390
1907	Monongah, USA	360
1908	Bockum, Germany	350

However, mining is a hazard not only for miners but also for the people living in the neighbourhood of the mines. Several accidents have shown such impacts by the uncontrolled release of toxic mining materials. For example, in April 1998 in Spain, a dam for the storage of colliery wastes broke. The wastes included chemical elements like arsenic, cadmium, thallium and other metals. When the dam broke, about four million cubicmeters of wastewater and mud was released. Although most of the mud deposited close to the dam, other parts flooded an area close to a national park and ran into the river Guardiama. The accident yielded to major ecological damage (Sjöstedt 2004).

In January 2000, the same type of accident occurred in a gold mine in Romania. Again, the dam for the colliery waste broke releasing, among other materials, about 120 tonnes of cyanide. The waste materials ran into the river Lapus, which is connected to the Theiss and the Danube. It was estimated that more than 1,000 tonnes of fish were poisoned.

An extremely tragic event was the slide of a waste site related to a coal mine in Aberfan in Wales in 1966. This slide claimed 144 victims, most of them children. In 1985, a dam broke in Trentino, Italy, killing 260 people and destroying more than 60 building (European Commission 2003).

However, mining is not the only profession with a certain degree of risk. Other professions also show higher risks as compared to the average profession value. For example, fishing and agricultural professions are exposed to such higher risks (DIRERAF 2007). The building industry also shows higher values. In the German federal state of Saxony, fatal, accidents in the construction industry account for 50% of all fatal industrial accidents. The protection against accidents at work is of major interest – not only for the companies alone but also for authorities. For example, in Germany, there exists the Federal Institute for Occupational Safety and Health (BAUA 2008), Association of German Safety Engineers (VDSI 2008) and the Association of Freelance Safety Engineers and Interplant Services (BFSI). This topic is also dealt with in occupational medicine and environmental medicine and includes the assessment of temperatures (Hausladen et al. 2002) and chemicals at the workplace (Böse-O'Reilly et al. 2001).

2.4.12 Information Technology

Many industries have existed for decades or centuries, for example, agriculture or mining. However, in such industries, new technologies emerge and are used. In fact, many of these new technologies in such traditional industries are based on new information technologies. Electronic information processing is of utmost importance and has today changed the average working conditions in over a few decades. The personal computer was introduced about 20 years ago. Nowadays, cars, airplanes, ships, elevators, washing machines and traffic lights are at least partly controlled by this technology. Some technologies have experienced an even wider penetration by information technology, like telecommunication and financing. In many of these products, information technology can decide about life and death of humans. So if this technology fails, it might impose a risk to humans. Therefore, in many cases, the so-called "redundant systems" are used, which still remain operable after some partial failure. Safety is understood in this field as the solution of a given problem in a given time.

However, problems are known. For example, the major public discussion about the Year 2000 problem should be mentioned. In the year 1999, computer failures probably costed about 100 billion dollars in the US alone.

Furthermore, information technology is corrupted intentionally through computer viruses. In 2001, computer viruses caused a damage of about 13 billion US dollars, in 2002 between 20 and 30 billion dollars and in 2003 about 55 billion dollars (c't 2004, Tecchanel 2004, PC Magazin 2004).

Software engineering, as a special part of the new information technology, was introduced in 1968. Some examples of project failure should be mentioned here. The German Ministry of Finance tried to develop a software called FISCUS, which was planned to be used in all 650 tax offices in Germany and, therefore, homogenise the tax declaration system. The development of the software had already costed about 900 million Euros, but was cancelled after 13 years in 2004, as the goal still remained inaccessible. It is quite interesting to note that the program failed partly due to bureaucracy.

Another example in Germany was the introduction of the web-based program 2All in 2005. This program was designed for the new unemployment regulations in Germany called Hartz-IV-reformation. The new regulations were to be built into the program, and about 2.5 million applications were supposed to be administered through this software. When the software was introduced, only a small bug yielded rather bigger problems: the account numbers were automatically right justified instead of being left justified. Therefore, the bank transfers did not work and a huge number of people did not receive their unemployment benefits (Aßmann et al. 2006).

A third example in Germany was the introduction of the distance-based toll for trucks. Although this is an extremely complicated system, here too some software failures occurred. For example, the automatic update function in the on-board units of the trucks did not function properly causing many problems with the toll registration (Aßmann et al. 2006).

2.4.13 Food

The introduction of factory farming and industrial production of food yielded to new risks in agriculture in the last decades. While in earlier centuries crop failure due to bad weather conditions was the major risk, new risks have emerged, for example, with artificial food additives. The population is extremely sensitive to risks related to food, since food is a general need for living. However, examples from the last decades show that indeed problems do occur (Table 2-74).

Table 2-74. Some food scandals in Europe (Dittberner 2004)

Year	Region or country	Problem
1971	Rhineland-Palatinate	Illegal amount of HCH (hexachlorocyclohexane) in milk
1972	Baden-Württemberg	Illegal amount of HCH in milk and vegetables
1976		Salmonellae in poultry
1977	Germany	Illegal amount of HCH in milk
1979	Hesse, Hamburg	Illegal amount of HCH in milk
1979	Nordrhein-Westfalen	Illegal amount of thallium in milk
1979	Hamburg	Illegal amount of dieldrin (pesticide) in 500 tonnes of Danish butter
1980	Germany	Detection of the synthetic hormone (diethylstilboestrol) in veal
1981	Spain	Illegal mixture of olive oil with rape oil that was designated for industrial usage. About 20,000 suffered toxication, some dead.
1982	Germany	Salmonellae in nearly 70% of all deep-frozen chicken detected
1984	Bonn	Illegal amount of HCH in milk
1985	Austria, Germany	Detection of anti-freeze agents in Austrian and German wines
1986	Ukraine, Europa	Radioactivity in food caused by emission of radioactive material from the nuclear power plant in Chernobyl
1986	Italy	Red wine mixed with methyl alcohol, about 30 people died.
1987	Germany	Detection of worms in sea fishes. The consumption of sea fishes fell sharply and new control procedures were imposed.
1987	U.K.	First publication of BSE
1988	Germany	Hormones found in about 70,000 calves
1993	Germany	Reports about spoilt meat in freezers
1994	Germany	Pesticide detected in baby food
1995	Bavaria, Baden-Württemberg	Antibiotics found in honey
1996	Germany	Application of toxic disinfection agents for cleaning of hen-coops
1997	Germany	Illegal import of beef from U.K.
1997–1999	Itaty, Belgium	Production of butter using suet and chemicals in Italy, distribution mainly in Belgium.
1999	Belgium	Animal feed mixed with dioxin containing industrial oil leading to, prohibition of eggs, butter and meat products

(Continued)

Table 2-74. (Continued)

Year	Region or country	Problem
2000	Spain	Illegal amount of pesticides found in spanish capsicum
2001	Germany, Austria	Application of hormones and vaccines for the fattening of pigs
2001	Europe	Import of shrimps from Asia treated with anti-biotics
2002	Netherlands, Germany	Import of veal loaded with chloramphenicol
2002	Sweden, Europe	Detection of acrylamid in some foods
2002	Thailand, Hesse	Import of Thai poultry containing nitrofuran
2002	Germany	Nitrofen (herbicide) found in wheat
2002	Italy, Germany	Import of turkey containing tetracyclin
2002	Germany	Spoiled poultry in trade
2002	Germany, Mecklenburg-Western Pomerania	Storage of wheat in the ground of a former military airport, wheat contained high amounts of lead
2003	Germany, Israel	Breast milk substitute contained insufficient amounts of vitamin B1, in Isreal two babies died
2003	Italy	Poison attack on mineral water and milk products

One of the greatest disasters in industrial food production was probably the BSE epidemic in Great Britain and Europe. It has been found that the BSE (bovine spongiform encephalopathy) originating from sheep infected with scrapie. Unfortunately, several years ago, the idea was born to recycle these sick sheep using their animal protein for ruminant feed production. Other theories consider the spontaneous jump of the disease from sheep to cows. Independent from that theory, sick animals were used for the production of ruminant feed, resulting in the illness of more than 200,000 cows. The number of precautionary slaughters may have run into millions.

The scrapie disease has been well known since about 200 years ago, and the old farmer rule, not to use meat from such sheep, may have prevented the transmission of BSE to humans in previous times. However, the transmission from cows to other mammals has been proven, and therefore the species barrier can be exceeded. A comparable disease is already known from epidemics in Papua New Guinea. Therefore, intensive scientific works were carried out to discuss the transmission of BSE to humans. Since the middle of the 1990s, a modified version of the Creutzfeld–Jakob disease has been observed. However, the proof of the causal relationship seems to be extremely difficult. Nevertheless, in contrast to the original Creutzfeld-Jakob disease, now even young people are hit by the disease.

Currently less than 200 people are infected with the disease. This is a rather low number, but considering an incubation time period of almost 20 years, numbers can still increase dramatically. Recent studies show that incubation time and the probability of a disease depends on the genotype (Dittberner 2004, Worth Matravers et al. 2000, Prusiner 1995, Carrell 2004).

The introduction of a new technology in farming was probably the cause of the epidemics. While most consumers did not realise a change of technology, with genetically modified food, consumers are much more aware of the change. Here, many people refuse to accept such food. However, one should not forget that genetically modified food has been used for decades. Not only is breeding a technology to change the genetic code, radioactive radiation has been also used for at least 30 years to cause mutations in plants and obtain species, with genetically altered properties. In contrast to this widely applied technology, genetic engineering permits the production of transgenic creatures with genetic code from other species (Uni-protokolle 2004, Spahl & Deichmann 2001).

Independent from new diseases and genetically modified food, many other risks are emerging in relation to food. The entire diet has changed in developed countries. This includes the effects of food, design of foods or current cultural development about the skinniness of females as well as the amount of time for meals. An overview about such impacts can be found in Grimm (2001) and Pollmer (2006).

2.5 Health Risks

Most people in developed countries die by health problems, not by natural or technical risks. Major causes are cardiovascular diseases and cancer. Then, further causes are ranked. However, in developing countries, the figures look different. Here, infections are of much concern. Figure 2-67 shows the causes of death in Germany and in developing countries. In this section, the health risks are discussed.

2.5.1 Cardiovascular diseases

About 50% of the population in developed countries die of cardiovascular diseases. Between 1975 and 1995, on average of more than 700,000 US Americans died from cardiovascular diseases per year (Parfit 1998).

The term cardiovascular disease, in general, can be defined as all morbid changes in the heart and the vascular system, but includes many different diseases. It includes, for example, coronary diseases, cardiac insufficiency, local asphyxia, (mostly) myocardial infarction, high blood pressure and strokes. The diagnosis is based on several techniques, such as anamneses, identifying family risks, diagnosis of the body, electrocardiogram, blood diagnosis or angiographies. Although it is the elderly who are generally exposed to this disease, some progress has been made over the last decades to decrease the incidence and mortality rates associated with this disease. For example, Finland, which experienced an over proportional burden of cardiovascular diseases, introduced a special preventive program in the 1980s. This program included the regular control of blood pressure, the prohibition of smokings, launching the subject "health and hygiene" in schools and the stimulation of the industry to provide low fat and low salt food. As a consequence of the program, the number of cardiovascular disease dropped (Walter 2004). Not only the incidence rate dropped, however, but also the mortality rate dropped through the improvement of medications like beta blockers, thrombolysis or angiotensin-converting enzyme inhibitors. It has been assumed that about 2/3 of the decrease in deaths is based on preventive measures, and about 1/3 is based on acute treatment. Therefore, an early identification of people at risk is necessary. Investigations to identify major risk groups include the:

- Framingham Heart Study, USA
- Seven Country Study, USA
- MONICA (Monitoring of Trends and Determinants of Cardiovascular Diseases), WHO
- ARIC (Atherosclerosis Risk in Communities), USA
- PROCAM (Prospective Cardiovascular Münster), Münster, Germany
- KORA (Cooperative Health Research in the region Augsburg)
- SHIP (Study of Health in Pomerania), Greifswald, Germany
- Heinz Nixdorf Recall Study (Risk Factors, Evaluation of Coronary Calcification, and Lifestyle), Essen, Germany

The studies have so far identified the so-called causal risk factors (Table 2-75), conditional risk factors and predisposing risk factors. The four major risk factors are hypertension, smoking, hypercholesteremia and diabetes. They can explain about 75–85% of all new cases. Risk calculators can be found on the internet, for example at Universität Münster (2007) or at BNK (2007). A good news about these risk factors is the fact that they are reversible or treatable. Further risk factors, which are still under discussion, include Hypertriglyceridemia, Lipoprotein-a, hyperhomocystinemia, hypercoagulability, c-reactive protein (CRP), processes of inflammation and coronary calcium deposit (WHO 2003, Schaefer et al. 2000, Fargeman 2003, Marmot & Elliott 1995, Keil et al. 2005, Kolenda 2005).

Table 2-75. Risk factors for cardiovascular diseases (Grundy 1999)

Risk factor	Categorical level
Cigarette smoking	Any current
Blood pressure	≥ 140 mm Hg systolic
	≥ 90 mm Hg diastolic
LDL cholesterol	≤ 160 mg/dL
HDL cholesterol	< 35 mg/dL
Plasma glucose	> 126 mg/dL (fasting)

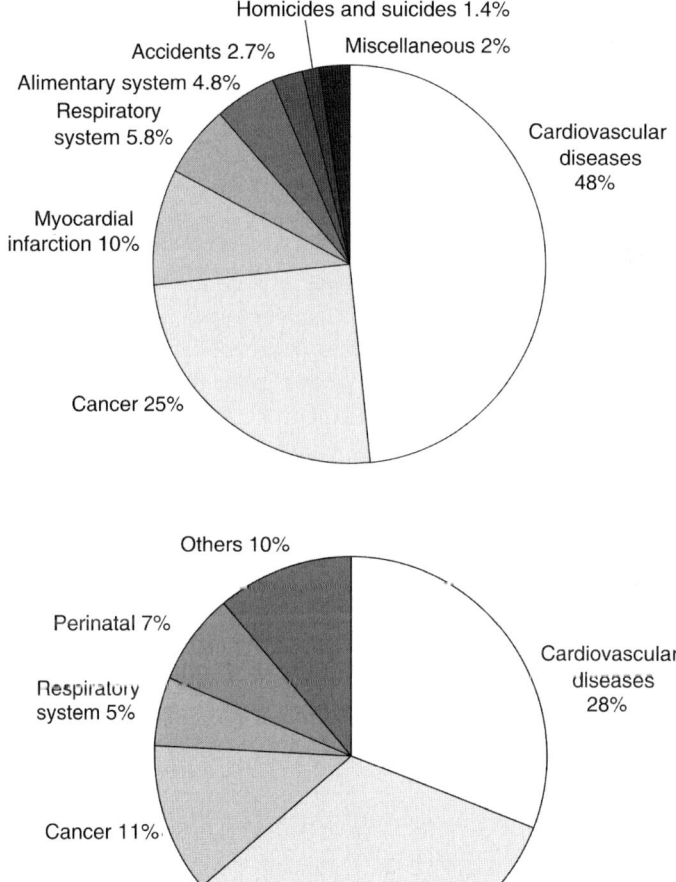

Fig. 2-67. Causes of death in developed (*top*) and developing countries (*bottom*) according to the UNO and the German Federal Statistical Office

2.5.2 Cancer

Cancer is the cause of about 25% of all deaths in developed countries. Cancer is usually described as a malignant tumour with extensive immaturity of cells, autonomy and invasive growth as well as the ability to build metastasis (Roche 1993). Tumour cells can appear at any stage during the entire lifetime of humans. The cause is usually multi-causal and does not depend on only one factor; even causal relationships for some circumstances have been found. In most cases, the tumour cells can be identified and destroyed by endogen defense. Especially, however, in older people, the defense tires and the number of erroneous cells increases. Therefore, at a certain age, the risk of cancer increases.

In 1997 in Germany, about 330,000 people became ill with some form of cancer. About 1/4 of the 330,000 patients were younger than 60, but the average age was 66 for men and 67 for women. People with cancer lose approximately 8 years from the average life expectancy. Fortunately, for some cancer types, the chance of survival has increased. This is mainly based on improved treatment techniques as well as on earlier diagnosis and some changes in the ratio of different cancer types. For example, the number of people with gastric cancer has decreased, while the number with large bowel cancer has increased. For the latter the chance of surviving is higher as compared to the former. (Cancer Register 1997).

The overall probability of getting cancer during one's entire lifetime is given in Table 2-76. Although the numbers are imposed for the US, they can be extended to most developed countries. It is quite interesting to note how high such lifetime values are. For example, one in every three people will be confronted with skin cancer at some point during their lifetime. Also, every tenth smoker will be hit by some cancer. It has been estimated that about 1/4 of all cancers are connected to smoking. Figures 2-68 and 2-69 show relations between cancer and smoking. Furthermore, chemicals seem to be able to cause cancer. About 2,000 chemicals have proven to be carcinogens in animal experiments (Henschler 1993). Relations with some special professions or special type of living have been found centuries ago (Table 2-77). Causes for cancer are shown in Fig. 2-70. However, since ionising radiation is often considered as cancer risk, in Figs. 2-71 and 2-72, the causes of radiation are given again to show how such causal chains can be extended to describe causes.

Table 2-76. Risk of cancer during lifetime (EPA 1991)

	Carcinoid situation or material	Cancer lifetime probability
1.	Sun (skin cancer)	3.3×10^{-1}
2.	Smoking (one package per day)	8.0×10^{-2}
3.	Natural radon concentration inside a house	1.0×10^{-2}
4.	Natural radiation outside of houses	1.0×10^{-3}
5.	Passive smoker	7.0×10^{-4}
6.	Artificial chemicals inside buildings	2.0×10^{-4}
7.	Air pollution in industrial areas	1.0×10^{-4}
8.	Chemicals in drinking water	1.0×10^{-5}
9.	Chemicals in food	1.0×10^{-5}
	(a) 60 gram peanut butter per week (natural aflatoxin)	8.0×10^{-5}
	(b) once per year a trout from lake Michigan	1.0×10^{-5}
10.	Leaking of chemicals from a dump site	$1.0 \times 10^{-4} – 1.0 \times 10^{-6}$

Table 2-77. Early medical investigations in cancer causes taken from Henschler (1993)

Year	Author	Type of cancer	Work or type of living
1743	Ramazzini	Breast cancer	Nun
1761	John Hill	Nose cancer	Snuff
1775	Percival Pott	Scrotum cancer	Chimney sweep
1795	Soemmering	Lip cancer	Pipe smoker
1820	Ayrton et al.	Skin cancer	Arsenic therapy
1874	Volkmann	Scrotum cancer	Brown coal tar
1895	Rehn	Cyst cancer	Aniline worker
1902	Frieben	Skin cancer	X-rays
1933	CIF	Nose cancer	Nickel extraction
1935	Lynch & Smith	Lung cancer	Asbestos
1940	Müller	Bronchial cancer	Cigarette smoking
1943	Wedler	Mesothelioma	Asbestos
1974	Creech & Johnson	Liver hemangiosarcoma	Vinyl chloride

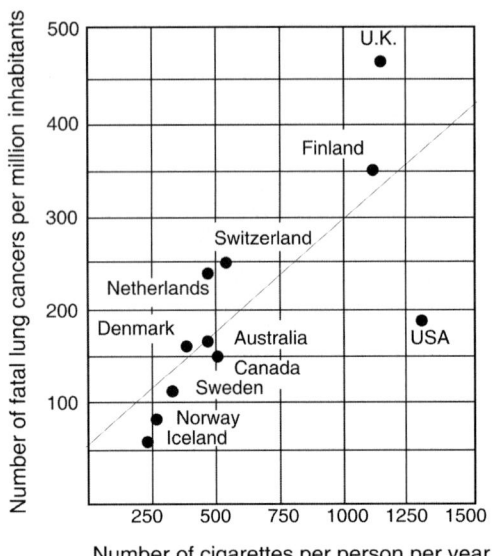

Fig. 2-68. Relationship between the average number of cigarettes smoked in a country and the number of fatal lung cancer cases taken from Henschler (1993)

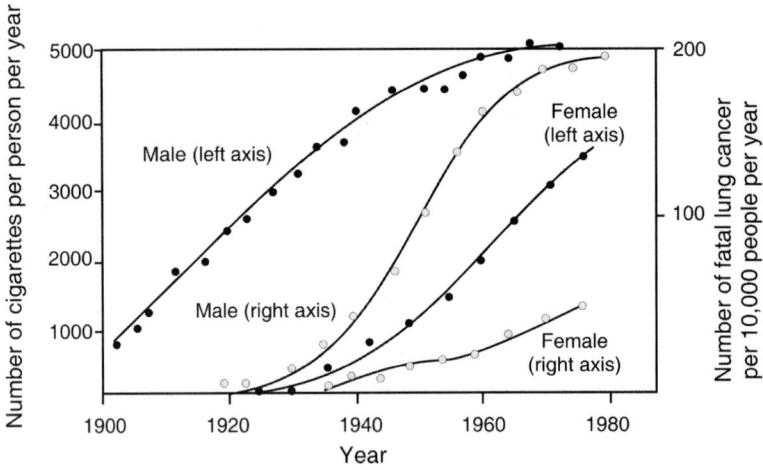

Fig. 2-69. Development of the number of cigarettes smoked over time and the number of fatal lung cancer cases taken from Henschler (1993)

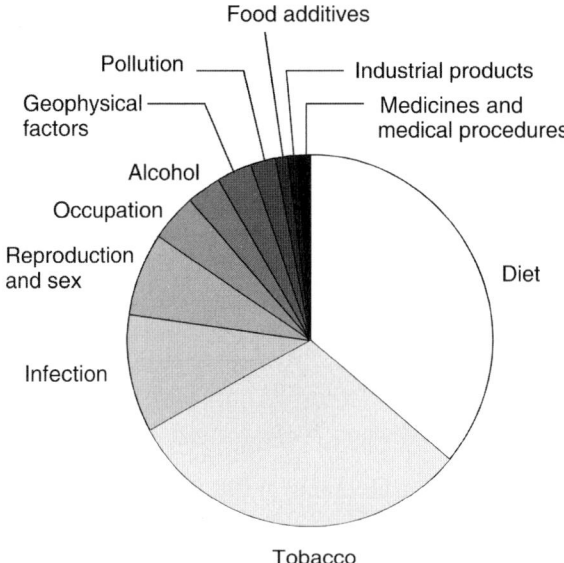

Fig. 2-70. Causes of cancer according to Doll & Peto (1981). Further works have been carried out, for example see Schmähl et al. (1989)

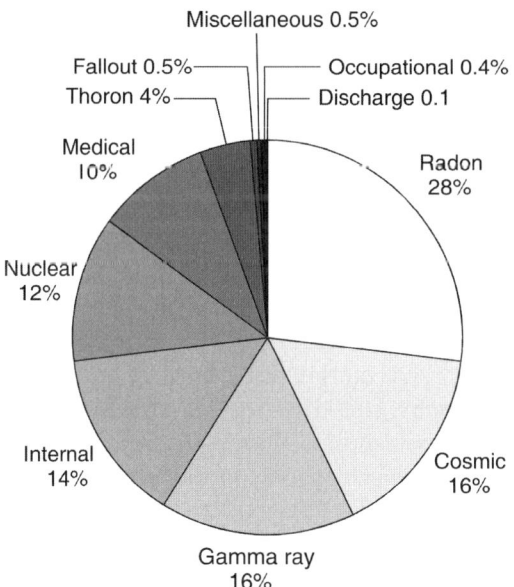

Fig. 2-71. Causes of radiation according to NRPB (1986)

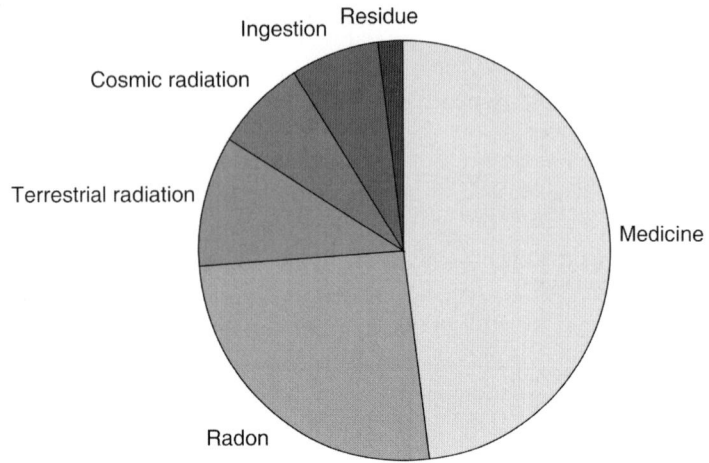

Fig. 2-72. Causes of radiation according to BfS (2007)

2.5.3 Birth

Even common natural processes include hazards. This becomes clearly visible during childbirth. Table 2-78 shows some numbers of infant and mother mortality rates. Here, again, the difference between lesser developed countries and developed countries becomes clearly visible. This becomes even more visible in Tables 2-79, 2-80, 2-81. It has been estimated that per year about 10.6 million children under the age of five die worldwide. This overly large number should be avoidable with current knowledge (Razum & Breckenkamp 2007, IFRC 2006). In some endogenous societies, the infant mortality rate reaches up to 30% (Schiefenhövel 2004). This is a value close to that of animals. For example, for tigers in the Bangladesh swamps, the pup mortality rate reaches 60% in the first three months.

Fortunately, the infant mortality rate in developed countries yields much better values. In many countries, values less than 1% are reached. The lowest values are about 0.4%. This progress is heavily related to some medical and hygienic improvements in the late 19th century in Europe and America. This progress does not only cover birth-related morality but also virtually all diseases known on earth.

Table 2-78. Infant mortality and mother mortality rates for different countries (Zwingle 1998)

Country	Infant mortality rate		
		Average infant mortality rate:	
Columbia	28×10^{-3}	Developed countries	8×10^{-3}
Brazil	43×10^{-3}	Developing countries	64×10^{-3}
Nicaragua	46×10^{-3}		
Mexico	28×10^{-3}	Average mother mortality rate during birth:	
USA	7×10^{-3}	Developed countries	1.0×10^{-4}
Russia	17×10^{-3}	Developing countries	50.0×10^{-3}
U.K.	6×10^{-3}		
Italy	6×10^{-3}		
Turkey	42×10^{-3}		
Germany	5×10^{-3}		
Mali	123×10^{-3}		
Egypt	63×10^{-3}		
Nigeria	63×10^{-3}		
Botswana	60×10^{-3}		
Iran	35×10^{-3}		
Saudi Arabia	29×10^{-3}		
India	72×10^{-3}		
China	31×10^{-3}		
Japan	4×10^{-3}		
Bangladesh	82×10^{-3}		
Papua New-Guinea	77×10^{-3}		
Australia	5×10^{-3}		

Table 2-79. Overall number of child deaths in different countries (Razum & Breckenkamp 2007)

S.No.	Country	Number of child deaths under the age of five
1	India	2,204,000
2	Nigeria	1,059,000
3	Congo	589,000
4	China	537,000
5	Ethiopia	515,000
6	Pakistan	482,000
7	Bangladesh	289,000
8	Uganda	203,000
9	Angola	199,000
10	Niger	194,000

Table 2-80. Number of child deaths per 1,000 births in different countries (Razum & Breckenkamp 2007)

S.No.	Country	Number of child deaths under the age of five
1	India	283
2	Nigeria	260
3	Congo	259
4	China	219
5	Ethiopia	205
6	Pakistan	204
7	Bangladesh	203
8	Uganda	200
9	Angola	197
10	Niger	194

Table 2-81. Spartial distribution of child deaths and ranking of the causes (Razum & Breckenkamp 2007)

	Africa	Southeast Asia	Eastern Mediterranean	Western Pacific	America	Europe
Absolute number of deaths under the age of five per year	4.4 Mio	3.1 Mio	1.4 Mio	1.0 Mio	0.4 Mio	0.3 Mio
Number of deaths per 1,000 births	171	78	92	36	25	23
Perinatal complications	26%	44%	43%	47%	44%	44%
Pneumonia	21%	19%	21%	13%	12%	12%
Diarrohea	16%	18%	17%	17%	12%	13%
Malaria	18%	0%	3%	0%	0%	0%
Measles	5%	3%	4%	1%	0%	1%
HIV/AIDS	6%	1%	0%	0%	1%	0%
Accidents	2%	2%	3%	7%	5%	7%
Others	5%	12%	9%	13%	25%	23%

Mio – million

2.5.4 Adverse Effects

The goal of medical treatment is the healing and support of ill people. Usually people suffering from some kind of illness are under increased risk. If errors are made during treatment, it might directly influence the survival probability of the patient and may result in a death, which may be avoidable.

There are different types of faults or errors during treatment. For example, a wrong diagnosis is made, a wrong treatment might be administered or the treatment might be carried out erroneously. There might also, however, be some known or unknown adverse side effects of the pharmaceutics. For example, a known adverse effect of chemotherapy during cancer treatment is a large decrease in the quality of life of the patient. In such a case, however, a gain in the quality of life after the treatment is assumed. In general, an adverse event is defined as disadvantaged but unintentional effect of a treatment, of course considering the right medical treatment and the recommended dose of the pharmaceutics.

The number of fatalities caused by adverse side effects is estimated between 44,000 and 98,000 per year in the USA. Comparable relative values exist in Canada, Australia, U.K. and New Zealand (Wreathall 2004, JCAHO 2004, Davis et al. 2001, Vincent et al. 2001, Wilson et al. 1995). Even though such values seem to be rather high, one should keep in mind that the number of fatalities without any treatment would result in much higher fatality rates. The consequence of these investigations showing such numbers should not be taken as a general criticism of the medical system, but should assist with the introduction of quality standards and risk management procedures. The number of avoidable cases is assumed to be approximately 40-70% (Wreathall 2004). The causes for such cases are human errors, miscommunication, technical failure or known side effects. However, as stated in chapter "Indetermination and risk", the complex system "humans" will show surprising reactions. This influences the outcomes of medical treatment. An example of this is spontaneous cancer healing, which is extremely rare.

One intensively and publicly discussed adverse side effect is in relation to vaccinations. Although modern types of vaccinations are beyond any comparison with historical types of vaccinations, which had very high incidence rates of adverse side effects, even nowadays adverse effects occur. For example, in Saxony, a federal state of Germany, nearly 9.2 million vaccinations were given between the years 2001 and 2004. From these 9.2 million vaccinations, 10 cases caused adverse effects. The distribution of these cases in relation to the different vaccinations is shown in Fig. 2-73. Of course, if one looks over a longer period, the ratio changes. From 1990 to 2000, there were about 22 million vaccinations in Saxony alone, yielding to 23 cases of harm. Here, the major contribution came from the BCG vaccination (Bigl 2007).

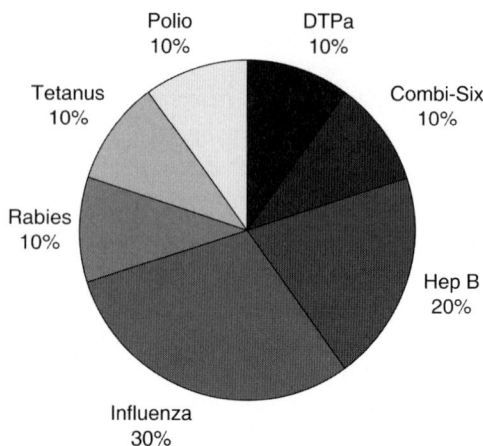

Fig. 2-73. Ratio of different vaccinations to overall number of undesirable effects in Saxony, based on Bigl (2007)

Probably the most used pharmaceutic worldwide is acetylsalicylic acid (aspirin). Aspirin has been proven to drop the probability of cardiac infarctions or strokes. On the other hand, it increases the probability of gastrorrhagia and cerebral apoblexy. Therefore, this pharmaceutic might be applied as a safety measure. However, it might also decrease the safety for patients in other areas. While advantages and disadvantages can be weighed against each other, it becomes more difficult when a combination of different pharmaceutics is taken, which and therefore might increase the risks to patients. Smith et al. (1966) showed an exponential growth of adverse effects of pharmaceutics related to the linear growth of drugs administered. The identification of such adverse effects based on the application of different pharmaceutic is very difficult, as shown in Table 2-82. Usually the number of investigated patients during the registration phase lies between 3,000 and 5,000, and during the limited registration of three years between 10,000 and 100,000. It should be mentioned that the registration of a new pharmaceutic costs about 500 million Euros. It has been estimated that about 6% of all hospitalisation is caused by such adverse effects. Adverse effects yield to about 0.1%–0.2% of all deaths. About 80% of these deaths may be avoidable (Fauler 2006).

The problems with identifying the causal relationships (AVP 2005, Oerlinghausen 2006) between pharmaceutics and adverse effects become visible through the dysmelia syndrome incident. Here, one tablet taken during the third and sixth week of pregnancy could yield to embryopathy. The product was later withdrawn. The product cycle of different pharmaceutics is shown in Figs. 2-74 and 2-75, giving the impression that more such adverse effects are found during use. However, as the next section shows, pharmaceutics have been successfully applied against many diseases.

Table 2-82. Number of patients treated for one, two or three adverse effects with a confidence interval of 95% related on some different incidences. It becomes clear that low incidences are extremely difficult to prove (WHO 1983)

Incidence of adverse effects	Number of patients treated for adverse effects		
	one	two	three
1/100	300	480	650
1/200	600	960	1300
1/1,000	3,000	4,800	6,500
1/2,000	60,00	9,600	13,000
1/10,000	30,000	48,000	65,000

Time in the market of various active ingredients prior to recall

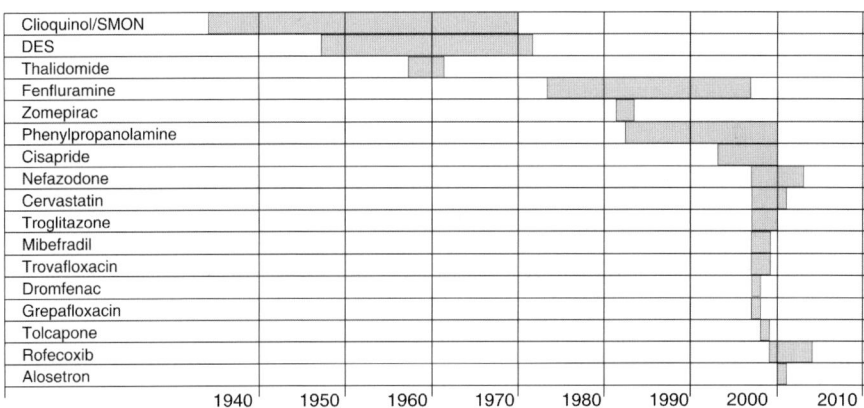

Fig. 2-74. Time in the market for various active ingredients prior to recall subject to the type of application (Munich Re 2005b)

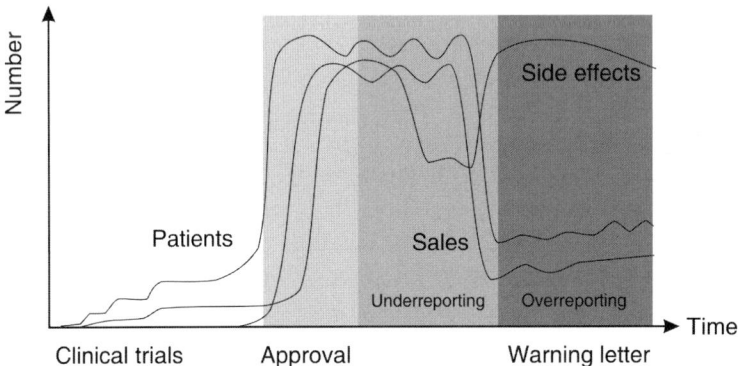

Fig. 2-75. Model for the development of patient numbers, sales numbers and numbers of side effects during the establishment of a drug (Munich Re 2005b)

2.5.5 Epidemics and Pandemics

Epidemics were great hazards during the development of mankind, especially before the Industrial Revolution. The Industrial Revolution yielded to a dramatic improvement in working and living conditions. Nevertheless, epidemics and pandemics did also occur after the Revolution.

While the term epidemic describes the massive occurrence of a disease in a limited area or over a limited time, a pandemic describes an epidemic that reaches major areas of a country or a continent. Throughout history, there have been several diseases that reached the status of an epidemic and/or pandemic. These diseases were mainly infectious diseases. Examples are given to illustrate the magnitude and severity of such pandemics.

Influenza in conjunction with bacterial superinfections quite often occur as an epidemic or pandemic. For example, in 1889 and from 1918 to 1919, major influenza pandemics were experienced, resulting in a large number of fatalities. The Spanish influenza in the years 1918 and 1919 emerged in four waves, claiming about 22 million lives. In Germany alone, somewhere between 225,000 and 300,000 people or 0.5% of the population died (Jütte 2006). It has been estimated that about 500 million people were infected with Spanish influenza. In recent decades, influenza pandemics have also occurred. For example, in 1957, there was an outbreak of Asian influenza with approximately 1 million fatalities, and in 1968 there was an outbreak of Hong Kong influenza resulting in about 700,000 fatalities. The latest influenza pandemic threat occurred in 2006 when a mutation of the H5N1 bird flu virus was thought to have occurred. The British government estimated that there could be up to 320,000 victims in Britain alone. However, even "normal" influenza seasons in Germany claim several thousand victims as shown in Table 2-83.

Since the 16th century, about 30 influenza pandemics have been identified. Probably well known, however, are the plague pandemics of the medieval times.

Table 2-83. Number of fatalities per influenza season in Germany (RKI 2007)

Influenza season	Fatalities
1995/96	32,000
1996/97	7,000
1997/98	6,000
1998/99	20,000
1999/00	12,000
2000/01	7,000
2001/02	4,000
2002/03	17,000
2003/04	6,000

2.5.6 Bubonic Plague

The Bubonic plague, also called the Black Death, was probably the greatest plague from the 14th to 18th century in Europe. It is a highly contagious disease, which can be found in rats and is transmitted through fleas. There exist several types of plague, such as the lung plague or the skin plague. The mortality rate depends on the type of plague, but has historically reached values of 95% for the lung plague, 75% for the bubonic plague or 100% for the sepsis. The plague usually originates from plague reservoirs located in the north of Asia, such as Siberia or Mongolia. Iran and Africa are sometimes mentioned as places of origin (Roche 1993).

The first historical data concerning the plague can be found in a description of the siege of Athens during the pelopennical wars in about 430 B.C. Approximately one-third of the Athens population died at this time by the plague. In 170 A.D., there were some plague epidemics in the eastern part of the Roman Empire. In 542, the Justinian plague destroyed all goals of the Emperor Justinian to re-establish the Roman Empire. In 630 A.D., a plague epidemic occurred in the Persian Empire (Leberke 2004, Grau 2004).

Probably the worst plague pandemic occurred from 1347 to 1352. During that time, presumably 25 million people were killed by the disease, which was approximately one-quarter of the entire European population. For example, in the German city of Lübeck, 90% of the entire population was lost. Even wealthier people died: 25% of all landlords as well as 35% of the town councilmen died. In the German city of Erfurt, half the population died, and in Frankfurt/Main, about 2,000 deaths per day were counted during the pandemic (Grau 2004).

The origin of this historic pandemic can be traced back to the peninsula Crimea. Here, some Genoese traders were locked in the city of Kaffa by Tatarian and Awarian soldiers. The city was beset for quite some time. It was unclear whether the disease started inside the city or in the lines of the besieger. After the beset was over, the Genoese traders returned to Genoa, halting at Constantinople, Messina and Naples. It was in these places that the plague occurred first. The pandemic then reached virtually the whole of Europe. First, Italy was hit, followed by Vienna in 1348 and then England and Scandinavia in 1350. Even Iceland was touched by the disease. Whether the information about the number of fatalities is correct remains uncertain. It seems rather certain, however, that many regions were almost completely depopulated and became deserted. In Italy alone, it took about a century to restore the population number to what it was before the plague.

When this plague ended in the middle of the 14th century, one may not think that the plague did not return. The German city of Lübeck was hit by the plague not only in 1350 but also in 1406, 1420, 1433, 1451, 1464, 1483, 1484, 1525–1529, 1537, 1548, 1550, 1564, 1565, 1625 and 1639. A plague pandemic occurred in Italy in 1630, in London from 1661 to 1666 and in Vienna from 1678 to 1679, each claiming about 100,000 lives. In 1720 and 1814, a plague epidemic occurred in France and Belgrade, respectively. Even at the end of the 19th century, a pest pandemic was observed in Asia, starting in 1894 in Hong Kong and reaching China, Japan and India in 1896. It has been estimated that about 15 million people died. By trade ships, this pandemic also reached Suez in 1897, South Africa in 1899, San Francisco in 1900 and Paris, in 1920. In Paris, only 20 cases were observed. The last plague epidemic was observed in 1911 in Manchuria and fought off successfully. Based on World Health Organisation data, about 1,000–3,000 cases still appear per year worldwide. In 1994 in India, an epidemic centre was identified. In general, the data show that the hazard posed by the plague has been permanently decreased over the last centuries. This development, however, is not valid for other infectious diseases, for example malaria.

2.5.7 Malaria

There exist different types of malaria: malaria quartana, malaria tertiana and malaria tropica. The disease malaria tropica can be deadly. All types of malaria are caused by a protozoan of the genus plasmodium (*Plasmodium vivax*, *Plasmodium ovale, Plasmodium falciparum*). Plasmodium are intracellular parasites. They are transmitted by Anopheles mosquitoes. The Roman philosopher Marcus Terentius Varro had assumed that malaria was caused by very small animals transmitted by mosquito bites (Köster-Lösche 1995). There is one development cycle of the plasmodium inside the mosquito and one inside the human. The incubation time, depending on the plasmodium, takes about 7 to 40 days.

The number of infected people worldwide has been estimated between 300 and 500 million. Fourty percent of the world population lives in malaria-contaminated areas. Every year, the disease claims between 1 and 3 million lives, most of them children under five years. The major areas where the disease is prevalent are Africa, South America and Asia (Fig. 2-76). Africa, however, has 90% of all infected people worldwide. Historically, malaria-contaminated areas reach much further north, for example in Europe.

There were some malaria epidemics in the 16th century (1557–1558), 17th century (1678–1682), 18th century and 19th century (1806-1811, 1845–1849, 1855–1860, 1866–1872). As a result of the increase in the cultivation of land and the heavy application of DDT (dichlorodiphenyltri-chloroethane), malaria was on the retreat. Unfortunately, however, the plasmodium became resistant. The most used pharmaceutics, chloroquine and mefloquine, to prevent malaria will now have to be applied in higher doses.

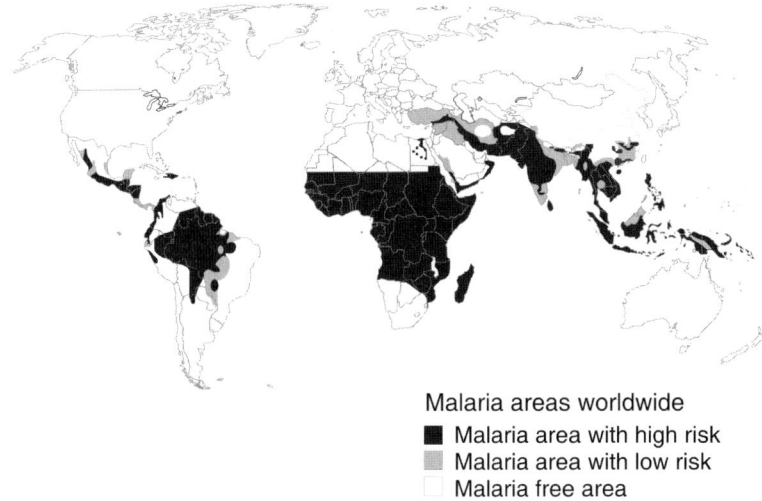

Malaria areas worldwide
■ Malaria area with high risk
　Malaria area with low risk
　Malaria free area

Fig. 2-76. Geographical distribution of malaria worldwide

2.5.8 AIDS

HIV (Human Immunodeficiency Virus)/AIDS (Acquired Immuno Deficiency Syndrome) is a rather new disease, and only scientifically known since 1980 when it was described for the first time in the US. The first patient was Gaetan Dugas; however, based on later investigations, AIDS can be traced back to diseases from chimpanzees and other primates. This means that the disease probably did not just start in 1980 but the pandemic started then (Köster-Lösche 1995).

After an incubation time of 2–6 weeks, for about 7-10 days, a first acute HIV infection may become appearent. This infection is usually not diagnosed as HIV as no antibodies are detectable. After the acute infection, a chronic infection starts running over for an average of ten years. During that time, the immune system is increasingly damaged and the number of CD4 cells (lymphocyte) drops from 1,000 to about 200 per µl blood. At that level, the immune system is so heavily damaged that so-called opportunistic infections occur. Based on the number of lymphocytes per µl blood, different stages of HIV infection are classified.

It was in the early 1980s about 150 people died from the disease, and about 500 people were identified as infected. By 1985, about 12,000 people were infected. According to some publications, between 34 and 46 million people are currently infected with HIV/AIDS. In 2003, between 4.2 and 5.8 million people were newly infected. This figure includes about 800,000 children. In some regions, the number of new infections is decreasing. In Germany in the 1980s, the new infection number was about 8,000, whereas currently the number is between 2,000 and 2,500. In other regions, however, the number is increasing dramatically. In some regions, however, such as southern Africa, more than 10% of the population is infected. In some Asian regions, the number of infected people is also growing significantly (Bloom et al. 2004, BfG 1997).

In 2003, about 2.3–3.5 million people died from AIDS. It has been estimated that over 20 million fatalities have been claimed since the beginning of the disease. Mathers & Loncar (2006) have estimated that the overall number of fatalities from AIDS for 2030 will be 117 million with a yearly number of victims of about 6.5 million. The strong influence of AIDS on mortality for South Africa is shown in Fig. 2-77. While in most countries worldwide, the number of deaths in the age cohorts between 15 and 40 is rather low, here a maximum is reached (van Gelder 2003).

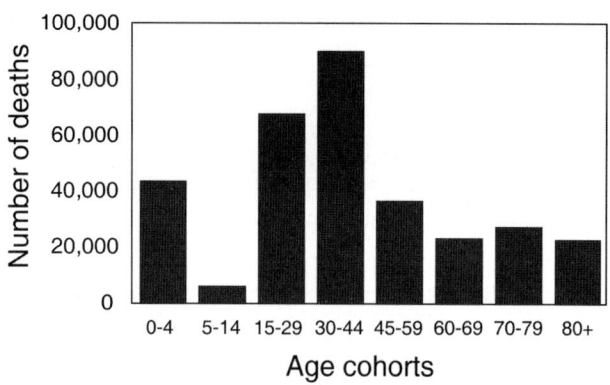

Fig. 2-77. Mortality diagram for different age cohorts in South Africa for AIDS according to van Gelder (2003)

Since the first description of AIDS and the discovery of the HIV by Robert Gallo and Luc Montagnier in 1983 and 1984, much research work has been carried. However, still the disease is not curable. In 1987, the first HIV medicine was introduced. Currently, 25 single or combination preparations, from four active substance classes are admitted for treatment.

2.5.9 Tuberculosis

Tuberculosis is an infectious disease that occurs in episodes. Between a third and a half of the entire world population is infected by this disease. Every year about 100 million people are newly infected by the pathogen *Mycobacterium tuberculosis*. The estimated number of deaths worldwide is between 2 and 3 millions. About 5,000 people worldwide die from tuberculosis every day, 95% of them in developing countries (RKI 2004).

The relationship between insufficient social conditions and the prevalence of the disease can be observed in Russia after the collapse of the Soviet Union. The radical social change yielded to the impoverishment of major parts of the population. This resulted in a dramatic increase in the number of people infected with tuberculosis. In only 10 years, the number of people infected with tuberculosis doubled. The mortality rate actually trippled. In addition, the interrupted treatment of certain people, for example prisoners, supported the development of treatment-resistant tuberculosis bacteria (RKI 2004).

2.5.10 Cholera

Historically, cholera was also a major plague. In 1817, about 600,000 Indians died from cholera. In the beginning, this dicase did not affect the British; therefore, it was not of concern to them. However, after 9,000 soldiers out of 18,000 died at the British headquarters they became alert (Köster-Lösche 1995). Figure 2-78 shows an example of the pattern of cholera during a London epidemic in 1854. Here, the relation with water quality becomes visible.

Fig. 2-78. Cholera pattern for London taken from Fragola & Bedford (2005)

2.5.11 Other Infectious Diseases and Outlook

The listing of diseases together with the number of victims could be extended arbitrarily. For example, smallpox claimed about 500 million lives between 1880 and 1980, and even nowadays measles claims about 900,000 victims worldwide. In 1995 alone, more than 50 million people died by infectious diseases (Bringmann et al. 2005). The list, however, would blast the frame of the book. For the interested reader, the UN report "Global burden of disease" is recommended.

Especially in many developing countries, the number of infectious diseases still remains high. Diseases such as malaria, amoebic dysentery, bilharziasis, leishmaniasis and sleeping sickness claim a huge number of victims. Pathogens are also becoming more and more resistant. This is true not only in developing countries but also in developed countries. For example, the pathogen *Staphylococcus aureus* has become methicillin-resistant; the enteroccocus pathogen has become vancomycin-resistant; and some phylum pathogens from staphylococcus have become vancomycin-intermediate-resistant. No new antibiotics have been developed since the 1970s with the exception of the pharmaceutic linezolid. On the other hand, many infectious diseases have been battled fairly successfully over history, for example cholera, smallpox, leprosy, syphilis and plague by pharmaceutics, and improved living conditions. This shows that not only the health risks but also social conditions or social risks cause the death.

2.6 Social Risks

2.6.1 Suicide

Suicide is the deliberate abortion of one's own life. The abortion can be done either in an active way by carrying out a certain dangerous action, for example jumping from a structure, or else by endangering life-supporting actions like avoiding medicine or addicted to drinking. The worldwide number of successful suicides has been estimated by about 1 million per year. However, this value shows geographical patterns as well as change over time. In Germany, the number of suicides has dropped from 18,700 in 1982 to about 12,000 in 2000 (Helmich 2004, Eichenberg 2002). Maximum suicide rates were reached in the region of the former USSR (Fig. 2-79). In comparing between genders, more males commit suicides than females. In Russia, males commit up to four times as many suicides as females, whereas in Germany the suicide rate for men is about double that of women.

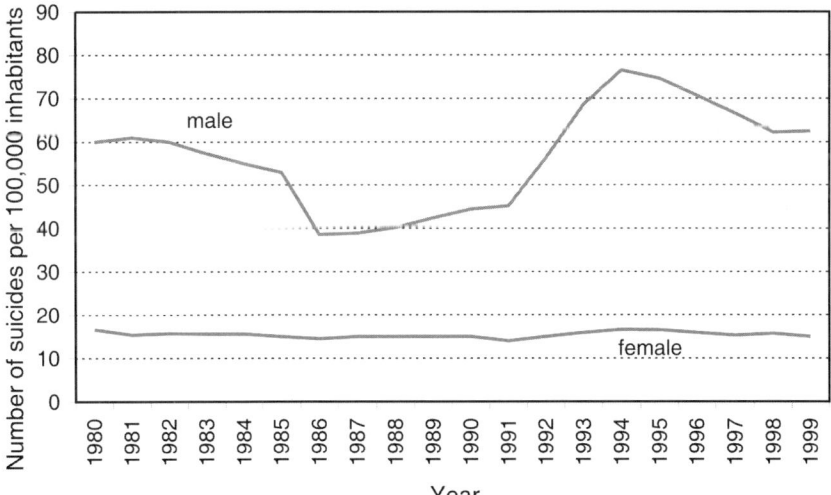

Fig. 2-79. Development of the suicide rate in Russia (Felber 2004)

The suicide rate is not constant acron life stages. In general, one can state that the suicide rate increases with increasing age. Every second women carrying out a suicide is older than 60 years. Suicide rates reach a peak in youth. In the age bracket of 15 and 29 years, suicide is the second highest cause of death (Dlubis-Mertens 2003).

In Germany, suicides contribute to 1.3% of all deaths. However, it has been estimated that the number of unreported cases is approximately 25% of the known suicide numbers. It is difficult to know how many drug misuses, car accidents and other accidents were caused intentionally. Suicide attempts are not recorded in Germany due to data protection restraints.

Media has a great impact on suicides. The internet has been used for organisation of collective suicides. Mass media reports causes sometimes called suicide links or copycat suicides. This is also called the "Werther effect", since the first reported suicide link happened after the publication of Goethes book "The Sorrows of Young Werther". Therefore, in several countries, there exist rules for reporting about suicides. One of the most famous examples is the ban on reporting about suicides that occur at the Golden Gate Bridge in June 1995, as the number of suicides at this structure was close to 1,000. In France, pictures from suicides are prohibited in the mass media. In Austria, reports about suicides using the underground were stopped due to a suicide link.

Examples of several preferred spots for suicides are given in Table 2-84. In Germany, some suicides happen at the Göltzschal Bridge, the biggest brick arch bridge in the world (Fig. 2-80).

Table 2-84. Some structures or locations with suicide numbers

Structures	Number of suicides
Golden-Gate-Bridge in San Francisco	~ 1,200
Eiffel Tower	~ 370
Skyway-Bridge in Florida	~ 81 since 1989
Arro-Secco-Bridge in Pasadena (California)	
Mount Mihara Vulcano in Japan	
Empire-State-Building in New York	
Space-Needle in Seattle	

One of the major causes of suicides is depression. It is estimated that in Germany about 4 million people (5% of the population) are affected by this disease. If this is true, then suicides could be classified as a health risk rather than as a social risk. However, in most cases, depression is caused by a single dramatic social event, such as the loss of job, the loss of a partner or the loss of a child. These are damages to the social network in which people live, and if the remaining social network is not strong enough to heal the injury, then the individual human might fail.

Fig. 2-80. Göltzschtal Bridge in Germany

2.6.2 Poverty

Poverty can be understood as a specific lifestyle. This lifestyle is defined by a lack of opportunities. The UNO have defined poverty as "deprivation of the basis of living". Poverty can be measured in different ways. The poverty parameters might be based on monetary income, poverty line, unsatisfied basic needs indicator or the poverty head count index (Coudouel et al. 2003).

Poverty can be found in all geographic regions of the world and affect persons of all ages. Based on the definition by the amount of available monetary resources, about one-fifth of the earth's population is poor, living with less than a dollar per day. This relates to 1.2 billion people. But, of course, poverty is not equally distributed geographically, even if it can be found worldwide. Developing nations, in particular, show a higher contribution to the number of poor people worldwide. Considering the limited access to even elementary health services in developing nations, nearly 60% of the population can be defined as living in poverty. In some countries, for example Mexico, values between 70% and 90% of the population are considered to be living in poverty.

Some parts of the population are more exposed to poverty than others. For example, it has been estimated that about 10.6 million children under five years of age die per year worldwide, mainly due to insufficient supply of nutrition or medical treatment (Razum & Breckenkamp 2007). A general relationship between the social status and the health of children was given by Lampert & Kurth (2007). The influence of unemployment and poverty on health has been investigated by Weber et al. (2007).

Several studies estimated the loss of life expectancy due to poverty even in developed countries. The range of lost life expectancy lies between 5 and 12 years in Germany. In other countries, a value of 7 years is mentioned (Cohen 1991). In developing countries, average life expectancy is heavily influenced by poverty, reaching an extremely low value of only 33 years in Zimbabwe. Poverty is often related to abnormal political conditions and war conditions.

2.6.3 War

Wars can yield to such a dramatic increase in mortality rates that they are often not considered a risk in the classical sense. On the other hand, wars happen regularly (Richardson 1944), and therefore values about the mortality rates should be included. It should be mentioned here that mortalities during the times of war are connected not only to military actions but also to the decrease in normal living conditions. For example, many people have died during wars by famine or insufficient medical treatment. Examples are given to stress about the loss of human lives under war conditions.

During the World War I, 34 countries with an overall population of 1.52 billion were in a state of war. About 67 million soldiers were mobilised. Approximately 13% of the soldiers died, about 30% were injured, and about 5% were disabled after the war. This gives a mortality rate of 3% (3×10^{-2}) per year for soldiers (Schenck 1965).

In the World War II, the mortality rate for German soldiers was about 12-15% per year. Between 1939 and 1945, about 5.3 million men were killed. The mobilisation rate (soldiers per total number of men living in a country) reached nearly 50%. Over the entire war period 25% of all men in Germany died (Overmans 1999).

With a total of 50 million fatalities, the World War II could be considered the biggest manmade disaster in human history. Some suggest, however, that the wars in the 20th century actually built a chain of wars that are all connected to each other. World War I was is considered the starting point of the chain of wars, yielding not only to the vast number of fatalities but also to the collapse of many different countries, including the civil wars that lasted over several years with an extraordinary loss of lives. Examples of civil wars are the ones that occurred in Russia and China. The economical consequences of World War I might also be in part responsible for the political success of the Nazi party in Germany.

As a consequence of the World War II, Europe was divided. In Asia, empires were forever changed. For example, the influence of Japan was heavily decreased, leaving Korea as an intermediate state. This was the basis for the Korean War as well as the Vietnam War. In the end, this chain of wars was the major contributor to a loss between 150 and 200 million lives in the 20th century. It has been estimated that between 4% and 5% of all deaths in the 20th century were a result of wars. Numbers of war-related deaths are given in Table 2-85.

Although the 20th century showed the highest number of war fatalities, wars have probably been a permanent companion of human development. Wars were lead during the old Egyptian time, during the time of the Greeks, during the Roman Empire, during the migration time and throughout the medieval ages. Table 2-86 gives the numbers of war fatalities for some major wars throughout the last five centuries.

In addition to numbers, some detailed remarks are given here. One war mentioned in Table 2-86 was the 30-year war in Europe. During that war, the loss of soldiers contributed only a minor portion to the overall loss of human life. The war was heavily marked by the violence of soldiers against the civilian population. Even worse though were the famines and epidemics. A detailed summary of victims was given by a royal Prussian doctor in the 19th century in the book "History of epidemics, famine and war during the time of the 30 year war". For example, the Mark Brandenburg lost about half its population. However, it should be mentioned that the population was often uprooted, not killed. Therefore, the number of fatalities might be lower. During that time, it was not permitted for farmers to move away from their villages. Many farmers used the chaotic situation during the war to escape to the growing cities. Hence, of the 30-year war yielded to an uprooting of the German population.

However, the huge loss of life remains if one looks at the overall population numbers. Before the 30-year war, the German empire had a population of about 22 million. After the war, the population was estimated to be approximately 13.5 million. The loss of the Netherlands and Bohemia is already included in the numbers.

Table 2-85. Number of war-related deaths over the centuries (Eckhardt 1991, Leger Sivard 1996)

Century	War death in millions	Death per 1,000 people
1st–15th	3.7	Not specified
16th	1.6	3.2
17th	6.1	11.2
18th	7.0	9.7
19th	19.4	16.2
20th	109.71	44.4

Table 2-86. Number of war-related deaths for select wars (Renner 1999, White 2007)

Conflict	Period	Number of people killed in millions	Civilian victims in percent
Peasants' War in Germany	1524–1525	0.17	57
Dutch independence War versus Spain	1585–1604	0.18	32
30-year War in Europe	1618–1648	4.00	50
Spanish Succession in Europe	1701–1714	1.25	–
7-year War (Europe, North America, India)	1755–1763	1.36	27
French Revolutionary and Napoleonic wars	1792–1815	4.19	41
Crimean War	1854–1856	0.77	66
US Civil War	1861–1865	0.82	24
Peruguay versus Brasil and Argentina	1864–1870	1.10	73
Franco-Prussian War	1870–1871	0.25	25
Congo Free State War	1886–1908	8.00	
US-Spanish War	1898	0.20	95
Mexican Revolution	1910–1920	1.00	
Russian Civil War	1917–1922	4.0–10.0	
China: Warlord and Nationalist Era	1917–1937	40.0	
World War I	1914–1918	26.0	50
Armenian massacres	1915–1923	1.00	
Stalin's regime	1924–1953	15-30	
World war II	1939–1945	53.5	60
Chinese Civil War	1945–1949	1.0–6.0	
Mao Zedong's regime including famine	1949–1975	40–45.0	
Korean War	1950–1953	2.7–2.9	50
Rwanda and Burundi	1959–1995	0.7–1.7	
Vietnam War	1960–1975	3.0	58
Nigeria	1967–1970	2.0	50
Cambodia	1970–1989	1.2	69
Bangladesh secession from Pakistan	1971	1.0	50
Afghanistan	1978–1992	1.5	67
Mozambican Civil War	1981–1994	1.0	95
Sudan Civil War	1984	1.5	97

In contrast to the historic wars, especially during the early ages and the medieval ages, the weapons in modern times have reached a much higher destruction power. The development of the destruction power is given in Table 2-87 in terms of explosive power and killing index. The explosive power is given in the mass equivalent of TNT. The killing index, which is used in some armies for qualification, gives the number of possible targets. Of course, this number is rather a theoretical value, which considers only the maximum gun speed. It does not consider any logistic problems with ammunition. Based on Table 2-87, the highest destruction capacity is found with the application of nuclear bombs (Albrecht 1985).

Table 2-87. Explosive force and killing index for weapons (Albrecht 1985)

Weapon	Explosive force in TNT	Killing index
Javelin		18
Sword		20
Bow and arrow		20
Crossbow		32
Drake, 12 pounds, 16th century		43
Flint with flintlock, 18th century		47
Muzzleloader rifle, middle 19th Century		150
Field gun, 12 pounds, 17th century		230
Breech loading fire arm, end of 19th century		230
Repeating rifle, World war I		780
Field gun type Gribeauval, 12 pound grenade, 18th century		4,000
Machine gun, World War I		13,000
Machine gun, World War II		18,000
Field gun, explosive grenade 75 mm, end 19th century		34,000
Tank, World War I (two machine guns)		68,000
Airplane, World War I (one machine gun, two bombs)		23,0000
Field gun, explosive grenade 155 mm, World war I		470,000
Howitzer, grenade with proximity fuze 155 mm, World War II		660,000
V-2-Rocket, World War II		860,000
Tank, World War II (one gun, two machine guns)		2.2 million
Fighter bomber, World War II (eight machine guns, two bombs)		3 million
Hiroshima nuclear bomb	20 kiloton	49 million
Short-range ballistic missile type Lance	0.05 kiloton per war head	60 million
Short-range ballistic missile type Lance	1 kiloton per war head	170 million
Howitzer Caliber 155 m, type M 109	0.1 kiloton per war head	680 million
Tactical rocket, French type Pluton	20 kiloton	830 million
Phantom-fighter bomber with bomb B-61	350 kiloton	6.2 billion
intermediate-range ballistic missile, French model M-20	1 megaton	18 billion
Intercontinental ballistic missile, Soviet type SS-18	25 megaton	210 billion
USA, 1954, test of nuclear fusion bomb	15 megaton	
USSR, 1960, test of biggest nuclear fusion bomb	60 megaton	

Nuclear bombs are military explosive devices that use nuclear fusion or fission as the explosion energy supply. As already mentioned, the explosive power is usually given in the mass equivalent of TNT. The nuclear fission energy supplied by 1 gram U-235 corresponds with 2×10^4 kg TNT or 8.2×10^{10} Joule. Nuclear fission is usually initiated by the bombardment of nuclides U-235 and Pu-239 with neutrons. Because nuclear fission produces free neutrons after it has started, a chain reaction can be designed. Such a chain reaction depends on the critical mass. A typical critical mass is 50 kg U-235. Because the theoretical critical load can be decreased by some technical measures, smaller bombs with less weight (two orders of magnitude) are possible. The bomb over Hiroshima used only 1 kg U-235, reaching an explosive power of about 2×20^7 kg TNT.

In contrast to the above, nuclear fusion bombs obtain their energy from nuclear fusion. While the size of bombs based on nuclear fission is limited due to the effect of critical masses, the explosive power of nuclear fusion bombs is infinite due to the possible unlimited production and storage of deuterium. In 1962, a H-bomb was ignited 400 km above the Pacific Ocean. Although the explosion of the 1.5×10^9 kg TNT equivalent bomb happened 1,500 km away from Honolulu, it yielded to a collapse in the energy supply there. The bomb with the highest explosive power was launched in 1960 by the USSR with an equivalent mass of 6×10^{10} kg TNT (Rennert et al. 1988).

Explosive power of the magnitude of 10^{10} kg TNT can destroy structures completely within a distance of 8 km from the explosion centre. Within a distance of 15 km, major structural damages can be observed. An explosive power of 10^7 kg TNT yields to major retina burns caused by the high glare effect (100 times higher than the sun). The inflame point of a 2×10^8 kg TNT breaches a distance of 4.5 km. Table 2-88 shows the transformation of the explosion energy into other energy forms (Rennert et al. 1988).

Table 2-88. Transformation of Energy forms (Rennert et al. 1988)

Energy	Percentage
Shock wave	50
Thermal radiation	35
Immediate ionising radiation	5
Late ionising radiation	10

The explosion of a nuclear bomb with a mass equivalent of 10^6 TNT would probably completely eliminate a city of the size New York. The bomb would be ignited about 2 km above the city (Tables 2-89 and 2-90). The thermal radiation of the explosion would yield to heavy fires in the city. The heavy radiation would be followed by a shock wave, which might partially extinguish some fires and cause new fires by transporting burning material. The fire would grow into a firestorm. The heat of the fire produces a wind with a speed of up to 160 km per hour. Such a wind would blast huge amounts of ash into the atmosphere. If one considers that perhaps hundreds of cities would experience such a tragedy, during a nuclear war it becomes clear that inconceivable amounts of dust and ash would be transported into the air. Such huge masses would heavily influence the shielding of the earth's surface from sunlight and would yield to a dramatic temperature drop. This is called nuclear winter. Table 2-91 shows different scenarios for nuclear world wars and the amount of ash and dust that would be released. Based on the assumed worldwide effects of such a nuclear world war, it becomes clear that no winner will exist, even if no direct hits occur (Turco et al. 1985).

Table 2-89. Fatality radius of a nuclear bomb explosion based on the shock wave (Rennert et al. 1988)

Explosive power in TNT equivalent	Fatality radius (km)
10^6	0.2
10^7	0.5
10^8	1.5
10^9	3.0
10^{10}	6.6

Table 2-90. Effects of radioactive radiation caused by a nuclear bomb explosion with the mass equivalent of 10^6 kg TNT (Rennert et al. 1988)

Distance from explosion (m)	Dose (Gy)	Time to death (days)
400	180	1–2
500	80	1–2 (5 min capable of acting)
640	30	5–7
760	6.5	5–7 (2 h capable of acting)

Table 2-91. Investigated scenarios of nuclear wars (Turco et al. 1985)

Scenario	Overall explosive power in mega tonnes	Inhabited or industrialised targets in percent explosive power	Explosive power of single warheads in mega tonnes	Overall number of explosions	Ash particles < 1 μm in million tonnes	Dust particle < 1 μm in million tonnes
Weak over ground explosions	5,000	33	0.1–1	22,500	300	15
Full exchange of nuclear strokes	10,000	15	0.1–10	16,160	300	130
Mean exchange of nuclear strokes	3,000	25	0.3–5	5,433	175	40
Limited exchange of nuclear strokes	1,000	25	0.2–1	2,250	50	10
Strike against military targets	3,000	0	1–10	2,150	0	55
Strike against hard targets (bunkers)	5,000	0	5–10	700	0	650
Strike against urban centers	100	100	0.1	1,000	150	0
World War	25,000	10	0.1–10	28,300	400	325

The current development of warfare suggests rather smaller, locally concentrated types of wars. Such wars give the public the impression that the risk for the population and the soldiers is dramatically lower as compared to historical wars. Nevertheless, the current wars in Afghanistan or Iraq does not show such expected low mortality rates. However, there might be some progress in a lower risk for American soldiers, but the mortality rate still remains high as compared to a peaceful life in developed countries. About 100,000 Iraqi fatalities were estimated for the second Iraq war in 2003 and 2004 (Roberts et al. 2004). With a population of 22 million, this gives a mortality rate of about $4–5 \times 10^{-3}$ per year.

In 2003, about 300 US soldiers died in battles. Here, the number of exposed soldiers was estimated to be 100,000. This corresponds to a soldier mortality rate of 3.0×10^{-3}. For comparison reasons, the yearly mortality rate of a German soldier in World War II was 0.1 per year. It is interesting to note that the mortality rate of US soldiers in the second Iraq war actually did not decrease after the war itself. The mortality rate remains high in the order of 3.0×10^{-3} per year. The overall number of losses for us reached nearly 4,000 by the end of 2007 with about 30,000 injured.

In general, the number of military actions worldwide has risen as shown in Table 2-92 (Ipsen 2005). Also, the length of wars has risen as some examples in Table 2-93 show. Both effects are in contrast to the observed developments in the 1990s due to a drop in military expenditures. After the collapse of communism, world military spending dropped significantly (Table 2-94). The latest developments, however, show rather alarming signs.

Table 2-92. Number of wars (Ipsen 2005)

Period	Number of armed conflicts	Increase in armed conflicts
1945–1969	5–15	10
1969–1975	15–30	15
1975–1992	5–52	17
Since 1992	≈30	

Table 2-93. Length of some wars during the 20th century (Ipsen 2005)

Country	Beginning of war	Duration until today in years
Burma	1949	56
Columbia	1965	40
Israel/Palestine	1967	38
Northern Ireland	1969	36
Philippines	1970	35
Cambodia	1975	30
Iraq/Kurdistan	1976	29

In 2005, world military expenditures reached about 1,100 billion US dollars. This represents 2.5% of the world gross domestic product. Since the middle of the 1990s, military expenditure has increased by about 30% (Table 2-94). Currently, the US is responsible for a major portion of military expenditure worldwide, mainly due to the ongoing military actions (Table 2-95). In other countries, military expenditure is increasing as well, for example Russia, China and India. This increase, however, mainly corresponds to the economic growth (SIPRI 2006). The highest military expenditure per capita is found in Israel (Table 2-96).

The high American military expenditure is strongly connected to the military actions in Afghanistan and Iraq. These military actions can be seen as a direct or indirect consequence of the terrorist attack against the World Trade Center on 11 September 2001 in New York. In addition, this shows a continuous trend from wars being fought between states to wars between a state and a non-state organisation. This development is strongly connected to the definition of terrorism.

Table 2-94. World military spending in 1986 and 1994 in billion US dollars (Conetta & Knight 1997, US ACDA 1996, IISS 1996)

	1986	1994	% Change
World	1297.0	840.3	−35.2
OECD	622.6	540.9	−13.1
Non-OECD World	674.4	299.4	−55.6
NATO	562.6	469.3	−16.6
Non-NATO World	734.4	371.0	−49.5
Non-NATO OECD	60.0	71.6	+19.3
US	365.3	288.1	−21.0
Non-US World	931.7	552.2	−40.7
Non-US OECD	257.3	252.8	−1.7
Non-US NATO	197.3	181.2	−8.2

However sometimes the so-called state terrorism occurs in wars. This has been observed by the installation of concentration camps in Germany during the Nazi times. The first concentration camps were introduced by the English in the Boer war between 1899 and 1902. The mortality rate in the camps reached up to 12% per year. That was a comparable value to some German concentration camps during the World War II. However, there were concentration camps in World War II, where the mortality rates reached more than 60% (Schenck 1965). Numbers of the systematical killing of Jews during the holocaust is, for example, given in Hewitt (1997).

Table 2-95. Current military expenditures in billion US dollars (CIA 2005)

Country	Current military expenditures
US	276.7
China	55.9
France	46.5
Japan	39.5
Germany	38.8
UK	31.7
Italy	20.2
Saudi Arabia	18.3
Brasil	13.4
Korea, South	13.1
India	11.5
Australia	11.4
Iran	9.7
Israel	8.9

Table 2-95. (Continued)

Country	Current military expenditures
Spain	8.6
Turkey	8.1
Canada	7.9
Taiwan	7.6
Netherlands	6.5
Greece	6.1
Korea, North	5.2
Singapore	4.5
Sweden	4.4
Argentina	4.3
Egypt	4.0

Table 2-96. Current military expenditures in US dollars per capita (CIA 2005)

Country	Amount
Worldwide average	311
Israel	1,487
Singapore	1,003
US	986
Brunei	977
Kuwait	931
New Caledonia	925
Qatar	911
Oman	893
Bahrain	801
Saudi Arabia	778
France	778
Norway	687
United Arab Emirates	654
Greece	574
UK	530
Sweden	495
Cyprus	482
Australia	475
Germany	466
Denmark	460
Netherlands	404
Taiwan	356
Italy	349
Switzerland	348
Finland	347

2.6.4 Terrorism

Terrorism can be seen as a specific type of violence and crime. Since 1983 in the USA, the most common definition of terrorism is: "The term terrorism means premeditated, politically motivated violence perpetrated against noncombatant targets by subnational groups or clandestine agents, usually intended to influence an audience." (DoS 2004)

Only within a few years after the World Trade Center terrorist attacks in 2001, the war on terror was launched. Terrorism, however, has a long history. Data of terror victims of the last few decades have been statistically investigated by Bogen & Jones (2006). Whether this war is a sufficient safety measure to mitigate terrorism will not be discussed here; in general, the number of worldwide terrorist attacks remains high. Of course, this depends very strongly on the definition used. For example, if all attacks in Iraq are considered terrorist attacks, this remains a centre of terrorism, but the incidents there can also be considered military actions.

Obviously a peak in fatalities as a result of terrorist attacks was reached in 2001 with the attacks on the World Trade Center in New York, on the Pentagon in Washington and the airplane hijacking in Pennsylvania. While in 2001, more than 3,000 lives were claimed, in other years the number was usually less than 1,000 and only a few hundreds worldwide. For example in 2002, about 700 terrorist victims were claimed. The number of terrorist attacks worldwide is between 200 and 400 per year (DoS 2004).

The terrorist attacks show clusters at different locations and at different times. Such location clusters are clearly related to some political troubles of mainly unresolved problems. For example, such a location cluster can be found in Israel, where intensive investigations into terrorism and fighting terrorism have been carried out. For a long time, Northern Ireland and Great Britain have formed a terror location cluster. The consequences, however, of the terrorist attack at the World Trade Center in 2001 reached a new level. These consequences are not only in terms of direct victims of the incident but also the indirect consequences affecting the lives of probably a billion people in terms of political changes and level of freedom.

Further major terrorist attacks can be found in Europe. For example, the attack on a train in March 2004 in Spain claimed about 200 lives and injured more than 1,000 people. The blasting of a Boeing 747 in 1988 above Lockerbie in Scotland claimed 280 victims. During the world soccer championship in Germany, the probability of a bomb attack was estimated as 0.38% (Woo 2006).

Examples of major terrorist attacks worldwide over the last 30 years are either based on the insured value of loss or on the number of victims as can be seen in Tables 2-97 and 2-98.

Table 2-97. Terror attacks with highest insured damages (Swiss Re)

Insured damage [1]	Fatalities	Date	Incident	Country
19,000	3,000	11.09.2001	Attack on the World Trade Center, Pentagon and other buildings	USA
907	1	24.04.1993	Bomb blasting in London close to the NatWestTower	U.K.
744	–	15.06.1996	Bomb blasting in Manchester	U.K.
725	6	26.02.1993	Bomb explosion at the underground parking lot of the World Trade Center in New York	USA
671	3	10.04.1992	Bomb explosion in the London financial district	U.K.
398	20	24.07.2001	Suicide bombing at Colombo International Airport	Sri Lanka
259	2	09.02.1996	Bomb attacks against London South Key Docklands	U.K.
145	166	19.04.1995	Bomb attack against a government building in Oklahoma City	USA
138	270	21.12.1988	Bomb explosion on PanAm Boeing 747 above Lockerbie	U.K.
127	0	17.09.1970	Blasting of three hijacked passenger airplanes in Zerqa	Jordan

[1] Million US-dollars 2001.

Table 2-98. Terror attacks with maximum victims (Swiss Re)

Insured damage [1]	Fatalities	Date	Incident	Country
19,000	3,000	11.09.2001	Attack on World Trade Center, Pentagon and other buildings	USA
–	300	23.10.1983	Bomb attack against US and French forces camp	Lebanon
6	300	12.03.1993	Series of 13 bomb attacks in Bombay	India
138	270	21.12.1988	Bomb explosion on PanAm Boeing 747 above Lockerbie	U.K.
–	253	07.08.1998	Two bomb attacks a against US embassy in Nairobi	Kenya
145	166	19.04.1995	Bomb attack against a government building in Oklahoma City	USA
45	127	23.11.1996	Hijacked Boeing 767-260 of Ethiopian Airlines crashed into the Indian Ocean	
–	118	13.09.1999	Bomb attack destroyed residential building in Moscow	Russia
–	100	04.06.1991	Arson in an arm cache in Addis Abeba	Ethiopia
6	100	31.01.1999	Bomb attack in Ceylinco House in Colombo	Sri Lanka

[1] Million US-dollars 2001.

As already mentioned, terrorist attacks quite often occur in local and temporal clusters. Example of this, are the bomb warnings and attacks in the UK during the 1980s and 1990s. Many readers will probably remember the warnings during Christmas time at shopping centres or subways in London. Here, the terror attacks were strongly related to the independence endeavours in Northern Ireland; such examples can also be found in other European countries, for example Spain. The author himself has heard bomb explosions in London and has experienced the checking of railway tunnels against bombs in Spain.

Even in other countries like Germany, terror clusters occurred during the 1970s and 1980s due to the Red Army. In Japan in 1995, an attack with poisonous gas was carried out in the subway of Tokyo, killing 5 and injuring about 5,000 people. In Russia, India and Indonesia, terrorist attacks were conducted.

The magnitude of terrorist attacks is not only defined by the number of victims but also by the impact on the social systems. Such impacts can be preliminarily defined by the requirement for help (medical staff, fire brigades, police and army). A catastrophic terrorist event can cause damages that yield to the rupture of normal life in a city, region or country. This was very visible after the World Trade Center attacks, when air traffic from and to the eastern part of the US was completely closed down. Such secondary effects very often cause a much higher damage when compared to the direct initial damage not only in terms of financial values but also in terms of trust in the society and changes in the system. For example, in 2005 when an attack using liquid bombs at London Heathrow was planned, the boarding regulations for all airports inside the EU region was changed, which affected millions of passengers. Therefore, some countries, for example the Netherlands, have introduced a policy to restore normal life, since complete safety measures against intelligent terrorist attacks are impossible under normal living conditions. For example, after a terrorist bomb attack, security checks could be strengthened not only at airports but also at railway stations, buses, schools and universities and even supermarkets. This would completely alter the standard living conditions.

The classification of disastrous and non-disastrous terrorist attacks is mainly separated by the number of casualties. Usually 300–500 casualties are considered and treated by regular emergency help, whereas incidents surpassing this number require additional emergency help from neighbouring areas. For example, incidents with 1,000 casualties require additional help from a distance of up to 200 km.

One has to be aware of the fact that during such events the medical attention is not comparable to normal treatment conditions. Some recommendations for an emergency state give about 2 min observation time per patient (Adams et al. 2004).

In general, in many countries, there are organisational structures to deal with terrorist acts of different kinds, since such acts might not only use bombs but also nuclear, biological and chemical weapons (Davis et al. 2003). Such weapons would heavily influence the social system. Therefore, special preparations for such cases are required by organisations such as the police, the secret service and also the medical systems. An example of special preparation is the storage of iodine tablets in case nuclear weapons are used either in a war or terrorist situations. Such tablets would prevent or at least lower the storage of radioactive material in the human body.

2.6.5 Crimes

Terrorist attacks are a special type of criminal acts. These might include murder and homicide; however, murder and homicide are not only committed as terrorist attacks. In general, murder is defined in laws as criminal offence against life, which is characterised by cattiness and base motives. The killing of people that does not fulfill these characteristics is called homicide.

In the US, about 22,000 people are killed per year by murder, homicide and police actions to prevent criminal acts (Parfit 1998). For comparison, Table 2-99 lists murder and homicide rates for different times and different cities (Remde 1995, Walker 2000 and BKA 2004).

Table 2-99. Murder and homicide rates for different countries or cities

Year	Country or city	Number of murders and homicides per 100,000 inhabitants per year
1981	USA	10.2
1989	USA	9.8
1998	USA	6.3
1993	Johannesburg, South Africa	155.0
1993	London, U.K.	2.5
1999	Dresden, Germany	0.8
1996	Dresden, Germany	1.9

Figure 2-81 shows the development of the rate of murder and homicide in the USA for nearly a century. There seems to be a relation between times of economical wellbeing and the number of murders. It also shows that the executions may have only a limited effect on the murder rate.

For Germany, the development is shown in Fig. 2-82. The figure includes not only committed but also attempted murders. Additionally, one has to consider that the data until 1993 only represent West Germany, whereas later data for unified Germany are given (Rückert 2004). Although the interpretation of the data is difficult due to the inhomogeneity, the data in general show a decrease in the murder rate since 1993.

Often such average numbers do not consider that the frequency of criminal acts is high in some specific areas of a city or a region. Police use the possibility to draw geographic maps including data about the frequency of criminal acts. Figure 2-83 shows a part of the city map from Philadelphia indicating regions of high crime density in the year 1998 (Fragola & Bedford 2005).

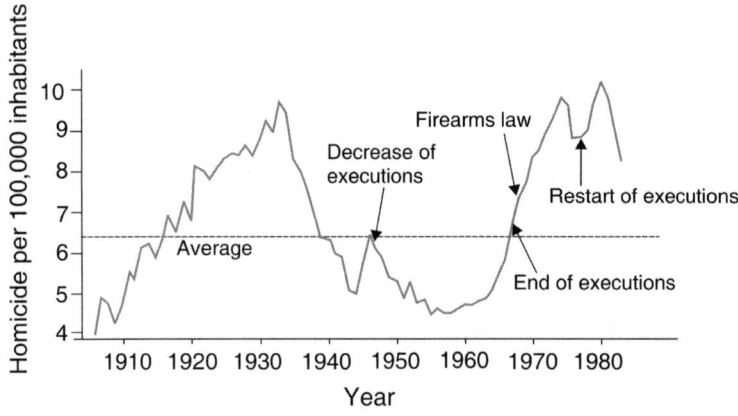

Fig. 2-81. Development of homicide rate in the US from 1900 to 1980

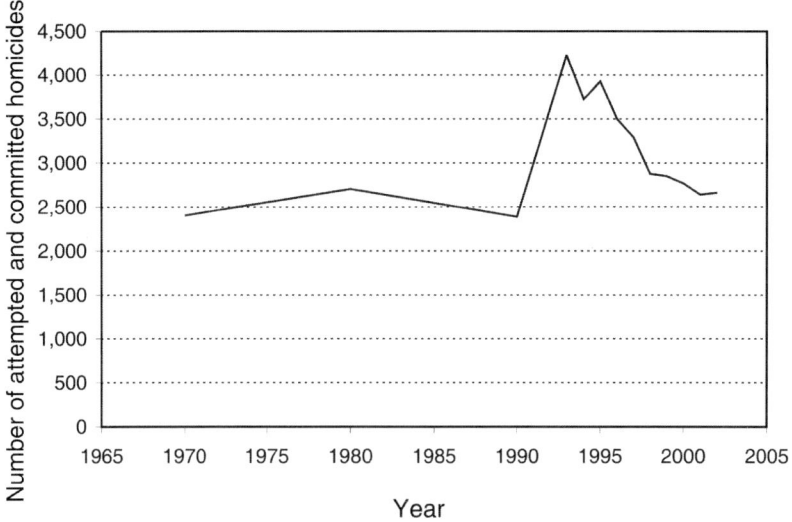

Fig. 2-82. Number of attempted and committed homicides in Germany (Rückert 2004)

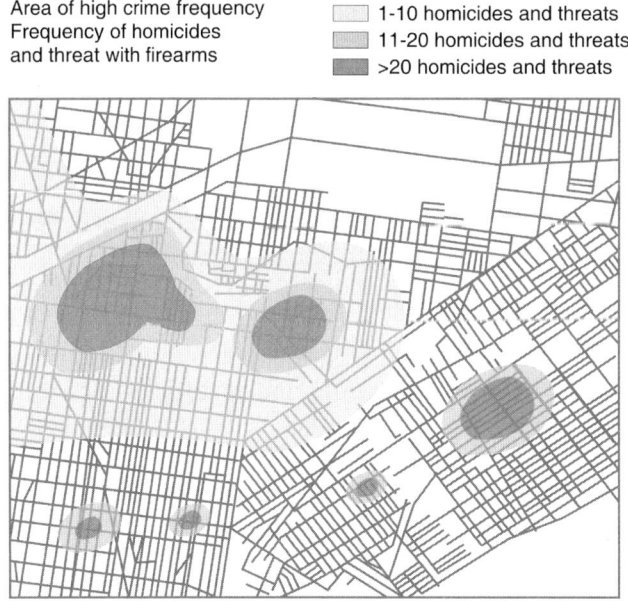

Fig. 2-83. Example of a GIS (Geographical Information System) for identification of crime hot spots (Fragola & Bedford 2005)

The crimes mentioned here are strongly related to violence against humans. The worldwide problem of violence has been intensively studied by the WHO (Krug et al. 2002).

In the last years, violence from teenagers in developed countries has grasped major media attention, for example massacres in schools. It is well known that morbid social conditions might yield to depressive behaviour. There exists a relationship between depressive behaviour and aggressive behaviour of teenagers. Many young people cloister themselves after serious abuse, which is often caused by disproportional development in young age. This might yield to extensive TV viewing, computer consumption and finally to extreme violent behaviours (Te Wildt & Emrich 2007).

2.6.6 Drug Abuse and Alcoholism

Drug-related crimes and the trade of illegal drugs are not only a criminal hazard but also a hazard to the customer. Drug addiction is one of the biggest killers in both developing and developed countries. In Germany, about 100,000 tobacco-related deaths per year are recorded. Drugs are rarely identified as a direct cause of death; however, they are quite regularly identified as an initial cause of illnesses such as cancer (43,000 deaths per year in Germany), cardiovascular diseases (37,000 deaths per year in Germany) and respiratory diseases (20,000 deaths per year in Germany). The probability of falling ill with lung cancer and laryngeal carcinoma is twenty times and ten times higher, respectively, for a smoker. Fortunately, over the last few years, the percentage of smokers has fallen from about 60% (1980) to about 46% (1999) for males and from 54%–34% for females. Still, it has been estimated that tobacco will claim about 8.3 million lives per year in 2030.

Next to smoking, the second biggest killer is alcohol. The identification of alcohol as the cause of death still remains difficult. The estimated number of victims in Germany per year lies between 40,000 and 50,000. Data for Switzerland is given in Table 2-100. The WHO has defined a critical alcohol consumption of 20 g per day for females and 60 g per day for males. In Germany, about 10–15% of all men reach that value and about 3–5% of all women. In total, about 170,000 people are addicted to alcohol, of these about 25,000 men and 6,000 women are treated in hospitals and a further 88,000 men and 10,000 women are treated ambulant. Indirectly alcohol causes about 33,000 car accidents in Germany with about 1,500 fatalities. That is about 20% of all motor vehicle fatalities (BMG 2000).

Alcoholism can be classified according to Table 2-101. Newer research projects distinguish only two groups of alcohol addicts. Type 1 begins the alcohol addiction after the age of 25. The addiction is characterised by rather low social complications in comparison to a type II addict. Type II is characterised by an early start of alcohol addiction in association with heavy social complications and the early observation of alcohol and drug misuse in the kinship at a young age.

Table 2-100. Death numbers, lost life-years and chronic diseases caused by alcoholism in Switzerland for people of age between 15 and 74 (Gutjahr & Gmel 2001)

	Male						Female					
	Chronic consequences		Acute consequences		Portion of alcohol-related death from all deaths		Chronic consequences		Acute consequences		Portion of alcohol-related death from all deaths	
Age	Number of fatalities	Lost life-years	Number of fatalities	Lost life-years	Mortality (%)	Lost life-years (%)	Number of fatalities	Lost life-years	Number of fatalities	Lost life-years	Mortality (%)	Lost life-years (%)
15–24	1	35	51	2,728	12.1	12.0	1	118	9	504	6.7	7.7
25–34	12	483	62	2,823	9.4	9.6	6	388	13	552	5.6	6.3
35–44	61	2,162	49	1,741	10.9	11.3	34	1,308	11	397	8.7	9.7
45–54	168	4,273	51	1,306	11.5	12.2	66	1,796	12	292	7.4	8.2
55–64	220	3,232	38	587	7.4	7.8	81	1,343	10	148	5.0	5.8
65–74	236	1,193	36	176	4.0	4.9	121	581	15	64	3.2	3.9
>74	235	–	68	–	1.8	–	412	–	70	–	2.0	–
Total	933	11,378	355	9,361	4.2	9.7	721	5,534	140	1,957	2.7	6.9

Table 2-101. Classification of alcoholism (Möller et al. 2001)

Type of alcoholism	Type	Addiction	Signs of addiction	Frequency (%)
Alpha	Conflict drinker	Only psychological	No loss of control, capability of abstinence	5
Beta	Occasional drinker	None	No loss of control, capability of abstinence	5
Gamma	Addictive drinker	First only psychological, later physical	Loss of control, but occasional capability of abstinence	65

(Continued)

Table 2-101. (Continued)

Type of alcoholism	Type	Addiction	Signs of addiction	Frequency (%)
Delta	Habitual drinker	Physical	Inability for abstinence, continuous low level of drunkenness	20
Epsilon	Episodical drinker	Psychological	Loss of control for several days during excessive drinking	5

The health consequences of alcohol addiction can be diverse. For example, a chronic consumption of about 60 g of pure alcohol by males and about 20 g of pure alcohol by females can cause a cirrhosis of the liver as well as psychological successor diseases. Such diseases are acute delirium, alcoholic hallucination, alcoholic jealousy mania, personality changes, dementia, Korsakowsyndrome or Wernicke cerebropathy (Möller et al. 2001).

In addition to legal drugs such as alcohol, there are illegal drugs that are misused. In Germany, probably cannabis is the most frequently used illegal drug. It has been estimated that about 2 million people in Germany consume cannabis at least once per year. That represents 2.5% of the entire population of 80 million. Ecstasy is used in Germany by at least 500,000 people once per year (6%) (BMG 2000).

In Germany, approximately 8,000 heroin consumption beginners, about 7,000 Amphetamine consumption beginners, 5,000 cocaine consumption beginners and 4,000 ecstasy consumption beginners are registered per year. Consumption beginners are people who have been identified in connection with the misuse of drugs. They include taster and first consumers and not necessarily addicted persons (BMG 2000).

The majority of drug related deaths are either due to multiple misuse of different drugs or to the long-term misuse of drugs over several years. In 1999 in Germany, about 1,800 drug-related deaths were registered (Fig. 2-84). A further topic is the drug misuse by children and teenagers (Stolle et al. 2007). Furthermore, in addition to illegal drugs, legal drugs are also misused; however, this misuse is difficult to record. Such drugs include painkillers, sleeping pills, tranquilisers, stimulants and purge. The misuse here is mainly assumed based on general consumption data of such drugs. On average, 20% of the female and 10% of the male population have taken such pharmaceutics in the last four weeks. Considering some age-dependant cohorts, higher values can be found. For example, in the group with age greater than 50 years, up to 30% of the females used the mentioned drugs in the last four weeks (BMG 2000).

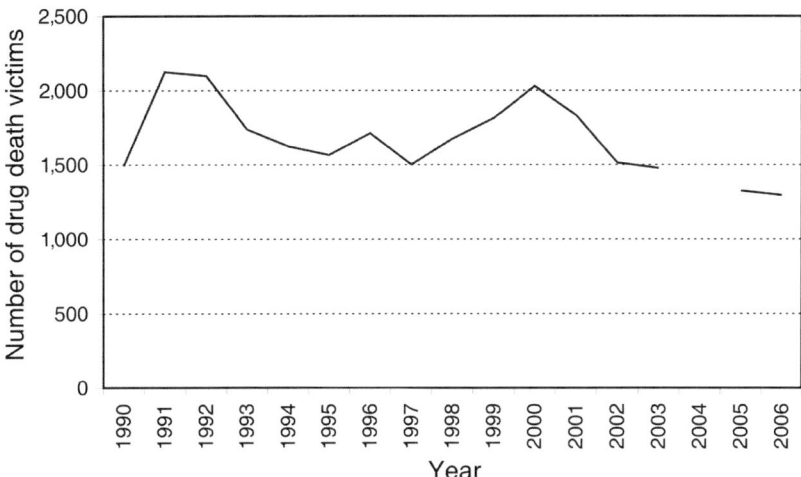

Fig. 2-84. Number of drug victims in Germany (DNN 2004)

2.6.7 Mountaineering

Obviously humans do not seek absolute safety, but they choose to make a trade-off between risks and advantages. A good proof of this statement is mountaineering. Here, humans accept a higher risk by attempting to achieve something favourable for them. Some examples to visualise such higher risks in mountain regions are given here.

At the beginning of the 1990s, about 40 people were killed during the ascent to the Pike Lenin in the Pamir Mountains (Häußler 2004). From 1975 to 2002, about 1,200 people had climbed Mount Everest. From this population, about 175 people died during attempts. In 1996 alone, 15 climbers were killed. This means that for every seven successful climbs, one climber gets killed. However, in the last years, the ratio has been improved, as shown in Fig. 2-85 (Klesius 2003). An overview of mortalities at different mountain peaks is shown in Table 2-102.

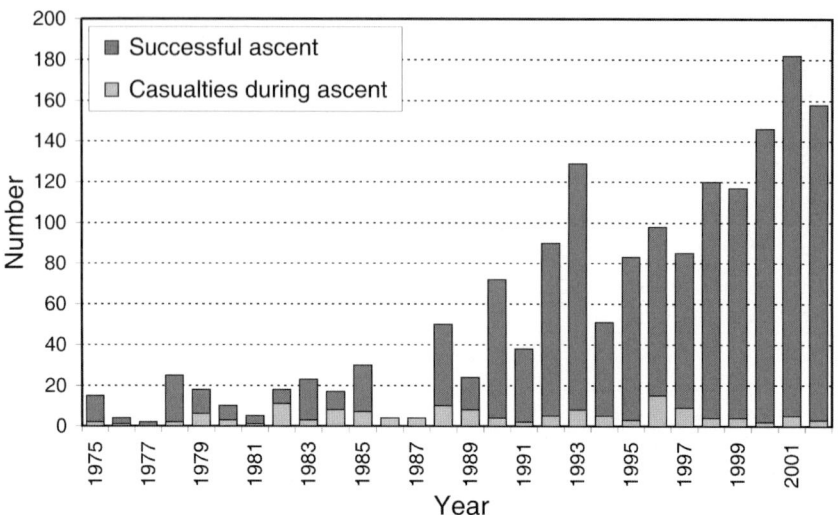

Fig. 2-85. Number of successful climbs of Mount Everest for different years including the number of fatalities (Klesius 2003)

2.6.8 Sport

It seems to be strange to consider sport as a risk. However, as shown in the section on adverse effects, pharmaceutics too can be considered a risk. The original way of living of humans did not require such organised and controlled physical activities like sport. Although such activities are required, they might impose some risks of injuries to humans. Table 2-103 shows different kinds of sporting activities and related sport accidents based on 100 hospitals in the US. However, independent from the number of accidents, such accidents are in most cases not life-threatening.

Table 2-102. Fatality rates for people climbing different mountain peaks (Höbenreich 2002)

	Height (m)	Number of people successfully reaching the summit	Percent of repetition	Numbers on summit	Overall number of fatalities	Number of fatalities during ascent	Ratio of fatalities to reached peak	Ratio of fatalities during descent to reached peak
Everest	8,850	1,173	299	874	165	40	1:7.1	1:29.3
K2	8,611	164	1	163	49	22	1:3.4	1:7.5
Kantschendzönga	8,586	153	7	146	38	7	1:4	1:21.9
Lhotse	8,516	129	1	128	8	2	1:16.1	1:64.5
Makalu	8,463	156	0	156	19	8	1:8.2	1:19.5
Cho Oyu	8,201	1,090	92	998	23	5	1:47.4	1:218
Dhaulagiri	8,167	298	8	290	53	5	1:5.6	1:59.6
Manaslu	8,163	190	1	189	51	3	1:3.7	1:63.3
Nanga Parbat	8,125	186	2	184	61	3	1:3.1	1:62
Annapurna	8,091	109	3	106	55	8	1:2	1:13.6
Gasherbrum I	8,068	164	3	161	17	3	1:9.7	1:54.7
Broad Peak	8,047	217	5	212	(+107)	18	(1.2)%	22.2%
Gasherbrum II	8,035	468	12	456	15	3	1:31.2	1:156
Shisha Pangma	8,027	167	2	165	(+434)	19	(0.3)%	10.5%
Sum	4,664	436	4,228	4,769	591	115		
Average	8,282	333.1	31.1	—	42.2	8.2	1:7.9	8.8

Table 2-103. Estimations of sporting accidents in the US for the year 1998 (NIIC 1999)

Sport	Number of accidents	Percent of accidents in age classes		
		0–4	5–14	15–24
Basketball	631,186	0.6	31.5	46.4
Bicycling	577,621	7.1	55.0	15.2
American Football	355,247	0.3	45.0	43.1
Baseball	180.,582	4.5	50.4	23.3
Soccer	169,734	0.5	45.7	37.6
Softball	132,625	0.3	19.2	30.1
Fitness equipment	123,177	0.4	13.9	26.3
In-line skating	110,783	0.7	61.1	18.7
Trampoline jumping	95,239	9.6	69.6	14.0
Skiing	81,787	0.5	14.2	15.9

2.6.9 Panic

Mass panic is a special human behaviour yielding to crowd stampede, which can result in fatalities. It has been observed at sport events. Some examples of panics are given.

During a memorial ceremony, about 14 children were killed in stampede in June 2000 in Addis Ababa. During the Troitsa Festival in Minsk, Belarus, on 30 May 1999, about 53 people were killed at the entrance of a subway station. A fire disaster occurred in West Warwick at Rhode Island on 20 February 2003. A fire in the club "The station" caused a panic, in that 99 people were trampled to death and 190 severely injured. On 6th December 1976, a panic occurred in Port-au-Prince, Haiti, at a World Cup soccer match between Haiti and Cuba. After a goal, a fan lighted a fire cracker in the stands. Other fans believing it was gun fire panicked and knocked down a soldier. The soldier's gun then went off and killed two children. The panic continued, and further people were trampled to death and the soldier committed suicide (CDL 2007).

The prediction of human behaviour in such situations is rather complex. There are many causes for such behaviour–like bad weather situation at the end of major open-air events, fire and smoke in closed rooms, complex building structures, rain, hail, storm, ice, time pressure and fear or hysteria –and are all contagious phenomena and possible causes. Furthermore, during such events, the individual freedom of action is limited and a new, collective behaviour emerges. This yields actually to some juristic problems, which are mentioned in Kretz (2005).

A simple recommendation such as providing more emergency exits might not help in many cases. Emergency exits are often not useful because most people try to leave the building the way they entered (Kretz 2005). Recently, some numerical models have been developed to predict crowd stampede caused by panic (Helbing et al 2000).

References

Abel M (1974) Massenarmut und Hungerkrisen im vorindustriellen Europa

Abele G (1974) Bergstürze in den Alpen. Wissenschaftlichen Alpenvereinshefte 25, München

Abramovitz J (2001) Unnatural Disasters. Worldwatch Paper 158, October 2001, Worldwatch Institute

Adams H-A, Vogt M & Desel H (2004) Terrorismus und Medizin – Versorgung nach Einsatz von ABC-Kampfmitteln. Deutsches Ärzteblatt. Jahrgang 101, Heft 13, 26. März 2004, pp. B 703–B 706

Ahrens R (2001) Wie viel Nutzen für wie viel Risiko. Dokumentation des Workshops. Frankfurt am Main, Februar 2001

Albrecht U (1985) Einführung. Rüstung und Sicherheit. Spektrum der Wissenschaft: Verständliche Forschung. Verlagsgesellschaft, Heidelberg 1985, pp. 7–16

Ale BJM (2003) Keynote lecture: Living with risk: a management question. In: Bedford T &, van Gelder PHAJM (Eds): Safety & Reliablity – (ESREL) European Safety and Reliability Conference 2003, Maastricht, Netherlande, Balkema Publishers, Lisse 2003, Vol. 1, pp. 1–10

Alén C, Johansson A, Bengtsson P-E, Johansson L, Sällfors G & Berggren B (1999) Landslide risk analysis in infrastructure planning. Application of Statistics and Probability (ICASP 8), (Hrsg) RE Melchers & MG Stewart, Sydney, Vol. 1, pp. 227–234

Allen PA (1997) Earth Surface Processes. Blackwell Science, Oxford

Altavilla A, Garbellini L & Spazio A (2000) Risk assessment in the aerospace industry. Proceedings – Part 2/2 of Promotion of Technical Harmonization on Risk-Based Decision-Making, Workshop, May 2000, Stresa, Italy

Alvarez LW, Alvarez W, Asaro F & Michel HV (1980) Extraterrestrial cause for the Cretaceous–Tertiary extinction. Science 208(4448) pp. 1095–1108

Ammann W, Buser O & Vollenwyder U (1997) Lawinen. Basel, Birkhäuser Verlag

Andrey J & Mills B (2003) Collisions, casualties, and costs: Weathering the elements on canadian roads. ICLR Research, Paper Series – No. 33

Andrey J, Mills B & Vandermolen J (2001) Weather Information and Road Safety, Paper Series – No. 15, August 2001, The Institute for Catastrophic Loss Reduction (ICLR)

Armstrong B & Williams K (1986) The Avalanche book. Fulcrum, Golden

Aßmann U, Demuth B & Hartmann F (2006) Risiken in der Softwareentwicklung. Wissenschaftliche Zeitschrift der Technischen Universität Dresden, 55, Heft 3–4, pp. 105–109

AVP (2005) Arzneimittelkommission der deutschen Ärzteschaft. Arzneiverordnung in der Praxis. Pharmavigilanz. Berlin

Bachmann H (1997) Erdbebensicherung der Bauwerke. In (Ed) Mehlhorn, G.: Der Ingenieurbau: Grundwissen, Teil 8: Tragwerkszuverlässigkeit, Einwirkungen. Verlag Wilhelm Ernst & Sohn, Berlin, 1997

Bachmann A (2003) Ein wirklichkeitsnaher Ansatz der böenerregten Windlastenauf Hochhäuser in Frankfurt/Main. Technischen Universität Darmstadt. Dissertation

Baladin RK (1988) Naturkatastrophen – Der Pulsschlag der Naturgewalten. 2. Auflage, Verlag MIR, Moskau und BSB G.G. Teubner Verlagsgesellschaft Leipzig

BAUA (2008) Federal Institute for Occupational Safety and Health, www.baua.de

Bell R (2007) Lokale und regionale Gefahren- und Risikoanalyse gravitativer Massenbewegungen an der Schwäbischen Alb. Dissertation. Rheinischen Friedrich-Wilhelms-Universität Bonn

Bergmeister K (2006) Beton unter hohen Temperaturen – eine Frage der Tunnelsicherheit, Beton- und Stahlbetonbau, 101(2), pp. 74–80

Bergmeister K (2007) Opening lecture: Innovative technologies to upgrade fire safety of existing tunnels. In: Taerwe L & Proske D (Eds) Proceedings of the 5th International Probabilistic Workshop, Ghent 27–28 November 2007, pp. 1–14

Bergmeister K, Curbach M, Strauss A & Proske D (2005) Sicherheit und Gefährdungspotentiale im Industrie und Gewerbebau. Betonkalender 2006, Teil II, Ernst & Sohn

Bergmeister K, Suda J, Hübl J, & Miklau FR (2007) Schutzbauwerke gegen Wildbachgefahren. Betonkalender 2008, Ernst und Sohn

BfG (1997) – Bundesministerium für Gesundheit: AIDS-Bekämpfung in Deutschland. Bonn 1999, 7. überarbeitete Auflage, Kölnische Verlagsdruckerei

BfS (2007) Federal office for radiation protection. www.bfs.de

Bieseke D (1986) Wenn Adenauer Hunde geschlachtet hätte. – Die Selbstverwirklichung des Hundes durch Beißen. Eine Aufzeichnung des Schreckens. Berlin: Ararat

Bieseke D (1988) Wie vor. Alle Hunde in den Himmel. – Bissiges zu einem Missbrauch. Böblingen: Tykve

Bigl S (2007) Undesirable effects after vaccination in the free state of Saxony. Ärzteblatt Sachsen, 3, 2007, pp. 128 –134 and 5, 2007, pp. 230

BKA (2004) Bundeskriminalamt: Polizeiliche Kriminalstatistik 1999 für die Bundesrepublik Deutschland, http://www.bka.de

Bloom DE, Bloom LR, Steven D & Weston M (2004) Business and HIV/AIDS: Who me? In: Talyor K & DeYoung P (Eds) A global review of the business response to HIV/AIDS. Joint United Nations Program on HIV/AIDS,

BMG (2000) Bundesministerium für Gesundheit. Drogen- und Suchtbericht 1999 der Drogenbeauftragten der Bundesregierung.

BNK (2007) http://www.bnk.de/transfer/euro.htm

Bogen KT & Jones ED (2006) Risks of Mortality and Morbidity fromWorldwide Terrorism: 1968–2004. Risk Analysis, Vol. 26, No. 1, pp. 45–59

Bolt BA, Horn WL, Macdonald GA & Scott RF (1975) Geological Hazards – Eartquakes – Tsunamis – Volcanoes – Avalanches – Landslides – Floods. Springer Verlag, Berlin, Heidelberg, New York

Böse-O'Reilly S, Kammerer S, Mersch-Sundermann V & Wilhelm M (2001) Leitfaden Umweltmedizin. Urbach & Fischer, München – Jena

Brandhofer P & Heitmann K (2007) REACH – Die neue Herausforderung für Ihr Unternehmen. Ecomed Sicherheit, Verlagsgruppe Hüthing Jehle Rehm GmbH, Landsberg

Breitsamer F (1987) Wenn Hunde Menschen töten – eine fachpolizeiliche Untersuchung für die Praxis – Naturbedingtes Fehlverhalten der Tiere oder vorwerfbares Schuldverhalten der Menschen? Die Polizei, 77, 8, pp. 267–271

Bringmann G, Stich A & Holzgrabe U (2005) Infektionserreger bedrohen arme und reiche Länder – Sonderforschungsbereich 630: Erkennung, Gewinnung und funktionale Analyse von Wirkstoffen gegen Infektionskrankheiten", BLICK, Forschungsschwerpunkt, pp. 22–25

Broughton E (2005) The Bhopal disaster and its aftermath: a review. Environmetnal Health. 4: p 6

Brüske-Hohlfeld I, Kreienbrock L & Wichmann H-E (2006) Inhalation natürlicher Strahlung: Lungenkrebs durch Radon. GSF – Forschungszentrum für Umwelt und Gesundheit GmbH in der Helmholtz-Gemeinschaft. Klemm C, Guldner H & Haury H-J (Eds) Strahlung. 18. Ausgabe 2006. Neuherberg. pp. 37–43

BSI (2003) PD 7974–1: 2003: Application of fire engineering principles to the design of buildings; Part 1: Initiation and development of fire within the enclosure of origin

Bub H, Hosser D, Kersen-Bradley M & Schneider U (1983) Eine Auslegungssystematik für den baulichen Brandschutz. Brandschutz im Bauwesen, Heft 4, Berlin, Erich Schmidt Verlag GmbH

Bürger M, Sedlag U & Zieger R (1980) Zooführer. Urania-Verlag, Leipzig

Busse N & Sattar M (2007) Die Renaissance der Kernkraft. Frankfurter Allgemeine Zeitung. 11. Januar 2007, Nr. 9, p 5

c't (2004) Milliarden Schaden durch Liebesbrief: http://www.heise.de/newsticker/meldung/9390

Cancer register (1997) Arbeitsgemeinschaft Bevölkerungsbezogener Krebsregister in Deutschland (Eds) Krebs in Deutschland – Häufigkeiten und Trends. Gesamtprogramm zur Krebsbekämpfung. Saarbrücken, 1997

Carlier P (2004) Nuclear energy and sustainable future. International Journal of Nuclear Power 18(2–3) pp. 39–40

Carrell RW (2004) Prion dormancy and disease. Science 306 (5702), pp. 1692–1693

Cassini P & Pons-Ineris P (2000) Risk assessment for the transport of goods through road tunnels. Proceedings – Part 2/2 of Promotion of Technical Harmonization on Risk-Based Decision-Making, Workshop, May, 2000, Stresa, Italy

CDL (2007) Crowd Dynamics Limited. http://www.crowddynamics.com/

Chapmann CR, Durda DD & Gold RE (2001) The comet/asteroid impact hazard: a system approach. Office of Space Studies, Southwest Research Institute, Boulder CO 80302 und Space Engineering and Technology Branch, Johns Hopkins University Applied Physics Laboratory, Laurel MD 20723, 24 February

Chlond B, Lipps O & Zumkeller D (1998) Das Mobilitätspanel (MOP) – Konzept und Realisierung einer bundesweiten Längsschnittbeobachtung. Universität Karlsruhe – Institut für Verkehrswesen. IfV – Report Nr. 98–2

CIA (2005) World Factbook, 28 July 2005

Coch NK (1995) Geohazards Natural and Human. Prentice Hall, New Jersey

Cohen BL (1991) Catalog of Risks extended and updated. Health Physics, Vol. 61, September 1991, pp. 317–335

Conetta C & Knight C (1997) Post-Cold War US Military Expenditure in the Context of World Spending Trends. Cambridge, MA, Commonwealth Institute, Project on Defense Alternatives Briefing Memo #10, January 1997. http://www.comw.org/pda/bmemo10.htm

Considine M (2000) Quantifying Risks in the Oil and Chemical Industry. Proceedings – Part 1/2 of Promotion of Technical Harmonization on Risk-Based Decision-Making, Workshop, May, 2000, Stresa, Italy

Costa JE (1984) Physical geomorphology of debris flows. In: Costa JE & Fleischer PJ (Eds): Developments and Applications of Geomorphology. Berlin, Springer: pp. 268–317

Coudouel A, Hentschel JS & Wodon QT (2003) Poverty: Measurement and Analysis. Worldbank

Coussot P, Laigle D, Arattano M, Deganutti A & Marchi L (1998) Direct determination of rheological characteristics of debris flow. Journal of Hydraulic Engineering 124(8) pp. 865–868

Crozier MJ (1998) Landslides. In: DE Alexander & RW Fairbridge (Eds.) Encyclopedia of Environmental Science. Kluwer Academic Publishers, Dordrecht, The Netherlands. pp. 371–374

Cruden DM & Varnes DJ (1996) Landslide types and processes, In: Turner AK & Schuster RL (Eds) Landslides: Investigation and Mitigation. TRB Special Report, 247, National Academy Press, Washington, pp. 36–75

Cui P, Chen X, Waqng Y, Hu K & Li Y (2005) Jiangia Ravine debris flows in south-western China. In: Jakob M & Hungr O (Eds) Debris-flow Hazards and Related Phe-nomena. Springer, Berlin, Heidelberg: pp. 565–594

DA (2007) – Deutsches Ärzteblatt. Von vielen Ärzten unterschätzt. Jg. 104, Heft 30, 27th July 2007, p B 1899

Daniels GG (Ed) (1982) Planet earth – vulcanos. Time-Life Books. E.V. Amsterdam

Davis LE, LaTourrette T, Mosher DE, Davis LM & Howell DR (2003) Individual Preparedness and Response to Chemical, Radiological, Nuclear, and Biological Terrorist Attacks. RAND: Public Safety and Justice, Santa Monica

Davis P, Lay-Yee R, Briant R, Schug S, Scott A, Johnson S & Bingley W (2001) Adverse events in new zealand public hospitals. Principal findings from a National Survey, No. 3 Occasional Papers. December 2001: Department of Public Health and General Practice, Christchurch School of Medicine and Health Sciences, University of Otago, Christchurch, New Zealand

Deiters S (2001) Tagish Lake Meteorit – Überbleibsel von der Entstehung des Sonnensystems. 28. August 2001. http://www.astronews.com/news/artikel/2001/08/0108-029.shtml

Deiters S (2002) Meteoriten – Teil der bayerischen Feuerkugel gefunden 31. Juli 2002. http://www.astronews.com/news/artikel/2002/07/0207-024.shtml

Diercke (2002) Weltatlas. Westermann Schulbuchverlag GmbH, Braunschweig 1988, 5. Auflage

Dilley M, Chen RS, Deichmann U, Lerner-Lam AL, Arnold M, Agwe J, Buys P., Kjekstad O, Lyon B & Yetman G (2005) Natural disaster hotspots – A global risk analysis. The World Bank, Hazard Mnagment Unit, Washington

DIRERAF (2007) Development of Public Health Indicators For Reporting Environmental and Occupational Risks Related To Agriculture And Fishery, http://www.direraf.com/

Dittberner K-H (2004) Europas wichtigste Lebensmittel-Skandale, http://earth. prohosting.com/khdit/BSE/Skandale.html#1982_1, http://bse.khd-research.net

DKKV (2002) Deutsches Komitee für Katastrophenvorsorge e.V.: Journalisten-Handbuch zum Katastrophenmanagement 2002. Erläuterungen und Auswahl fachlicher Ansprechpartner zu Ursachen, Vorsorge und Hilfe bei Naturkatastrophen. 7. überarbeitete und ergänzte Auflage, Bonn: 2002

Dlubis-Mertens K (2003) Suizidforen im Internet – Ernstzunehmende Beziehungen. Deutsches Ärzteblatt, Jahrgang 99, Heft 3, März 2003, p 118

DNN (2004) – Dresdner Neueste Nachrichten: Zahl der Drogentoten sind auf Tiefststand. Drogenbeauftragter der Bundesregierung, Freitag, 23th April, p 1

DNN (2005) – Dresdner Neueste Nachrichten: Zugfahrt in den Tod, 4. January 2005, p 3

Doerr A, Guerke G & Ruebarsch D (2002) The german risk-based explosive safety code. 30th DOD Explosive seminar 2002

Doll R & Peto R (1981) The causes of cancer: quantiative estimates of avoidable risks of cancer in the U.S. toady. Journal of the National Cancer Institute, 66, pp. 1191–1308

DoS (2004) US Department of State. Patterns of Global Terrorism 2003

EANM (2007) www.eanm.org

EAS (2008) – European Avalanche Services. www.avalanches.org

Eastlake K (1998) A century of sea disasters. Brown Partworks Limited, German Version, Gondrom Verlag Blindlach

Eckhardt W (1991) War-Related Deaths Since 3000 BC, Bulletin of Peace Proposals, December 1991

Egli T (2005) Objektschutz gegen gravitative Naturgefahren. Vereinigung Kantonaler Feuerversicherungen. Bern

Eichenberg C (2002) Suizidprophylaxe. Deutsches Ärzteblatt, Jahrgang 99, Heft 8, August 2002, p 366

EM-DAT (2004) The OFDA/CRED International Disaster Database, Université catholique de Louvain, Brussels, Belgium. 2004, http://www.cred.be/emdat/profiles/natural/germany.htm

EPA (1991) US Environmental Protection Agency: Environmental Risk: Your Guide to Analysing And Reducing Risk. Publication Number 905/9–91/017, October 1991

Erismann TH & Abele G (2001) Dynamics of Rock Slides and Rock Falls. Springer, Berlin, Heidelberg, New York

European Commission (2002) Vorschlag für eine Richtlinie des Europäischen Parlamentes und des Rates zur Eisenbahnsicherheit in der Gemeinschaft und zur Änderung der Richtlinie 95/18/EG des Rates über die Erteilung von Genehmigungen an Eisenbahnunternehmen und der Richtlinie 2001/14/EG über die Zuweisung von Fahrwegkapazität der Eisenbahn, die Erhebung von Entgelten für die Nutzung von Eisenbahninfrastruktur und die Sicherheitsbescheinigung, Brüssel, 23.1.2002

European Commission (2003) Entwurf einer Richtlinie über die Bewirtschaftung von Abfällen aus der mineralgewinnenden Industrie. Brüssel, 2. Juni 2003, http://wko.at/up/enet/bergbauabfrl.pd

European Commission (2007) ExternE – Externalities of enery. http://www.externe.info/

Evans AW (2004) Railway Risks, Safety values and Safety costs. Imperial College London, Centre for Transport Studies, August 2004

Falk M, Brugger H & Adler-Kastner L (1994) Avalanche survival chances. Nature 368, 21 (03 March 1994), p 482

FAO (2003) Food and Agriculture Organisation (Hrsg.) The State of Food Insecurity in the World 2003. Monitoring Progress Towards the World Food Summit and Millennium Development Goals Year. Rom

Fargeman O (2003) Coronary Artery Disease. Genes, Drugs and the agricultural connection, Elsevier, Amsterdam

Fauler J (2006) Risiko Arzeimittel. Wissenschaftliche Zeitschrift der Technischen Universität Dresden, 55, Heft 3–4, pp. 79–83

Felber W (2004) Suizidraten in den Ländern der ehemaligen Sowjetunion und in Sachsen. Technische Universität Dresden, Klinik und Poliklinik für Psychatrie und Psychotherapie

Fell R. & Hartford D (1997) Landslide Risk Management. In: Cruden D & Fell R (Eds) Landslide Risk Assessment. Proc. Of the IUGS Working Group on Landslides Workshop on Landslide Risk Assessment, Honolulu, Feb. 19–21, A.A. Balkema, Rotterdam

FEMA-NIBS (1999) Federal Emergency Management Agency and National Institute of Buidling Sciences: Earthquake Loss Estimation Methodoly, HAZUS 99, Technical Manual, Washington

Fernández-Steeger TM (2002) Erkennung von Hangrutschungssystemen mit Neuronalen Netzen als Grundlage für Georisikoanalysen. Dissertation, Fakultät Bio- und Geowissenschaften, Universität Karlsruhe

Fileppo E, Marmo L, Debernardi ML, Demetri K & Petusio R (2004) Fire prevention in underground works: software modelling applications. (Hrsg) Spitzer C, Schmocker U & Dang VN: International Conference on Probabilistic Safety Assessment and Management 2004, Berlin, Springer Verlag, London, Vol. 2, pp. 726–731

Fischer D (2003) Jahrhundert-Hochwasser oder drastische Klimaveränderung. In: Fischer D & Frohse J (Eds) Als dem Löwen das Wasser bis zum Rachen stand. Elbhang-Photo-Galerie, Dresden 2003, pp. 71–75

Flemmer S, Willing M & Brehlo A (1999) Brandkatastrophen – Die verheerendsten Brände des 20. Jahrhunderts. Tosa Verlag, Vienna

Foss MM (2003) LNG Safety and security. Center for energy economics. Texas, October 2003, www.utexas.edu/energyecon/lng

Fragola JR & Bedford T (2005) Identifying emerging failure phenomena in complex systems through engineering data mapping. Reliability Engineering and System Safety, 90, pp. 247–260

Fraser C (1966) The avalanche enigma. John Murray Publishers Ltd., London

Fraser C (1978) Avalanches and snow safety. John Murray Publishers Ltd., London

Fréden C (1994) Geology. National Atlas of Sweden. First Edition

Fricke H (2006) Modellierung von Öffentlichen Sicherheitszonen um Verkehrsflughäfen und deren wirtschaftliche Konsequenzen. Wissenschaftliche Zeitschrift der Technischen Universität Dresden, 55, Heft 3–4, pp. 123–130

Friedl JF (1998) Hauptursachen für Brände. Brandschutz, Beilage DBZ, 1

Füller H (1980) Das Bild der modernen Biologie. Urania-Verlag. Leipzig

Garrick BJ (2000) Nonradioactive Waste Disposal. Proceedings – Part 1/2 of Promotion of Technical Harmonization on Risk-Based Decision-Making, Workshop, May, Stresa, Italy

Gauer P, Lied K & Kristensen K (2008) On avalanche measurements at the Norwegian full-scale test-site Ryggfonn. Cold Regions Science and Technology 51, pp. 138–155

Geipel R (2001) Zukünftige Naturkatastrophen in ihrem sozialen Umfeld. Zukünftige Bedrohungen durch (anthropogene) Naturkatastrophen. Hrsg. Volker Linneweber. Deutsches Komitee für Katastrophenvorsorge e.V. (DKKV) pp. 31–41

Gellert W, Gärtner R, Küstner H & Wolf G (Eds) (1983) Kleine Enzyklopädie Natur. VEB Bibliographisches Institut, Leipzig, 21. durchgesehene Auflage

Geophysical Institute (2007) www.gedds.alaska.edu/AuroraForecast/

Gero D (1996) Aviation Disasters. Motorbuch Verlag, Stuttgart (in German)

Gore R (2000) Wrath of the gods. Centuries of upheaval along the Anatolian fault. National Geographic, July 2000, pp. 32–71

Graf H (2001) Klimaänderungen durch Vulkane, Max-Planck-Institut für Meteorologie. Hamburg http://www.mpimet.mpg.de/institut/jahresberichte/jahresbericht-2002.html

Grau G (2004) Der schwarze Tod. Wochenpost Nr. 30/1988, http://home.eplus-online.de/jmct/interess/pest.html

Grimm H-U (2001) Aus Teufels Topf – die neuen Risiken beim Essen. Knaur Taschenbuchverlag: München

GRS (1999) Gesellschaft für Anlagen- und Reaktorsicherheit mbH. Zur Sicherheit des Betriebs der Kernkraftwerke in Deutschland. July 1999: Köln

Grundy SM (1999) Primary prevention of coronary heart disease: integrating risk assessment with intervention. Circulation. 100, pp. 988–998

Grünthal G, Mayer-Rosa D & Lenhardt WA (1998) Abschätzung der Erdbebengefährdung für die D-A-CH-Staaten – Deutschland, Österreich, Schweiz. Bautechnik, 10, pp. 753–767

Gucma L & Przywarty M (2007) Probabilistic method of ships navigational safety assessment on large sea areas with consideration of oil spills possibility, In: Taerwe L & Proske D (Eds), Proceedings of the 5th International Probabilistic Workshop, Ghent 27–28 November 2007

Gucma L (2005) Keynote lecture: Methods of probability assessment of ship accidents on restricted water areas. In: Kolowrocki K (Ed) Advances in Safety and Reliability, Taylor & Francis Group, London, pp. 717–730

Gucma L (2006) Restricted water area optimization with risk consideration, Proceedings of the 4th International Probabilistic Symposium, 12–13 October 2006, Berlin, pp. 289–299

Gucma M (2006) Multi-factor MANOVA method for determination of pilot system interface, Proceedings of the 4th International Probabilistic Symposium, 12–13 October 2006, Berlin, pp. 367–376

Gürke G (2002) Risikoanalyse für die munitionstechnische Sicherheit; Abschlussbericht. Fraunhofer Institut für Kurzzeitdynamik, Ernst-Mach-Institut; E 19/02; Januar 2002

Gutenberg B & Richter CF (1954) Seismicity of the Earth. Princeton University Press, Princeton

Gutjahr E & Gmel G (2001) Die sozialen Kosten des Alkoholkonsums in der Schweiz Epidemiologische Grundlagen 1995-1998. Lausanne: Schweizerische Fachstelle für Alkohol- und andere Drogenprobleme. SFA/ISPA

Hauptmanns U, Herttrich M & Werner W (1991) Technische Risiken. Springer-Verlag, Berlin

Haury H-J (2001) Die Zahl der Todesopfer von Tschernobyl in den deutschen Medien – ein Erklärungsversuch. GSF-Forschungszentrum für Umwelt und Gesundheit. April 2001

Hausladen G, de Saldanha M, Nowak W & Liedl P (2002) Bauklimatik und Energietechnik für hohe Häuser. Betonkalender 2003, Ernst und Sohn, Berlin 2003, pp. 303–364

Häußler O (2004) Am blutigen Berg im Himmelsgebirge. Frankfurter Allgemeine Zeitung, Dienstag 10th August 2004, Nr 184, p. 9

Heidelberg (1999) Haushaltsbefragung zum Verkehrsverhalten in der Region: http://www.heidelberg.de/rathaus/publik/haushaltsbefragung-verkehr.pdf

Helbing D, Farkas I & Vicsek T (2000) Simulating dynamical features of escape panic, Nature 407, pp. 487–490

Helmich P (2004) Selbstmord – Ein Wort, das es nicht geben sollte. Deutsches Ärzteblatt, Jg. 101, Heft 23, 4. Juni 2004, pp. B 1374–B 1375

Heneka P & Ruck B (2004) Development of a strom damage risk map, In: Mahlzahn D & Plapp T (Eds) A review of strom damage functions. Disasters and Society, Logos Verlag, pp. 129–136

Henschler D (1993) Krebsrisiken im Vergleich – Folgerungen für Forschung und politisches Handeln. GSF – Mensch und Umwelt. Ein Magazin des GSF-Forschungszentrum für Umwelt und Gesundheit. 8. Ausgabe, März 1993, pp. 65–73

Hewitt K (1997) Regions of risk: a geographical introduction to disasters. Addison Wesley Longman Limited, London

Hintermeyer H (1998) Schiffskatastrophen – von der Spanischen Armada bis zum Untergang der Pamir. Pietsch Verlag, Stuttgart

HMUE (1992) Hessisches Ministerium für Umwelt, Energie und Bundesangelegenheiten: Hochsicherheitsdeponie-Konzepte. Entwicklung und Planung eines Modellvorhabens für eine Hochsicherheitsdeponie als Sonderabfallager, Ergebnisse einer Studie. Erich Schmidt Verlag, Berlin 1992

Höbenreich C (2002) Todesrisiko Achttausender – Trockene Zahlen, nüchterne Fakten. Berg & Steigen. 1/2002, pp. 29–32

Hoffmann HJ (2000) When Life nearly came to an end. The Permian Extinction. National Geographic. Number 3, September 2000, pp. 100–113

Höffner R, Niemann H-J, Hölscher N & Hubert W (2005) Sturmsicherheit: Den Spielraum immer wieder ausloten. Rubin 2/05, pp. 14–24

Holdgate R (2001) A mandate for sprinklers? Fire prevention and fire engineer Journal. September 2001. Fire Protection Association, London

Holicky M & Markova J (2003) Reliability analysis of impacts due to road vehicles. Applications of Statistics and Probability in Civil Engineering, Der Kiureghian A, Madanat S & Pestana JM (Eds), Millpress, Rotterdam, pp. 1645–1650

Holub M (2007) Studienmaterial Wildbach- und Lawinenverbauung. University of Natural Resources and Applied Life Sciences Vienna

Hosser D., Keintzel E & Schneider G (1991) Seismische Eingangsgrößen für die Berechnung von Bauten in deutschen Erdbebengebieten. Abschlußbericht zum Forschungsvorhaben Harmonisierung europäischer Baubestimmungen. Eurocode 8 – Erdbeben. Universität Karlsruhe

Hubert E, Debray B & Londicke H (2004) Governance of the territory around hazardous industrial plants: decision process and technological risk. (Hrsg) C Spitzer, U Schmocker & VN Dang: International Conference on Probabilistic Safety Assessment and Management 2004, Berlin, Springer Verlag, London, 2004, Vol. 3, pp. 1258–1263

Hübl J (2006) Vorläufige Erkenntnisse aus 1:1 Murenversuchen: Prozessverständnis und Belastungsannahmen; In: FFIG, G Reiser (Hrsg), Geotechnik und Naturgefahren: Balanceakt zwischen Kostendruck und Notwendigkeit, Institut für Geotechnik, BOKU Wien, Geotechnik und Naturgefahren, 19.10.2006, Wien.

Hübl J (2007) Skriptum Wildbach- und Lawinenverbauung, Institut für Alpine Naturgefahren, Universität für Bodenkultur Wien (unveröffentlicht)

Huet P & Baumont G (2004) Lessons learnt from a Mediterranean Flood (Gard, September 2002) (Hrsg) Spitzer C, Schmocker U & Dang VN: International Conference on Probabilistic Safety Assessment and Management 2004, Berlin, Springer Verlag, London, 2004, Vol. 2, pp. 638–643

Hungr O, Evans SG, Bovis MJ & Hutchinson JN (2001) A review of the classification of landslides of the flow type. Environmental & Engineering Geoscience, VII(3), pp. 221–238

IAEA (1991) The International Chernobyl Project: Assessment of Radiological Consequences and Evaluation of Protective Measures. Technical Report by an International Advisory Committee, Vienna

IAEA (2007) International Atomic Energy Association http://www.iaea.org/Publications/Factsheets/English/ines.pdf

ICLR (2008) Institute for Catastrophic Loss Reduction. www.iclr.org

IFRC (2006) International Federation of Red Cross and Red Crescent Societies. World Disasters Report – Focus on neglected crises. Geneva

IISS (1996) International institute for strategic studies. The Military Balance, Oxford University Press, London

Inhaber H (2004) Risk analysis applied to energy systems. Encyclopedia of Energy, Elsevier, Amsterdam

IPCC (2001) Intergovernmental Panel on Climate Change. Climate Change 2001: The Scientific Basis, Cambridge University Press, Cambridge

Ipsen, K. (2005) The eternal war – capitulation the international law for the reality? (in German). In: Rubin 2/2005. Ruhr-University Bochum, pp. 26–31

IPSN (1996) Institut de Protection et de Sûreté Nucléaire (IPSN): Bilanz über die gesundheitlichen Folgen des Reaktorunfalls von Tschernobyl, http://www.grs.de/products/data/3/pe_159_20_1_ipsn_d.pdf

IRR (2008) Institute of Risk Research. http://www.irf.univie.ac.at

Issler D, Lied K, Rammer L, Revol P, Sabot F, Cornet ES, Bellavista GF, Sovilla B (1998) European Avalanche Test Sites - Overview and Analysis in View of Coordinated Experiments. SAME – Avalanche Mapping, Model Validation and Warning Systems. Reports, on CD, Fourth European Framework Programme Environment and Climate

ISL (2007) Institute of Shipping Economics and Logistics. www.isl.org

Iversion MR & Denlinger RP (2001) Flow of variably fluid-ized granular masses across three-dimensional terrain: 1. Coulomb mixture theory. Journal of Geophysical Research 106(B1): pp. 537–552

Iverson RM (1997) The physics of debris flows. Reviews of Geophysics 35(3): pp. 245–296

Jablonski D (2002) Survival without recovery after mass extinctions. Proceedings of the National Academy of Science of the United States of America 99, pp. 8139–8144

Jacob P (2006) 20 Jahre danach: Der Unfall von Tschernobyl. GSF – Forschungszentrum für Umwelt und Gesundheit GmbH in der Helmholtz-Gemeinschaft. C Klemm, H Guldner & H-J Haury (Eds) Strahlung. 18. Ausgabe 2006. Neuherberg. pp. 46–54

JCAHO (2004) Joint Commission on Accreditation of Healthcare Organizations: Patient Safety Standard LD.5.1 and LD.5.2

Jelenik A (2004) Ghana – Mit Kräutern gegen Malaria und Aids. Deutsches Ärzteblatt, Jg. 101, Heft 23, 4. Juni 2004, pp. B1381–B1382

Jensen PH (1994) The Chernobyl accident in 1986 – Causes and Consequences. Lecture at the Institute of Physics and Astronomy, University of Aarhus, 30. November 1994

Johnson AM (1970) Physical Processes in Geology. Freeman and Cooper, San Francisco

Jonkman SN (2007) Loss of Life Estimation in Flood Risk Assessment. Theory and application. Delft cluster

Jorigny M, Diermanse F, Hassan R & van Gelder PHAJM (2002) Correlation analysis of water lelves along dike-ring areas. In: Hassanizadeh SM, Schotting RJ, Gray WG & Pinder GF (Eds), Volume 2, Proceedings of the XIVth International Conference on Computational Methods in Water Resources (CMWR XIV), June 23–28, 2002, Delft, Elsevier Science, Developments in Water Science, 47, pp. 1677–1684

Julien PY & O'Brien JS (1997) Selected Notes on Debris Flow Dynamics. In: Armanini A & Michiue M (Eds) Recent Developments on Debris Flows. Springer, Berlin, pp. 144–162

Junkert A & Dymke N (2004) Strahlung – Strahlenschutz. Eine Information des Bundesamtes für Strahlenschutz. Braunschweig

Jütte R (2006) Geschichte der Medizin – Verzweifelter Kampf gegen die Seuche. Deutsches Ärzteblatt, Jg. 103, Heft 1–2, 9. January 2006, pp. A32–A33

Kääb A, Huggel C & Fischer L (2006) Remote-sensing technologies for monitoring Climate Change impacts on Glacier- and Permafrost-related Hazards. In: Nadim F, Pöttler R, Einstein H, Klapperich H and Kramer S (Eds), Proceedings of the 2006 ECI Conference on Geohazards, Lillehammer, Norway, Bepress, 2007

Kaatsch P, Spix C, Schulze-Rath R, Schmieded S & Blettner M (2008) Leukaemia in young children living in the vicinity of German nuclear power plants. International Journal of Cancer. 1220, pp. 721–726

Kafka P (1999) How safe is safe enough? In: Schuëller GI & Kafka P (Eds) An unresolved issue for all technologies. Safety and Reliability. Balkema, Rotterdam, pp. 385–390

Kantha L (2006) Time to Replace the Saffir-Simpson Hurricane Scale? Eos, 87(1), 3 January 2006 p. 3

Karger C R (1996) Wahrnehmung und Bewertung von „Umweltrisiken". Was können wir aus Naturkatastrophen lernen? Arbeiten zur Risiko-Kommunikation. Heft 57. Programmgruppe Mensch, Umwelt, Technik, Forschungszentrum Jülich GmbH, Jülich: März 1996

KBA (2003) Kraftfahrzeugbundesamt, http://www.kba.de

Keil U, Fitgerald AP, Gohlke H, Wellmann J & Hense H-W (2005). Risikoabschätzung tödlicher Herz-Kreislauf-Erkrankungen. Die neuen SCORE-Deutschland-Tabellen für die Primärprävention. Deutsches Ärzteblatt. Jg. 102, Heft 25, pp. B1526–B 1530

Kellerer A M (2006) Von der Dosis zum Risiko. GSF – Forschungszentrum für Umwelt und Gesundheit GmbH in der Helmholtz-Gemeinschaft. Klemm C, Guldner H & Haury H-J (Eds) Strahlung. 18. Ausgabe 2006. Neuherberg. pp. 23–36

Khanduri A & Morrow G (2003) Vulnerability to buildings to windstorms and insurance loss estimation. Journal of Wind Engineering and Industrial Aerodynamics, 91, pp. 455–467

Kichenside G (1998) Katastrophale Eisenbahnunfälle – Die schwärzesten Tage. Bechtermünz Verlag, Augsburg 1998 (Great Train Disasters. The Word's Worst Railway Accidents. Parragon 1997)

Kiefer D (1997) Sicherheitskonzept für Bauten des Umweltschutzes. DAfStb, Heft 481, Beuth Verlag GmbH, Berlin.

Kingery CN & Bulmash G (1984) Airblast Parameters from TNT Spherical Air Burst and Hemispherical Surface Burst, US Army Ballistic Research Laboratory, ARDC; April 1984

Klesius M (2002) The state of the Planet. National Geographic, September 2002, pp. 103–115

Klesius M (2003) Everst's Greatest Hits. National Geographic. May 2003, pp. 2–71

Kloas J & Kuhfeld H (2002) Stagnation des Personenverkehrs in Deutschland. DIW-Wochenberichte 42/02, Deutsches Institut für Wirtschaftsforschung Berlin

Knobloch B (2005) Euro Finance Week – Initiative Finanzstandort Deutschland am 25.10.2005, Frankfurt am Main, EuroHypo AG.

Knoflacher H (2001) Stehzeuge. Der Stau ist kein Verkehrsproblem, Böhlau Wien

Koeberl C (2007) Meteorite impact cratering on Earth: Hazards, and geological and biological consequences. Departmentkongress Bautechnik und Naturgefahren. Berlin: Ernst & Sohn. pp. 75–77

Koeberl C & Virgil LS (2007) Terrestrial Impact Craters Slide Set. Lunar and Planetary Institute, http://www.lpi.usra.edu/publications/slidesets/craters/

Kolenda K-D (2005) Sekundärprävention der koronaren Herzkrankheit: Effizienz nachweisbar. Deutsches Ärtzeblatt, Jg. 102, Heft 26, pp. B1596–B1602

Kolymbas D (2001) Introduction to landslides and debris flow phenomena.

Konersmann R (2006) Die Vorteile der QRA bei der Abschätzung des externen Risikos der Flughäfen. 44. Tutzing Symposium "QRA – Quo Vadis?", 12.-15.3.2006 in der evangelischen Akademie Tutzing

Korotikin IM (1988) Seeunfälle und Katastrophen von Kriegsschiffen. Militärverlag der Deutschen Demokratischen Republik. 4. unveränderte Auflage. Berlin

Köster-Lösche K (1995) Die großen Seuchen – Von der Pest bis Aids. Insel Taschenbuchverlag, Frankfurt am Main und Leipzig

Kretz T (2005) Der Fußgängerverkehr – Theorie, Experiment, Anwendung. XIV Heidelberger Graduiertenkurs Physik, 8.4.2005

Kröger W & Høj NP (2000) Risk Analyses of Transportation on Road and Railway. Proceedings – Part 2/2 of Promotion of Technical Harmonization on Risk-Based Decision-Making, Workshop, May 2000, Stresa, Italy

Kroker E & Farrenkopf M (1999) Grubenunglücke im deutschsprachigen Raum. Katalog der Bergwerke, Opfer, Ursachen und Quellen. Bochum

Krom A & de Wit S (2006) Reliability analysis of the fire protection in the high speed train tunnel "Groene Hart". In: Proske D, Mehdianpour M & Gucma L (Eds) Proceedings of the 4th International Probabilistic Workshop, Berlin 21–13 October 2006, pp. 161–176

Krug EG, Dahlberg LL, Mercy JA, Zwi AB & Lozano R (2002) World report on violence and health. World Health Organization, Geneva

Krystek R & Zukowska R (2005) Time series – the tool for traffic safety analysis. In: Advances in Safety and Reliability, Kolowrocki (Ed) Taylor and Francis, London, pp. 1199–1202

Kunz M (2002) Simulation von Starkniederschlägen mit langer Andauer über Mittelgebirgen. Dissertation. Universität Fridericiana Karlsruhe

Kurmann F (2004) Hungerkrisen. http://www.lexhist.ch, 2004

Labrousse E (1944) La crise del l'economie francaise à la fin de l'ancien régime et au début de la Révolution

Lampert T & Kurth B-M (2007) Sozialer Status und Gesundheit von Kindern und Jugendlichen. Deutsches Ärzteblatt 2007, 104(43), pp. A2944–A2949

Langenhorst F (2002) Einschlagskraft auf der Erde – Zeugen der kosmischen Katastrophen. Sterne und Weltraum, Juni 2002, pp. 34–44

Lanius K (1988) Mikrokosmos – Makrokosmos. Das Weltbild der Physik. Urania Verlag, Leipzig

Leberke M (2004) Mit dem Tod tanzen. http://tms.lernnetz.de/religion2.htm

Leger Sivard R (1996) World Military and Social Expenditures 1996, Washington, DC: World Priorities

Leicester R, Aust M & Reardon G (1976) A statistical analyses of the structural damage by cyclone Tracy. Civil Engineering Transactions 2, pp. 50–54

Lieberwirth P (2003) Ein Beitrag zur Wind- und Schneelastmodellierung. In: Proske D (Ed) Proceedings of the 1st Dresdner Probabilistik-Symposium. Fakultät Bauingenieurwesen, Technische Universität Dresden. pp. 123–138

LIN (2003) Lexas Information Network. Flugsicherheit, http://www.aviationinfo.de/flugsicherheit.html

Lind N & Hartford D (1999) Probability of human instability in flooding: A hydrodynamic model. Application of Statistics and Probability (ICASP 8), Sydney, 2, pp. 1151–1156

Lomnitz C & Rosenblueth E (1976) Seismic Risk and Enginnering Decisions. Elsevier, Amsterdam

Maier A (2006) Feuer – vom Risikofaktor zum Risikomanagement. Schadenspiegel 2/2006. Munich Re. pp. 21–25

Mann G (Edr) (1991) Propyläen Weltgeschichte – Eine Universalgeschichte. Propyläen Verlag Berlin – Frankfurt am Main

Marmot M & Elliott P (1995) Coronary Heart Disease Epidemiology. From Aetiology to Public Health. Oxford Medical Publications

Mason BG, Pyle DM & Oppenheimer C (2004) The size and frequency of the largest explosive eruptions on Earth. Bulletin of Volcanology, 66, pp. 735–748

Mathers CD & Loncar D (2006) Updated projections of global mortality and burden of disease, 2002–2030: data sources, methods and results. Evidence and Information for Policy Working Paper, Evidence and Information for Policy, World Health Organization, October 2005, Revised November 2006

Matoussek M & Schneider J (1976) Untersuchungen zur Struktur des Sicherheitsproblems, IBK-Bericht 59, ETH Zürich

Mattmüller M (1982) Die Hungersnot der Jahre 1770/71 in der Basler Landschaft. Gesellschaft und Gesellschaften. Hrsg N Bernard &, Q Reichen, pp. 271–291

Mattmüller M (1987) Bevölkerungsgeschichte der Schweiz, Teil 1, 1987, pp. 260–307

McAneney KJ (2005) Australian Bushfire: Quantifying and Pricing the Risk to Residential Properties, Proceedings of the Symposium on Planning for Natural Hazards – How Can We Mitigate the Impacts? University of Wollongong, 2–5 February 2005

McClung D & Schaerer P (1993). The Avalanche Handbook. (1011 SW Klickitat Way), Seattle, Washington 98134. The Mountaineers, USA

McDougall PR & Riedl C (2005) US-amerikanische Verordnung will Gefahren durch Weltraummüll verhindern. Schadenspiegel Heft 2(48), pp. 2–8

Mears A (2006) Avalanche dynamics. www.avalanche.org

Mechler R (2003) Natural Disaster Risk Management and Financing Disaster Losses in Development Countries. Dissertation. Universität Fridericiana Karlsruhe

Mehlhorn G (Ed) (1997) Bauphysik and Brandschutz. In: Der Ingenieurbau: Grundwissen in 9 Bänden. Verlag Ernst und Sohn, Berlin

Melosh HJ (1989) Impact Cratering. A Geologic Process. Oxford University Press, Oxford

Merz T (2001) Müll. In Böse-O'Reilly S, Kammerer S. Mersch-Sundermann V. Wilhelm M. (Hrsg.): Leitfaden Umweltmedizin. 2. Auflage, Urban & Fischer, München & Jena

Mielke H (1980) transpress Lexikon: Raumfahrt. 6. bearbeitete Auflage, Transpress VEB Verlag für Verkehrswesen, Berlin

Ministry of Public Works (2001). Decree 9th May 2001: "Minimum safety requirements regarding land use planning for areas around major hazard installations" (in Italian), issued on "Supplemento Ordinario" n. 151 to the "Gazzetta Ufficiale Italiana n. 138", 16th June 2001

Minoura K, Imamura F, Kuran U, Nakamura T, Papadopuoulos GA, Sugawara D., Takahashi T & Yalciner AC (2005) A Tsunami generated by a possible submarine slide: evidence for slope failure triggerd by the North Anatolian fault movement. Natural Hazards 36, pp. 297–306

Möller HJ, Laux G & Deisler A (2001) Psychiatrie und Psychotherapie. 2. Auflage, Thieme Verlag

Montanari A & Koeberl C (2000) Impact Stratigraphy: The italian record. Lecture Notes in Earth Sciences, Vol. 93, Springer Verlag, Heidelberg,

Morell V (1999) The Sixth Extinction. National Geographic Magazin, National Geographic Society, 195(2), pp. 43–59

Moser S (1987) Wie sicher ist Fliegen? Orell Füssli Verlag Zürich und Wiesbaden, 2. Auflage 1987

Munich Re (2000). Die Welt der Naturgefahren. CD-Programm

Munich Re (2001) Weltkarte der Naturkatastrophen 2000

Munich Re (2003) Topics Geo: Annual Review: Natural Catastrophes 2003

Munich Re (2004a) NatCatSERVICE

Munich Re (2004b) Topics Geo: Annual Review: Natural Catastrophes 2003

Munich Re (2005a) Achtung Stürme. Topics Geo 2004, München

Munich Re (2005b) Topics 2/2005, pp. 23–24

Munich Re (2006a) Hurricans, stronger, more frequent, more costly. Hurrikane – stärker, häufiger, teurer. Assekuranz im Änderungsrisiko. München

Munich Re (2006b) Hurrikansaison 2005: Zeit zum Umdenken. Topics Geo 2005. München

Munter W (1999) 3 mal 3 Lawinen, Bergverlag Rother, München

NaDiNe (2007) – Natural Disasters Networking Platform. GFZ Potsdam

NASA (1999) Impact Hazard.: http://liftoff.msfc.nasa.gov/Academy/SPACE/SolarSystem/Meteors/ImpactHazard.html

NASA (2003) Trail of black holes and neutron starts points to ancient collision. 12.8.2003. http://www.nasa.gov

National Research Council – Committee on Space Debris (1995) Orbital Debris: A technical assessment, Washington D.C., National Academy Press

NATO (1997) Manual of NATO Principles for the Storage of Military Ammunition and Explosives-AASTP-1. Document AC/258-D/453; August 1997

Newhall CG & Self S (1982) The volcanic explosivity index (VEI). An estimate of explosive magnitude for historical volcanism. Journal of Geophysical Research 87, pp. 1231–1238

Newson L (2001) The Atlas of the World's Worst Natual Disasters. Dorling Kindersley, London

NGI (2007) Tsunamies. http://www.geohazards.no/projects/tsunamis.htm

NGS (1998a) National Geographic Society. Biodiversity – Millenium in Maps, Dezember 1998, Washington, D.C.

NGS (1998b) National Geographic Society: Physical Earth – Millenium in Maps, March 1998, Washington, D.C.

NGS (1999) – National Geographic Society. Universe – Millenium in Maps, October 1999, Washington, D.C.

NIIC (1999) The National Injury Information Clearinghouse. Product Summary Report, Injury Estimates for Calendar Year 1998, and National Electronic Injury Surveillance System (NEISS). U.S. Consumer Product Safety Commission, Washington, D.C. http://www.cpsc.gov

NN (1997) Exhibition at the Lippischen Landesbibliothek 4.8.-12.9.1997: Brand und Katastrophe – Alte Bücher zum Feuerlöschwesen

NN (2003) http://www.geocities.com/extinct_humans/meteor.htm

NN (2004) http://www.naturgewalt.de/duerrechronologie.htm

NRPB (1986) National radiological protection board. Living with radiation. London. HMSO

NZZ (2004) Der Hitzesommer 2003 im 500-jährigen Vergleich: Zwei Grad wärmer als 1901 bis 1995. Neue Züricher Zeitung – Internationale Ausgabe. 5. März 2004, Nr. 54, p 43

NZZ (2004) Über Frauen, die den Hunger bekämpfen. Neue Züricher Zeitung – Internationale Ausgabe. 6./7. März 2004, Nr. 55, p 9

O'Neill R (1998) Natural Disasters. Parragon Books, Bristol

Oerlinghausen BM (2006) Kausalitätsbewertung hepatotoxischer Reaktionen. Deutsches Ärzteblatt, Jg. 103, Heft 36, 8th September 2006, pp. B2000–B2001

Overmans R (1999) Deutsche militärische Verluste im Zweiten Weltkrieg. Beiträge zur Militärgeschichte Band 46. R. Oldenbourg Verlag, München

Paffrath G (2002) Altlasten: Erkundung-Risikobewertung-Sicherheit-Techniken, Lecture material

Paffrath G (2004) Die Anwendung des Multibarrierenkonzepts zur Erhöhung der Sicherheit bei Deponien und behandelten Altlasten, Fachhochschule Darmstadt, Fachbereich Chemie- und Biotechnologie

Paretzke HG, Oeh U, Pröhl G & Schneider K (2007) Radioactivity in the population by nuclides in the environment (in German), 13 March 2007, Leipzig

Parfit M (1998) Living with Natural Hazards. National Geographic, 194(1), pp. 2–39

Paté-Cornell ME & Fischbeck PS (1994) Risk Management for the tile of the Space Shuttle. The Institute of Management Sciences. Stanford University, Californien. Interfaces 24: 1 January–February 1994, pp. 64–86

PC Magazin (2004) 55 Milliarden US-Dollar Schaden durch Viren, 19. Januar

Peden M et al. (2004) World Report on Road Traffic Injury Prevention: Summary, World Health Organization. Geneva

PELEM (1989) Panel of Earthquake Loss Estimation Methodology: Estimating Losses from Future Earthquakes. National Academy Press

Petak W & Atkisson A (1992) Natural Hazard Risk Assessment and Public Policy. Springer. New York

Pierson TC (1986) Flow behavior of channelized debris flows, Mount St. Helens, Washington. In Abrahams AD (Ed) Hillslope Processes. Allen and Unwin, Boston, pp. 269–296

Pilkington M & Grieve RAF (1992) The Geophysical Signature of Terrestrial Impact Craters.Reviews of Geophysics, 30, pp. 161–181

Plate EJ (2003) Regional consultation Europe – Report for EWC II Second International Conference on Early Waring. 28–29 July 2003, Potsdam

Pohl R (2004) Talsperrenkatastrophen. University of Technology Dresden, Professor für Hydromechanik

Poisel R (1997) Geologische-geomechanische Grundlagen der Auslösemechanismen von Steinschlag, in: Tagungsband "Steinschlag als Naturgefahr und Prozess", Institut für Wildbach und Lawinenschutz (Hrsg.); Universität für Bodenkultur-Wien

Pollmer U (2006) Wohl bekomm's! Prost Mahlzeit. Kiepenheuer & Witsch, Köln

Prasse HG (1983) Untersuchungen der Trümmer- und Splitterwirkung bei Exp-losionsereignissen in Munitionslagerhäusern; Fraunhofer Institut für Kurz-zeitdynamik, Ernst-Mach-Institut; E 16/83; Juni

Preuß E (1997) Reise in Verderben – Eisenbahnunfälle der 90er Jahre. Transpress Verlag Stuttgart

Promat tunnel (2008). http://www.promat-tunnel.com/en/tunnel-fires.aspx

Prokop A (2007) The application of terrestrial laser scanning for snow and avalanche research. PhD-thesis, Department of Civil Engineering and Natural Hazards, University of Natural Resources and Applied Life Sciences, Vienna

Proske D (2003) Beitrag zur Risikobeurteilung von alten Brücken unter Schiffsanprall, PhD work, University of Technology Dresden

Proske D, Lieberwirth P & van Gelder P (2005) Sicherheitsbeurteilung historischer Steinbogenbrücken. Dirk Proske Verlag: Wien Dresden

Prusiner SB (1995) Prionen-Erkrankungen. Spektrum der Wissenschaft, März, pp. 44

PTM (2004) Pacific Tsunami Museum Inc. Tsunami Photographs http://www.tsunami.org/archivespics.htm, 2004

Pudasaini SP & Hutter K (2007) Avalanche dynamics. Dynamics of rapid flows of dense granular avalanches. Springer

QG (2004) Queensland Government, State Counter Disaster Organization http://www.disaster.qld.gov.au/disasters/landslides_history.asp

Rackwitz R & Streicher H (2002) Optimization and Target Reliabilities. JCSS Workshop on Reliability Bades Code Calibration. Zürich, Swiss Federal Institute of Technology, ETH Zürich, Switzerland, March 21–22

Rackwitz R (1997) Einwirkungen. in (Hrsg.) Mehlhorn G: Der Ingenieurbau: Grundwissen, Teil 8: Tragwerkszuverlässigkeit, Einwirkungen. Verlag Wilhelm Ernst & Sohn, Berlin

Rahikainen J & Keski-Rahkonen O (2004) Statistical Determination of Ignition Frequency of Structural Fires in Different Premises in Finland; Fire Technology 40; pp. 335–353

Razum O & Breckenkamp J (2007) Kindersterblichkeit und soziale Situation: Ein internationaler Vergleich. Deutsches Ärzteblatt 2007, 104(43), pp. A2950–A2956

Remde A (1995) Afrikas Süden, Namibia-Botswana-Zimbabwe-Südafrika Richtig Reisen, Köln: DuMont

Renner M (1999) Ending Violent Conflict. Worldwatch Paper 146, April 1999, Worldwatch Institute

Rennert P, Schmiedel H & Weißmantel C (Eds) (1988) Kleine Enzyklopädie Physik. Hrsg. 2. Auflage, Leipzig: VEB Bibliographisches Institut

Richardson LF (1944) The distribution of wars in time. Journal of the Royal Statistical Society, 107(¾), pp. 242–250

Rickenmann D (1999) Empirical relationships for debris flows. Natural Hazards. 19(1). pp. 47–77

Riegler A (2007) Beeinträchtigung der Lebensqualität durch die Rückkehr von Großraubtieren (Bär, Luchs und Wolf). Studienarbeit, Universität für Bodenkultur Wien

RKI (2004) Robert Koch Institut: Tuberculose. http://www.rki.de

RKI (2007) Robert Koch Institut: Influenza. http://www.rki.de

Roberts L, Lafta R, Garfield R, Khudhairi J & Burnham G (2004) Mortality before and after the 2003 invasion of Iraq: cluster sample survey. The Lancet, published online October 29, 2004 http://image.thelancet.com/extras/04art10342web.pdf (access 5th November 2007)

Roche Lexikon Medizin (1993) Hoffmann – La Roche AG and Urban & Schwarzenberg (Eds). 3rd Edition, Urban & Schwarzenberg. München, Wien, Baltimore

Romeike F (2005) Preis der Erkenntnis. Risknews 3/2005, p 3

Rosenthal W (2004) MaxWave Rogue waves – Forecast and impact on marine structures, GKSS Forschungszentrum GmbH, Germany, http://w3g.gkss.de/projects/maxwave/, 2004

Rückert S (2004) Tatort-Analyse. Die Zeit, Nr. 16, 7. April 2004, pp. 15–16

Ruge B (2004) Risk Matrix as Tool for Risk Assessment in the Chemical Process Industries. (Hrsg) Spitzer C, Schmocker U & Dang VN: International Conference on Probabilistic Safety Assessment and Management 2004, Springer Verlag, Berlin, London, Vol. 5, pp. 2693–2698

Ruppert A (2000). Application of the Term „Risk" from the viewpoint of the German Chemical Industry. Proceedings – Part 1/2 of Promotion of Technical Harmonization on Risk-Based Decision-Making, Workshop, May, 2000, Stresa, Italy

Rütz N (2004) Versicherungsprodukte und Umwelthaftungsrecht unter besonderer Berücksichtigung von Öko-Audit und ISO 14001. BTU Cottbus, Fakultät Umweltwissenschaften und Verfahrenstechnik. http://www.tu-cottbus.de/BTU/Fak4/Umwoek/Publikationen/AR_4_01.pdf

Savage SB & Hutter K (1989) The motion of a finite mass of granular material down a rough incline. Journal of Fluid Mechanics 199, pp. 177–215

SBA (2005) Statistisches Bundesamt, ifo Institut für Wirtschaftsforschung e. V.: Die volkswirtschaftliche Bedeutung der Immobilienwirtschaft, Sonderausgabe, p 47

SBA (2006) Statistisches Bundesamt: Unfallgeschehen im Straßenverkehr 2005, Statistisches Bundesamt Wiesbaden, Deutschland

Schaefer H, Jentsch G, Huber E, & Wegener B (2000). Herzinfarkt-Report, Urban & Fischer

Scheer J (2000) Versagen von Bauwerken. Band I: Brücken. Ernst & Sohn Verlag für Architektur und technische Wissenschaften GmbH, Berlin

Schenck EG (1965) Das menschliche Elend im 20. Jahrhundert – Eine Pathographie des Kriegs-, Hunger- und politischen Katastrophen Europas. Nicolaische Verlagsbuchhandlung Herford, Herford

Schenk C, Beitz S & Buri P (2005) Zwischenbilanz des deutschen Beitrages zum Wiederaufbau. Ein Jahr nach der Flutkatastrophe im Indischen Ozean. Bundesministerium für wirtschaftliche Zusammenarbeit und Entwicklung (BMZ), November 2005, Bonn

Schiefenhövel W (2004) Fertilität zwischen Biologie und Kultur. Traditionelle Geburtenkontrolle in Neuguinea. Neue Züricher Zeitung, 13./14. März, Nr. 61, p 57

Schild M (1982) Lawinen. Dokumentation für Lehrer, Skilager- und Tourenleiter. Lehrmittelverlag des Kantons Zürich

Schlatter H-P, Fermaud C & Wigley P (2001) Safety Assessment of Structures Exposed to Railway Impacts. Safety, Risk and Reliability – Trends in Engineering, Malta, Liaison Committee

Schmähl D, Preussmann R & Berger MR (1989). Causes of cancer – an alternative view to Doll and Peto (1981). Klin Wochenschr. 1989 Dec 4; 67(23). pp. 1169-1173

Schneider U & Lebeda C (2000) Baulicher Brandschutz. Verlag W. Kohlhammer. Stuttgart

Schraube H (2006) Höhenstrahlung: die Exposition beim Fliegen. GSF – Forschungszentrum für Umwelt und Gesundheit GmbH in der Helmholtz-Gemeinschaft. Klemm C, Guldner H & Haury H-J (Eds) Strahlung. 18. Ausgabe 2006. Neuherberg. pp. 14–15

Schröder H (2004) Sturmfluten an der ostfriesischen Küste. http://home.t-online.de/home/Heiner.Schoeder.html, April 2004

Schröder UG & Schröder BS (2005) Strahlenschutzkurs für Mediziner. Georg Thieme Verlag KG, Stuttgart

Sedlag U (1978) Die Tierwelt der Erde. Urania-Verlag, Leipzig

Selby MJ (1993) Hillslope Materials and Processes, Oxford University press, Oxford

SGU (2004) Sveriges Geologiska Undersökning, http://www.sgu.se/geologi/jord/skred/skred_e.htm

Shiu YK & Cheung WM (2003) Case studies on assessment of global landslide risk in Hong Kong. In: Der Kiureghian A, Madanat S & Pestana JM (Eds) Applications of Statistics and Probability in Civil Engineering, Millpress, Rotterdam, pp. 1393–1400

SIPRI (2006) Yearbook 2006 – Armaments, Disarmament and International Security, Published in June 2006 by Oxford University Press on behalf of Stockholm International Peace Research Institute

Sjöstedt J (2004) Entwurf eines Berichtes über die Mitteilung der Kommission über die Sicherheit im Bergbau: Untersuchung neuerer Unglücke im Bergbau und Folgemaßnahmen. Ausschuß für Umweltfragen, Volksgesundheit und Verbraucherpolitik, (KOM(2000) 664 – C5-0013/2001 – 2001/2005(COS)), http://www.europarl.eu.int

Smith JW, Seidl LG & Cluff LE (1966) Studies on the epidemiology of adverse drug reactions. V. Clinical factors influencing susceptibility. Ann. Intern. Med. 65, 4, pp. 629–640

Smith K (1996). Environmental Hazards – Assessing Risk and Reducing Disaster. Routledge: London

Smolka A & Spranger M (2005) Tsunamikatastrophe in Südostasien. Topics Geo: Jahresrückblick Naturkatastrophen 2004, Munich Re, München, pp. 26–31

Smolka A (2007) Vulkanismus – Neuere Erkenntnisse zum Risiko von Vulkanausbrüchen. Münchner Rück. Schadenspiegel 1/2007, München, pp. 34–39

Sornette D (2006) Critical Phenomena in Natural Sciences. Chaos, Fractals, Selforganisation and Disorder: Concepts and Tools. Second Edition. Springer: Berlin Heidelberg

Sovilla B, Schaer M & Rammer L (2008). Measurements and analysis of full-scale avalanche impact pressure at the Vallée de la Sionne test site. Cold Region Science and Technology 51, pp. 122–137

Spahl T & Deichmann T (2001) Das populäre Lexikon der Gentechnik – Überraschende Fakten von Allergie über Killerkartoffeln bis Zelltherapie. Eichborn Verlag, Frankfurt am Main

Sparks S & Self S (2005) Super eruptions: global effects and future threats. Report of a Geological Society of London, Working Group

Spiegel (2003) China: 200 Menschen starben bei Erdgasunglück. 25.12.2003. http://www.spiegel.de

Stolle M, Sack P-M & Thomasius R (2007) Drogenkonsum im Kindes- und Jugendalter. Deutsches Ärzteblatt, Jg. 104, Heft 28-29, 16th Juli 2007, B1819-B1827

Stone R (2007) Too late, earth scans reveal the power of a killer landslide. Science, 311, pp. 1844–1845

Swisdak MM (2001) The Determination of Explosion Yield/TNT Equivalence From Airblast Data. Indian Head Division/Naval Surface Warfare Center; March 2001

SwissRe (2002) Natur- und Man-made Katastrophen 2001. Sigma Nr. 1, SwissRe, Zürich

Synolakis CE (2007) A hydrodynamics perspective for the 2004 Megatsunami. In: Nadim F, Pöttler R, Einstein H, Klapperich H and Kramer S (Eds) Proceedings of the 2006 ECI Conference on Geohazards, Bepress, Lillehammer, Norway

Synolakis CE, Bardet JP, Borrero J, Davies H, Okal E, Silver E, Sweet J & Tappin D (2002) Slump origin of the 1998 Papua New Guinea tsunami, Proceedings of the Royal Society of London, Series. A, 458, pp. 763–789

Tadele F (2005) Ethiopia: University to offer undergraduate degree in disaster risk management. Disaster Reduction in Africa – ISDR Informs, Issue 6, 6th December 2005, pp. 14–16

Te Wildt T & Emrich HM (2007) Die Verzweiflung hinter der Wut. Deutsches Ärzteblatt, Jg. 104, Heft 10, 9th March 2007, B554–B556

Tecchanel (2004) Code Red . 2,6 Milliarden US-Dollar Schaden, http://www.tecchannel.de/news/allgemein/6449/

Toroyan T & Peden M (2007) Youth and Road Safety, World Health Organization, Geneva

Totschnig R & Hübl J (2008) Historische Ereignisdokumentation. University of Natural Resources and Applied Life Sciences, Institute of Mountain Risk Engineering, Vienna

Tri Data Corporation (1998) Fire in the United States 1986–1995. Federal Emergency Management Agency. United States Fire Administration, National Fire Data Center, FA-183, August 1998

Trouillet, P (2001) Truck Impacts on French Toll-Motorways Bridge's Piers. Safety, Risk and Reliability – Trends in Engineering, Malta 2001, on CD

Turco RP, Toon OB, Ackermann TP, Pollack JB & Sagan C (1985) Die klimatischen Auswirkungen eines Nuklearkrieges. Spektrum der Wissenschaft: Verständliche Forschung. Verlagsgesellschaft, Heidelberg, pp. 52–64

Tyagunov S, Grünthal G, Wahlström R, Stempniewski L, Zschau J & Münich JC (2006) Erdbebenrisiko-Kartierung für Deutschland. Beton- und Stahlbetonbau 101, Heft 10, pp. 769–782

Uguccioni G (2004) The criteria for compatibility between industrial plants and land use in Italy. ESRA Newsletter July 2004, ESRA, pp. 2–5

Union Carbide (2007) Bhopal Information Center. http://www.bhopal.com/irs.htm

Uni-protokolle (2004) http://www.uni-protokolle.de/Lexikon/Gentechnik.html

Universität Münster (2007) http://chdrisk.uni-muenster.de/calculator.php

US ACDA (1996) World Military Expenditures and Arms Transfers 1995, US Government Printing Office, Washington D.C.

USGS (2001) http://wwwneic.cr.usgs.gov/neis/eqlists/eqsmosde.html

USGS (2008) http://volcanoes.usgs.gov/Products/Pglossary/vei.html

US-NRC (2004) United States Nuclear Regulatory Commission: Fact Sheet: The Accident at Three Mile Island. http://www.nrc.gov, 2004

van Breugel K (2001) Establishing Performance Criteria for Concrete Protective Structures fib-Symposium: Concrete & Environment, Berlin, 3–5. October 2001

Van der Heuvel M (2006) Strahlenquelle Atomkern: Ionsierende Strahlung. GSF – Forschungszentrum für Umwelt und Gesundheit GmbH in der Helmholtz-Gemeinschaft. Klemm C, Guldner H & Haury H-J (Eds) Strahlung. 18. Ausgabe 2006. Neuherberg. pp. 6–13

Van der Hoven I (1957) Power Spectrum of Horizontal Wind speed in the Frequency Range from 0.0007 to 900 Cycles per Hour. Journal of Meteorology, 14, pp. 160–164

van Gelder PHAJM (2003) Cost Benefit Analysis of Drugs against AIDS in South Africa using a Life-Quality Index, Chapter 2 in: Proceedings of the Workshop on Advanced Models in Survival Analysis, Bloemfontein, January 13 2003

Varnes DJ (1978) Slope Movements and Types and Processes, In: Schuster RL & Krizek J (Eds) Landslides – Analysis and Control. Transportation Research Board Special Report 176, National Academy of Sciences, Washington D.C.

VDSI (2008) Association of German Safety Engineers, http://www.vdsi.de

Vesilind PJ (2004) Chasing Tornadoes. National Geographic Magazine, April 2004, pp. 3–37

VGB (2001) Allgemeine Wohngebäude-Versicherungsbedingungen

Vincent C, Neale G & Woloshynowych M (2001) Adverse events in Bristol hospitals: preliminary retrospective record review. British Medical Journal, 322, 2001, pp. 517–519

VMKUG (2004) Vierte Ministerielle Konferenz Umwelt und Gesundheit. Reaktionen des Gesundheitswesens auf extreme Wetter- und Klimaereignisse, Arbeitspapier, EUR/04/5046267/13 28. April 2004, Budapest, Ungarn, 23.–25. Juni 2004

Walker D (2000) Death penalty has slim effect on murder rate, Guardian News Service, The Jakarta Post, 5. p 5

Walter U (2004) Gesündere Lebensmittel müßten billiger werden. Gesundheit – Das Magazin aus Ihrer Apotheke. Januar 2004, pp. 16–17

Weber A (1964) Wildbachverbauung, Kapitel XIII, in: "Taschenbuch landwirtschaftlicher Wasserbau" Uhden, Otto (Hrsg.), Franckh'scher Verlagsbuchhandlung, Stuttgart, pp. 483–528

Weber A, Hörmann G & Heipertz W (2007) Arbeitslosigkeit und Gesundheit aus sozialmedizinischer Sicht. Deutsches Ärzteblatt 2007, 104(43), pp. A2957–A2962

Webster PJ, Holland GJ, Curry, JA & Chang H-R (2005) Changes in tropical cyclone number, duration, and intensity in a warming environment, Science, 309, pp. 1844–1846

Wegener A (1915) Die Entstehung der Kontinente und Ozeane. Nachdruck 2005: Alfred-Wegener Institut für Polar und Meeresforschung, Bremerhaven

Weidl T & Klein G (2004) A new determination of air crash frequencies and its implications for operation permissions. (Hrsg) Spitzer C, Schmocker U & Dang VN: International Conference on Probabilistic Safety Assessment and Management 2004, Berlin, Springer Verlag, London, Vol. 1, pp. 248–253

Weilert A & Hosser D (2007) Probabilistic Safety Concept for Fire Safety Engineering based on Natural Fires. In: Taerwe L & Proske D (Eds), 5th International Probabilistic Workshop, Ghent, 2007, pp. 29–42

Wenz W, Mönig H, Flemming K, Gehring D, Hoffmann G, Konermann G, Prütz W, Reinwein H & Wannenmacher (1980) Radiologie, Springer Verlag Berlin Heidelberg

White M (2007) Twentieth Century Atlas – Worldwide Statistics of Death Tolls. http://users.erols.com/mwhite28 (access 2nd November 2007)

WHO (1983) Safety requirements for the first use of new drugs and diagnostic agens in man. COMS.

WHO (2004) World Helath Organisation. Neglected Global Epidemics: Three Growing Threats. The World Health Report 2003 – Shaping the Future, Chapter 6, http://www.medicusmundi.ch/mms/services/bulletin/bulletin200401/kap02/14who.html

Williams M (2006). Snow Hydrology: Avalanches http://snobear.colorado.edu/Markw/SnowHydro/avalanches.html

Williams M, Dunkerley D, De Decker P, Kershaw P & Chappell J (1998) Quaternary Environments, Second Edition. Arnold (Hodder Headline Group), London New York Sydney

Williams R (1999) After the deluge – Central America's Storm of the Century. National Geographic. 196(5), pp. 108–129

Wilson N (1998) Great Sea Disasters. Parragon 1998, German Version, Bechtermünz Verlag, Weltbild Verlag, Augsburg 1998

Wilson RCL, Drury SA & Chapman JL (1999) The Great Ice Age: Climate Change and Life, Routledge: London

Wilson RM, Runciman WB, Gibberd RW, Harrison BT, Newby L & Hamilton HD (1995) The quality of Australian health care study. The Medical Journal of Australia, 163, pp. 458–471

WLV (2006) Jahresbericht 2005 des Forsttechnischen Dienst für Wildbach- und Lawinenverbauung. Bundesministerium für Land- und Forstwirtschaft, Umwelt und Wasserwirtschaft, Sektion Forst. Wildbach- und Lawinenverbauung. Wien

Weidl T & Klein G (2004) A new determination of air crash frequencies and its implications for operation permissions. C Spitzer, U Schmocker & VN Dang (eds), International Conference on Probabilistic Safety Assessment and Management 2004, Berlin, Springer Verlag, London, 2004, Volume 1, pp. 248–253

Woo G (2006) Terror attacks during world soccer championship 2006. Tageszeitung Österreich, 17 May 2006, p. 2

Wörner J-D (1997) Grundlagen zur Festlegung und Beurteilung von Dichtheitsanforderungen für Anlagen mit wassergefährdenden Stoffen (Grundlagen 1994). DAfStb Heft 481, Anlage, Beuth Verlag, Berlin

Woronzow-Weljaminow BA (1978) Das Weltall. Urania-Verlag Leipzig Jena Berlin, 2. durchgesehene Auflage

Worth Matravers P, Bridgeman J & Ferguson-Smith M (2000) The BSE Inquiry Report. http://www.bseinquiry.gov.uk/report

Wreathall J (2004) PRA, Patient Safety and Insights for Quality Improvement in Healthcare. In: Spitzer C, Schmocker U & Dang VN (Eds) International Conference on Probabilistic Safety Assessment and Management 2004, Berlin, Springer Verlag, London, Vol. 4, pp. 2206–2211

Yeh H, Imamura F, Synolakis CE, Tsuji Y, Liu P & Shi S (1995) The Flores Island Tsunamis, EOS, Transactions, American Geophysical Union, 74(33), Seite 369, pp. 371–373

Zebrowski E (1997) Perils of a Restless Planet. Scientific Perspectives on Natural Disasters. Cambridge University Press, Cambridge

Zhang S (1993) A comprehensive approach to the observation and prevention of debris flow in China. Natural Hazards, 7, pp. 1–23

Ziervogel G (2005) Understanding resilient, vulnerable livelihoods in South Africa, Malawi, Zambia. Disaster Reduction in Africa – ISDR Informs, Issue 6, 6th December 2005, pp. 19–23

3 Objective Risk Measures

3.1 Introduction

As discussed in chapter "Indetermination and risk", introducing the language of mathematics is a trial to improve objectivity. As shown, this can only be done partially; nevertheless, the application of parameters might be useful for risk- informed decisions (Arrow et al. 1996).

This chapter mainly discusses different types of risk parameters. First, some simple parameters are introduced, which permit a rough risk estimation. Subsequently, the risk parameters are discussed from simple ones to those more sophisticated ones. Such parameters are mortalities, fatal accident rates, family of F-N diagrams and lost life- years. The most sophisticated risk parameter – the quality of life – is then discussed in chapter "Quality of life – the ultimate risk measure".

3.2 Rough Risk Measures

The computation of risks usually requires advanced models and high efforts of computation. But, sometimes, such computations give only the impression of high quality results, for example, if data are missing. Under such conditions, it might be adequate to use simple risk computations. Here, the strength of rough risk measures or matrices becomes visible. In an easy and understandable way, risks are estimated using mainly only tables or diagrams.

3.2.1 NASA Risk Assessment Table

As shown in the following tables, frequencies of events (Table 3-1) and damages of events (Table 3-2) are simply approximated. The values taken from the table are then added up to create a risk management table

(Table 3-3), which can be compared with goal values (Table 3-4). With this, a quick risk assessment can be carried out.

Table 3-1. Hazard probability rank

Level	Description	Component	Fleet or entire construction
A	Frequent	Presumably frequent	Continuously found
B	Likely	Several times during lifetime	Widespread
C	Occasional	Presumably one time per lifetime	Several times
D	Remote	Unlikely during lifetime	Unlikely but has to be considered
E	Unlikely	So unlikely that it can be excluded	Unlikely but possible

Table 3-2. Hazard severity categories

Level	Description	Scenario and details
I	Disaster	Death and system loss
II	Critical	Serious accident, severe system damage
III	Insignificant	Minor accident, minor system damage
IV	Neglectable	Damage less than that in minor accident

Table 3-3. Risk management matrix

Probability of hazard	Severity			
	Disaster	Critical	Insignificant	Neglectable
Likely	1	3	7	13
Probable	2	5	9	16
Occasional	4	6	11	18
Rare	8	10	14	19
Unlikely	12	15	17	20

Table 3-4. Risk acceptance index

Goal value	Category
1–5	Unacceptable
6–9	Undesirable
10–17	Acceptable with further assessment
18–20	Acceptable without further assessment

3.2.2 Australian Standard AS/NZS 4360

Another example of the same procedure is shown in AS/NZS 4360 standard (Schmid 2005). Again, the frequency (Table 3-5) and the intensity of events (Table 3-6) are approximated, summarised (Table 3-7) and compared with goal values (Table 3-8).

Table 3-5. Qualitative measure of likelihood

Level	Description	Scenario and detail	Probability
16	Very likely	Will happen under virtually all conditions	> 85%
12	Highly likely	Will happen under most conditions	50–85%
8	Fairly likely	Will happen quite often	21–49%
4	Unlikely	Will happen sometimes	1–20%
2	Very unlikely	Not expected	<1%
1	Almost impossible	Possible but very surprising	<0.01%

Table 3-6. Qualitative measure of consequence

Level	Description	Scenario and detail
1,000	Disaster and catastrophe	Fatalities, release of poison with considerable effects on the environment, bankruptcy
100	Major accident	Serious injuries, release of poison from production, serious hazard for the business
20	Average accident, substantial damage	Medical care required, release of poison but not outside the production area or without any effects on the environment, but substantial loss of profit
3	Minor accident	First aid required, released poison is immediately bound, low effects on business
1	Neglectable, non- significant	No injuries, no effects on environment, neglectable effects on business

Table 3-7. Risk management matrix

Likelihood	Seriousness				
	Neglectable	Minor	Substantial	Major	Disaster
Very likely	16	48	320	1,600	16,000
Highly likely	12	36	240	1,200	12,000
Fairly likely	8	24	160	800	8,000
Unlikely	4	12	80	400	4,000
Very unlikely	2	6	40	200	2,000
Almost impossible	1	3	20	100	1,000

Table 3-8. Risk acceptance index

Value	Category
> 1,000	Non-acceptable
101–1,000	Not desired
21–100	Acceptable
< 20	Neglectable

3.2.3 Kent Scale

The Kent Scale has been used by the US Defense Intelligence Agency in 1980 to transfer terms into frequencies (Table 3-9). The Kent Scale has been also adapted for the estimation of geological risks in oil search (Table 3-10).

Table 3-9. Measure of likelihood

Term	Synonym	Rank	Probability
Highly certain	Apparently certain, highly probable	5	91–100%
Probable	Chance is high, people believe in it	4	61–90%
Equal chance	Balanced	3	41–60%
Improbable	Not probable	2	11–40%
Impossible	Low chance, very doubtful	1	1–10%

Table 3-10. Measure of consequence

Term	Synonym	Rank	Probability
Proofed	True	8	98–100%
Virtually certain	Convincible	7	90–98%
Highly likely	Strong belief, highly likely	6	75–90%
Likely	Presumably true, good chances	5	60–75%
Balanced	Less good chances, balanced	4	40–60%
Presumably not true	Unlikely, bad chances	3	20–40%
Possible but unlikely	Very low chance, highly unlikely	2	2–20%
Invalidity proofed	Impossible	1	0–2%

3.2.4 Australia's Paper's Risk Score

The Australia's Paper's risk score permits in a fast way a rough estimation of a risk. In Fig. 3-1, the risk score is shown. First, the probability of an event is estimated. Here, the example from Burgmann (2005) has been taken. The probability is chosen between unusual and remotely possible. In the second step, the first consequence is estimated, the exposure. So if an accident happens, how often is it connected to an uncontrolled exposure. This gives the frequency timeline. From there, the consequence in terms of fatalities is estimated (line B). Connecting the points of line A and line B gives the risk score lying in a certain region.

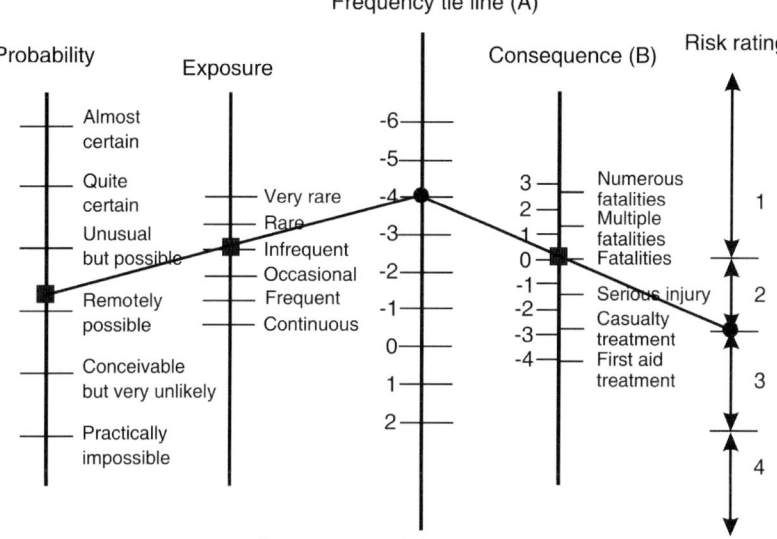

Fig. 3-1. Australia's Paper's risk score (Burgmann 2005)

3.2.5 Hicks Scale

Taken from Shortreed et al. (2003), the Hicks tables permit a rough estimation of the risks. Again here, tables are used for the estimation of frequency (Table 3-11), consequence estimation (Table 3-12) and evaluation
(Table 3-13).

Table 3-11. Event frequency classification system

Weight	Possibility
Frequent (5)	1 or more events per year
Probable (4)	12 or more events in 10 years
Occasional (3)	1 or more events per 30 years
Remote (2)	1 or more events per 200 years
Improbable (1)	Less than 1 event per 200 years

Table 3-12. Consequence rating for health risk evaluation

Consequence	Public health consequences
Catastrophic (100)	Multiple fatalities and injuries
Major (60)	Single fatality, permanent total disability
Serious (25)	Major injuries, partial injury or longer term injury
Moderate (10)	Minor injuries, medical aid and low severity impairment
Minor (2)	Slight injury, illness, first aid not required

Table 3-13. Risk score and levels (Event frequency × Consequence rating)

Risk score	Risk level	Action required
>400	Extreme risk	Intolerable, immediate action necessary to reduce risk
100–400	High risk	Unacceptable for long-term, risk controls have to be implemented
30–100	Moderate risk	Undesirable, evaluate risk reduction measures in long-term
<30	Low risk	No mitigation necessary, periodic evaluation to maintain at low level

3.2.6 Risk Diagram from Hoffmann-La Roche AG

A further example of a rough risk measure is taken from the company Hoffmann-La Roche AG (Hungerbühler et al. 1998). This procedure is diagram-based as shown in Fig. 3-2. Here again, risks are assessed based on some rough estimations of the frequency of an event and of the damage.

3.2.7 Further Rough Measures

There exist many further rough risk measures. Examples of applications can be found in GTZ (2004) or Greminger et al. (2005) with the program RiskPlan. In general, rough measures permit the user to quickly estimate risks based only on some rough or incomplete information about frequencies and damages. Although it does not give the impression of an objective risk measure, in many cases, it is the best that can be achieved in terms of objectivity. Furthermore, if people use such simple procedures, they might be more willing to understand and apply the concept of risk assessment.

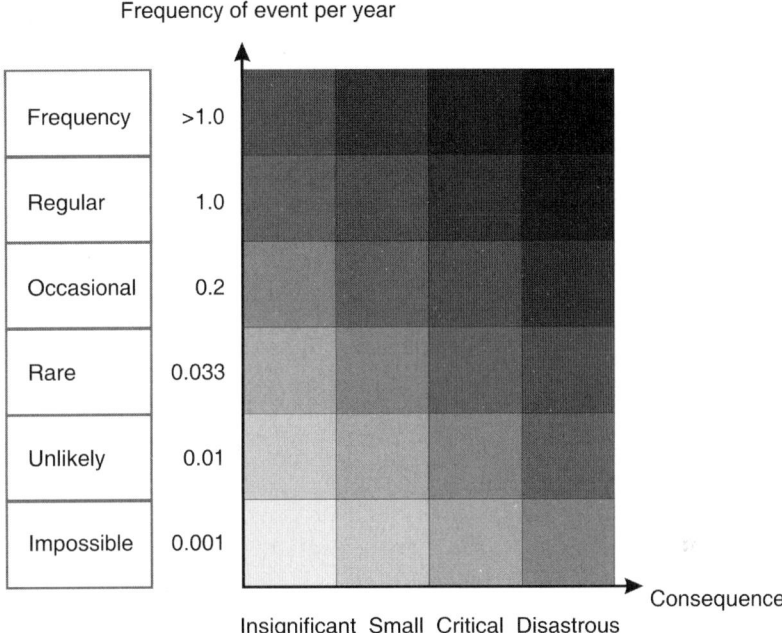

Frequency of event per year

Frequency	>1.0				
Regular	1.0				
Occasional	0.2				
Rare	0.033				
Unlikely	0.01				
Impossible	0.001				

Consequence

Insignificant Small Critical Disastrous

Human	Human	Minor injuries	Injuries	Injuries	Fatalities
	Vicinity	Distur-bance	Irritation	Evacua-tion
Environment		Only system affected	System and area affected	Reversible damages	Irre-versible damages
Tangible assets	SFrancs	<1 mio	1-2 mio	2-10 mio	>10 mio
	Breakdown of system	Days	Weeks	Months	Years

mio- million and SFrancs- Swiss Francs

Fig. 3-2. Risk diagram from Hoffmann-La Roche AG (taken from Hungerbühler et al. 1998)

3.3 Specialised Risk Measures

3.3.1 Introduction

While the rough risk measures have found wide applications in many fields, including natural hazards, chemical industries and aerospace industry, special measures have also been developed in other fields. Such measures cannot be introduced here completely, but some examples will be given from two fields: toxicology and medicine. Schütz et al. 2003 give an overview about the diversity of such parameters, which might be used for further studies.

3.3.2 Toxicology

3.3.2.1 Concentration-Based Risk Measures

Dose or concentration of toxic substances can be used as an indicator for risks to creatures. Different kinds of dose measures are known, which differ in the consequences and in the percentage of affected creatures to the exposed ones. The effective concentration (EC) considers different kinds of effects of toxins, whereas the lethal concentration (LC) only considers the percentage of dead creatures. Other measures are the NOEC and the LOEC, which are explained in short in the following list:

- NOEC – No observed effect concentration
- LOEC – Lowest observed effect concentration
- EC_{50} – Median effective concentration
- LC_{10} – 10% quantile lethal concentration
- LC_{50} – Median lethal concentration

Figure 3-3 shows the sorting of the mentioned measures in a log concentration scale of a toxic substance. Another point of interest is the exposure time, usually 96 h. Furthermore, Table 3-14 gives some example values of LD_{50}.

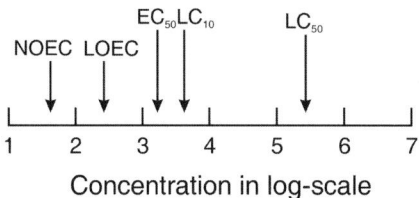

Concentration in log-scale

Fig. 3-3. Toxic effects as a function of concentration (Burgmann 2005)

Table 3-14. Approximate acute LD_{50} values for animals for select chemicals (Burgmann 2005)

Chemical	Dose (LD_{50}) in mg per kg body weight
Sodium chloride	4,000
Ferrous sulfate	1,500
Strychnine sulfate	2
Nicotine	1
Dioxin (TCDD)	0.001
Botulinum	0.00001

3.3.2.2. Unit Risk Value

The unit risk value gives the estimated number of additional cancer risk cases based on the inhalation of a substance over 70 years in a concentration of 1 $\mu g/m^3$ air. Examples are given in Table 3-15 and Table 3-16. This value can then be adapted to the current conditions in terms of time and concentration. This value is widely used for emission protection laws in Germany and the EU.

Table 3-15. Unit risk values for a few substances (LUW 2005)

Substance	Unit risk value
Arsenic	4×10^{-3}
Asbestos [1]	2×10^{-5}
Benzole	9×10^{-6}
Cadmium	1.2×10^{-2}
Diesel soot [2]	10×10^{-5}
PAK (BaP)	7×10^{-2}
2,3,7,8-TCDD	1.4

[1] Related to 100 fibres per m^3.
[2] Measurement of particle mass.

Table 3-16. Risk based on unit risk values (LUW 2005)

Substance	1:1,000	Risk [1] 1:2,500	1:5,000	Concentration
Arsenic	13	5	2.5	ng/m^3
Asbestos	220	88	44	$fibres/m^3$
Benzole	6.3	2.5	1.3	$\mu g/m^3$
Cadmium	4.2	1.7	0.8	ng/m^3
Diesel soot	2.8	1.1	0.6	$\mu g/m^3$
PAK (BaP)	3.2	1.3	0.6	ng/m^3
2,3,7,8-TCDD	39	16	7.8	fg/m^3

[1] 1:1,000 means one additional case of cancer and so on.

3.3.3 Health Care System

3.3.3.1 Relative Risk

The human exposure to substances is a problem not only of eco-toxicology but also for the health system. Here, many new substances are continuously developed as medication. To identify whether such substances are indeed helping humans or increase a risk, risk measures are applied. One such risk measure is the relative risk (RR). It can be computed based on Schütz et al. (2003):

$$RR = \frac{A/(A+B)}{C/(C+D)} \tag{3-1}$$

The variables consider the number of negative affected and non-negative affected persons (Table 3-17). Negative affected here is understood as appearance of a disease.

Table 3-17. Meanings of the variables according to Eq. (3-1)

	Negative affected	Non-negative affected
To substance exposed	A	B
Not to substance exposed	C	D

3.3.3.2 Risk Difference

A further health- related risk measure is the risk difference (RD). The risk difference is computed as:

$$RD = A/(A+B) - C/(C+D) \tag{3-2}$$

The parameter is also known as absolute risk reduction. A value of $RD = 0$ indicates no change, whereas $RD < 0$ shows that the risk has decreased, and $RD > 0$ indicates that the substance actually increases the risk to patients (Schütz et al. 2003).

3.3.3.3 Pooled Risk Difference

In many cases, not only the investigation of one author shall be considered in risk estimations, but further publications should also be considered. Here, the so-called pooled risk difference (PRD) can be used.

$$PRD = \frac{\sum_{i=1}^{k}\left(A_i \cdot \left(\frac{C_i + D_i}{N_i}\right) - C_i \cdot \left(\frac{A_i + B_i}{N_i}\right)\right)}{\sum_{i=1}^{k}\left(\frac{(A_i + B_i) \cdot (C_i + D_i)}{N_i}\right)} \tag{3-3}$$

with $N_i = A_i + B_i + C_i + D_i$

The three different measures discussed in this section do not explicitly describe the negative effects of some treatment or pharmaceutics. However, in many cases, the negative effect is a loss of human lives. This yields to the term mortalities.

3.4 Mortalities

3.4.1 Introduction

Every day, about 400,000 humans are born worldwide. In the same time, about 200,000 people die. Many of the people had a long and satisfying life, but others die much too early. In developed countries, the causes of deaths are registered. For example, someone may die in connection with a road accident, an airplane crash or a heart attack. At the end of a year, the data can be added up and the causes can then be ranked according to the number of people killed by each cause. This gives a good impression about the size of hazards, but does not permit the comparisons with other regions or countries. This can be achieved by the introduction of relative values. Such values can easily be computed by the ratio of the absolute number to the entire population. Then, the comparison can be extended to a world-wide level, at least to all regions where such data are available and published on a regular basis.

Consider an example of approximately 5,000–7,000 road accident fatalities in Germany per year over the last few years. In contrast, about 50 people drown per year. Assuming a population of 80 million, the ratio can be calculated:

$$\frac{\text{Number of road accident fatalities}}{\text{Population}} = \frac{5{,}000...7{,}000}{80{,}000{,}000} \qquad (3\text{-}4)$$

$$= 6.25 \cdot 10^{-5}...8.75 \cdot 10^{-5} \approx 1.0 \cdot 10^{-4}$$

$$R_H = \frac{N_o}{N_G} \leq \max R_H \qquad (3\text{-}5)$$

The result of such calculation is the mortality or frequency of deaths per year for a certain region or country. Sometimes, people use the term probability of death. The term probability is used here based on the assumption of extrapolation of historical data into the future. Whether that is valid or not is another question.

3.4.2 Choice of Population and Representative Year

First, the focus should be more on the choice of population. The above equation has assumed that the entire population uses the road transport system. This indeed is quite true, since more than 90% of the population are road users in one way or another. But, how is it with motorcyclists?

Assume about 2,000 fatalities in connection with motorbike accidents in Germany per year. Based on the entire population, this gives

$$\frac{\text{Number of motor bike accident fatalities}}{\text{Population}} = \frac{2,000}{80,000,000} = 2.25 \cdot 10^{-5} \qquad (3\text{-}6)$$

But, if not the entire population joins the group of motorcyclists, the computed mortality might be wrong. Assuming that only 5 million people are motorcyclists, the mortality changes to

$$= \frac{2,000}{5,000,000} = 4.0 \cdot 10^{-3} \qquad (3\text{-}7)$$

This shows that one has to take care in choosing the population size. This is valid not only for the population but also for the time period. There might, for example, be a year with a strong winter and, therefore, the traffic volume has been rather low, yielding to a lower number of fatalities. Such discussion might look here artificial, but there has been a strong discussion in Bavaria, Germany, whether the rate of children with leukemia in the neighbourhood of nuclear power plants has increased or not. One of the major points of discussion was the choice of the representative time period. For example, there might be a few years in statistics showing a significant increase, whereas other years behave completely normal.

Another example for time dependency is the killing of people by lightening. Between 50 and 100 people were killed in Germany by lightening 50 years ago; current numbers are between 3 and 7 fatalities per year (Zack et al. 2007).

3.4.3 Different Units

The next critical point is the choice of the unit "per year". This is not necessarily a must; for some means of transport, other units of mortality might be more appropriate, for example distance-based units. In addition, for other hazards, such as chemicals, a distance unit might not be applicable. The following list gives an impression about the diversity of such mortality units (adapted from Slovic 1999).

- Death per year
- Death per million people in the population
- Death per million people within x miles of the source of exposure
- Death per unit of concentration
- Death per facility
- Death per tonne of toxic air released

- Death per tonne of toxic air absorbed by people
- Death per tonne of chemical produced
- Death per million dollars of product produced
- Death per trip
- Death per km travelled
- Death per energy produced

Femers & Jungermann (1991) have given a more extended table of risk units. Halperin (1993) compares different risk parameters for travel as listed in Table 3-18.

Table 3-18. Different risk parameters for travel according to Halperin (1993)

Type of risk parameter	Definition
Mileage Death Rate	Annual fatalities per 100 million vehicles per year
Registration Death Rate	Annual fatalities per 100,000 people in the population
Population Death Rate	Annual fatalities per 10,000 vehicles
Trip Fatality Risk	Risk of dying for a specific trip by a specific type of traveller with a specific means of transport and transport vehicle
Aggregate Fatality Risk	Lifetime risk of dying from a specific cause
Route Fatality Risk	Risk of dying for a specific route

3.4.4 Causes

The computation of mortalities and epidemiological surveys assumes that the causes are given correctly. Nevertheless, that can be distrusted in many cases. A quality investigation in postmortem examinations showed that about 1,000 non-natural deaths are assessed as natural deaths in Germany. Such differences are known worldwide between death certificate, mortality statistics and autopsies (Bratzke et al. 2004, Vennemann et al. 2006, Schelhase & Weber 2007).

Sometimes, there can be trends, just like in the fashion world, in some scientific fields, which influence the mortality statistics, for example car-induced dust or cot death.

3.4.5 Examples

Keeping such drawbacks in mind, the ranking on mortalities might still be a good indicator for fatal risks. Figure 3-4 gives a rough estimate about the diversity of mortalities.

An extensive list of mortalities is given in Table 3-19. Here many numbers for different technological, natural, health or social risks can be found.

Furthermore, Table 3-20 gives inverse values by mentioning actions that increase the mortality by one in a million. This value of increased mortality by one in a million is of great importance since it is included in many regulations as "permitted maximum additional risk" or as so-called "de minimis risk." However, the origin of this value remains uncertain; even some indications for the development in the 1960s have been found (Kelly 1991). Interestingly, this value is also applied as goal value for the probability of failure of structural elements (Spaethe 1992).

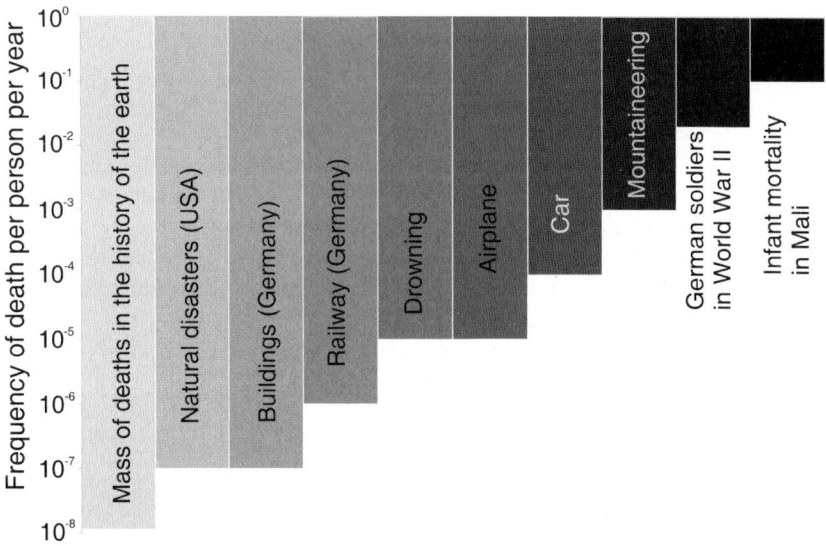

Fig. 3-4. Mortalities for different actions

Table 3-19. Mortalities for different causes. Most values are taken from Paté-Cornell (1994), Parfit (1998), Kafka (1999), Mathiesen (1997), Spaethe (1992), Ellingwood (1999), James (1996), Proske (2004), DUAP (1997), Ditlevsen (1996), Overmans (1999) and Zwingle (1999). A complete list of references for the data can be found in Proske (2004). Same causes can have different values based on different reference times and different reference regions

Causes of death	Mortality per year
Children in jungle in the first two years in Irian Jaya	$2.5 \cdot 10^{-1}$
Infant mortality in Mali	$1.2 \cdot 10^{-1}$
German soldiers in World War II	$7.0 \cdot 10^{-2}$
Infant mortality in the developing world	$6.4 \cdot 10^{-2}$
Storebælt link bridge (<19 fatalities) computation	$2.0 \cdot 10^{-2}$
Motorcyclists (USA)	$2.0 \cdot 10^{-2}$
Males between age 54 and 55 in the GDR 1988	$1.0 \cdot 10^{-2}$
Females between age 60 and 61 in the GDR 1988	$1.0 \cdot 10^{-2}$

Table 3-19. (Continued)

Causes of death	Mortality per year
Loss of shuttle during mission (NASA 1989)	$1.0 \cdot 10^{-2}$
General mortality in the US	$9.0 \cdot 10^{-3}$
General mortality in the US 1999	$8.6 \ 10^{-3}$
Infant mortality in the developed world	$8.0 \cdot 10^{-3}$
Cancer (US 1999)	$5.7 \ 10^{-3}$
Heart disease (US 1999)	$5.7 \ 10^{-3}$
Maternal mortality during birth in the developing world	$5.0 \cdot 10^{-3}$
Trapeze artist (US)	$5.0 \cdot 10^{-3}$
Acceptable risk in the British heavy industry (old value)	$4.0 \cdot 10^{-3}$
Smoking (US 1999)	$3.6 \ 10^{-3}$
US soldiers in the second Iraq war 2003	$3.0 \ 10^{-3}$
Smoking (US)	$3.0 \ 10^{-3}$
Heart disease in the US (1975–1995)	$2.9 \cdot 10^{-3}$
Cancer every age (UK)	$2.8 \cdot 10^{-3}$
Mountaineering (international)	$2.7 \cdot 10^{-3}$
Parachuting (USA)	$2.0 \cdot 10^{-3}$
Spaceman (ESA Crew Recovery Vehicle)	$2.0 \cdot 10^{-3}$
Acceptable risk in the British heavy industry (new value)	$2.0 \cdot 10^{-3}$
Canvey Island (UK)	$2.0 \cdot 10^{-3}$
Workers in heavy and construction industry (UK 1990)	$1.8 \cdot 10^{-3}$
Deep-sea fishing	$1.7 \cdot 10^{-3}$
Violent crimes in Johannesburg in 1993	$1.5 \cdot 10^{-3}$
Alpine climbing (50 h per year)	$1.5 \cdot 10^{-3}$
Underground mining in Germany in 1950	$1.3 \cdot 10^{-3}$
Workers on off-shore platform (UK 1990)	$1.3 \cdot 10^{-3}$
Flying (crew)	$1.2 \cdot 10^{-3}$
Smoking (death caused by cancer)	$1.2 \cdot 10^{-3}$
Workers in oil and gas production	$1.0 \cdot 10^{-3}$
Males between age 17 and 18 in the GDR in 1988	$1.0 \cdot 10^{-3}$
Females between age 35 and 36 in the GDR in 1988	$1.0 \cdot 10^{-3}$
Mountaineering (US 1999)	$1.0 \ 10^{-3}$
Maximum tolerable risk for workers	$1.0 \ 10^{-3}$
Acceptable risk for medical surgeries	$1.0 \cdot 10^{-3}$
Acceptable risk for British off-shore platform	$1.0 \cdot 10^{-3}$
Acceptable risk for Norwegian off-shore platform	$1.0 \cdot 10^{-3}$
Smoking (400 h per year)	$1.0 \cdot 10^{-3}$
Deep-sea fishing	$8.4 \cdot 10^{-4}$
Underground mining (US 1970)	$8.4 \cdot 10^{-4}$
Firefighters (US)	$8.0 \cdot 10^{-4}$
Hang-gliding (US)	$8.0 \cdot 10^{-4}$
Underground mining (UK 1950)	$7.4 \cdot 10^{-4}$
Working in a coal mine (US)	$6.3 \cdot 10^{-4}$
Underground mining in Canada in 1970	$6.2 \cdot 10^{-4}$
Underground mining in Germany in 1980	$5.9 \cdot 10^{-4}$
Failure of dams	$5.0 \cdot 10^{-4}$

(Continued)

Table 3-19. (Continued)

Causes of death	Mortality per year
Unexpected death (US)	$3.7 \cdot 10^{-4}$
Farmers (US)	$3.6 \cdot 10^{-4}$
Coal mining	$3.3 \cdot 10^{-4}$
Lung cancer in Germany	$3.2 \cdot 10^{-4}$
Underground mining (UK 1970)	$3.0 \cdot 10^{-4}$
Coal mining (1500 h per year)	$3.0 \cdot 10^{-4}$
Motor vehicle road accidents (US 1967)	$2.7 \cdot 10^{-4}$
Unexpected death (Australia)	$2.5 \cdot 10^{-4}$
Car accident (US)	$2.4 \cdot 10^{-4}$
Policemen (US)	$2.2 \cdot 10^{-4}$
Car accident	$2.2 \cdot 10^{-4}$
Car accident (US 1999)	$2.0 \ 10^{-4}$
AIDS (US 1995)	$2.0 \cdot 10^{-4}$
Car accident (300 h per year)	$2.0 \ 10^{-4}$
Construction workers	$1.7 \cdot 10^{-4}$
Swimming (50 h per year)	$1.7 \cdot 10^{-4}$
Acceptable risk according to Ford in the 1970s	$1.6 \cdot 10^{-4}$
Workers in forestry	$1.5 \cdot 10^{-4}$
AIDS (US 1996)	$1.5 \cdot 10^{-4}$
Construction workers (2,200 h per year)	$1.5 \cdot 10^{-4}$
Mining workers	$1.4 \cdot 10^{-4}$
Container ship crew	$1.3 \cdot 10^{-4}$
Flying (passenger)	$1.2 \cdot 10^{-4}$
Motor vehicle accidents in Germany in 1988	$1.2 \cdot 10^{-4}$
AIDS worldwide	$1.2 \cdot 10^{-4}$
Boating (80 h per year)	$1.2 \cdot 10^{-4}$
Failure of bridges	$1.1 \cdot 10^{-4}$
Housework	$1.1 \cdot 10^{-4}$
Holiday (UK 1990)	$1.0 \cdot 10^{-4}$
Construction workers	$1.0 \cdot 10^{-4}$
Acceptable risk for old buildings	$1.0 \cdot 10^{-4}$
Maximum tolerable risk for the public	$1.0 \ 10^{-}$
Accidents at home (US 1999)	$1.0 \ 10^{-4}$
Fall (US 1967)	$1.0 \cdot 10^{-4}$
Household	$1.0 \cdot 10^{-4}$
Violent crime (US 1981)	$1.0 \cdot 10^{-4}$
Girls under age of 14 years (the Netherlands)	$1.0 \cdot 10^{-4}$
Frequency of cancer with actions required	$1.0 \cdot 10^{-4}$
Maternal mortality at birth in the developed world	$1.0 \cdot 10^{-4}$
Violent crime (US 1981)	$9.8 \cdot 10^{-5}$
Railway men	$9.6 \cdot 10^{-5}$
Road accidents (UK)	$9.1 \cdot 10^{-5}$
Fatalities during police actions in the US	$8.6 \cdot 10^{-5}$
Fall (Germany 1988)	$8.1 \cdot 10^{-5}$
Shipping (scheduled services)	$8.0 \cdot 10^{-5}$

Table 3-19. (Continued)

Causes of death	Mortality per year
Workers in agricultural industry	$7.9 \cdot 10^{-5}$
Air traffic (scheduled service 10,000 miles per year)	$6.7 \cdot 10^{-5}$
Violent crime (US 1998)	$6.3 \cdot 10^{-5}$
Nuclear power plant failure in Germany (early/late fatalities)	$5.8 \cdot 10^{-5}$
Workers in metal industry	$5.5 \cdot 10^{-5}$
Industrial plants in West Australia	$5.0 \cdot 10^{-5}$
Boating (US)	$5.0 \cdot 10^{-5}$
Industrial plants in New South Wales, Australia	$5.0 \cdot 10^{-5}$
Working in factory	$4.0 \cdot 10^{-5}$
Fire and explosions in US in 1967	$3.7 \cdot 10^{-5}$
Storebælt link bridge (20–200 fatalities) computation	$3.0 \cdot 10^{-5}$
Joining a rodeo (US)	$3.0 \cdot 10^{-5}$
Hunting (US)	$3.0 \cdot 10^{-5}$
Drowning in US in 1967	$2.9 \cdot 10^{-5}$
Fire (US)	$2.8 \cdot 10^{-5}$
Workers in energy production	$2.5 \cdot 10^{-5}$
Acts of violence in London in 1993	$2.5 \cdot 10^{-5}$
Dam failure with fatalities in the US	$2.5 \cdot 10^{-5}$
Flights (20 h per year)	$2.4 \cdot 10^{-5}$
Drugs in Germany in 1999	$2.2 \cdot 10^{-5}$
Workers in the chemical industry	$2.1 \cdot 10^{-5}$
Airplane crash in USA in 1999	$2.0 \ 10^{-5}$
Civil engineers	$1.9 \ 10^{-5}$
Average above all workshops	$1.9 \ 10^{-5}$
Average above all industries	$1.8 \ 10^{-5}$
Railway (200 h per year)	$1.5 \cdot 10^{-5}$
Working accidents (UK)	$1.4 \cdot 10^{-5}$
Acceptable risk by General Motors in the 1990s	$1.2 \cdot 10^{-5}$
Acceptable risk	$1.1 \cdot 10^{-5}$
Failure of shopping centres and sport centres (West Australia)	$1.0 \ 10^{-5}$
Sport and open air structures (New South Wales, Australia)	$1.0 \ 10^{-5}$
Storages and office structures (New South Wales, Australia)	$1.0 \ 10^{-5}$
Maximal acceptable risk in known situations (Netherlands)	$1.0 \ 10^{-5}$
Fire in USA in 1999	$1.0 \ 10^{-5}$
Failure of building constructions	$1.0 \cdot 1^{-5}$
Acceptable cancer risk for a substance (FDA)	$1.0 \cdot 1^{-5}$
Acceptable risk for new structures (the Netherlands)	$1.0 \cdot 10^{-5}$
Acceptable risk (the Netherlands)	$1.0 \cdot 10^{-5} – 1.0 \cdot 10^{-6}$
Flights in USA in 1967	$9.0 \cdot 10^{-6}$
Electrical engineer	$8.0 \cdot 10^{-6}$
Road traffic (10,000 miles per year, careful driving)	$8.0 \cdot 10^{-6}$
Fire of structures	$8.0 \cdot 10^{-6}$
Fire	$8.0 \cdot 10^{-6}$
Gas poisoning in USA in 1967	$7.9 \cdot 10^{-6}$

(Continued)

Table 3-19. (Continued)

Causes of death	Mortality per year
Average over all service industries	$7.0 \cdot 10^{-5}$
Railway traffic in USA in 1967	$5.0 \cdot 10^{-6}$
Railway traffic in Germany in 1988	$4.4 \cdot 10^{-6}$
Storebælt link (> 200 fatalities computed)	$3.0 \cdot 10^{-6}$
Freezing in USA in 1967	$1.6 \cdot 10^{-6}$
Natural disasters in the US	$1.4 \cdot 10^{-6}$
Air traffic in Germany in 1988	$1.2 \cdot 10^{-6}$
Residential houses and hotels (New South Wales, Australia)	$1.0 \cdot 10^{-6}$
Maximal acceptable risk for new situations (the Netherlands)	$1.0 \cdot 10^{-6}$
Maximal permitted mortality	$1.0 \cdot 10^{-6}$
Acceptable risk	$1.0 \cdot 10^{-6}$
De minimis Risk (term explained in chapter "Law and risk")	$1.0 \cdot 10^{-6}$
Drought USA (1980–2000)	$1.0 \cdot 10^{-6}$
Acceptable probability of cancer by a substance (EPA)	$1.0 \cdot 10^{-6}$
Risk by nuclear power plants (USNRC)	$1.0 \cdot 10^{-6}$
Acceptable risk	$1.0 \cdot 10^{-6}$
Hunger, thirst, exhaustion in USA in 1967	$9.7 \cdot 10^{-7}$
Natural disasters (earthquakes, floods, etc.) in USA in 1967	$8.2 \cdot 10^{-7}$
Offshore shipping (USA)	$8.0 \cdot 10^{-7}$
Airplane accident (USA)	$7.5 \cdot 10^{-7}$
Floods (USA)	$6.0 \cdot 10^{-7}$
Flooding in the USA (1967–1996)	$5.4 \cdot 10^{-7}$
Railway traffic (USA)	$5.1 \cdot 10^{-7}$
Failure of hospitals, schools (New South Wales, Australia)	$5.0 \cdot 10^{-7}$
Failure of hospitals, schools (West Australia)	$5.0 \cdot 10^{-7}$
Lightening (USA)	$5.0 \cdot 10^{-7}$
Lightening in USA in 1967	$4.4 \cdot 10^{-7}$
Hurricanes (USA 1967–1996)	$3.7 \cdot 10^{-7}$
Lightening (USA 1967–1996)	$3.2 \cdot 10^{-7}$
Bite and scratch by animals in USA in 1967	$2.2 \cdot 10^{-7}$
Structural failure	$1.0 \cdot 10^{-7}$
De minimis risk for workers	$1.0 \cdot 10^{-7}$
Accidents on the way to work using public transport	$1.0 \cdot 10^{-7}$
De minimis risk	$1.0 \cdot 10^{-7}$
Lightening (UK)	$1.0 \cdot 10^{-7}$
Structural failure	$1.0 \cdot 10^{-7}$
High and low atmospheric pressure in USA in 1967	$6.5 \cdot 10^{-8}$
Earthquake (1990–2000)	$5.1 \cdot 10^{-8}$
Hailstorm (USA 1990–2000)	$3.1 \cdot 10^{-8}$
Volcano eruption (USA 1990–2000)	$2.2 \cdot 10^{-8}$
Mass extinction in geology	$1.1 \cdot 10^{-8}$
De minimis risk for the public	$1.0 \cdot 10^{-8}$
Acceptable risk for cancer at the end of 1950s	$1.0 \cdot 10^{-8}$
Meteorite impact	$6.0 \cdot 10^{-1}$

Table 3-20. Actions that increase the chance of death by one in a million (Viscusi 1995, McBean & Rovers 1998, Covello 1991 – original source Wilson 1979)

Activity	Cause of death
Smoking 1.4 cigarettes	Cancer, heart disease
Drinking 0.5 liter of wine	Cirrhosis of liver
Spending 1 h in a coal mine	Black lung disease
Spending 3 h in a coal mine	Accident
Living 2 days in New York or Boston	Air pollution
Travelling 6 min by a canoe	Accident
Travelling 10 miles by a bicycle	Accident
Travelling 300 miles by a car	Accident
Flying 1,000 miles by airplane	Accident
Flying 6,000 miles by jet	Cancer from cosmic radiation
Living 2 months in Denver	Cancer from cosmic radiation
Living 2 months in a average masonry building	Cancer from natural radiation
One chest X-ray taken in a good hospital	Cancer from radiation
Living 2 months with a cigarette smoker	Cancer, heart disease
Eating 40 tablespoons of peanut butter	Cancer caused by aflatoxin B
Drinking Miami drinking water for 1 year	Cancer caused by chloroform
Drinking 30 × 12 oz cans of diet soda	Cancer caused by Saccharine
Living 5 years at side boundary of a typical NPP	Cancer from radiation
Drinking 1,000×1 oz soft drinks from plastic bottles	Cancer from acrylonitril monomer
Living 20 years near a polyvinyl chloride plant	Cancer from vinyl chloride
Eating 100 charcoal broiled steaks	Cancer caused by benzopyrene
Living 150 years within 20 miles of a NPP	Cancer from radiation
Living 50 years within 5 miles of a NPP	Cancer from radiation

3.5 Life Expectancy

If the mortalities can be understood as a risk measure, then the sum of all causes of mortality can also be considered a risk measure: this would be the life expectancy of humans. If average life expectancy increases, then according to the understanding of mortalities, the sum of risk decreases.

Indeed the average life expectancy in developed countries has increased dramatically over the last centuries. Figures 3-5 and 3-6 show this development. Additionally, for the year 2000, the diversity of this measure for about 170 countries is included in Fig. 3-5. There has been a strong discussion whether this permanent growth of life expectancy per year of about 2–3 months will continue or not (Weiland et al. 2006). Oeppen & Vaupel (2002) have suggested that this growth will continue the next decades.

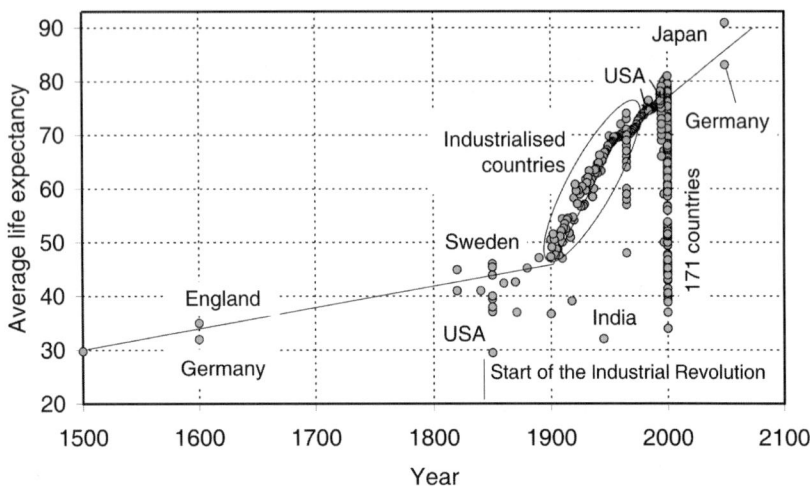

Fig. 3-5. Development of average life expectancy in years using data from Cohen (1991), NCHS (2001), Rackwitz & Streicher (2002), Skjong & Ronold (1998), Easterlin (2000), Becker et al. (2003) and IE (2004)

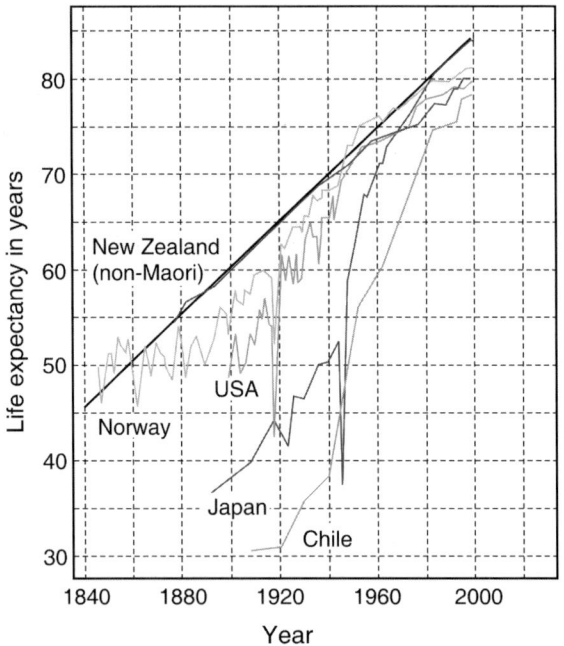

Fig. 3-6. Development of maximum life expectancy for select countries according to Oeppen & Vaupel (2002)

However, sooner or later, this growth will meet new boundaries. As mentioned in the beginning of the book, humans are time-limited beings. Therefore, the human body experiences ageing. Since the 1950s, ageing has been described by development and reaction of ROS (reactive oxygen species) and RNS (reactive nitrate species) in organisms. Further theories describe ageing by (Kleine-Gunk 2007):

- Theory of development of AGE (advanced glycosylation endproducts)
- Theory that aging is a development of lack of hormones
- Theory of ageing by Hayflicks division constant
- Chronical inflation processes

The question remains whether such processes are controllable with a certain amount of resources spent. Here, the change of death, causes might indicate such future barriers. Figure 3-7 shows the causes of death in Germany as example, of a developed country as well as the distribution for a developing country. Clearly, the distribution of causes has changed. This change is also visible in Fig. 3-8 taken from Adams (1995). A nice visualisation is given in Fig. 3-9 by Müller (2003).

However, the question then is, how reliable is the data. Figure 3-10 is a summary of different references about the human population of the western hemisphere in 1492, just as an example. Although one might consider modern times using better data, very often this problem changes to ontological problems as mentioned in chapter "Indetermination and risk".

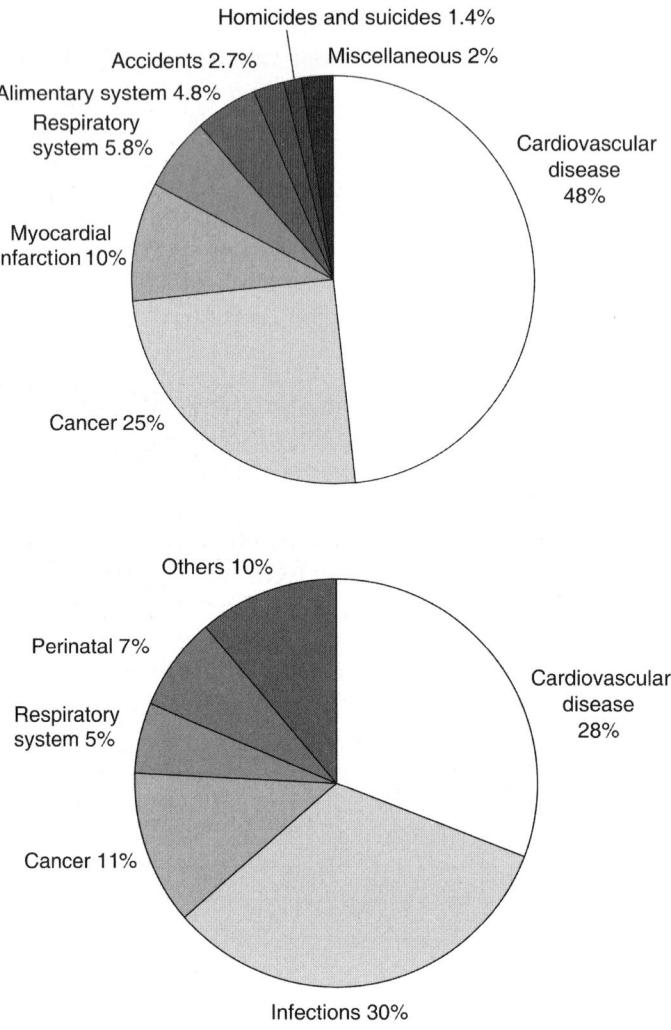

Fig. 3-7. Causes of death for a developed country (*top*) and causes of death for a developing countries (*bottom*) (GFSO 2007 and Bringmann et al. 2005)

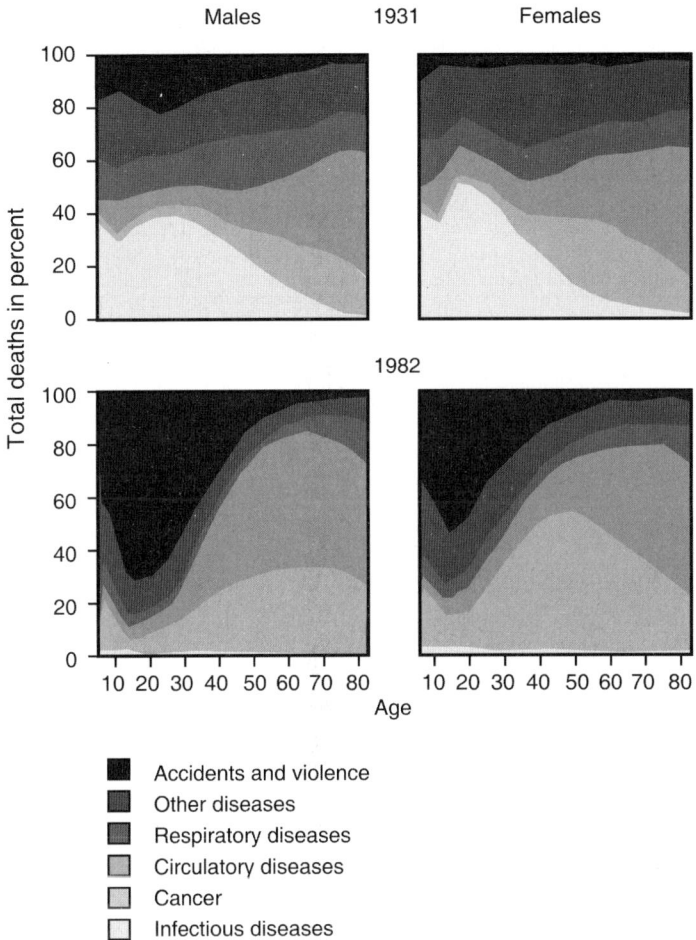

Fig. 3-8. Development of causes of death over time (1931 and 1982) according to Adams (1995)

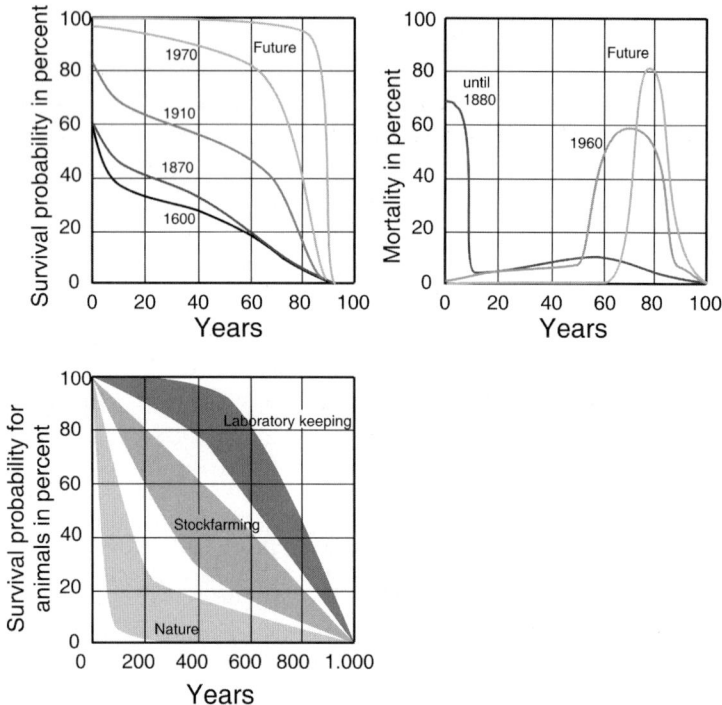

Fig. 3-9. Cumulative survival probabilities for different times, mortality curves for different times and survival curves for animals in nature, stockfarming and laboratory keeping (Müller 2003)

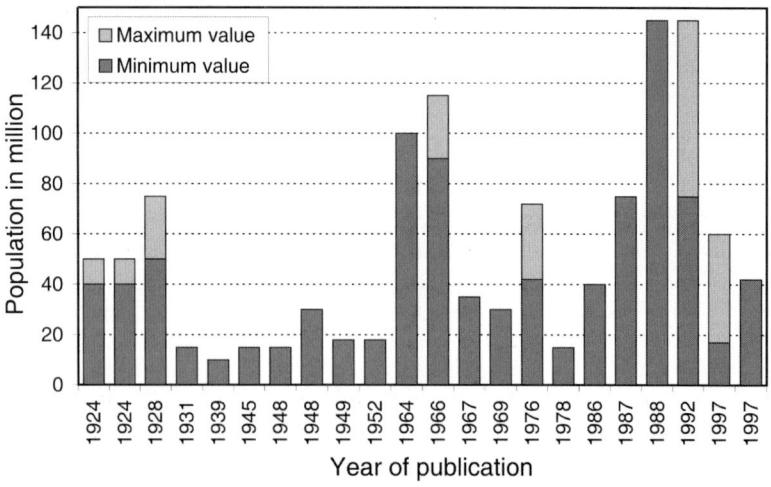

Fig. 3-10. Estimation of population of the western hemisphere in 1492 (White 2003)

3.6 Fatal Accident rates

Since mortality and life expectancy are risk parameters not related to the time of exposure, they might give biased results (Table 3-21). For example, a person could live for one year in a house and could use a motorbike for 2 h per year; both could cause the same mortality. To improve comparison, a standardised basic time of exposure has to be introduced. The parameter Fatal Accident Rate (FAR) is based on this idea. Here, the time of exposure is defined with 10^8 h or 11,415.5 years (Bea 1990). The long time period was chosen to give result numbers without the application of exponents. Sometimes, a slightly different parameter is used with a standard exposure time of 10^3 h (Jonkman et al. 2003). Additionally, there are also FARs related to distances. In general, the FAR can be given as follows:

$$FAR = \frac{N_{ot}}{N_{Gt}} \cdot \frac{t_N}{t_R} \leq \max FAR \qquad (3-8)$$

The application of this parameter will be shown for the risk of flying. First, it is assumed that the yearly mortality of a flight crew member is 1.2×10^{-3}. The yearly flight time is assumed with 1,760 h. Then, the mortality per flight hour is related to the flight hours:

$$\frac{1.2 \cdot 10^{-3}}{1,760} = 6.82 \cdot 10^{-7}/h \qquad (3-9)$$

Putting the standardised exposure time of 10^8 h into the relation, it yields to

$$6.82 \cdot 10^{-7} \cdot 10^8 = 68.2 \ \text{fatalities} \qquad (3-10)$$

Table 3-22 lists several risks in terms of this risk parameter. For flying, values in the range of 120–250 are found. Additionally, some other references give one fatality per 588,000 flight hours. Relating this time to the standardised exposure time of 10^8 h, the following value can be computed.

$$\frac{10^8}{588,000} = 170.1 \ \text{fatalities} \qquad (3-11)$$

This value fits quite well to the data given in Table 3-22. In general, the values in Table 3-22 reach from 0.0002 to 50,000. If one considers the highest value of 50,000, one can compute the average time between accidents. Dividing the 10^8 h by 50,000 yields 2,000 h. That is the average survival time for a person working as a jockey at the national British hunting race. For comparison, the average survival time for a person working in a household is about 5×10^7 h.

Table 3-21. Death rate of actions considering their exposure time (Melchers 1999)

Activity	Approximate death rate ($\times 10^{-9}$ death per exposure hour)	Typical exposure in hours per year	Typical risk of death ($\times 10^{-6}$ per year)
Alpine climbing	30,000–40,000	50	1,500–2,000
Boating	1,500	80	120
Swimming	3,500	50	170
Cigarette smoking	2,500	400	1,000
Air travel	1,200	20	24
Car travel	700	300	200
Construction work	70–200	2,200	150–440
Manufacturing	20	2,000	40
Building fires	1–3	8,000	8–24
Structural failures	0.02	6,000	0.1

Table 3-22. Fatal accident rates for different activities according to Hambly & Hambly (1994), Bea (1990), Camilleri (2001), Melchers (1999) and Haugen et al. (2005)

Activity	
Jockey at the national steeplechase	50,000
Plague in London in 1665	15,000
Professional boxer	7,000
Mountaineering and climbing	4,000
Rock climbing while on rock face	4,000
Alpine mountaineering	3,000–4,000
Alpine mountaineering (international)	2,700
Fireman in London air-raids 1940	1,000
Travel by canoe	1,000
Diving air in maritime oil production	685
Travel by motorcycle	660
Travel by helicopter	500
Swimming	350
Travel by scooter	310
Travel by motorcycle and moped	300
Driving moped	260
Travel by airplane (crew)	250
Smoking	250
Travel by airplane	240
Travel by boat	150
Travel by airplane (crew and passengers)	120
Diving saturation in oil production	97

Table 3-22. (Continued)

Activity	
Travel by bicycle	96
Travel by car	70
Police officer in Northern Ireland	70
Construction worker	67
Travel by car	60
Deep- sea fishing	59
Travel by car	57
Travel by car	56
Railway shunter	45
Smoking	40
Coal mining	40
Marine operations – anchor handling	37.4
Fishing industry	35
Lifting offshore and onshore	26.8
Coal mining	21
Walking beside a road	20
Offshore oil exploration	20
Marine operations – supply	18.1
Travel by air	15
Travel by car	15
Demolition – onshore	12.3
Prefabrication and construction onshore	10.4
Coal mines	8
Average man in the 30s from accidents	8
Average man in the 30s from disease	8
Construction industry	7.7
Construction industry	7–20
Marine activities – diving support	7.5
Lung cancer in Merseyside, England	7
Scaffolding	5.5
Travel by train	5
Travel by train	5
Construction industry	5
Average lung cancer (UK)	5
Deconstruction operations offshore	4.1
Metal manufacturing	4
Travel by train	4
Chemical industry	3.5
Staying home	3
Travel by bus	3
Homework	2.1
Influenza	2
Factory work	2
Travel by bus	1
Accident at home	1

(Continued)

Table 3-22. (Continued)

Activity	
Run over by car	1
Leukemia	0.8
Management and administrative activities	0.4
Off- duty time for workers in offshore ind.	0.2
Building fire	0.15
Fire	0.1–0.3
Radon gas natural radiation (UK average)	0.1
Terrorist bomb attack in London	0.1
Building fire in canton Bern	0.0620
Building fire in canton Zurich	0.0510
Contraceptive pill	0.0200
Malta earthquake MM-VII	0.0077
Building collapse	0.0020
Bite of poisonous animal	0.0020
Lightening	0.0010
Explosion of pressure tank for public	0.0006
Transport of dangerous goods for public	0.0005
Hit by airplane falling down	0.0002
Malta earthquake MM-VIII	0.0007
Contaminated landfill	0.0001

Of course, for this risk parameter, some goal values have been introduced and can be used for proof of safety. In the oil industry, an acceptable value of FAR of 15 per 10^8 has been introduced (Randsaeter 2000). Aven et al. (2005) mention a FAR of 10 for all persons on an installation based on the NORSOK Z-103 code. FAR values of 0.04–0.12 have been introduced in the field of fire risks in buildings for the public in Switzerland. In Norway, such values reach from 0.05 to 0.3 according to the SINTEF code (Maag 2004).

3.7 Family of *F-N* Diagrams

3.7.1 Introduction

The representation of risks only using mortality either based on the normal time or on exposure time does not permit the consideration of the cruelty of a single disaster. However, in chapter "Subjective risk judgement" it will be discussed that the dread of a single event might have considerable effects on the acceptance of risks. Therefore, the significance of the parameters mortality and FAR might be limited, and further improvements of the parameters are required. This can be easily visualised by pressing an apple through a wire, where the taste of apple remains but the shape is lost.

Mortality gives an average value over a certain time period but cannot describe some further properties of a single event.

This limitation has been described by the introduction of individual and social risks (Fig. 3-11). Mortalities give individual risks and do not consider any effects of a certain risk to the functioning of the society. However, people consider risks with a low frequency and a high number of fatalities as higher than risks with a high frequency and a low number of fatalities, even if both risks have the same mortality. However, other parameters might be able to consider such effects and are, therefore, appropriate for the expression of social risks. Here, one has to take care to distinguish social risks like poverty and social risks in terms of the overall number of people killed in a society by an accident or other event.

Fig. 3-11. Difference between individual risk (*top*) and social risk (*bottom*) (Jonkman et al. 2003)

One parameter that is able to consider the cruelty of a single accident or event is the *F-N* Diagrams. These diagrams use logarithmic scale *x*- and *y*-axes. On the *x*-axis, the consequence of an accident is shown, whereas on the *y*-axis, the frequency (*F*) or probability of such accidents or events is given. The consequence can be given as the number of fatalities (*N*) as well as in monetary units, energy or time. Therefore, sometimes, *F-N* diagram is also known as *F-D* diagram, where *D* stands for damage. Such damage parameters might be able to consider not only short-term damages like fatalities but also long-term effects.

The diagrams can either be cumulative or non-cumulative. The latter are called *f-N* diagrams. The development of such curvatures will be explained in detail. An example is taken from Ball & Floyd (2001). First of all, frequency and severity data about different types of accidents have to be known. A list of such possible data is given in Table 3-23. How such data can be computed is described in the following section.

It can be noted that the number of fatalities is not a natural number. This can happen during the computation of accidents, but is sometimes discussed. In the next step, the data are sorted based on the number of fatalities (Table 3-24). Table 3-24 can already be visualised as *f-N* diagram. However, widely used are the *F-N* diagrams. Here, the probability *F* of *N* or more fatalities is used. Therefore, the data have to be rearranged summing up the probabilities of different accidents with *N* or more fatalities. Also, Table 3-25 can be visualised (Fig. 3-12).

Table 3-23. Original accident data including frequencies and number of fatalities

Event no.	Number N of fatalities	Probability f of events per year
1	12.1	4.8×10^{-3}
2	123	6.2×10^{-6}
3	33.4	7.8×10^{-3}
4	33.2	9.1×10^{-4}
5	29.2	6.3×10^{-3}
6	15.6	7.0×10^{-4}
7	67.3	8.0×10^{-5}
8	9.5	4.0×10^{-3}
9	52.3	1.2×10^{-6}
10	2.7	3.4×10^{-4}

Table 3-24. Sorted accident data including frequencies and number of fatalities

Number N of fatalities	Probability f of events per year	Event no.
2.7	3.4×10^{-4}	10
9.5	4.0×10^{-3}	8
12.1	4.8×10^{-3}	1
15.6	7.0×10^{-4}	6
29.2	6.3×10^{-3}	5
33.2	9.1×10^{-4}	4
33.4	7.8×10^{-3}	3
52.3	1.2×10^{-6}	9
67.3	8.0×10^{-5}	7
123	6.2×10^{-6}	2

Table 3-25. Rearranged accident data

Number N of fatalities	Probability N of events per year	Event no.
1 or more	2.49×10^{-2}	1–10
3 or more	2.46×10^{-2}	1–0
10 or more	2.06×10^{-2}	1–7.9
30 or more	8.80×10^{-3}	2–4.7.9
100 or more	6.20×10^{-6}	2
300 or more	–	–

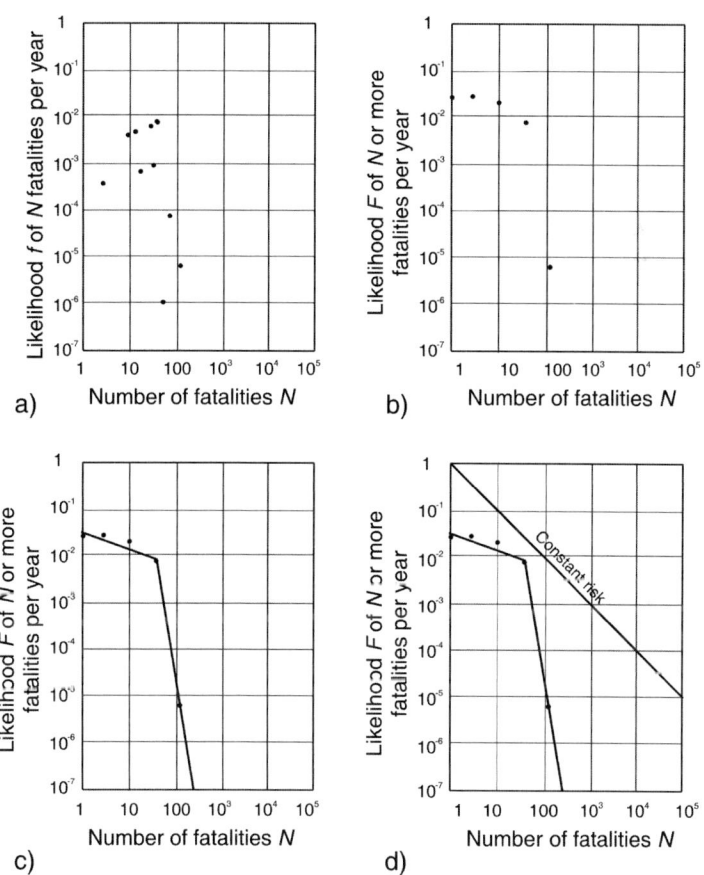

Fig. 3-12. Visualisation of the data from Table 3-25

The axes in the diagrams shown in Fig. 3-12 are with a logarithmic scale. The time scale for the frequency is usually years. It is easier to understand a return period in years than in hours. An example would be a return period of 100 years or 854,400 h. Figure 3-12 (a) is simply a visualisation of the accident data. In Fig. 3-12 (b), the data are already cumulated as it becomes clear in the axis lable. In Fig. 3-12 (c), the points are connected by straight lines to give a function between the number of fatalities and the frequency of accidents. There are also other techniques where the points are not directly connected but the lines are rather kept constant. Then, a staircase- like function arises. The centre of gravity of the area under the function represents the average number of fatalities (Jonkman et al. 2003). A line with a constant slope of 45° represents a constant risk, since an accident with a return period of 10 years multiplied with 100 fatalities gives the same result as an accident with a return period of 1 year with 1,000 fatalities. Finally, the function can be compared to proof functions. It should be mentioned that such diagrams are only valid for regions or time periods (Fig. 3-13). If the data change, the diagrams have to be changed as well.

The types of risks can be classified into four groups using such diagrams (Fig. 3-14). Risks of type K1 are well described using statistical procedures. There are rather high frequencies, but the severity is rather low. On the other hand, accidents with high number of fatalities are extremely low. Such risks show a strongly decreasing function in the *F-N* diagrams. Some technical risk such as road traffic belong to this type. Risks of type K2 show only a low dependency between the frequency and the number of fatalities. These risks show a low falling slope or nearly horizontal function. Here, natural risks such as earthquakes or flooding might fit. Risks of type K3 are only theoretically known. They are below the horizon of experience, and no statistical data are available. Risks of type K4 have a damage size higher than the size of the current population. There are no cases known yet. (van Breugel 2001).

In the last decades, the shape of the functions changed. With the increase of the world population and the increased application of mitigation measures, the functions started to rotate towards more cruel accidents.

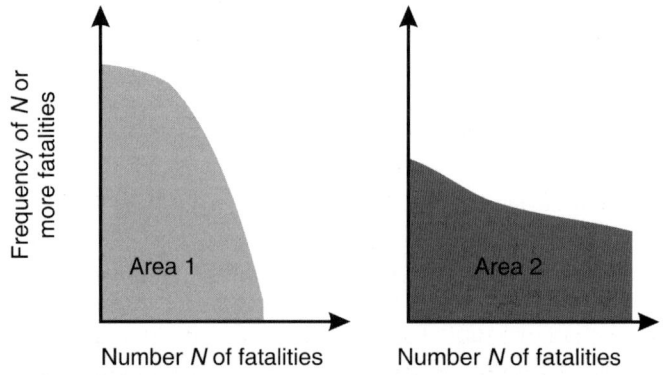

Fig. 3-13. Change of functions in *F-N* diagrams over time

F-N diagrams can be found in countless publications, including Rackwitz (1998), Larsen (1993), Hansen (1999), FHA (1990) and BZS (1995, 2003). Sometimes, they are used only for special problems, but even general applications are known. They are an excellent tool for the comparison of different technical risks and natural risks.

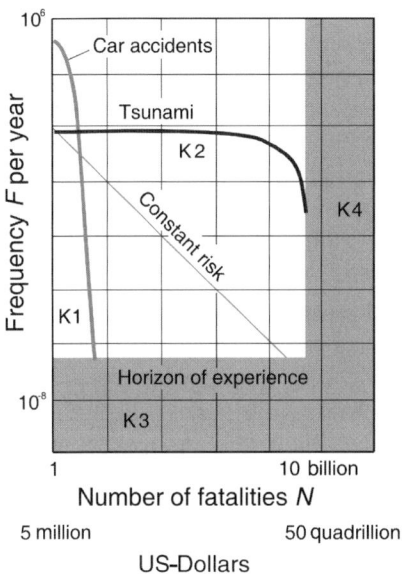

Fig. 3-14. Classification of risks according to van Breugel (2001)

3.7.2 Examples

The earliest known application of *F-N* diagrams was in the field of nuclear power plants. Originally nuclear power equipment was developed for military purposes. The first civil nuclear power plants were extensions of nuclear power engines from submarines and had only a small performance with less than 100 mW. They were introduced in the 1950s and 1960s. The containment was also an idea from the military submarines, and later extended to the protection of the public. In general, safety requirements were fulfilled by redundant systems. The design was mainly based on major accidents, such as earthquakes. The damage of the core was at that time not part of the design process. Although some risk assessment techniques were known in the 1950s, they were applied only qualitatively. However, the British scientist Farmer gained a breakthrough in risk assessment by introducing the first types of *F-N* diagrams. This tool, which was developed at the end of the 1960s, permitted a more objective judgement of the risk of nuclear power plants. This yielded to the WASH 1400 study (USAEC 1975) or the so-called "Rasmussen report" from the Massachusetts Institute of Technology charged by the US Atom Energy Commission. This study took about 3 years and attempted to quantify the safety of nuclear power plants in the US by modelling two plants, Surry and Peach Bottom. The safety assessment yielded to a comparison of nuclear power plants with other technical or natural risks. Even today, the study is widely quoted (Garrick 2000).

At the time of its publication in 1975, this study was seen as ambivalent. There was also a review of this report published in 1977 and known as the "Lewis report" (State University of California). In general, the Lewis report supported the application of risk parameters like the *F-N* diagrams but criticized the hidden uncertainty in computation. As a consequence, the "Rasmussen report" was withdrawn. This general attitude only changed after the Three Mile Island accident on 28th March 1979. Although the "Rasmussen report" had not exactly predicted this accident, however, a comparable scenario was included in the report. Therefore, probabilistic safety investigations were increasingly used. From 1979 to 1983, several nuclear power plants in the US were investigated using such techniques. Parallel to the development in the US, probabilistic safety assessments and risk assessments were also increasingly applied in other countries, such as Great Britain, the Netherlands, Hong Kong and Germany after the work by Farmer (1967). The development of goal values for risk assessment in the three countries, Great Britain, the Netherlands and Hong Kong, is shown in Fig. 3-15. It is not surprising that countries with a high population density were strongly involved in this process. In the middle of the 1980s, risk assessment using *F-N* diagrams was a common procedure worldwide for nuclear power plants (Garrick 2000). At the end of the 1980s, the NASA had used modern risk assessment tools for the Space Shuttle (Garrick et al. 1987). Since then, the variations of *F-N* diagrams can be found in many different fields. It appears useful to prove this with some examples. Figure 3-16 shows some examples from the WASH-1400 study. As mentioned earlier, the critical points of this study were the lack of information about uncertainty in these types of diagrams. Therefore, the diagrams in Fig. 3-17 show the confidence lines. It becomes clear that, in some regions, no statements can be given.

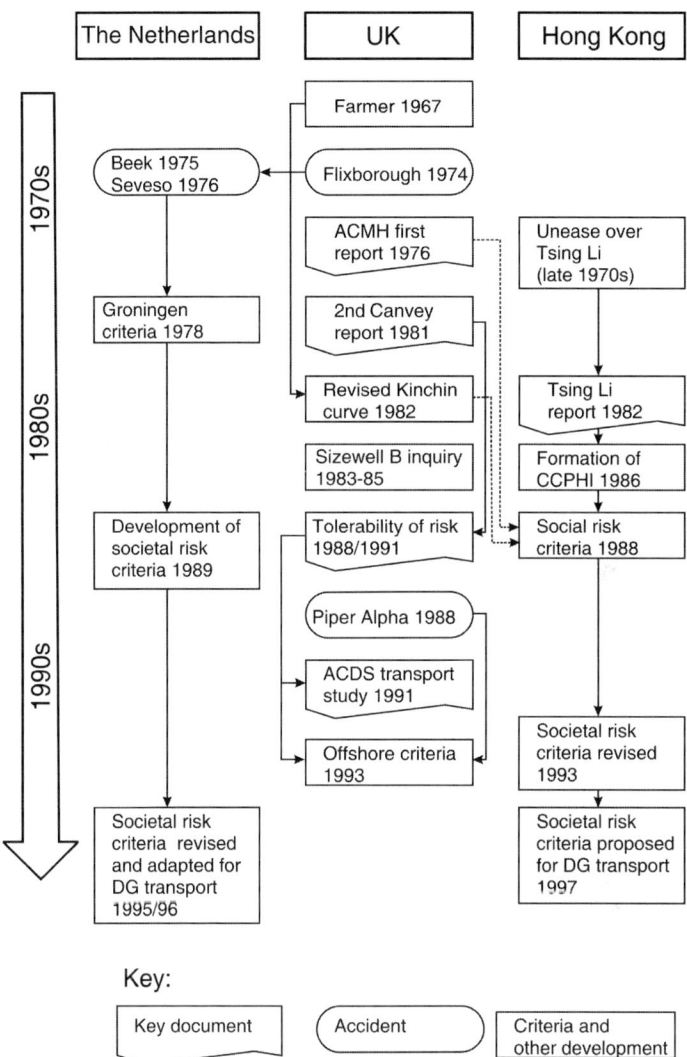

Fig. 3-15. Development of social risk criteria in three different countries according to Ball & Floyd (2001)

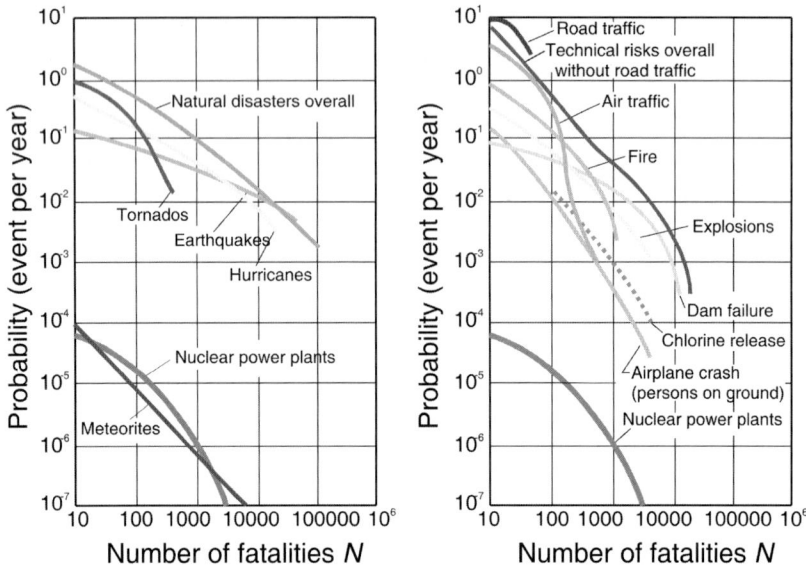

Fig. 3-16. *F-N* diagrams from the WASH 1400 report

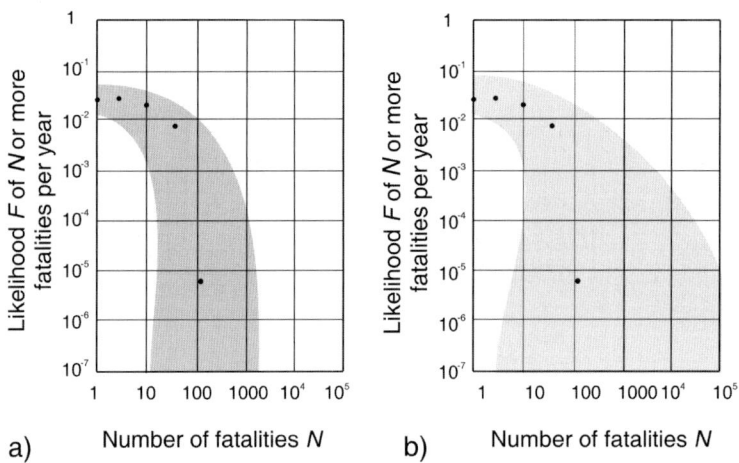

Fig. 3-17. *F-N* diagram including confidence intervals according to Ball & Floyd (2001)

Figure 3-18 shows examples from different industries. The term industry here includes different industries, for example the chemical industry and the shipping industry. Figure 3-18 permits the comparison between the chemical industry (Canvey Island) and a nuclear power plant. For the nuclear power plant, there is a distinction between early and delayed death. Figure 3-19 shows a diagram for the design and layout of a road in Alpine conditions. The road might be either in a tunnel improving the safety against some natural hazards like avalanches, but fire might be introduced as a new hazard, whereas the normal road might be exposed to natural hazards, but other hazards like fire might not be of considerable impact. Additionally, Fig. 3-19 shows a *F-N* diagram for an airport. The application of the shipping industry is shown in Fig. 3-20. Here, diagrams are given for different ship types. Finally, in Fig. 3-21, the storage of hazardous materials is considered. These examples are by no means complete. However, they should give a rough impression about the great diversity of the applications of *F-N* diagrams.

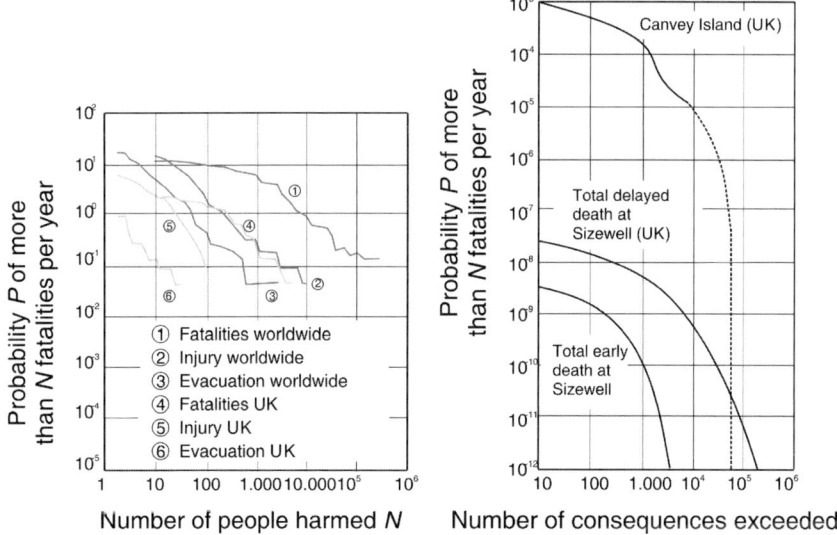

Fig. 3-18. *F-N* diagram for industry accidents (Kafka 1999)

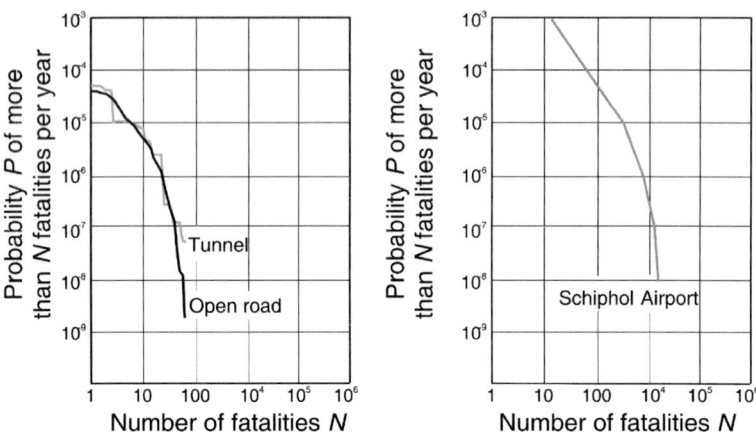

Fig. 3-19. *F-N* diagram for the comparison of a road design in an Alpine region (Kröger & Høj 2000) and a diagram for an airport (Vrijling et al. 2001)

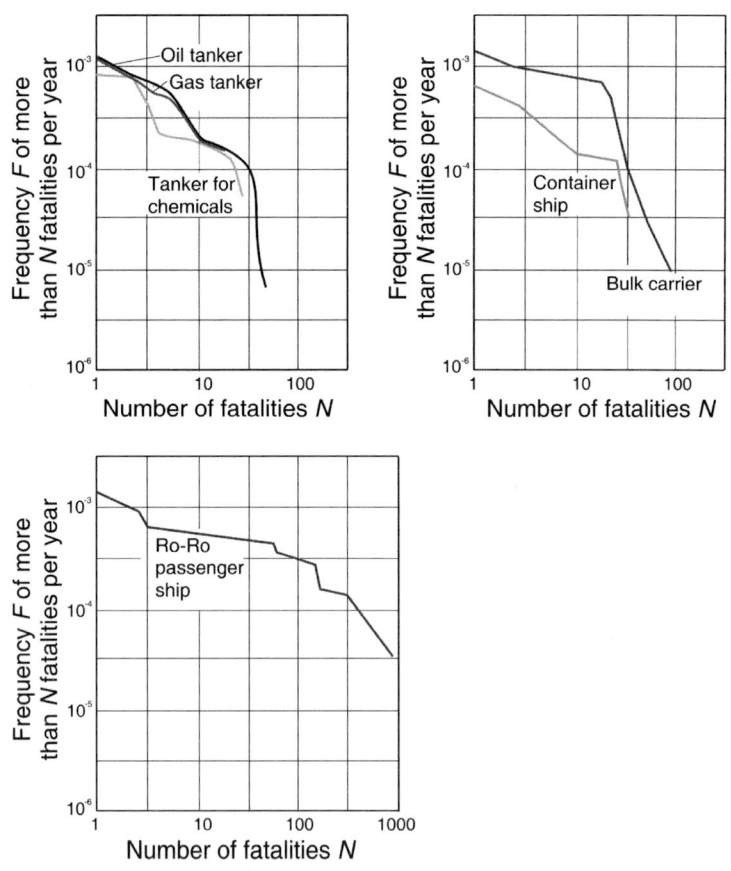

Fig. 3-20. *F-N* diagrams for different ship types (IMO 2000)

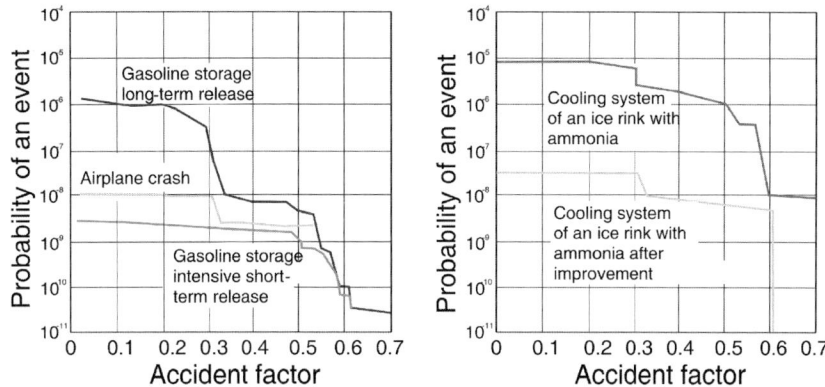

Fig. 3-21. *F-N* diagrams for the storage of hazardous materials (Gmünder et al. (2000)

3.7.3 Goal Curvatures for *F-N* Diagrams

So far, only diagrams including curvatures that describe a certain system or situation have been shown. However, no discussion about goal functions has taken place. Of course, if such risk parameters shall be successfully applied for risk assessment, one needs the criteria to decide whether a risk is acceptable or not. Indeed, in the last 30 years, several such proofs or goal curvatures have been developed for certain situations. Sometimes, the introduction of such curvatures intensified the discussion about the acceptance of fatalities. Indeed, the curvatures show acceptable values of fatalities, but on the other hand, since all actions include some risk, there are no activities without the possibility of negative consequences. To avoid living in the open air during a cold winter, one can build a house. However, a house can fail and can kill the owner. Therefore, two situations have to be compared and a better solution is identified. This can be done by risk parameters such as the *F-N* diagrams.

If such goal functions have to be introduced, one has to first understand the possible procedures to control the shape of such goal lines inside *F-N* diagrams. Usually, it is defined as

$$F \cdot N^a = k \qquad (3\text{-}12)$$

with *k* as constant and *a* as a factor to consider subjective risk judgement mainly described as risk aversion factor. Figure 3-22 shows the changes of curvature subject to changes of *k* and *a*. While the change of the factor *a* changes the slope of the line in the *F-N* diagram, the factor *k* shifts the line (Ball & Floyd 2001).

Usually the goal curvatures are not single lines but built up on several linear segments. Therefore, the described technique can be used to construct single linear segments. However, the construction of the lines can either be based on a single point and a given slope or by giving two points. Since 1976, the first procedure has been used: defining the frequency of a certain accident and defining the permitted change. For example, the ACMH (Advisory Committee for Major Hazards) had suggested that the probability of an accident with about 10 fatalities was less than 10^{-4} per year. This so- called anchor point can be used as start point for the construction of the line. In other cases, different values were chosen for the anchor point, for example in the Canvey Island investigation, an anchor point of 500 casualties and 2×10^{-4} frequency were suggested. Another investigation about the safety of nuclear power plants in the UK by the HSE (Health & Safety Executive) suggested 100 fatalities and 10^{-4} as probability. In the Netherlands an anchor point of 10 fatalities and 10^{-5} as probability were chosen (Ball & Floyd 2001).

After the choice of the anchor point, the direction has to be configured. Here, the risk aversion factor has to be declared. There have been many discussions about this value, and it is usually assumed in a range between 1 and 2 (Ball & Floyd 2001).

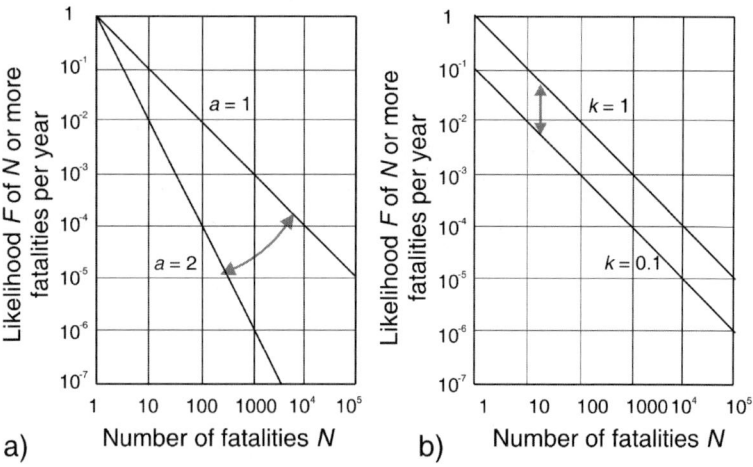

Fig. 3-22. Construction of lines in F-N diagrams based on slope (**a**) and shift of the lines (**b**)

After a complete line has been constructed, the area of the diagram can be divided into several areas. First, there will be one area under the line and one above it. Usually the area above the line is considered not acceptable. However, sometimes, several lines will exist in such diagrams forming several areas. Here, there might be areas with unacceptable risks, area with acceptable risks and areas with risks requiring further investigations. Such areas are sometimes called ALARP (As Low As Reasonable Practicable), meaning that the mitigation measures should be considered in relation to their efficiency. Figure 3-23 shows a typical example of such a diagram (Ball & Floyd 2001).

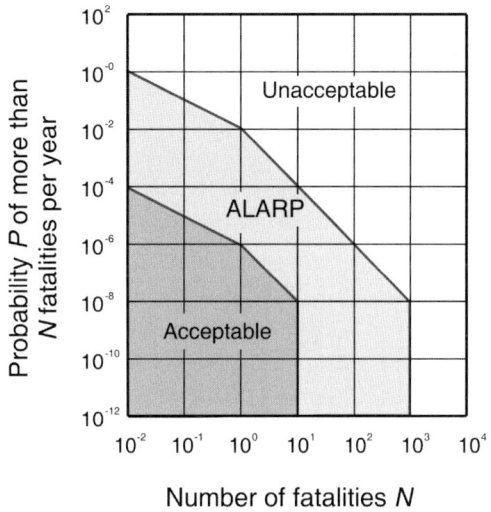

Number of fatalities N

Fig. 3-23. Example of proof lines for F-N diagrams

As already seen in the introduction about possible choices of an anchor point, the goal or proof lines might show some diversity. Indeed Figs. 3-24 and 3-25 show a number of such lines from different organisations and for different hazards. Figure. 3-24 shows (a) the Groningen curve (1978), (b) the Farmer curve (1967), (c) revised Kinchin curve (1982), (d) a curve from Hong Kong (1988), (e) a curve from the Netherlands (1980s), (f) a curve from ACDS from Great Britain (1991), (g) a curve for the safety of off-shore installations in the UK (1991) and (f) a curve from Hong Kong (1993). Figure 3-25 continues in (a) with again a curve from Hong Kong for the transport of hazardous material (1997), (b) a curve from Hong Kong for the transport of chlorine, (c) a curve from the Netherlands (1996), (d) again a curve from the Netherlands for the transport of dangerous goods (DG) (1996) and finally (e) a curve from Switzerland for the transportation of hazardous materials (1991–92). Since the diagram in Fig. 3-25 (e) uses an accident factor, the influences on this factor are shown in Fig. 3-26. A summary of all curves is shown in Fig. 3-27 to highlight the differences between goal curves (Ball & Floyd 2001).

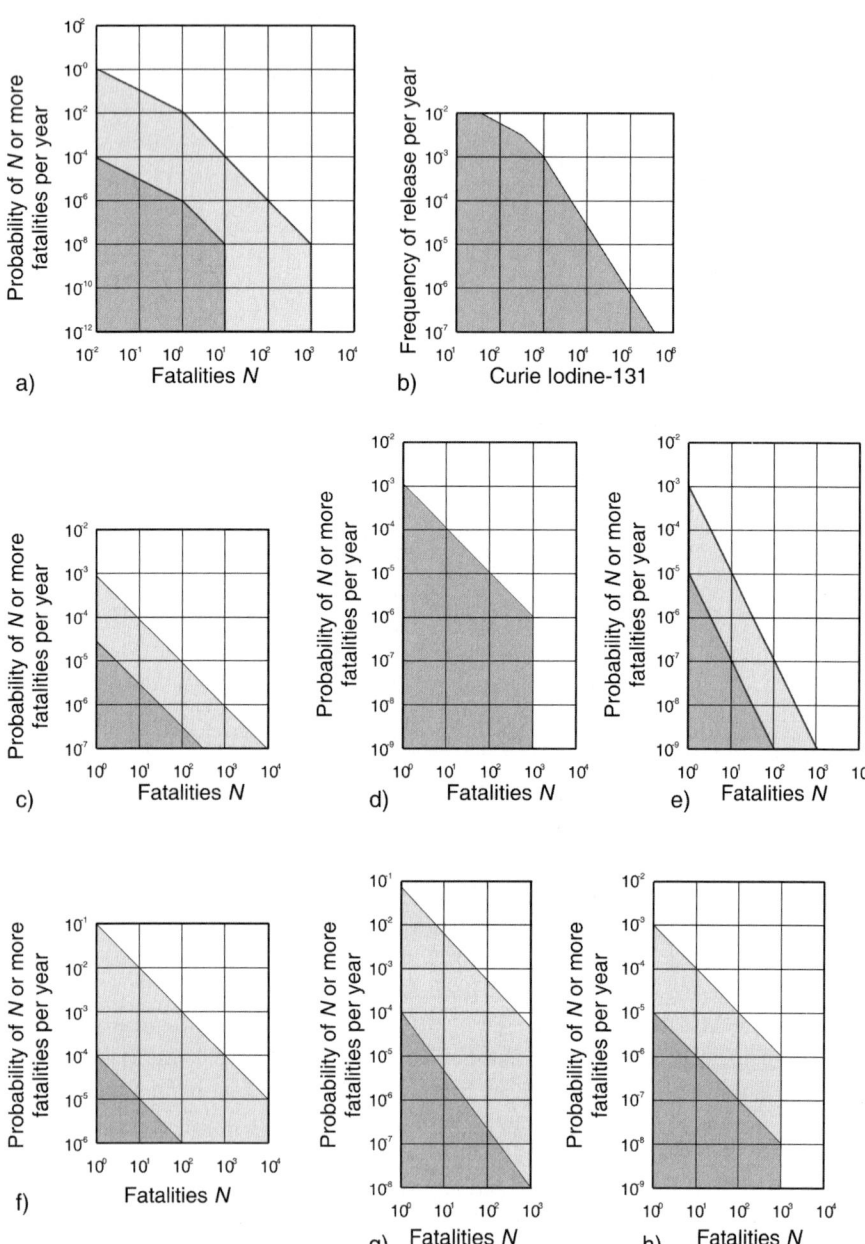

Fig. 3-24. Examples of proof lines for *F-N* diagrams (Ball & Floyd 2001)

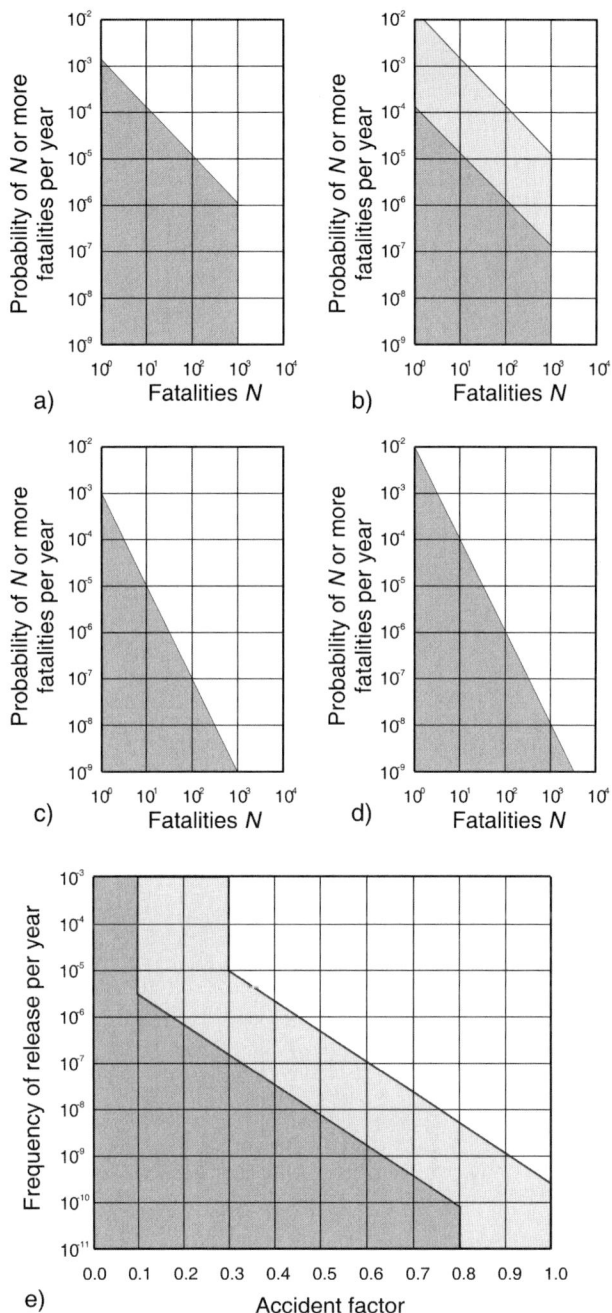

Fig. 3-25. Examples of proof lines for *F-N* diagrams (Ball & Floyd 2001)

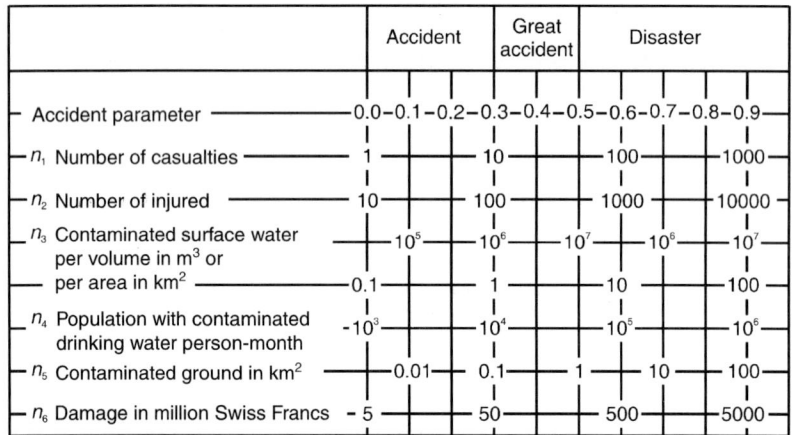

	Accident		Great accident	Disaster		
Accident parameter	0.0-0.1-0.2-0.3		-0.4-0.5	-0.6-0.7	-0.8-0.9	
n_1 Number of casualties	1		10	100	1000	
n_2 Number of injured	10		100	1000	10000	
n_3 Contaminated surface water per volume in m³ or per area in km²	10^5 0.1		10^6 1	10^7 10	10^6 100	10^7
n_4 Population with contaminated drinking water person-month	10^3		10^4	10^5	10^6	
n_5 Contaminated ground in km²	0.01		0.1	1 10	100	
n_6 Damage in million Swiss Francs	5		50	500	5000	

Fig. 3-26. Construction of the accident factor

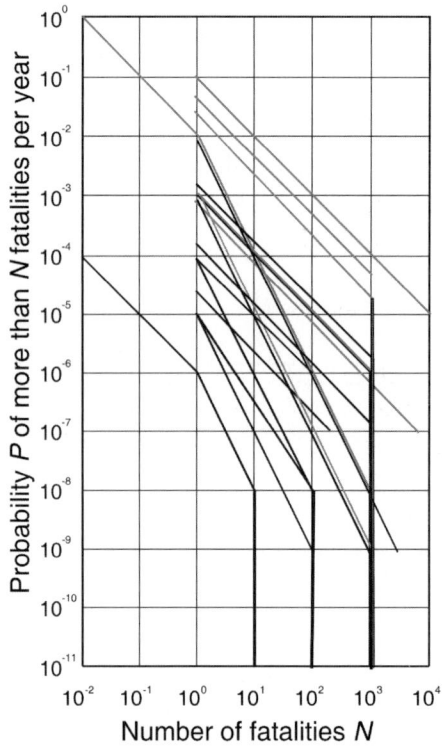

Fig. 3-27. Summary of all proof lines in one

A very easy application way by adapting *F-N* diagrams has been shown by Vrijling et al. (2001). This technique considers the following relationship as common for *F-N* diagrams:

$$P \leq \frac{C}{N^2} \tag{3-13}$$

This equation is transformed in

$$1 - f(N) < \frac{C_i}{N^2} \tag{3-14}$$

using

$$C_i = \left(\frac{\beta \cdot 100}{k \cdot \sqrt{N}} \right)^2 \tag{3-15}$$

Based on this approach, a simple formula can be developed:

$$E(N) + k \cdot \sigma(N) < \beta \cdot 100 \tag{3-16}$$

with
$E(N)$ — Average number of fatalities
$\sigma(N)$ — Standard deviation of fatalities
k — Confidence area, usually 3
β — Policy factor
N — Number of fatalities

This concept has been successfully applied for the estimation of the probability of dike failure in the Netherlands (Vrijling et al. 2001). About 40% of the area of the Netherlands is under sea level and is protected by dikes. This protected land is divided into 40 polders (N_A). For simplification reasons, it is assumed that about 1 million people live (N_P) in every polder. During the heavy flood in 1953, about 1% of the affected population died ($P_{d|i}$). Based on those numbers, one gets (Vrijling et al. 2001):

$$E(N) = N_A \cdot P_f \cdot P_{d|i} \cdot N_P = 40 \cdot P_f \cdot 0.01 \cdot 10^6 \tag{3-17}$$

$$\sigma(N) = N_A \cdot P_f \cdot (1 - P_f) \cdot (P_{d|i} \cdot N_P)^2 = 40 \cdot P_f \cdot (1 - P_f) \cdot (0.01 \cdot 10^6)^2 \tag{3-18}$$

$$E(N) + k \cdot \sigma(N) < \beta \cdot 100$$
$$= 40 \cdot P_f \cdot 0.01 \cdot 10^6 + 3 \cdot 40 \cdot P_f \cdot (1 - P_f) \cdot (0.01 \cdot 10^6)^2 < \beta \cdot 100 \tag{3-19}$$

This can be converted into

$$P_f + \frac{3 \cdot \sqrt{P_f - P_f^2}}{\sqrt{N_A}} = P_f + \frac{3 \cdot \sqrt{P_f - P_f^2}}{\sqrt{40}} = \frac{\beta \cdot 100}{N_A \cdot P_{d|i} \cdot N_P} = \frac{\beta \cdot 100}{40 \cdot 0.01 \cdot 10^6} \qquad (3\text{-}20)$$

The task to choose β remains. Based on Table 3-26 for different conditions in the Netherlands, the value can be chosen. In Table 3-26, lines 6 to 9 show computation examples, whereas lines 1 to 5 show empirical data. Figure 3-28 gives some indications for the choice of β.

Table 3-26. Data for different means of transport and the chosen policy value in the Netherlands (Vrijling et al. 2001)

Means of transport	N_A	P_{fi}	N_P	$E(N)$	$\sigma(N)$	$E(N)$ $+k \cdot \sigma(N)$	β
Airport Schiphol	$1.9 \cdot 10^6$	$5 \cdot 10^{-7}$	50	4.5	15.0	49.5	0.5
Airplane	$1.8 \cdot 10^6$	$5 \cdot 10^{-7}$	200	18.0	60.0	198.0	2.0
Car driving	$4 \cdot 10^6$	0.1		972.0	30.8	1.064.4	10.6
Hazardous material transport							0.1
Dam	40	0.01	10^6			1.937.0	19.4
Car driving (A.–A.[1])				7.1	2.7	15.0	0.15
Airplane (A.–A.[1])				0.3	4.1	13.0	0.13
Train (A.–A.[1])				0.05	0.4	1.3	0.013
High speed train				0.03	0.4	1.3	0.013

[1] A.–A.: Track Amsterdam–Antwerpen.

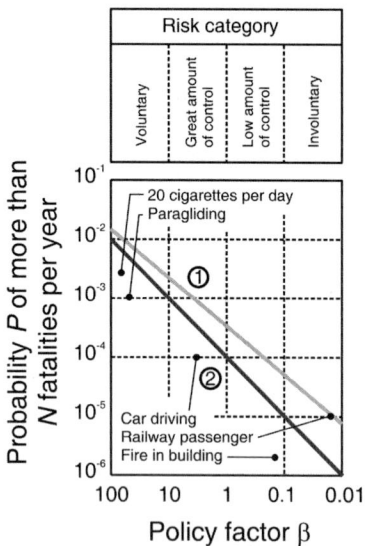

Fig. 3-28. Determination of the policy factor based on mortalities and degree of voluntariness (*line* ① according to Bohnenblust (1998) and *line* ② according to TAW (1988)). See also Jonkman et al. (2003), Merz et al. (1995) and BZS (2003)

It could be shown that goal mortalities and goal probabilities of failure of technical products can be considered as a special case of *F-N* diagrams. Therefore, they can be explained by the diagrams. On the other hand, while *F-N* diagrams might be used for some explanation of mortality values, they experience some disadvantages. *F-N* diagrams perform very well for the comparison of technical and natural risks, but are limited in their application for health or social risks.

To avoid problems with this risk indicator, they have been further advanced. One such extension is the usage of PAR values for the consequence axis. PAR (People At Risk) considers not only fatalities but also people affected in a different way. They might provide more information for some types of risks. Another modification might be the use of environmental damage indicators, for example time the environment needs to recover from an accident. Also, energy to recover might be used. Here, it has been attempted to transfer injuries and fatalities into lost energy. The equivalent energy of a human life has been expressed by 800 billion Joule (Jonkman et al. 2003). The family of the *F-N* diagrams with possible units for consequences of damages is shown in Fig. 3-29.

Besides the extension of the risk parameter *F-N* diagram, alternative risk parameters might be considered to overcome the disadvantages of the *F-N* diagrams.

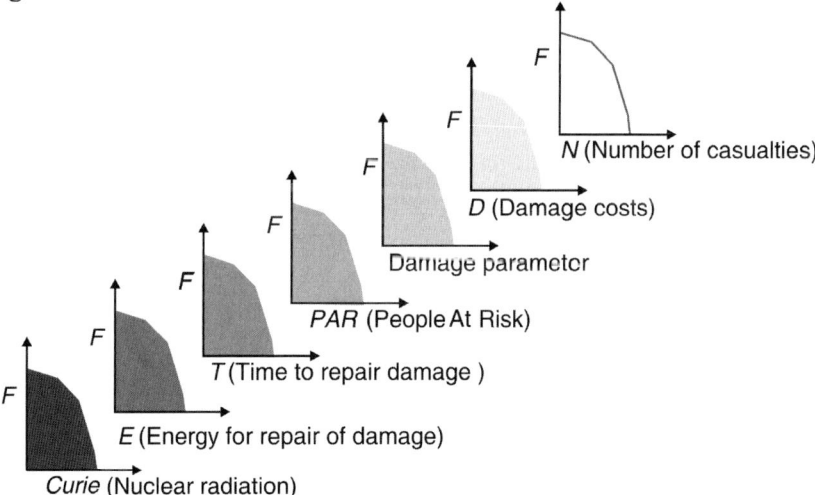

Fig. 3-29. Visualisation of the family of *F-N* diagrams with different damage units

3.8 Lost Life-Years

The risk parameters used so far do not consider any information about at which age a human dies. This might not be considered important at first impression, but that death at a certain age seems to be "natural". However, if young people die by a car accident, it can be considered a preventable risk. Therefore, the age information has to be considered for risk comparison. Since this is not possible with the risk parameter F-N diagrams, further improvements are required. Here, a parameter heavily used in the field of medicine might be practicable: the concept of lost life-years. The numerical formulation is rather simple:

$$LLY = e' - e \qquad (3\text{-}21)$$

where e′ is average life expectancy without a certain risk and e is average life expectancy with a certain risk. In the concept of lost life-years (LLY), other terms like years lost life (YLL) or lost life expectancy (LLE) are used as shown in Fig. 3-30.

Furthermore, the concept permits the introduction of morbidity into the risk parameter. This is quite often required since not all risks cause fatalities only in short term. Figure 3-31 shows the development of casualties over time for different types of risk.

Figure 3-30 shows that not only the difference of the life expectancy be numerically considered but also times of diseases and impairment. In this figure, the health profile of a human over lifetime is visualised (dark grey). Some accidents or diseases yield to a drop in the health profile, for example at age 15 or 24. The integral of the dark grey area gives the quality adjusted life-years (QALY), whereas the light grey area gives the disability adjusted life-years (DALY). These losses can be numerically added to the lost life years (Hofstetter & Hammitt 2001).

Some numerical examples of disability values are given in Table 3-27 for some diseases, and in Table 3-28 for some environmental stressors yielding to different diseases. Considering some 15 million life-years in the Netherlands, some 400,000 life-years are lost by environmental stressors like ozone, lead, noise and air pollution (Table 3-28). Such concepts have already been introduced in the 1970s by Kaplan et al. (1976).

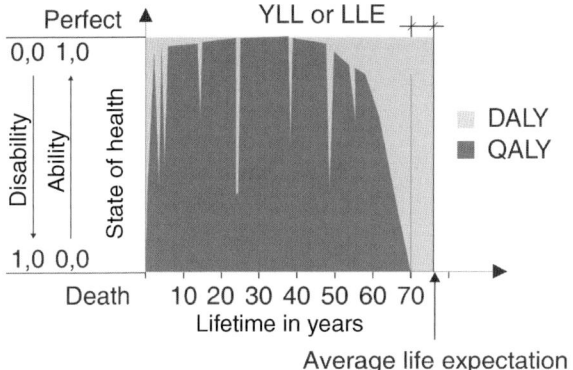

Fig. 3-30. Illustration of a human health profile over time and its numerical presentation by quality adjusted life-years (QALY) and disability adjusted life-years (DALY) adapted from Hofstetter & Hammitt (2001)

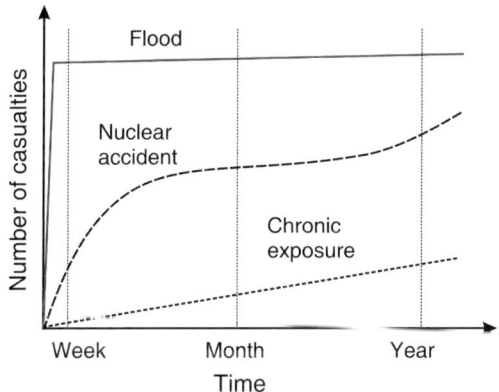

Fig. 3-31. Development of the number of casualties over time for different types of risk according to Jonkman (2007)

Table 3-27. Degree of disability according to Murray & Lopez (1997) and Perenboom et al. (2004)

Type of disability	Degree of disturbance
Vitiligo in face	0.020
Watery diarrhoea	0.021–0.120
Severe sore throat	0.021–0.120
Radius fracture in stiff cast	0.0121–0.240
Rheumatoid arthritis	0.0121–0.240
Below knee amputation	0.241–0.360
Deafness	0.241–0.360

(Continued)

Table 3-27. (Continued)

Type of disability	Degree of disturbance
Mental retardation	0.361–0.500
Unipolar major depression	0.501–0.700
Blindness	0.501–0.700
Paraplegia	0.501–0.700
Active psychosis	0.701–1.000
Dementia	0.701–1.000
Severe migraine	0.701–1.000
Quadriplegia	0.701–1.000
Major problems with performing daily-life activities	0.11–0.65

Table 3-28. Health consequences for five environmental risk factors evaluated by different health metrics for the Netherlands (adapted from Hofstetter & Hammitt, 2001)

Risk factors	Health effects	Incidence or prevalence cases per year	Duration	Disability weight	Δ QALY in years	QALY in %
Air pollution	Mortality total	7,114	10.9	1	77,543	19.28
	Mortality cardio-pulmonary	8,041	8.2	1	65,936	16.40
	Mortality lung cancer	439	13	1	5,707	1.42
	Chronic respiratory symptoms, children	10,138	1	0.17	1419	0.35
	Chronic bronchitis, adults	4,085	1	0.31	572	0.14
	Total				151,177	37.59
O$_3$	Mortality respiratory	198	0.25	0.7	50	0.01
	Mortality coronary heart disease	1,946	0.25	0.7	487	0.12
	Mortality pneumonia	751	0.25	0.7	188	0.05
	Mortality others	945	0.25	0.7	236	0.06
	Hospital admission, respiratory	4,490	0.038	0.64	75	0.02
	ERV, respiratory	30,840	0.033	0.51	519	0.13
	Total				1,554	0.39
Lead	Neuro-cognitive de-velopment (1-3 IQ-points)	1,764	70	0.06	7,409	1.84
Noise	Psychosocial effects: severe annoyance	1,767,000	1	0.01	159,030	39.54
	Psychosocial effects: sleep disturbance	1,030,000	1	0.01	82,400	20.49

Table 3-28. (Continued)

Risk factors	Health effects	Incidence or prevalence cases per year	Duration	Disability weight	Δ QALY in years	QALY in %
	Hospital admissions IHD	3,830	0.038	0.35	64	0.02
	Mortality IHD	40	0.25	0.7	10	0.00
	Total				241,504	60.05
Ozone deple-	Melanoma morbidity	24	6.9	0.1	50	0.01
tion	Melanoma mortality	7	23	1	161	0.04
	Basal	2,150	0.21	0.053	24	0.01
	Squamous	340	1.5	0.027	14	0.00
	Other mortality	13	20.2	1	263	0.07
	Total				511	0.13
Total					402,155	1
Mortality						37.44
Morbidity						62 56

On a global scale, about 1.4 billion life- years are lost by diseases, which represents about 259 life years per 1,000 human life-years. This is quite a high value (about 25%). Of course, geographically, the values differ. While in developed countries, the ratio is about 117 DALY lost for 1,000 inhabitant years, in China, 178 DALY are lost for 1,000 inhabitant years, in India it is about 344 and in some African countries it is 574. The DALY can be related not only to some geographic regions but also to certain diseases and other causes. For example, AIDS yields to 30 million lost life-years, yielding to about 2.2% of the global burden of disease, and TBC causes a loss of 46 million life-years representing 3.4% of the global burden of disease (Lopez et al. 2004).

However, besides diseases, other causes for the loss of life might be expressed using this risk measure. Table 3-29 gives some data and Fig. 3-32 visualises parts from this data.

Table 3-29. Lost life- years according to Cohen (1991), Cohen & Lee (1979), James (1996) and Covello (1991)

Cause	Lost life-years
Alcoholism	4,000
Poverty (France)	2,555–3,650
Being unmarried (male)	3,500
4% overweight	3,276
Cigarette smoking (male)	2,250
1 packet cigarettes per day	2,200

(Continued)

Table 3-29. (Continued)

Cause	Lost life-years
Heart disease	2,100
Cardiovascular disease	2,043
Low social relationships	1,642
Cardiovascular disease	1,607
Being unmarried (female)	1,600
Being 30% overweight	1,300
Cancer	1,247
Being a coal miner	1,100
Cancer	980
35% overweight	964
Being 30% overweight	900
< 8th grade education	850
Loss of parents during childhood	803
Cigarette smoking (female)	800
25% overweight	777
Low socioeconomic status	700
Stroke	520
Unemployment	500
Living in an unfavourable state	500
Pneumonia (Ghana)	474
Malaria (Ghana)	438
Army in Vietnam	400
Accidents (1988)	366
Accidents (1990)	365
Diarrhoea (Ghana)	365
Lung cancer	343
Cigar smoking	330
Work accident in agricultural industry	320
15 % overweight	303
Dangerous jobs (accidents)	300
Cancer of digestive organs	269
Cerebral vascular diseases	250
Work accident in construction industry	227
Pipe smoking	220
Increasing food intake 100 calories per day	210
Car accidents (1988)	207
Motor vehicle accidents	207
Car accidents (1990)	205
Tuberculosis (Ghana)	182
Work accident in mining industry	167
Chronic lung disease	164
Work accident in transport industry	160
Accidents other than car accidents (1990)	158
Pneumonia (influenza)	141
Alcohol (US average)	130

Table 3-29. (Continued)

Cause	Lost life-years
Suicide	115
Loss of one parent during childhood	115
Cancer of bladder	114
Cancer of genital area	113
Mountaineering (frequent)	110
Breast cancer	109
Pneumonia	103
Accidents at home	95
Suicide	95
Diabetes	95
Homicide	93
Cancer (Ghana)	91
Homicide	90
Legal drug misuse	90
Car accidents with collisions	87
Diabetes	82
Liver diseases	81
Accidents at home	74
Average job (accidents)	74
Car accidents without collisions	61
Work accident	60
Accidents in pubic	60
Jogging	50
Leukemia	46
Kidney infection	41
Drowning	41
Work accidents in production industry	40
Job with radiation exposure	40
Falls	39
Accidents to pedestrians	37
Car accident with pedestrian	36
Emphysema	32
Safest job (accidents)	30
Fall	28
Suffocation	28
Work accidents in service industry	27
Fire (burns)	27
Hang gliding	25
Parachuting	25
Arteriosclerosis	24
Drowning	24
Energy and petrol savings (small cars)	24
Generation of energy	24
Cancer in mouth	22
Poisoning with solid and fluid poisons	20

(Continued)

Table 3-29. (Continued)

Cause	Lost life-years
Fire	20
Illicit drugs (US average)	18
Fire in houses	17
Poison (solid, liquid)	17
Poisoning with solid and fluid poisons in houses	16
Car accidents against solid objects	14
Falling in flats	13
Suffocation	13
Ulcer	11.8
Asthma	11.3
Firearms accidents	11
Mountaineering for the entire population	10
Suffocation in flats	9.1
Sailing	9
Professional boxing	8
Natural radiation	8
Bronchitis	7.3
Diving (amateur)	7
Poisonous gases	7
Accidents with firearm	6.5
Accidents with machines	6.5
Killed from falling objects	6
Medical X-rays	6
Coffee	6
Car accident with pedestrian	5.7
Oral contraceptives	5
Accidents with bicycles	5
Tuberculosis	4.7
Inflammation of the gall bladder	4.7
Electric shock	4.5
Drowned in flats	4.2
Poisoning with gaseous poisons	4
Accidents with firearms in flats	3.8
Airplane	3.7
Food	3.5
All catastrophes combined	3.5
Hepatitis	3.3
Shipping	3.3
Poisoning with gaseous poisons in houses	2.6
Car accident with trains	2.5
Flu	2.3
Accidents with firearms in public	2.2
Snowmobile	2
Diet drinks	2
Reactor accidents (UCS)	2

Table 3-29. (Continued)

Cause	Lost life-years
Weather-related car accidents	1.8
Explosion	1.6
Railway traffic	1.3
Appendicitis	1.2
Extreme cold	1.0–2.1
Storm and flooding	0.9
Walking	0.9
Tornados	0.8
Lightening	0.7–1.1
Accidents with knives and razorblade	0.7
Extreme heat	0.6–0.7
American Football at university	0.6
Injuries from animals	0.6
Poisonous animals and plants	0.5
Race skiing	0.5
Drinking water in Florida	0.5
Flooding	0.4
Poisonous insects like bees, wasp and hornet	0.4
Poisonous animals and plants	0.4
Hurricane	0.3
American Football (high school)	0.3
Earthquake	0.2
Tsunami	0.15
Earthquake and volcano	0.13
Roasted meat	0.125
Bite from dogs	0.12
Poisonous animals like spiders and snakes	0.08
Nuclear power plant	0.05
Reactor accidents (NRC)	0.02
PAP test	–4
Air bags in car	–50
Safety improvements 1966–1976	–110
Mobile coronary care units	–125

If one looks carefully at the major causes of loss of lifetime, many of them are related to social risk, for example living in poverty, being unmarried, early school dropout or childhood without parents. Additionally, there is alcohol addiction, which might be simply a consequence of social imbalance as well. The first diseases in Table 3-29 might be connected to a virtual homogenisation of death causes due to the higher edges, but might also indicate social causes.

In general, summarising the results from lost life-year's data, one can conclude that, how people live defines their risk. In other terms, considering the quality of life is a must for risk evaluations. However, this "quality of life" term has a strong subjective component. Therefore, firstly, the next chapter will discuss subjective risk judgement, whereas the subsequent one will continue with the attempts to present quality of life in measures.

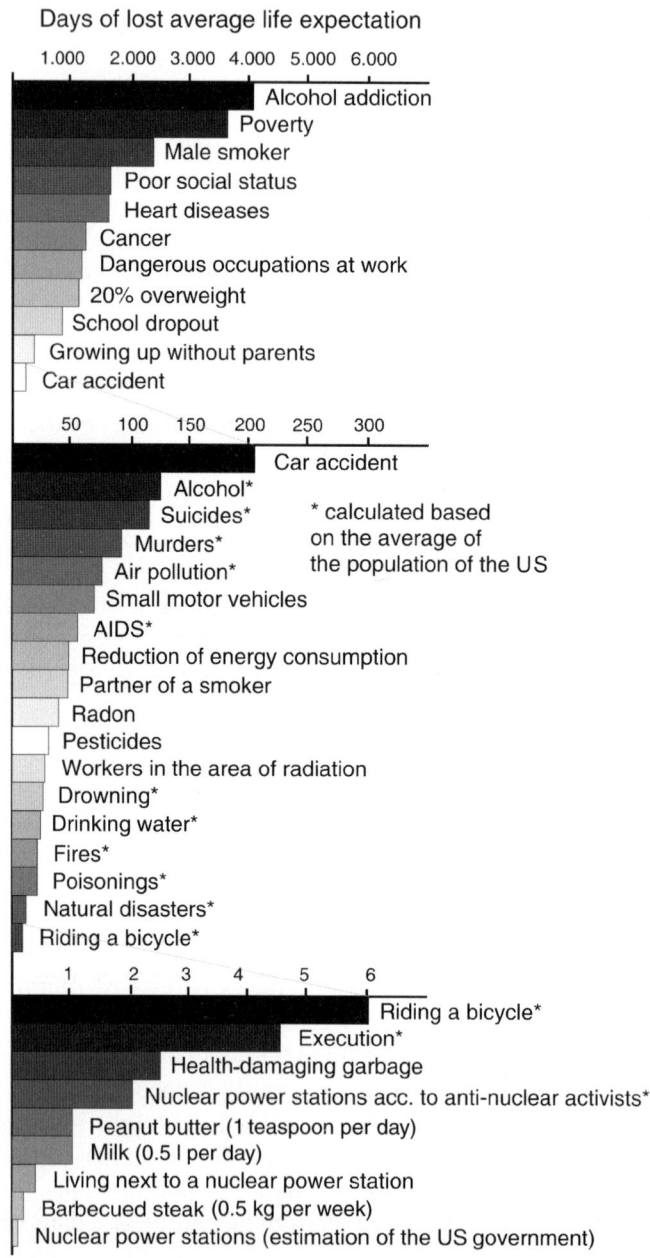

Fig. 3-32. Lost life-days according to Cohen (1991)

References

Adams J (1995) Risk. University College London Press

Arrow KJ, Cropper ML, Eads GC, Hahn RW, Lave LB, Noll RG, Portney PR, Russell M, Schmalensee R, Smith VK & and Stavins RN (1996) Is there a role for benefit-cost analysis in environmental, health and safety regulations. Science, 272, pp 221–222

Aven T, Vinnem JE & Vollen F (2005) Perspectives on risk acceptance criteria and management for installations – application to a development project. In: Kolowrocki (ed.) Advances in Safety and Reliability. Taylor & Francis Group, pp 107–114

Ball DJ & Floyd PJ (2001) Societal Risks. Final Report. School of Health, Biological/Environmental Sciences. Middlesex University. London

Bea RG (1990) Reliability criteria for New and Existing Platforms. Proceedings of the 22nd Offshore Technology Conference 7.-10. May 1990, Houston, Texas, pp 393–408

Becker GS, Philipson TJ & Soares RR (2003) The Quantity and Quality of Life and the Evolution of World Inequality. May 10, 2003

Bohnenblust H (1998) Risk-based decision making in the transportation sector; In: RE Jorissen, PJM Stallen, (eds.), Quantified societal risk and policy making. Kluwer academic publishers

Bratzke J, Parzeller M & Köster F (2004) Deutsches Forensisches Sektionsregister startet. Deutsches Ärzteblatt, Jahrgang 101, Heft 18, 30. April 2004, pp A1258–A1260

Bringmann G, Stich A & Holzgrabe U (2005) Infektionserreger bedrohen arme und reiche Länder - Sonderforschungsbereich 630: "Erkennung, Gewinnung und funktionale Analyse von Wirkstoffen gegen Infektionskrankheiten", BLICK, Forschungsschwerpunkt, pp 22–25

Burgmann M (2005) Risks and decisions for conservation and environmental management. Ecology, Biodiversity and Conservation. Cambridge University Press, Cambridge

BZS (1995) Bundesamt für Zivilschutz. KATANOS – Katastrophen und Notlagen in der Schweiz, eine vergleichende Übersicht, Bern

BZS (2003) Bundesamt für Zivilschutz. KATARISK – Katastrophen und Notlagen in der Schweiz. Bern

Camilleri D (2001) Malta's Risk Minimization to Earthquake, Volcanic & Tsunami Damage, Safety, Risk, Reliability – Trends in Engineering, Malta 2001

Cohen BL & Lee L (1979) Catalog of risk. Health Physics 36, pp 707–722

Cohen BL (1991) Catalog of Risks extended and updated. Health Physics, Vol. 61, September 1991, pp 317–335

Covello VT (1991) Risk comparisons and risk communications: Issues and problems in comparing health and environmental risks. In: RE Kasperson (ed.), Communicating Risk to the Public. Kluwer Academic Publishers, pp 79–124

Ditlevsen O (1996) Risk Acceptance Criteria and/or Decision Optimisation. Structural Engineering in consideration of Economy. 15 IABSE, Copenhagen

DUAP (1997) – Department of Urban Affairs and Planning: Risk criteria for land use safety planning. Hazardous Industry Planning Advisory, Paper No. 4, Sydney

Easterlin RA (2000) The Worldwide Standard of Living since 1800. Journal of Economic Perspectives. Vol. 14, No. 1, Winter 2000, pp 7–26

Ellingwood BR (1999) Probability-based structural design: Prospects for acceptable risk bases. Application of Statistics and Probability (ICASP 8), Sydney, Band 1, pp 11–18

Ellingwood BR (1999) Probability-based structural design: Prospects for acceptable risk bases. Application of Statistics and Probability (ICASP 8), Sydney, Vol. 1, pp 11–18

Farmer FR (1967) Siting criteria: a new approach. Nuclear Safety 8, pp 539–548

Femers S & Jungermann H (1991) Risikoindikatoren. Eine Systematisierung und Diskussion von Risikomassnahmen und Risikovergleichen. In: Forschungszentrum Jülich; Programmgruppe Mensch, Umwelt, Technik (Hrsg.): Arbeiten zur Risiko-Kommunikation, Heft 21. Jülich 1991.

FHA (1990) Department of Transportation, Federal Highway Administration: Guide Specification and Commentary for Vessel Collision Design of Highway Bridges, Volume I: Final Report, Publication Nr. FHWA-RD-91-006

Garrick BJ (2000) Invited Expert Presentation: Technical Area: Nuclear Power Plants. Proceedings – Part 2/2 of Promotion of Technical Harmonization on Risk-Based Decision-Making, Workshop, May 2000, Stresa, Italy

Garrick BJ et al. (1987) Space Shuttle Probabalistic Risk Assessment, Proof-of-Concept Study, Auxiliary Power Unit and Hydraulic Power Unit Analysis Report. Prepared for the National Aeronautics and Space Administration. Washington, D.C.

GFSO (2007) - German Federal Statistical Office. http://www.destatis.de

Gmünder FK, Schiess M & Meyer P (2000) Risk Based Decision Making in the Control of Major Chemical Hazards in Switzerland – Liquefied Petroleum, Ammonia and Chloride as Examples. Proceedings – Part 2/2 of Promotion of Technical Harmonization on Risk-Based Decision-Making, Workshop, May, 2000, Stresa, Italy

Greminger P et al. (2005) Bundesamt für Umwelt, Wald und Landschaft. RiskPlan, RiskInfo

GTZ (2004) – German Technical Co-operation. Risikoanalyse – eine Grundlage der Katastrophenvorsorge. Eschborn

Halperin K (1993) A Comparative Analysis of Six Methods for Calculating Travel Fatality Risk, Franklin Pierce Law Center,

Hambly EC & Hambly EA (1994) Risk evaluation and realism, Proc ICE Civil Engineering, Vol. 102, pp 64–71

Hungerbühler K, Ranke J & Mettier T (1999) Chemische Produkte und Prozesse – Grundkonzept zum umweltorientierten Design. Springer Verlag Berlin Heidelberg

Hansen W (1999) Kernreaktorpraktikum. Vorlesungsmitschriften. Institut für Energietechnik, Technische Universität Dresden

Haugen S, Myrheim H, Bayly DR & Vinneman JE (2005) Occupational risk in decommisining/removal projects. In: Kolowrocki (ed.), Advances in Safety and Reliability. Taylor & Francis Group, pp 807–814

Hofstetter P & Hammitt JK (2001) Human Health Metrics for Environmental Decision Support Tools: Lessons from Health Economics and Decision Analysis. National Risk Management Research Laboratory, Office of Research and Development, US EPA, Cincinnati, Ohio, September 2001

IE (2004) http://www.internationaleconomics.net/research-development.html

IMO (2000) – International Maritime Organisation: Formal Safety Assessment: Decision Parameters including Risk Acceptance Criteria, Maritime Safety Committee, 72nd Session, Agenda Item 16, MSC72/16, Submitted by Norway, 14. February 2000

James ML (1996) Acceptable Transport Safety. Research Paper 30, Department of the Parliamentary Library, http://www.aph.gov.au/library/pubs/rp/1995-96/96rp30.html

Jonkman SN (2007) Loss of life estimation in flood risk assessment – Theory and applications. PhD thesis, Rijkswaterstaat – Delft Cluster, Delft

Jonkman SN, van Gelder PHAJM & Vrijling JK (2003) An overview of quantitative risk measures for loss of life and economic damage. Journal of Hazardous Materials A 99, pp 1–30

Kafka P (1999) How safe is safe enough? – An unresolved issue for all technologies. In: Schuëller GI & Kafka P (eds.), Safety and Reliability. Balkema, Rotterdam, pp 385–390

Kaplan RM, Bush JW & Berry CC (1976) Health Status: Types of Validity and the Index of Well-being. Health Services Research 11, 4, pp 478–507

Kelly KE (1991) The myth of 10-6 as a definition of acceptable risk. In Proceedings of the 84th Annual Meeting of the Air & Waste Management Association, Vancouver, B.C., Canada, June 1991

Kleine-Gunk B (2007) Anti-Aging-Medizin – Hoffnung oder Humbug? Deutsches Ärzteblatt, Jg. 104, Heft 28-29, 16th Juli 2007, B1813-B1817

Kröger W & Høj NP (2000) Risk Analyses of Transportation on Road and Railway. Proceedings – Part 2/2 of Promotion of Technical Harmonization on Risk-Based Decision-Making, Workshop, May 2000, Stresa, Italy

Larsen OD (1993) Ship Collision with Bridges, The Interaction between Vessel Traffic and Bridge Structures. IABSE (International Association for Bridge and Structural Engineering), Zürich

Lopez AD, Mathers CD, Ezzati M, Jamison DT & Murray CL (2004) Global burden of disease and risk factors. World Health Organization

LUW (2005) – Landesumweltamt Nordrhein-Westfalen. Beurteilungsmaßstäbe für krebserzeugende Verbindungen. http://www.lua.nrw.de/luft/immissionen

Maag T (2004) Risikobasierte Beurteilung der Personensicherheit von Wohnbauten im Brandfall unter Verwendung von Bayes'schen Netzen. Insitut für Baustatik und Konstruktion, ETH Zürich, vdf Hochschulverlag AG an der ETH Zürich, IBK Bericht 282, März 2004 Zürich

Mathiesen TC (1997) Cost Benefit Analysis of Existing Bulk Carriers. DNV Paper Series No. 97-P 008

McBean EA & Rovers FA (1998) Statistical Procedures for Analysis of Environmental Monitoring Data & Risk Assessment. Prentice Hall PTR Environmental Management & Engineering Series, Vol. 3, Prentice Hall, Inc., Upper Saddle River

Melchers RE (1999) Structural Reliability Analysis and Prediction, John Wiley

Merz HA, Schneider T & Bohnenblust H (1995) Bewertung von technischen Risiken – Beiträge zur Strukturierung und zum Stand der Kenntnisse, Modelle zur Bewertung von Todesfallrisiken. Polyprojekt Risiko und Sicherheit: Band 3, vdf Hochschulverlag AG an der ETH Zürich

Müller U (2003) Europäische Wirtschafts- und Sozialgeschichte II: Das lange 19. Jahrhundert. Lehrstuhl für Wirtschafts- und Sozialgeschichte der Neuzeit. Vorlesung. Europäische Universität Viadrina, Frankfurt (Oder)

Murray C & Lopez A (1997) Regional patterns of disability-free life expectancy and disability-adjusted life expectancy: Global burden of disease study. The Lancet 349, pp 1347–1352

NASA (1989) Independent Assessment of Shuttle Accident Scenario Probabilities for the Galileo Mission. Vol. 1, Washington DC

NCHS (2001) – National Centre for Health Statistics: National Vital Statistics Report, Vol. 48, No. 18, 7. February 2001

Oeppen J & Vaupel JW (2002) Demography. Broken limits to life expectancy. Science 2002, 296, pp 1029–31

Overmans R (1999) Deutsche militärische Verluste im Zweiten Weltkrieg. Beiträge zur Militärgeschichte Band 46. R. Oldenbourg Verlag, München

Parfit M (1998) Living with Natural Hazards. National Geographic, Vol. 194, No. 1, July 1998, National Geographic Society, pp 2–39

Paté-Cornell ME(1994) Quantitative safety goals for risk management of industrials facilities. Structural Safety, 13, pp 145–157

Perenboom R, van Herten L, Boshuizen H & van den Bos G (2004) Trends in disability-free life expectancy. Disability and Rehabilitation 26(7), pp 377–386

Proske D (2004) Katalog der Risiken. Dirk Proske Verlag, Dresden

Rackwitz R & Streicher H (2002) Optimization and Target Reliabilities. JCSS Workshop on Reliability Bades Code Calibration. Zürich, Swiss Federal Institute of Technology, ETH Zürich, Switzerland, March 21–22

Rackwitz R (1998) Zuverlässigkeit und Lasten im konstruktiven Ingenieurbau. Vorlesungsskript. Technische Universität München

Randsaeter A (2000) Risk Assessment in the Offshore Industry. Proceedings – Part 2/2 of Promotion of Technical Harmonization on Risk-Based Decision-Making, Workshop, May 2000, Stresa, Italy

Schelhase T & Weber S (2007) Die Todesursachenstatistik in Deutschland. Bundesgesundheitsblatt-Gesundheitsforschung-Gesundheitsschutz, 7, 25. Juni 2007, DOI 10.1007/s00103-007-0287-6, pp 969–976

Schmid W (2005) Risk Management Down Under. Risknews 03/05, pp 25–28

Schütz H, Wiedemann P, Hennings W, Mertens J & Clauberg M (2003) Vergleichende Risikobewertung: Konzepte, Probleme und Anwendungsmöglichkeiten. Abschlussbericht zum BfS-Projekt StSch 4217.

Forschungszentrum Jülich GmbH, Programmgruppe „Mensch, Umwelt, Technik"

Shortreed J, Hicks J & Craig L (2003) Basic framework for Risk Management – Final report. March 28, 2003, Network for environmental risk assessment and management. Prepared for the Ontario Ministry of the Environment

Skjong R & Ronold K (1998) Societal Indicators and Risk acceptance. 17th International Conference on Offshore Mechanics and Arctic Engineering, 1998 by ASME, OMAE98-1488

Slovic P (1999) Trust, Emotion, Sex, Politics, and Science: Surveying the Risk-Assessment Battlefield, Risk Analysis, Vol. 19, No. 4, pp 689–701

Spaethe G (1992) Die Sicherheit tragender Baukonstruktionen, 2. Neubearbeitete Auflage, Wien, Springer Verlag

TAW (1988) Technische Adviescommissie voor de Waterkeringen; Some considerations of an acceptable level of risk in the Netherlands

USAEC (1975) United States Atomic Energy Commission. Reactor Safety Study, WASH-1400

van Breugel K (2001) Establishing Performance Criteria for Concrete Protective Structures fib-Symposium: Concrete & Environment, Berlin 3-5. Oktober 2001

Vennemann MMT, Berger K, Richter D & Baune BT (2006) Unterschätzte Suizidraten durch unterschiedliche Erfassung in Gesundheitsämter. Deutsches Ärzteblatt, Jg. 103, Heft 18, pp A 1222–A 1226

Viscusi W (1995) Risk, Regulation and Responsibility: Principle for Australian Risk Policy. Risk. Regulation and Responsibility Promoting reason in workplace and product safety regulation. Proceedings of a conference held by the Institute of Public Affairs and the Centre for Applied Economics, Sydney, 13 July 1995. http://www.ipa.org.au/ Conferences/viscusi.html

Vrijling JK, van Gelder PHAJM, Goossens LHJ, Voortman HG, Pandey MD (2001) A Framework for Risk criteria for critical Infrastructures: Fundamentals and Case Studies in the Netherlands, Proceedings of the 5th Conference on Technology, Policy and Innovation, "Critical Infrastructures", Delft, The Netherlands, June 26–29, 2001, Uitgeverrij Lemma BV

Weiland SK, Rapp K, Klenk J & Keil U (2006) Zunahme der Lebenserwartung – Größenordnungen, Determinanten und Perspektiven. Deutsches Ärzteblatt, Jg. 103, Heft 16, 21. April 2006, pp A 1072–A 1077

White M (2003) Twenthieth Centuray Atlas – Worldwide Statistcs of Death Tolls. http://users.erols.com/mwhite28

Wilson R (1979) Analyzing the Daily Risks of Life, Technology Review 81, February 1979, pp 40–46

Zack F, Rothschild MA & Wegener R (2007) Blitzunfall – Energieübertragungsmechanismen und medizinische Folgen. Deutsches Ärzteblatt, Jg. 104, Heft 51-52, 24th Dezember 2007, pp B3124–3128

Zwingle E (1998) Women and Population. National Geographic. Number 4, October 1998, pp 36–55

4 Subjective Risk Judgement

4.1 Introduction

The introduction of stochastic-based risk parameters has been a major step forward in providing effective safety measures; however, the decisions made by individual humans and human societies are only partly based on such numerically expressed risk measures. As we have seen in chapter "Indetermination and risk", this is quite good since scientifically based numerical models are also subject to individual and cultural assumptions. If stochastic-based risk parameters are only *one* basis for individual and social decisions, then the question arises: What else influences such decisions?. This chapter investigates these factors. The title of the chapter itself is somehow prone to discussion. In the literature, one finds not only the term subjective risk judgement but also risk perception or perceived safety.

In general, the author assumes that risk is more a rational concept, whereas safety always includes subjectivity, and therefore the term perceived safety might be misleading. However, independent from these considerations, the factors influencing the perception of safety will be investigated in this chapter.

To reach this goal, the chapter includes two major parts, namely individual subjective risk judgement and social risk judgement. First of all, the statement will have to be proven: that common people have a different perception of risks compared with stochastic-based numerical risk measures, and therefore showing that, other effects have to be considered.

4.2 Subjective Risk Judgement and Objective Risk Measures

Figure 4-1 shows a comparison of the statistical number of fatalities versus the subjective estimation of the number of fatalities for different causes. The subjective estimation is investigated by surveys, whereas the statistical number is computed on historical data. If both estimates yield identical results, the points in the diagram would follow the linear rising line.

Instead, there is a curvature; that is, for a few causes, the number of fatalities is overestimated (left part of the diagram), but for other causes, the number of fatalities is underestimated.

Other examples of subjective risk judgement are shown in Fig. 4-2 and Table 4-1. In the first, the results from a survey are shown giving the percentage of people avoiding a certain living region due to some risks. Here clearly, high crime rates are dominating. As mentioned in section "Lost life-years", social risks are the highest risks, and people are very much concerned about the functionally of the social system. Table 4-1 gives the results of a survey where people were asked to rank different risks. Again, the highest risks are connected to the failure of social systems or control mechanisms, even though the mortality numbers are low compared to other risks. Interestingly, some natural hazards are ranked higher than some technical risks like flying.

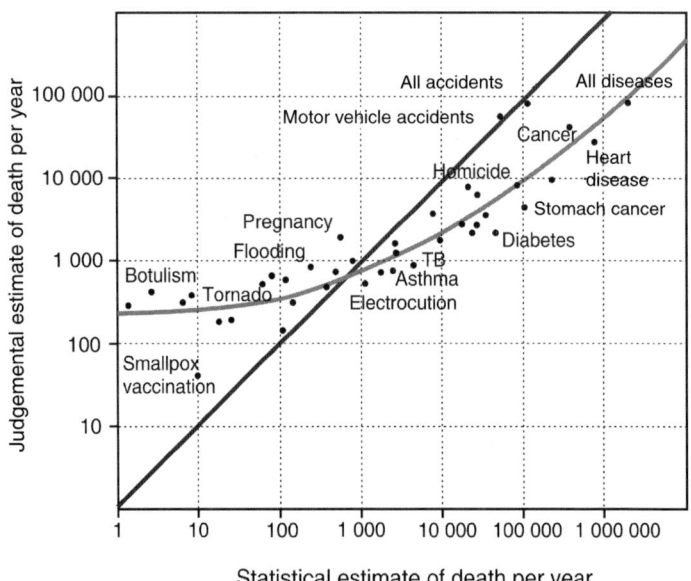

Fig. 4-1. Statistical versus subjective estimate of deaths per year for a certain region and a certain time period (taken from Viscusi 1995, Fischhoff et al. 1981)

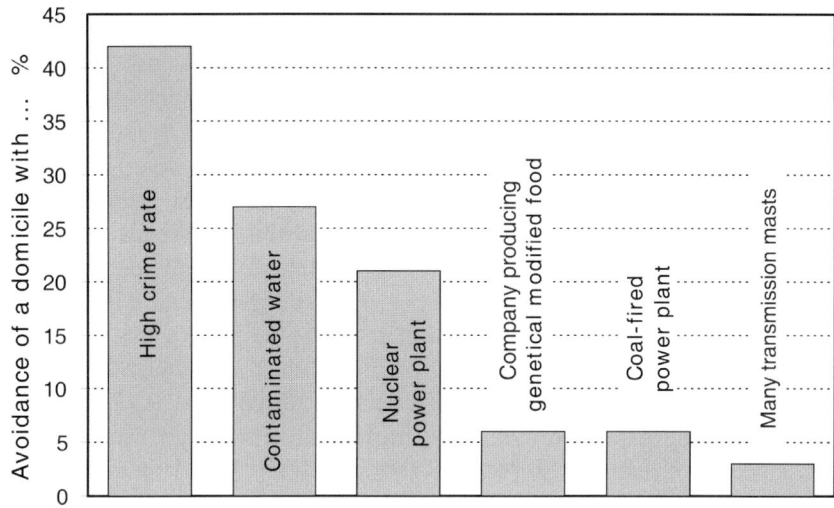

Fig. 4-2. Subjective risk judgement in terms of avoidance of domiciles (Zwick & Renn 2002)

Table 4-1. Subjective risk judgement: 100 corresponds to the highest risk, 0 to the lowest risk (Plapp & Werner 2002)

	Mean value	Median	Standard deviation
AIDS	77.6	83.5	22.5
Damage of ozone layer	71.4	74.5	20.4
Environmental pollution	67.5	70	20.9
Smoking	66.4	72	23.7
House fire	65.3	67	25.3
Earthquake	63.7	69	23.6
Volcanic eruption	60.2	67	28.4
Car driving	55.9	55	22.7
Nuclear power plants	55.8	58	28.9
Floods	52.9	57	23.7
Storm	47.9	48	20.9
Economic crisis	47.2	45	21.8
Genetically modified food	41.6	39.5	24.5
Alcohol	37.8	36	22.9
Non-ionised radiation	36.0	31	24.4
Skiing	35.8	33	20.7
Flying	33.4	31	18.6

4.3 Levels of Subjective Risk Judgement

Discrepancies between subjective judgement and statistically based measures are not only known in the field of risk assessment (Watzlawik 1985, von Förster & Pörksen 1999). Discrepancies can be explained with different theories. Such theories consider not only individual preferences but also social and cultural conditions. The weighting of the psychological, personal, cultural and social elements on subjective judgement is still under discussion. Douglas & Wildavsky (1982) suggested that culture is only able to describe about 5% of the variance of the subjective risk judgement. Furthermore, Table 4-2 summarises the influential weights of the different elements according to Schütz & Wiedemann (2005) and ILO (2007). As seen from that data, individual psychological and social aspects are of greater concern than cultural aspects (see also Dake 1991).

Table 4-2. Contribution of aspects to the subjective risk judgement according to Schütz & Wiedemann (2005) and ILO (2007)

Aspects	Contribution
Psychological and social aspects	80–90%
Personal aspects	10–20%
Cultural aspects	5%

The concept that social conditions might yield to illogical human reactions is well known. Even if humans know that what they are doing is wrong, they show the context-ruled behaviour. One of the best-known examples is the dollar auction game. Here, a dollar is auctioned. This may first sound simple, but the rule is that the second highest bidder also has to pay that highest amount but does not receive any money. This yields to the paradox situation caused by the context, in that people bid much higher than one dollar for one dollar. The interesting thing is that people know the dollar is worth only a dollar, but the auction conditions make them react in such a way (Teger 1980, Shubik 1971).

Another example is the well-known experiment by Milgram, where he asked people to punish other people with electric shocks. Even though information on the machine indicated that the shocks might be life-threatening, only 1/3rd of the participants stopped administering the shocks (Milgram 1974, Milgram 1997). To this class of experiments belongs the prison test by Zimbardo (Haney et al. 1973).

What people actually do is continue conforming to social assumptions and requirements. Latest research has proven this fact (Grams 2007, Hedström 2005, Ajzen 1988, Axelrod 1984, but see also Flood 1952).

4.4 Time-Independent Individual Subjective Risk Judgement

If there exists a strong bias between subjective risk judgement and statistical data, it should be of major interest to identify influences on the subjective judgement to predict risk acceptance for some actions or new technologies. Although social and individual aspects of subjective risk judgement cannot be completely separated, it is quite helpful to divide the influences in such a way.

Additionally, the identification of single influences on subjective judgement remains difficult. Therefore, in the beginning, groups of influences are introduced. Figure 4-3 is based on a survey introducing two influence parameter groups. It illustrates the perception of risks based on the degree of dread and degree of knowledge. In Fig. 4-3, there are four quadrants visible. The top right quadrant illustrates risks, which are considered both dreadful and unknown. In this quadrant, DNA technology or radioactive wastes are found. In most cases, the public does not accept such risks. Risks located at the bottom left quadrant are actually not considered risks at all, for example using a bicycle or playing football. People simply accept them. The risks in the other two quadrants might be accepted or refused by further parameters.

Here, it might be useful to identify further components of the dread of risk and the degree of knowledge. The components are listed in Table 4-3.

Dread might also be related to the term fear. Fear is defined as an extreme negative emotion about some conditions or objects. It mainly functions as a strong resource-controlling mechanism.

Table 4-3. Components of the degree of knowledge and the degree of dread of risk (Slovic et al. 1980)

Unknown risk	Dread risk
Not observable	Uncontrollable
Unknown to those exposed	Dread
Delayed effect	Global catastrophe
New risk	Fatal consequences
Risk unknown to science	Not equitable
	High risk to future generations
	Not easily reduced
	Risk increasing
	Involuntary
	Personal effect

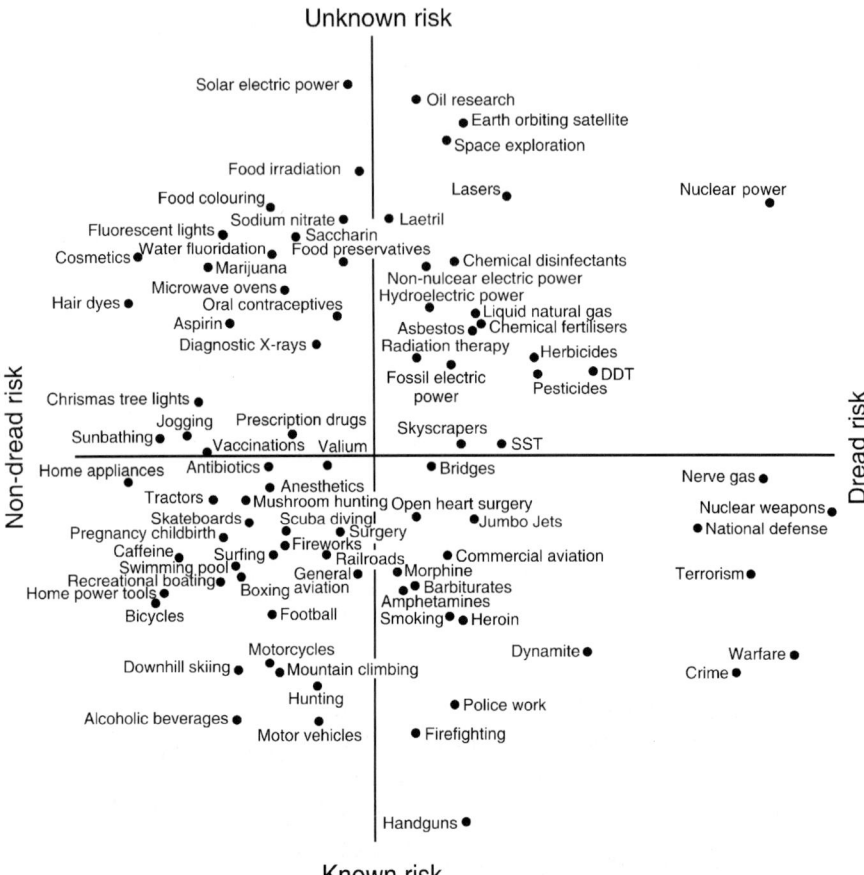

Fig. 4-3. Subjective risk judgement based on knowledge and dread of risk (Slovic, et al. 1980), further works by Sparks & Shepherd (1994) or Balderjahn & Wiedemann (1999) for the change of important risk judgement factors related to different professions

The effect of the degree of knowledge on subjective judgement is also illustrated in Figs. 4-4 and 4-5. Here, several hazards or risks are ranked according to the size of the hazard or risk (Fig. 4-4) and the degree of knowledge surrounding that hazard (Fig. 4-5). It is quite interesting to note that smoking does not fit into the theory of state of knowledge. Probably here, further effects like addiction or voluntariness become visible.

Look again at Fig. 4-3. The quadrant at the bottom left shows some risks that are actually not considered risks. Instead such events are considered as part of everyday living. Skiing is mainly seen by the major part of the population as sport and recreation–and not as a risk and consists of a high degree of voluntariness. Implicit inside Fig. 4-3 is an influence of voluntariness, even if it cannot be found on the axes.

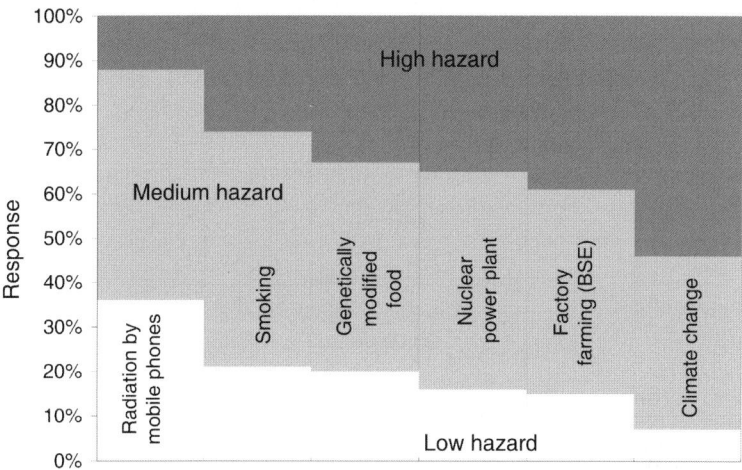

Fig. 4-4. Subjective estimation of hazards (low, medium or high) for different risks (Zwick & Renn 2002)

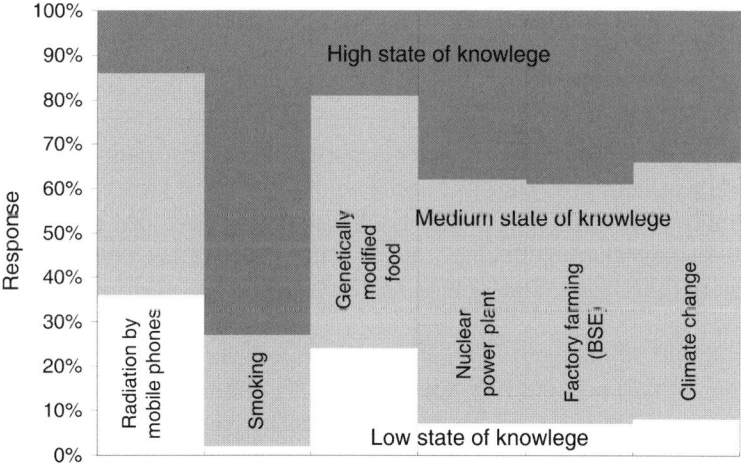

Fig. 4-5. Subjective estimation of knowledge for different items (Zwick & Renn 2002)

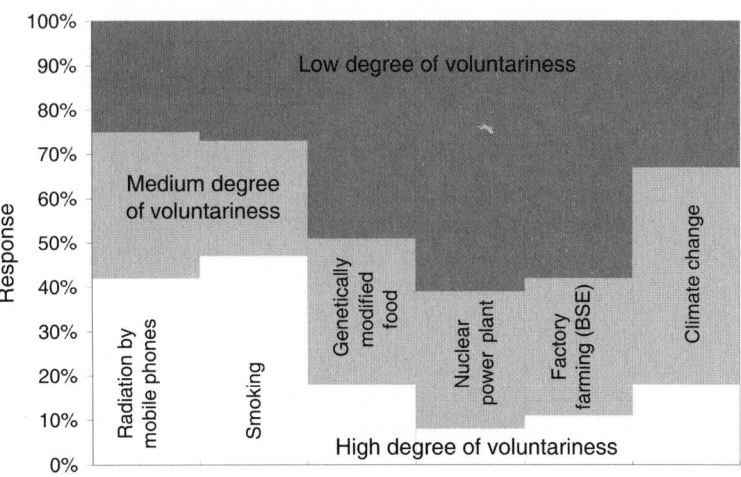

Fig. 4-6. Subjective estimation of voluntariness for different items (Zwick & Renn 2002)

Figure 4-6 shows the subjective estimated degree of voluntariness for the risks shown in Fig. 4-4. One should be careful about the estimation of voluntariness for smoking, since most people are not aware that they are addicted to it. Most people are confident of being able to stop smoking at any time, but just do not want to.

This gives an idea about the difficulties in identifying the causes of subjective risk judgement. Very often, different results occur in the ranking of risks with different questionnaires. Therefore, such investigations have to be done very carefully. Figure 4-2 and Table 4-1 show a variety of different subjective risk rankings. While the risks in Fig. 4-2 are completely connected to individual wellbeing, the risks in Table 4-1 are understood as risks to mankind. These different views have to be incorporated into the discussion.

An initial list about possible individual factors influencing subjective risk judgement can be found by splitting the diagram axis units from Fig. 4-3 as shown in Table 4-2. Further factors are listed in Table 4-4. Some publications mention up to 27 factors influencing subjective judgement (Covello 2001). For example, besides the factors listed in Table 4-4, Slovic (1999) also mentions the importance of sex and age for subjective judgement. For most types of risks, women are more risk averse than men, and older people are more risk averse than young people (Simon et al. 2003).

Table 4-4. Risk perception factors according to Covello et al. (2001) – see further Gray & Wiedemann (1997), Schütz et al. (2000) and Slaby & Urban (2002)

Voluntariness	Involuntary risks are perceived as greater risks as compared to voluntary risks.
Controllability	Risks under the control of others are perceived as greater risks as compared to risks under the individual's control.
Familiarity	Unfamiliar risks are perceived as greater risks as compared to familiar risks.
Equity	Unevenly distributed risks are perceived as greater risks as compared to evenly distributed risks.
Benefits	Risks with an unclear benefit are perceived as greater risks as compared to risks with a clear benefit.
Understanding	Risks difficult to understand are perceived as greater risks as compared to clearly understandable risks.
Uncertainty	Unknown risks are perceived as greater risks as compared to known risks.
Dread	Risks that create strong feelings such as fear are perceived as greater risks as compared to risks that do not create such strong feelings.
Trust	Risks connected to persons or institutions with low credibility are perceived as greater risks as compared to risks connected with trustful persons or organisations.
Reversibility	Risks with irreversible effects are perceived as greater risks as compared to risks without such effects.
Personal stake	Risks at a personal level are perceived as greater risks as compared to more impersonal risks.
Ethical and moral nature	Risks connected to low ethical or moral conditions are perceived as greater risks as compared to risks connected to high ethical or moral conditions.
Human versus natural origin	Man-made risks are perceived as greater risks as compared to natural risks.
Victim identity	Risks with identifiable victims are perceived as greater risks as compared to risks with only statistical victims.
Catastrophic potential	Risks creating spatial or temporal concentrated victims are perceived as greater risks as compared to risks that are diffuse over time and space.

Another approach to describe effects of subjective judgement is the comparison of risks with some characteristics of Greek-antic persons taken from Erben & Romeike (2003) and Klinke & Renn (1999). The main idea here is not the identification of single items of the subjective judgement but the identification of comparable situations. The concept is shown in relation to the traditional risk formula in Fig. 4-7, and examples are given in Fig. 4-8.

Damocles type of risks are defined by hazards and risks during a period of luck and prosperity. This type of risk was very well described by the Damocles sword, which was hanging over Damocles, head when he was living as Dionysus, a rich emperor. Such risks can be handled through mitigation measures. For example, Damocles could have worn a helmet (Fig. 4-9). The risk is additionally described as having a major public focus with the advantages often being forgotten, for example nuclear power plants. Cyclops type of risks are described as having lack of information about the return period of an event and high damages during an event. Examples of these risks are earthquakes or meteorite impacts. Pythia type of risks are identified by high uncertainty of the return period or probability of an event on well as the amount of damage. Such risks can be found in new technologies. Pandora type of risks have a high spreading of damages in both spatial and temporal dimensions. Additionally, the damages are mainly irreversible. Such risks are usually of global scale, such as nuclear wars, nuclear waste or chlorofluorocarbon. Cassandra type of risks are where the return period and the damages are widely known, but the time or space between cause and reaction is great. Therefore, humans do not really bother about such risks, for example the extinction of animals or plants. Medusa type of risks have a known amount of damage and return period and moderate spreading of damages over time and space, but humans and societies overreact to such risks.

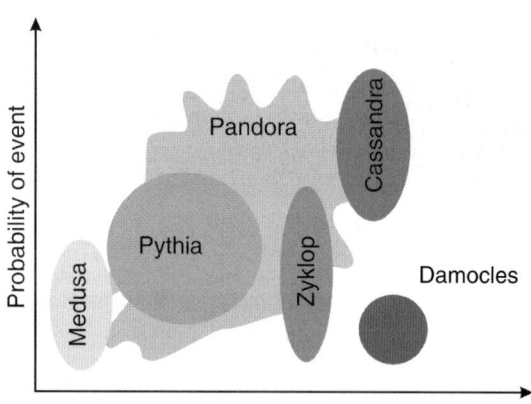

Fig. 4-7. Mythical-antic classification of risks (WBGU 1998)

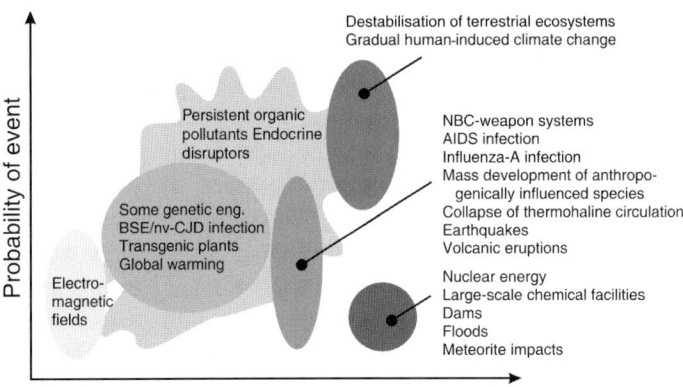

Fig. 4-8. Classification of certain technologies in the context of mythical-antic classification of risks (WBGU 1998)

Fig. 4-9. Be prepared!

Since the discussion of all individual factors influencing subjective judgement would go beyond the goal of this book, only a few factors will be discussed. The choice of factors is taken from a work of Wojtecki & Peters (2000). The major reason for choosing their factor classification is the specification of influence numbers. In their list, the most important factor is trust.

4.4.1 Trust

According to Wojtecki & Peters (2000), trust can yield to a change of risk perception by a factor 2,000. Based on system theory, trust represents an extension of our system. Therefore, individuals can dramatically improve their amount of resources. If one trusts a person, one can have easy access to their resources like time, money or power. One of the definitions of disasters is negative resources. The definition of danger is limited freedom of resources. So if one can increase the resources, one can better handle accidental events. This can also be done in conflict situations. The best examples probably are movies, where a protagonist tells people that they can trust him and do not have to worry. This is nothing else than an invitation to access his resources.

Mountaineering can also be considered within this context. Whether going alone or in a group, in most cases a subjective risk judgement, shows a dramatic bias. This bias is caused by the strong impression that one has access to the resources of other members of the group in terms of knowledge, experience, force, material and so on, leading to an overestimation of one's own resources. This last point can be easily destroyed by the experience of an accident, since trust can be more easily destroyed than built. There exists a rule of thumb which says, it takes about three times the effort to build trust than to destroy it (see also Figs. 4-10 and 4-11).

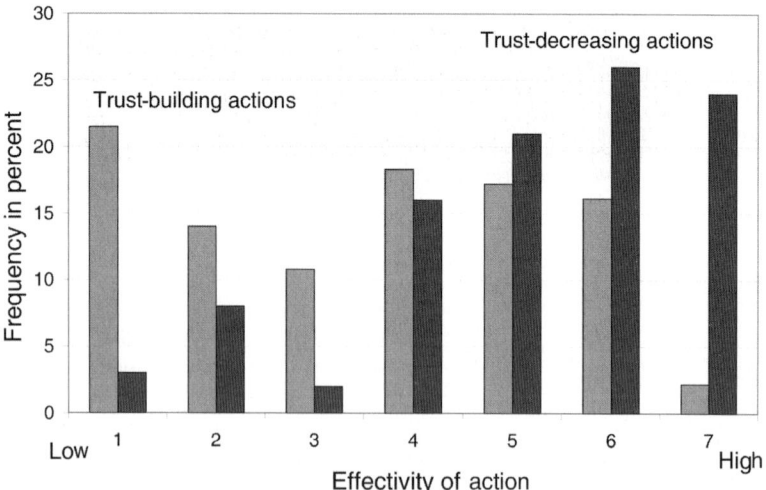

Fig. 4-10. Distribution of effectiveness of trust-building and trust-decreasing actions (Slovic 1993 and Slovic 1996)

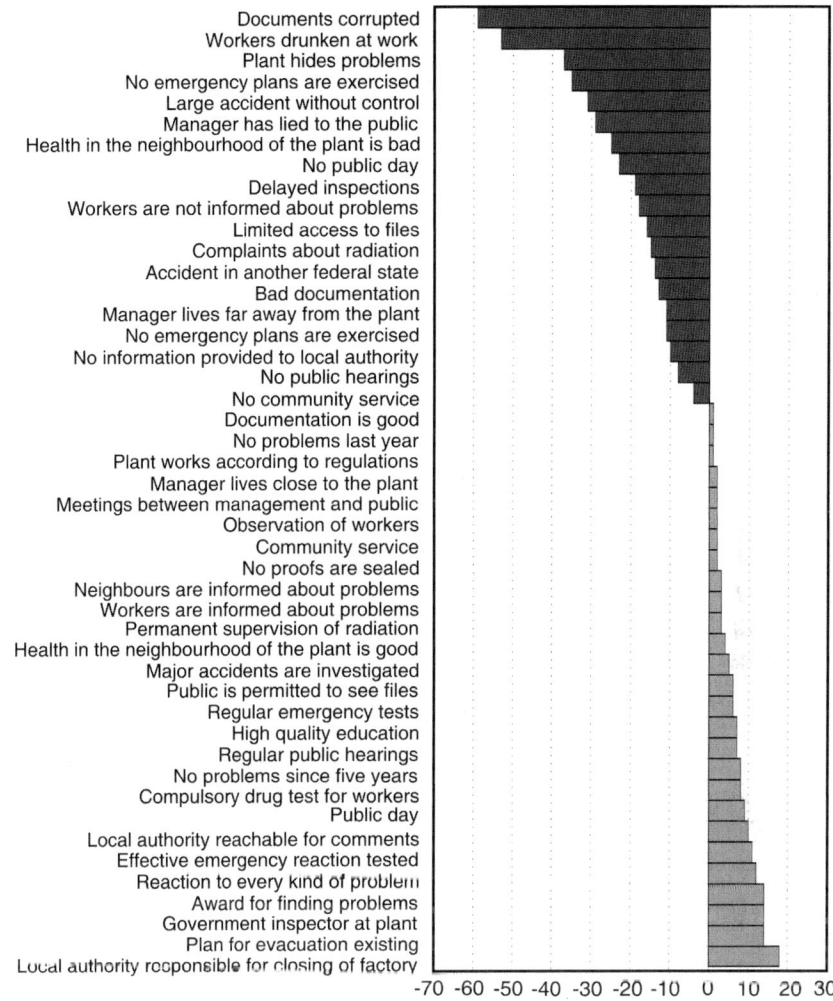

Fig. 4-11. Effectiveness of trust-building and trust-decreasing actions (Slovic 1993, 1996)

Especially for companies dealing with risky materials, such as nuclear power or genetically modified food, the described effect of trust-building is of major importance. Trust has to be developed over a long period; then it might pay off in accident situations. If an event that happened has heavily damaged trust, it might need strong efforts to re-establish this trust. Figure 4-12 shows some of the major factors contributing to the formation of trust. One of the possible ways to re-establish trust is to give the impression of control and voluntariness.

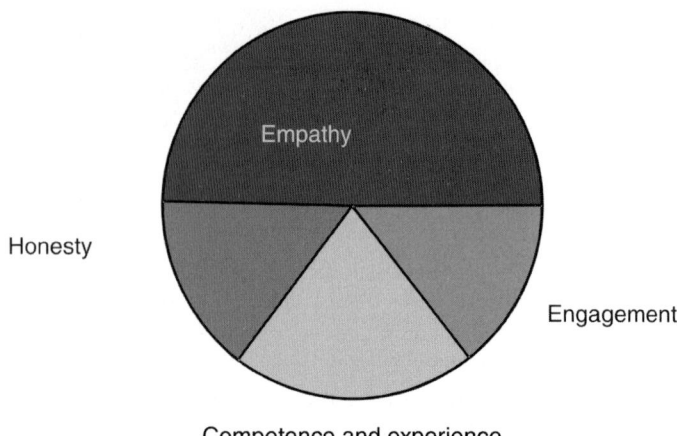

Competence and experience

Fig. 4-12. Components of trust (Wojtecki & Peters 2000)

The first documented evidence broaching the subject of trust and credibility was that by Aristotle. Kasperson introduced the following sub-items of trust: perception of competence, absence of bias, caring and commitment. Later, the sub-items were extended to commitment to a goal, competence, caring and predictability. Renn and Levine later introduced five sub-items:competence, objectivity, fairness, consistency and faith. Covello offered four sub-items: caring and empathy, dedication and commitment, competence and expertise, and honesty and openness (Fig. 4-12) (Peters et al. 1997).

A definition of trust was given by Rousseau et al. (1998): "Interpersonal trust is a psychological state that involves a reliance on another person in a risky situation based upon positive expectations of their intentions or behavior."

A good example of a trust-increasing action is the safety information offered on airplanes. Just before the airplanes take off, passengers are informed about the use and availability of life vests, the emergency escape routes and the proper behaviour in pressure loss situations. The purpose of such information stems mainly from the desire to build up trust rather than to decrease the number of fatalities during an emergency situation. Comparable to this are advertisements about the safety measures in cars. Here, often in an amusing way, the performance of cars in a crash situation is described. There has been, for example, an advertisement of a car using a smiling dummy.

Trust influences not only the acceptance of risk but also the functioning of economical systems. For people live in many countries, social conditions motivate them to produce further wealth. This seems to be valid for all developed countries in the world. However, this is not a concrete behaviour of humans, since one can also find countries where the social conditions motivate people to cabbage wealth from other people instead of producing the wealth (Knack & Zack 2001, Knack & Keefer 1997, Welter 2004). Such countries can be easily described as having a lack of trust in the social system. Here, the state as a representative of society is unable to fulfill the duty of safety, which is the major task of a state (Huber 2004).

Many psychological tests have shown that people mainly check the trustworthiness of other people through non-verbal information, and that the content of their communication (verbal information) is less important. Probably the most well-known tests were from Naftulin et al. (1973), where an actor gives a presentation to specialists about a certain topic. Most specialists did not realise that the presenter had no knowledge about the topic and was talking total nonsense. The ways in which the actor engaged the audience and emphasized certain points made the audience trust the information being presented. The actor used humour as well as references to other (non-existing) scientific works.

Ekman at the University of California has carried out newer studies. These studies showed that, under low-trust and high-concern situations, non-verbal information makes up to 75% of the entire message content. Therefore, non-verbal information is extremely important, often influencing the audience negatively.

Usually under normal conditions, the social system destroys trust. This becomes obvious in law cases where one expert speaks out against another expert and the public does not know whom to trust. Koren & Klein (1991) also showed that when two news articles were written, one stating that no additional cancer risks are observed close to nuclear power plants and another article stating that, in the Oak Ridge National Laboratory, there is an increase of blood cancer risk, the second article attracts much greater interest as compared to the first.

4.3.2 Control and Voluntariness

As already mentioned, humans tend to overestimate their capabilities and resources. People run into debt completely convinced that they might solve that problem in the future. People might use cars in a way they completely lose control. This general effect is called optimism bias. Figure 4-13 shows this effect for car driving. People were asked how well they judge their driving capabilities as compared to an average driver. If they consider themselves better than an average driver, they should give a positive value, if equal they should give a zero, and if worse they should give a negative value. The results showed that the participant assumes he/she drives better than an average driver. This effect can shift a risk by a factor of 1,000.

Such optimism bias has already been recognised in the cost estimation for public projects. The British Department of Transport (2004) has published numbers for the underestimation of structural costs for big projects (Flyvbjerg 2004) as shown in Table 4-5.

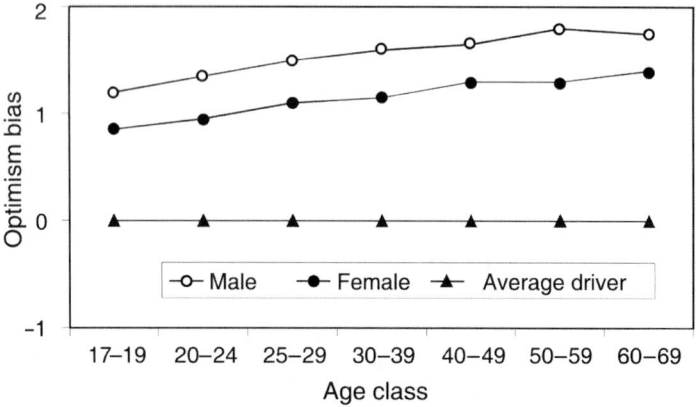

Fig. 4-13. Optimism bias for car drivers (Job 1999)

Table 4-5. Underestimated costs in percent of big public structural projects by optimism bias (Flyvbjerg 2004)

Project type	Optimism bias			
	Work duration		Capital expenditure	
	Upper	Lower	Upper	Lower
Standard buildings	4	1	24	2
Non-standard buildings	39	2	51	4
Standard civil engineering	20	1	44	3
Non-standard civil engineering	25	3	66	6
Equipment/development	54	10	200	10
Outsourcing	–	–	41	0

Optimism bias can be related to voluntariness and control. A good experiment has shown the influence of voluntariness in 1981. In a medical test, participants were required to take medication either with a radioactive coating, a bacterial coating or a heavy metal coating. Half the subjects were free to choose the coating, the other half were not. The discomfort level was double for the group that was not allowed to choose the type of coating. (Please note: There was no coating at all.)

Even if Fig. 4-14 does not consider voluntariness and control, it shows quite impressively this shift of risk by a factor of 1,000. The figure shows the observed number of fatalities per person-hour of exposure (FAR) versus the average annual benefit per person involved in terms of dollars. Even if the numbers are outdated, the relationships are still interesting. It shows that, leaving aside trust, control and voluntariness, one of the major factors that influence subjective risk judgement is benefit. This topic will be discussed again later, but it is interesting to note that people never make a sole risk judgement; they prefer to make a trade-off between possible loss (risk) and possible benefit (chance). Actually, psychologists have shown that people might have general problems in comparing risks because those are two negative options. Humans behave much better in comparing packages of risks and chances. This is not only because it looks easier here but also because humans are striving for advantages and benefits, not for disadvantages. This discussion will be continued in the chapter "Quality of life – the ultimate risk measure".

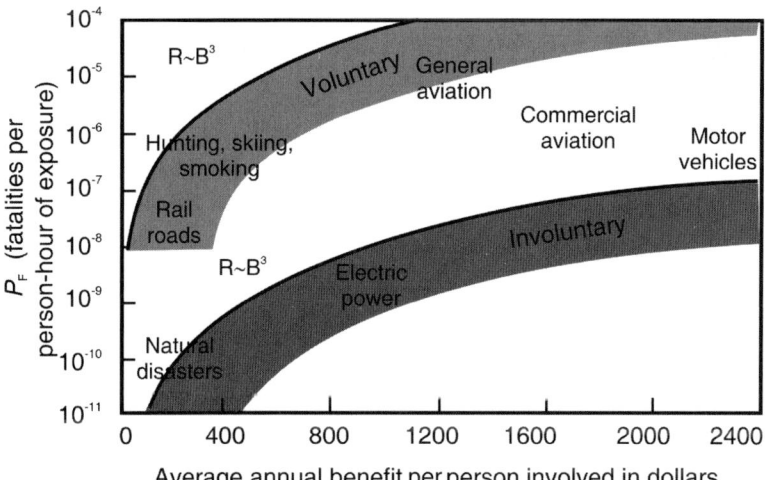

Fig. 4-14. FAR versus average annual benefit per person including separation of voluntary and involuntary actions (Starr 1969)

4.4.2 Benefit

As shown in Fig. 4-14, there exists a strong relationship between accepted risks and awareness of benefits. As seen from Fig. 4-14, this is not a linear relationship.

Figure 4-15 additionally supports the theory of non-linear behaviour in some regions of the diagram. What is quite interesting is the benefit awareness of smoking. The benefit awareness from car driving or flying is quite understandable. A car can be used in everyday life, is comfortable and easy to use, while flying takes one virtually around the world in only a few hours. Any other means of transport would take days or weeks. Nevertheless, such an awareness of benefit has to develop, as shown in Fig. 4-16. As mentioned in the first chapter, many new technologies were first due to insufficient awareness of their benefits.

Figure 4-17 tries to visualise that. New technologies might decrease a certain risk a to a lower level (risk b), but they might also impose a new risk (risk c). The drop from risk a to risk b would be a benefit.

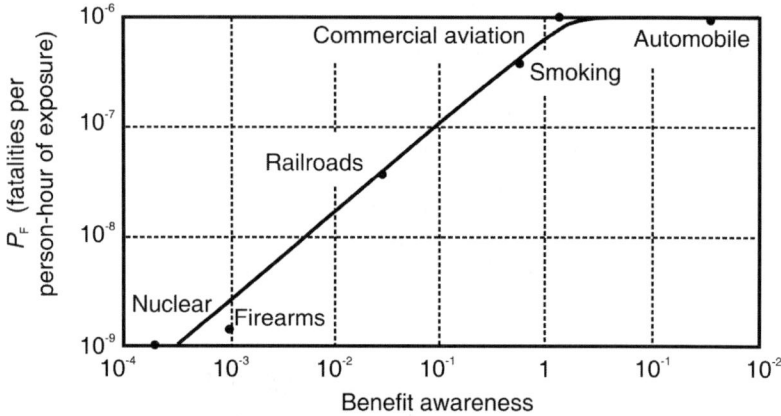

Fig. 4-15. FAR versus benefit awareness (Starr 1969)

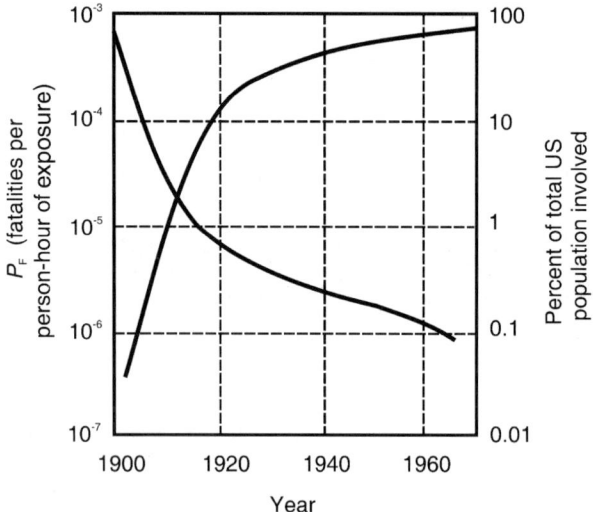

Fig. 4-16. Development of FAR over time and development of acceptance of air traffic technology (Starr 1969)

Very often this benefit is neglected. For example, consider the risk of structural collapse of a building. People might not consider it as acceptable under some conditions; however, living outside the structure in winter in some regions would impose a much higher risk.

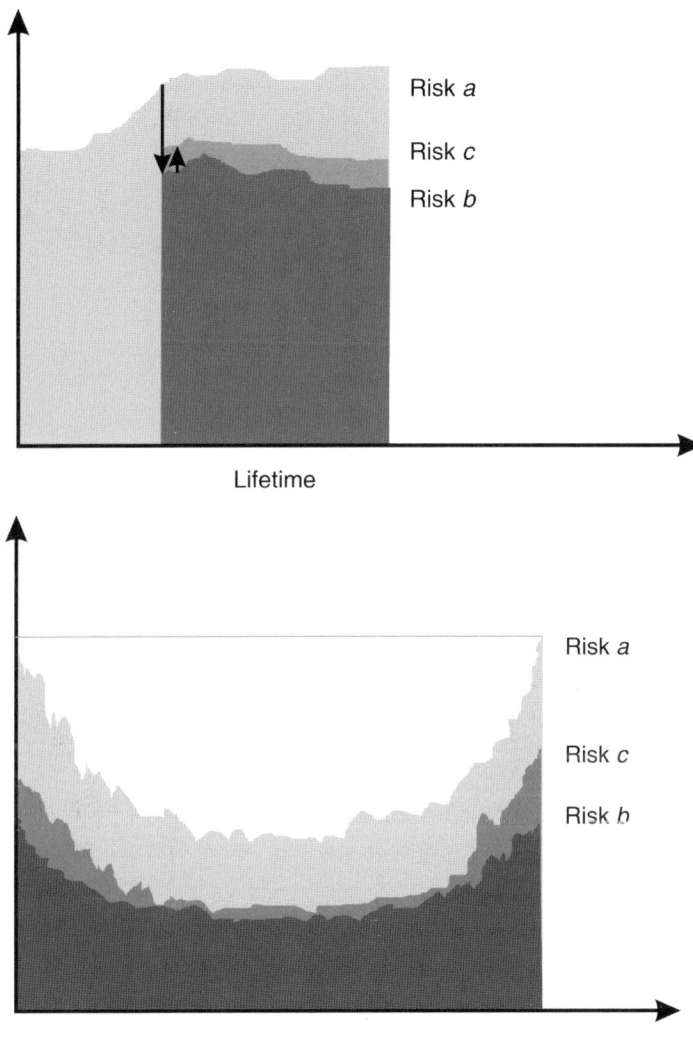

Fig. 4-17. The diagram at top shows the development of an initial risk (risk *a*) over a period of time, which is reduced by a certain technology or protection measure to risk *b*. The protection measure itself, however, causes a new risk, increasing risk *b* to risk *c*. The diagram at bottom shows this over a lifetime. It should be noted that, under some conditions, risk *c* can be higher than risk *a,* and therefore the intended protection measure does not really act as a protection measure

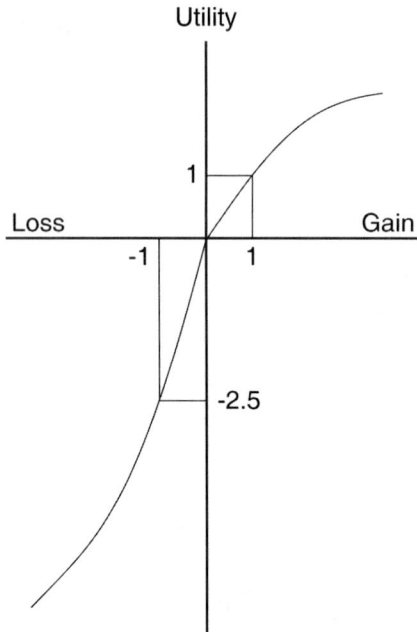

Fig. 4-18. Utility function according to Tversky & Kahnemann (1974), Tversky & Kahnemann (1981)

The work by Tversky & Kahnemann (1974, 1981) about the limit of the gain awareness function fits this consideration well. This is sometimes called the Easterlin paradox (Easterlin 1974), which states that the happiness of people rises with income to a certain threshold value, but not beyond. See also Fromm (1968, 2005) for a discussion. The same is valid for the loss; here the so-called reframing can be observed.

Interesting to note, however, are the different slopes of the utility function shown in Fig. 4-18: the slope of the loss curve is much steeper. A general rule of thumb says, a loss of 1,000 Euro hurts more than twice as much as the pleasure of winning a 1,000 Euro. The avoidance of negative values is based on the preservation of an already received major gain: the life. Further works about the non-linearity of the utility functions are given by Eisenführ & Weber (1993) and Jungermann et al. (1998).

Not only the intensity but also the choice of factors considering the benefit of some actions is difficult. Figure 4-19 and 4-20 show the influences on the decision-making process whether something brings advantages or disadvantages. Clearly, the so-called optimal solutions are very difficult to obtain under such a variety of influences. This statement should be kept in mind when thinking about optimal mitigation measures. Schütz et al. (2003) estimated that the choice and the weighting of factors is more important for risk assessment than the utility function.

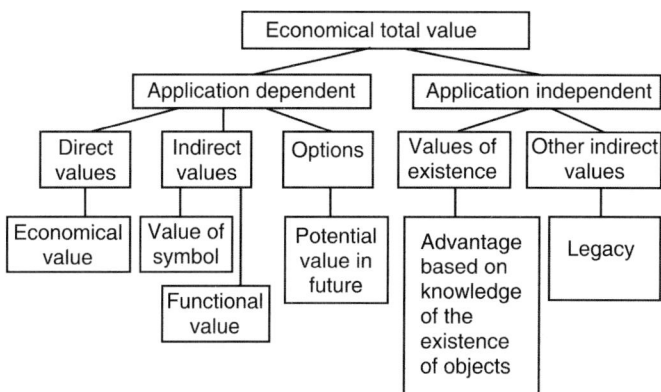

Fig. 4-19. Composites of the total economic values of the biosphere (WBGU 1999)

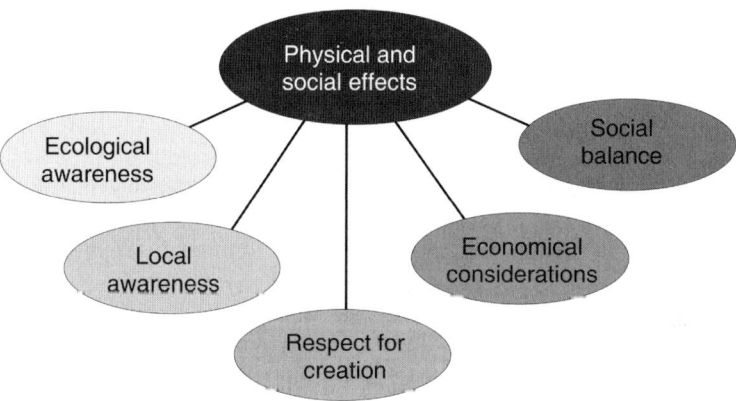

Fig. 4-20. Possible effects influencing physical and social wellbeing (WBGU 1999)

Additionally pure optimal solutions might not be adequate for survival, since Newman et al. (2002) have shown that optimal solutions do not protect from rare and very disastrous events. Unfortunately, such events might lead to non-recoverable effects, such as fatalities. This situation is comparable to a player who wins a gamble, but if he loses, he loses everything (Newman et al. 2002). Also, the data are processed in a non-rational way. Subjective probabilities are not directly used in subjective risk judgements; instead, a weighting function is used like the one shown in Fig. 4-21. In this function, very small probabilities are neglected, small probabilities are overestimated, high probabilities are underestimated and very high probabilities are overestimated. Figure 4-21 can be compared to Fig. 4-1. Prospect theory considers further that decisions are based on anchor points. Such anchor points might come from experience. Also, neuronal models confirm non-linear number representation in the brain (Shepard et al. 1975).

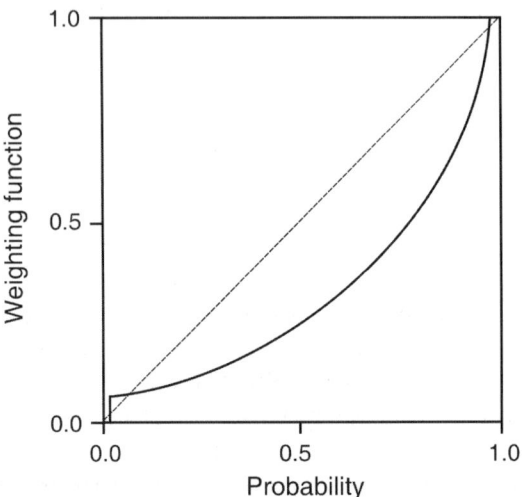

Fig. 4-21. Weighting function according to Tversky & Kahnemann (1974, 1981)

An example of such a non-linear weighting function is the Weber–Fechner law. It states that subjective power of sensation behaves logarithmically to the objective intensity of the stimulus. Especially for the perception of changes, this law is interesting. So, only a 3% change in hand-shake force, a 1–2% change in light intensity or a 10– 20% change of taste is noticeable. The law is given as:

$$E = c \cdot \ln \frac{R}{R_0} \qquad (4\text{-}1)$$

where R_0 is liminal stimulus, R is previous stimulus and c is a constant depending on the type of sense. The formula is not valid for all senses. For example, only some parts of the audio frequency band follow this law. An extension of this formula is Stevens power function (Dehaene 2003, Krueger 1989).

In general, the theory of human decisions is a wide field, and many theories have tried to deal with this issue, such as rational choice theory, Fishbein–Ajzen theory, game theory, frame selection theory and so on. For the concept of benefit and advantage, the same holds true as for the concept of risk and disadvantage: the difficulty to assessing values (Becker 1976, Feather 1990, Ajzen 1988, Davis 1973).

4.4.3 Fairness

Fair distribution of advantages and disadvantages of social decisions is of utmost importance for the functionability of societies. Traditional welfare economic concepts deal with this issue in terms of consequences of income distribution and individual utilities. However, in the field of risk assessments, such distribution examinations have to consider not only income distribution but also other terms like the distribution of longevity (Pliefke & Peil 2007).

However, the assessment of fairness is strongly limited by bounded rationality (Cook & Levi 1990, Dawes 1988) and context-based decisions. As already discussed in chapter "Indetermination and risk", any definition of terms includes indetermination. This is even more valid for some human concepts like fairness and justice. The final realisation of such terms is impossible, thus it calls for a decision maker outside the system, e.g. outside of the human society. Therefore, decisions are usually carried out by an extension of the models reaching a maximum degree of independence, as shown in Fig. 4-22.

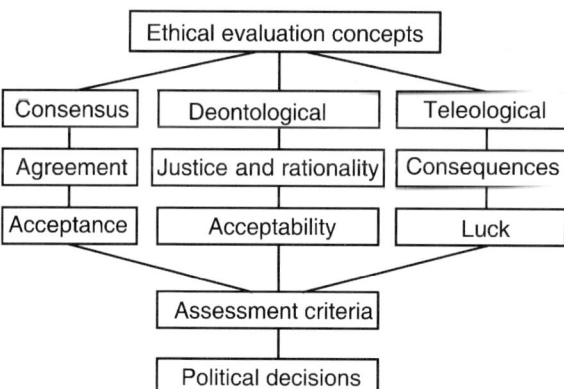

Fig. 4-22. Decision tree considering a wide range of criteria (WBGU 1999)

4.4.4 Alternatives

As already mentioned, humans have a good ability to compare packages of advantages and disadvantages. This statement already includes the assessment of risks or disadvantages based on other opportunities. A risk cannot only be assessed by itself since the boundary conditions have to be known, which is nothing other than alternatives in a decision-making process.

4.4.5 Origin of Risk

Lübbe (1989) gave some relationships for perception of risk and uncertainty based on socialcultural behaviour:

- The increase of human elements in the conditions of living decreases the acceptance of risks. This fits very well to Luhmann's explanation of risks (Luhmann 1997).
- Risk perception increases with more extensive control over natural and social processes (consider weather).
- Loss of experience increases awareness of uncertainty (no experience with rivers, etc.).
- The increase in civilisation-based complexity and uncertainty increases the awareness of uncertainty (time-limited working contracts).
- The awareness of uncertainty increases with decreasing social support (for example trust).
- The desire for safety increases with increasing technical and social safety levels.

4.4.6 Knowledge About Risk

People in general avoid situations where information about uncertainty is missing. Considering a situation with two options – in one, the probability information is given, and in the other, such information is missing, most people prefer the option with probability information. This has been evaluated during many experiments, and is now known as the ambiguity effect.

For a better understanding, consider this example. To play a lottery game, two boxes are given. In the first box, there are 100 balls, 50 black and 50 white. In the second box, there are 100 balls, but no information about the distribution of the colours is given. Under such conditions, most people tend to play with the first box (Rode et al. 1999).

4.5 Time-Dependent Individual Subjective Judgement

4.5.1 Personal Development and Subjective Risk Judgement

Riemann (1961) in his book describes the relation of people to risk, based on their early upbringing. The book mainly focusses on the development of different types of psychological characters, but also indirectly describes their attitude to risk. For example, hysterical people are quite open to risks, whereas obsessive people tend to spend an extreme amount of resources to fighting conscious risks.

Wettig (2006) and Beinder (2007) gave the same results. They state that infant stress caused by negative relationship experiences generates in the human brain the same type of circuit like a physical pain. Such scars in the brain yield to overreactions against certain risks or hazards. They are comparable to an open wound. Under such conditions, people spend more resources for the prevention of hazards. Such spending behaviour might be considered non-objective if the disturbances in early childhood are not considered.

A very special case is described in Wahle et al. (2005). They suggest that there are critical periods of molecular brain plasticity, where environmental conditions during an infant's first ten days decide the stress treatment behaviour that occurs when the infants become adults. Another example about psychological preferences subject to risk is shown in Fig. 4-23 by Renn (1992).

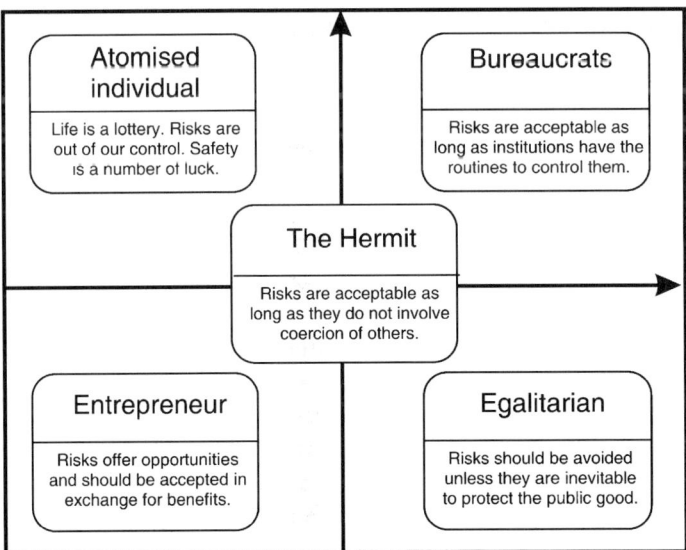

Fig. 4-23. Individual strategies to cope with risks taken from Renn (1992)

Not only the genetic core of a human and the individual history but also current environmental conditions influence the judgement of risk . Under situations causing high stresses, humans generally do not behave rationally and are mainly controlled by strong feelings that are shortcuts to assess situations. The capability for higher brain functions is preliminarily lost, and therefore information processing can drop up to 80% (Laney & Loftus 2005, McNally 2005, Shors 2006 and Koso & Hansen 2006). Even worse, if people have assessed a situation as negative and risky, they assume the worst and do not trust other people (Covello et al. 2001). Therefore, Sandman (1987) has defined risk as hazard + outrage.

The concepts of bounded rationality and cognitive heuristics and biases are widely discussed in other fields; they have strong effects on the risk judgement as well. It is of much importance since subjective risk judgement yields to real actions independent from objective conditions.

Such knowledge can be used to approach subjective risk judgement closer to more rational risk estimations by safety education. Geller (2001) has stated several points concerning the development of safety behaviour by humans:

- Self-perception is defined by behaviour.
- Direct persuasion has limited impact.
- An indirect approach is more likely to influence self-persuasion.
- Self-persuasion is a key to long-term behavioural change.
- Large incentives can hinder self-persuasion and lasting change.
- Mild threats influence self-persuasion more than severe threats.
- The more obvious the external control, the less the self-persuasion.
- Self-efficacy is a key to empowerment and long-term participation.
- Response efficacy is a key to empowerment and long-term participation.
- Motivation to act comes from outcome expectancy.

Several of these points fit very well to the Yerkes–Dodson curvature mentioned in chapter "Indetermination and risk". Self-perception is most important and relates very well to the issue of learned helplessness, where people just refuse responsibility and get used to regular help (GTZ 2004).

4.5.2 Individual Information Processing

Even if subjective judgement is neglected, the question arises whether humans are able to carry out rational decisions by information processing. Therefore, this section discusses the information processing capability of humans. The human brain carries out this information processing. Table 4-6 shows some main functions of different sides of the human brain.

Table 4-6. Main functions of the brain halves

Right brain	Left brain
Body language and imagery	Language, reading, calculation
Intuition and feelings	Reason, logic
Creativity and spontaneity	Laws, regulations
Curiosity, playing, risk	Concentration of one point
Syntheses	Analysis, detail
Art, dancing, music	Time sense
All-embracing relationships	
Feeling of space	

The human brain receives about one million bits of audio information per second and about 50 billion bits of visual information per second. See also Koch et al. (2006) for information transmission from the eye to the brain. The brain filters the incoming data heavily. Filtering is very strongly connected to the initial genetic program, the cultural background and the personal history of the individual (Schacter 1999, Ehlers 1996). Sometimes, it is called the individual access to the world. Filtering is carried out during processing and storing of data. The memory as information storage is usually divided into three parts (Planck 2007):

- Ultra short-term memory can retain 180–200 bits per second. Here, information can be kept only for up to 20. This memory builds up the consciousness range. It can only take a maximum number of 10 items but most people use only 3 or up to 7 items (Planck 2007).
- Short-term memory can retain 0.5–0.7 bits per second. This memory can keep information for hours and/or days.
- Long-term memory can retain 3 bits per minute. This memory can keep information for years or an entire lifetime. But, it keeps less than $1/10^{12}$ part of the original information. However, the overall storage capacity has been estimated up to 10^{15} bits (Ruggles 1996). For comparison purpose, in 2006, mankind has produced and stored data in the size of $3–5 \times 10^{15}$ bits (Keim 2002).

Further information about the computing power of the human brain can be found in Ruggles (1996) and Moravec (1998). However, the brain actually does not constantly process information. Instead, there exists a so-called psychological moment, which means only events with a time distance between them equal to or greater than 1/16th of a second can be identified as occurring one after another by the brain. Assuming that this time equals one bit, the number of bits per second depends on the division of seconds. For 1/16th of a second time step, this would yield 80 bits per second. Since for children the time steps per second are bigger, there is less information retained (Frank & Lobin 1998).

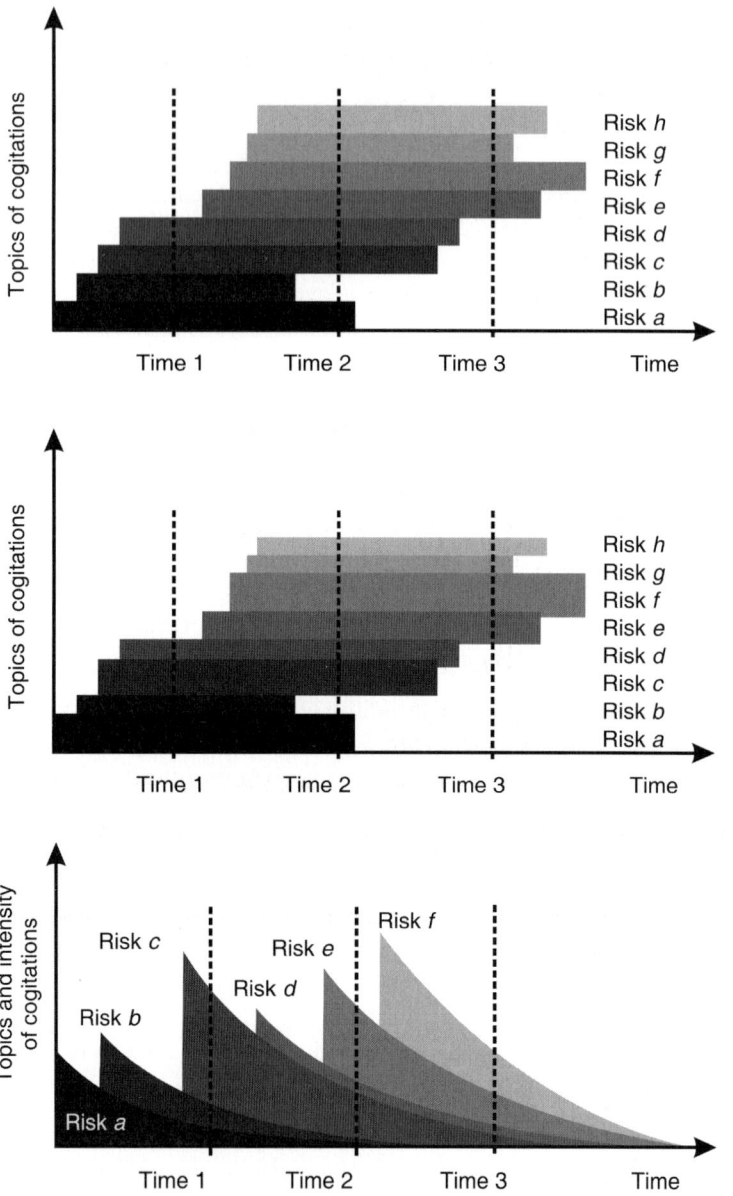

Fig. 4-24. The diagrams show topics of cogitations (only risks). The top diagram considers only time items, the middle diagram considers the intensity and the bottom diagram considers the Ebbinghaus memory (Ebbinghaus 1913) function

Based on those considerations, it is not surprising that humans filter and forget risks according to their preferences. Especially the amount of attention plays an important role for the judgement of risks. Figure 4-24 shows the modelling of risks in the brain over time. While in the top figure, only the persistence of perception of risks is visualised in a simple rectangle with the same intensity for all risks, in the middle figure, the intensity is different, and at the bottom, the Ebbinghaus memory curvature is applied (Ebbinghaus 1913). Empirical works by IKSR (2002) and Lustig (1996) confirm such models. The results of the IKSR are summarised in Fig. 4-25, giving the rule of thumb that 7 years after a disaster the awareness of this risk has dropped to the level as before. However, training can maintain the awareness. The loss of preparedness as shown in Fig. 4-26 is slower.

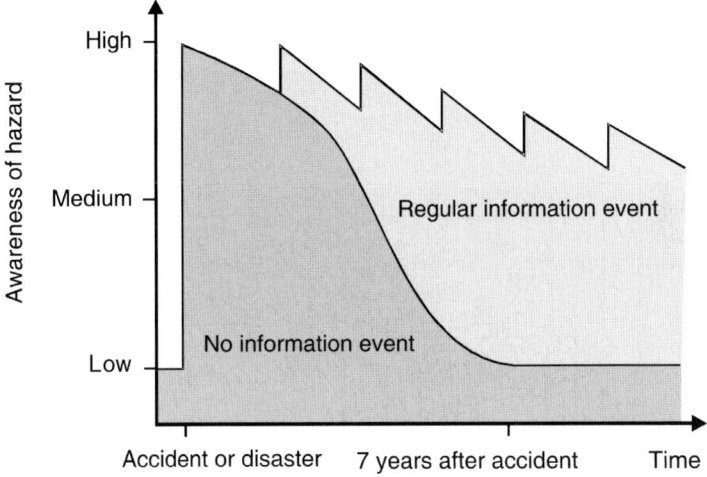

Fig. 4-25. Development of risk and hazard awareness over time after a disaster (IKSR 2002)

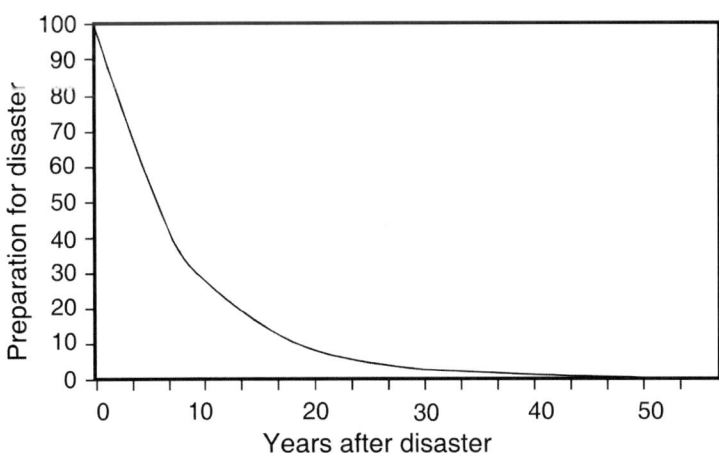

Fig. 4-26. Development of disaster preparation of a community after a disaster (Lustig 1996)

4.6 Social Subjective Risk Judgement

4.6.1 Basic Concepts

Humans, as stated in the preface of this book, are social beings. Therefore, subjective risk judgement cannot be fully understood without the consideration of the social systems (Krimsky & Golding 1992, Pidgeon et al. 2003). Theories about social systems show a great diversity and, as declared in the chapter "Indetermination and risk", they are highly complex systems (Coleman 1990).

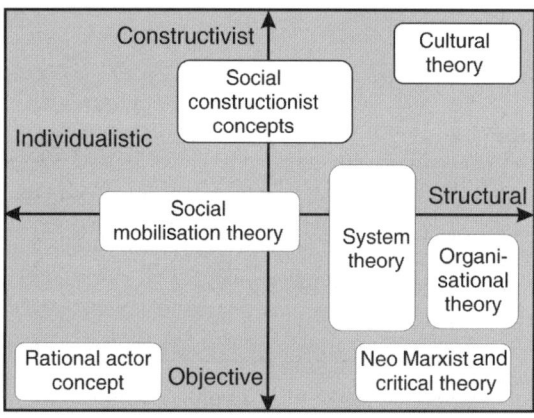

Fig. 4-27. Social theories in the context of risk (Renn 1992)

There exists a great variety of social theories describing the judgement of risks, as shown in Fig. 4-27. Even further, a clear division between individual and social elements is virtually impossible. Figure 4-28 shows possible procedures for risk estimation either on social or individual level. Such systems can permanently change the borders of the system (Fig. 4-29). This is very well known under critical conditions. If elements of a system are endangered, the system will spend resources in a short-time non-optimal way, just defending the integrity of the system. Only if the hazard exceeds a certain time or damage to the elements or the system exceeds a certain degree of seriousness, then the behaviour might shift back considering economical considerations. This has been heavily investigated among parent animals protecting their children. While under normal conditions, they attack even stronger animals and, therefore, endangering their own life, under extreme situations (starvation, drought), they leave their children without care. Similarly the, rules for military rescue actions quite often endanger more people to rescue a small number of military staff. Based on a theoretical optimisation analysis, this action should never be carried out, but under real-world conditions, it is of overwhelming importance that people keep their trust in the system. In contrast, people or societies might show the so-called learned helplessness, where they simply refuse to consider themselves as a self-sufficient system.

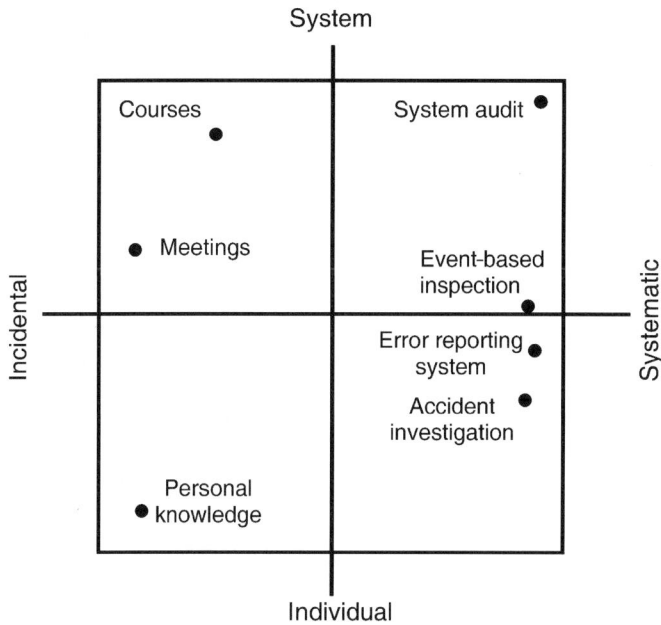

Fig. 4-28. Individual versus systematic estimation of risk (Wiig & Lindoe 2007)

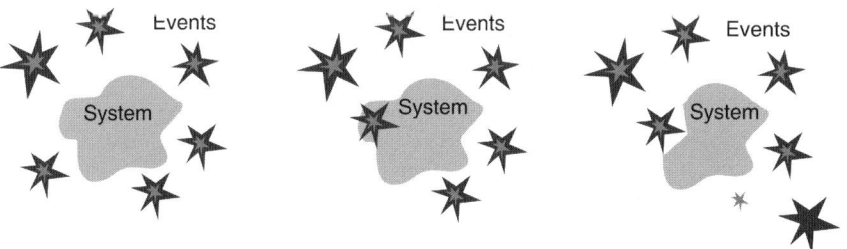

Fig. 4-29. Change of borders of a system under disastrous conditions

4.6.2 Long-Term Development of Social Subjective Judgement

As stated, not only individual behaviour but also social or cultural behaviour influence risk judgement. As already mentioned, the information flow process inside a society influences the individual risk judgement. Metzner (2002) has given a fine example of the iteration process between risk awareness and objective risks. Let us assume in Fig. 4-30 a new technology is introduced. First, the new technology increases the risk for the public, but people do not realise it. The new technology might be very fashionable or very useful, just like mobile phones. Then, there is an increasing awareness of risks caused by the new technology. The line thus bends upward becoming more vertical than horizontal. Therefore, subjective risk awareness is growing at a greater rate as compared to objective risks. Finally, at a certain level, regulators will react to the public's risk awareness by imposing new laws. Then, there remains a high level of subjective risk awareness, but the objective risk starts to fall. Finally, subjective awareness also falls. Whether both values end up being lower than the start value and whether there exists an objective risk are other questions. Figure 4-31 shows the positions of some hazards.

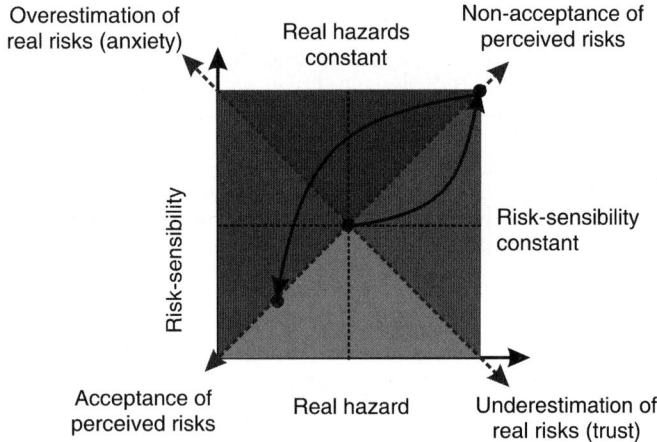

Fig. 4-30. Development of risk awareness and objective risk in societies (Metzner 2002)

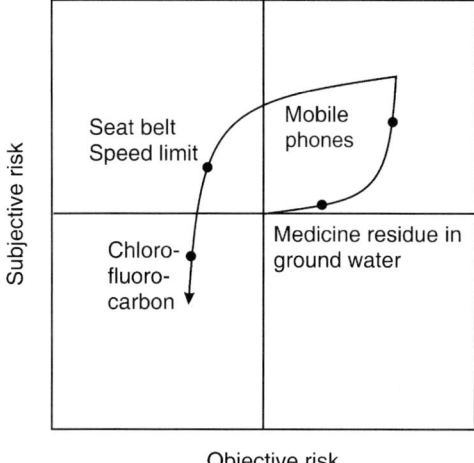

Fig. 4-31. Development of social risk judgement for different man-made risks

4.6.3 Short-Term Development of Social Subjective Judgement

Not only individual information processing but also social information processing influences risk judgement. Social information processing in developed countries is mainly carried out by the media. Nevertheless, media is not an objective risk informer. Sandman (1994) has found that:

- The amount of risk and hazard coverage is unrelated to the seriousness of the risk. Much more important for the coverage are timeliness, human interest, visual impression, proximity, prominence and drama.
- Most of the coverage focusses on social reactions, such as blame, fear, anger and outrage.
- Technical information inside a risk story is not really commendable by the audience. See also Johnson et al. (1992).
- Warning about risks is more common in media than reassuring. Since media acts as an early warning system, this is quite understandable. It is surprising that media tends to calm people in disaster situations.
- The interpretation of whether some information is assuring or disturbing depends very much on the individual reporter. Information that might be calming for experts does not necessarily have this effect on the public.

- For hazard and risk reporting, official sources are the most important.
- Frightening stories are more important for the career of journalists than calming stories. Therefore, media is more attracted to hazard stories than the reassuring ones.

However, in disaster situations, this behaviour changes and is, therefore, dependent on the time stage of a disaster. As Dombrowski (2006) and Wiltshire & Amlang (2006) have shown, disasters usually have indicators as they are related to processes, which might be released in a certain event. Figure 4-32 shows the development of such a disaster. However, like the term safety, disaster is strongly related to perception and, probably even more, to communication, since a disaster is mainly described by the failure of a social system, and therefore information about the status of such a system is required. Figure 4-33 shows the issue of risk communication in the development of a disaster. This is also called "hazard, risk and crisis chain" (Dombrowski 2006). The shape of such chains depends of the characteristics of the hazard and the reaction of the social system, as shown in Fig. 4-34.

Usually of major interest here are not only mitigation measures against hazards but also the capability of a society to deal with a crisis or a disaster. Here, Dombrowski has disintegrated such crisis management into a warning and reaction chain (Fig. 4-35). Additionally, the term warning has been separated from alarming, as shown in Table 4-7. After a disaster, usually a recovery period starts, which might last for months or years as shown in Fig. 4-36. This will yield to the term integral risk management, which does not only consider actions during the recovery, but also the possibility of a new disaster. Therefore integral risk management considers recovery actions, protections actions and disaster management actions.

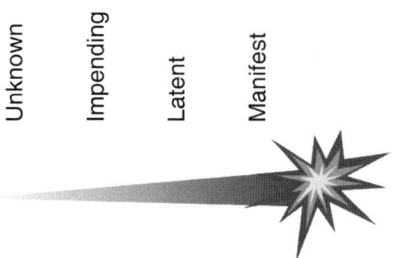

Fig. 4-32. Development of hazard perception (Dombrowski 2006)

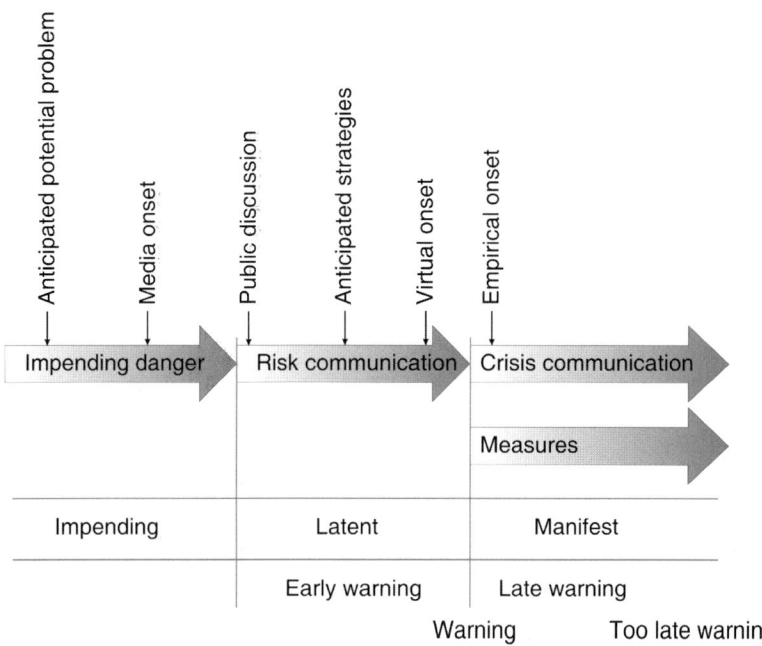

Fig. 4-33. The hazard, risk and crisis chain according to Dombrowski (2006)

Table 4-7. Differences between warning and alarming according to Dombrowski (2006)

Warning	Alarming
Social relation	Function relation
Advance of perception	Advance of time
Advance of assessment	Advance of technology
Advance of social	
Mainly indeterminate	Clear and outlined

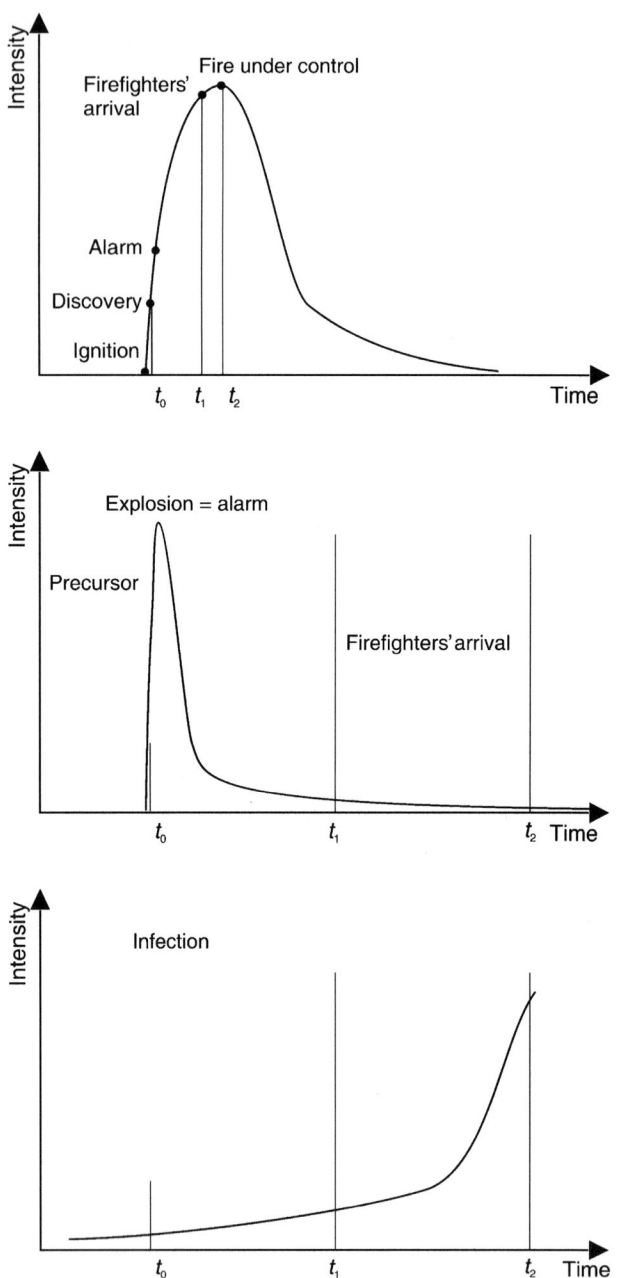

Fig. 4-34. Examples of crisis chains for fire, explosion and infections according to Dombrowski (2006)

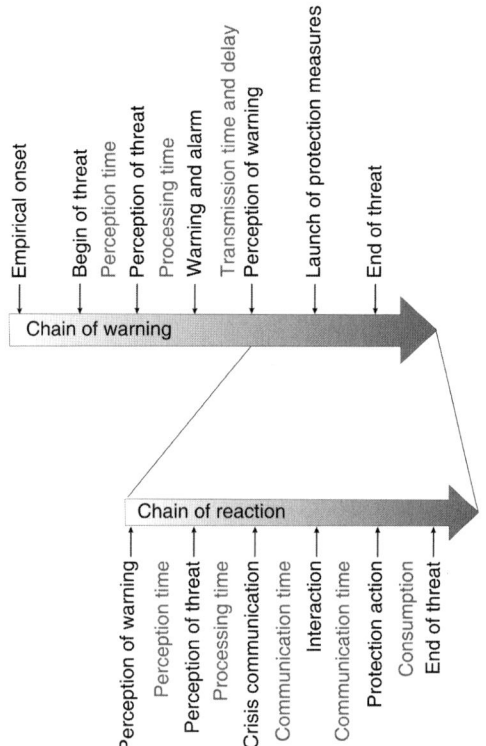

Fig. 4-35. Warning and reaction chain according to Dombrowski (2006)

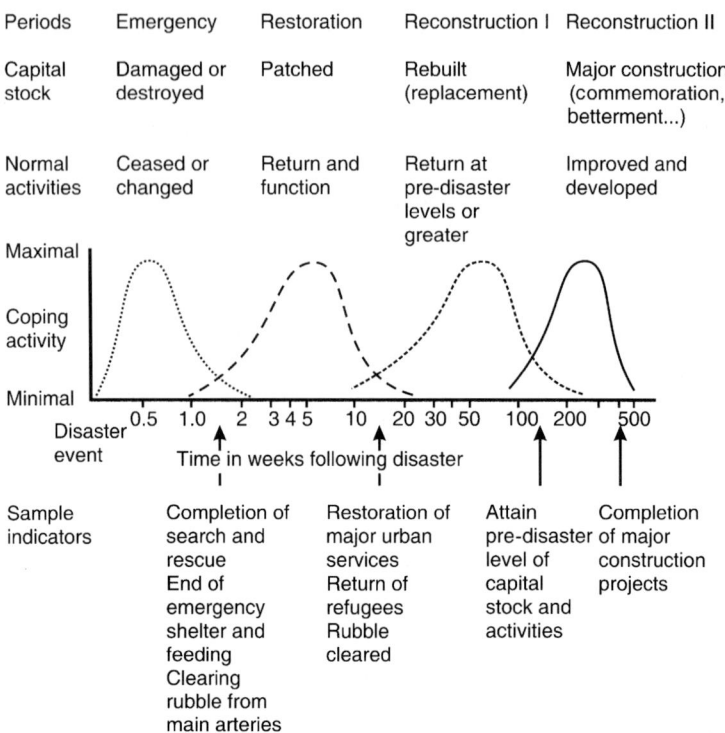

Periods	Emergency	Restoration	Reconstruction I	Reconstruction II
Capital stock	Damaged or destroyed	Patched	Rebuilt (replacement)	Major construction (commemoration, betterment...)
Normal activities	Ceased or changed	Return and function	Return at pre-disaster levels or greater	Improved and developed
Sample indicators	Completion of search and rescue End of emergency shelter and feeding Clearing rubble from main arteries	Restoration of major urban services Return of refugees Rubble cleared	Attain pre-disaster level of capital stock and activities	Completion of major construction projects

Fig. 4-36. Time scales for different time periods after disaster (Smith 1996)

4.7 Communication of Risk

The awareness of risks depends strongly on social background. Especially the communication of risks is, therefore, of major importance. William Leiss from the University of Calgary suggested: "One dollar for risk assessment and at the same time one dollar for risk communication, otherwise, one will easily pay one dollar for risk assessment and 10 or 100 dollars for risk communication." In the last few decades, intensive research has been carried out in this field, as many companies face the problem of communication risks.

In the beginning, it has been assumed that, by the introduction of objective risk measures, the public can be convinced about the acceptance of certain risks. Nevertheless, that concept left aside the limitations of the so-called objectivity in science. Scientists are also under pressure to fulfill performance measures in their respective field of science. Such performance measures reflect in the number of publications, amount of teaching hours or number of research contracts awarded. Especially in the field of medicine, this has yielded to the rule that researchers must provide information about the economic relationship between themselves and the producer of the pharmaceutics they investigate. Wiedemann (1999) has given a summary about certain effects influencing such risk communication problems (Figs. 4-37 and 4-38). Therefore, the interests of different participants in the decision-making process have to be considered.

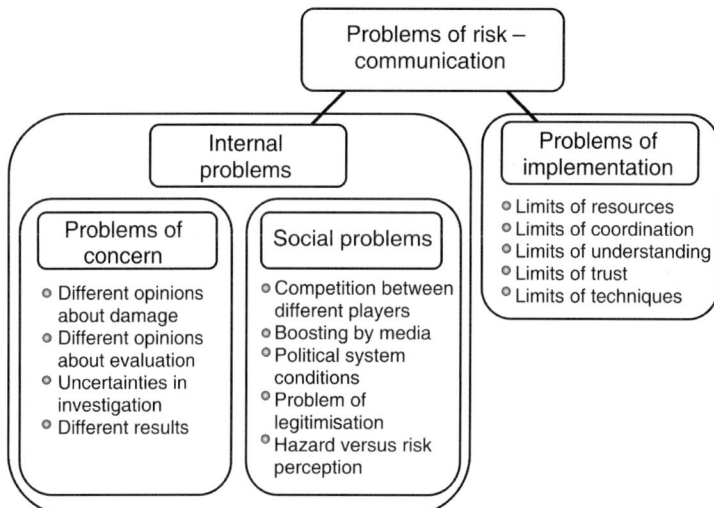

Fig. 4-37. Summary of different problems in risk communication (Wiedemann 1999)

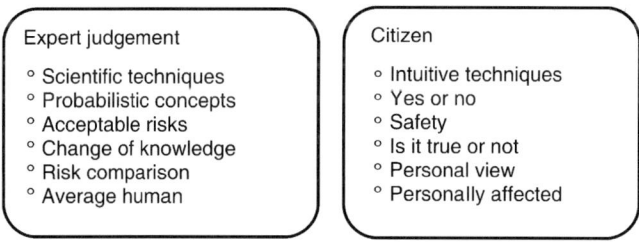

Fig. 4-38. Summary of different major assumptions for experts and citizens (Wiedemann 1999)

In the end, there is an iteration process about the acceptance of risks, which is only partly influenced by risk measures (Fig. 4-39). Metzner (2002) has given in Fig. 4-40 a classification of risk communication problems and their possible solutions. Note that in case D, there is no known solution. This case can be found quite often in real-world conditions, especially with the increasing dynamics in the social, economical and cultural systems. As already mentioned in chapter "Indetermination and risk", the numerical prediction of social systems currently cannot be done. These affect in a strong way the decision processes in society. One of the possible solutions is to keep the discussion going and refusing an ultimate solution. This leads to the concept of integral risk management, where future developments are already considered as future risk measures (Fig. 4-41). This concept can be applied to different types of disadvantages or different scales (Fig. 4-42).

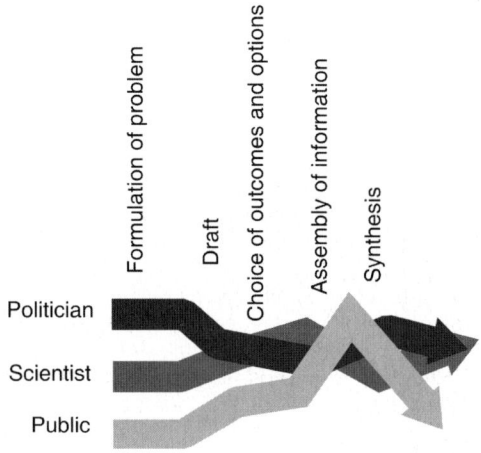

Fig. 4-39. Example of development of a decision-making process concerning a risk including contributions from politicians, scientists and the public (Wiedemann 1999)

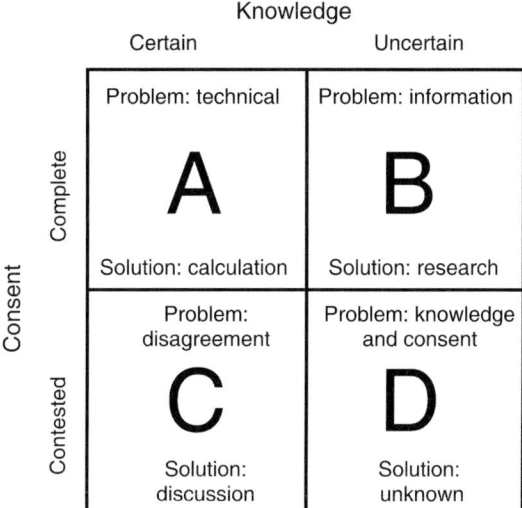

Fig. 4-40. Types of solutions for different types of problems including amount of knowledge (Metzner 2002)

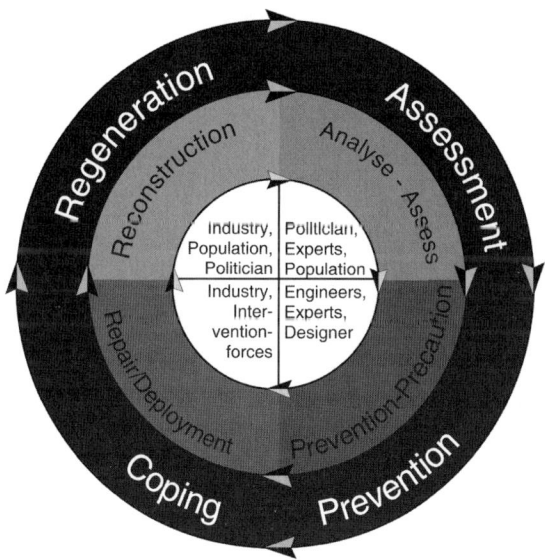

Fig. 4-41. Integral risk management cycle as a coping strategy with risk. Here, no final decision is made, instead after every major event, advantages are considered and implemented (Kienholz et al. 2004)

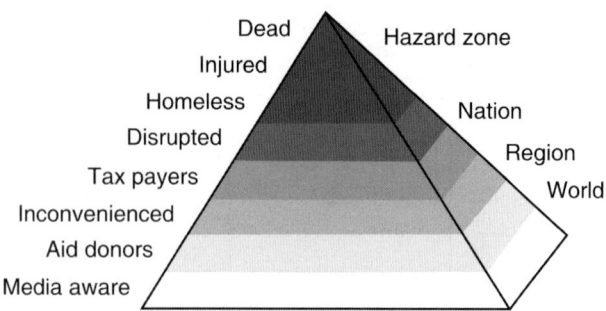

Fig. 4-42. Integral risk management implemented at different scales (Smith 1996)

However, returning to the problem of risk communication, strategies differ very much based on certain conditions. For example, based on the knowledge about information processing capability, recommendations for risk communication in critical situations give the following rules. There exists in critical situations a time window of only up to nine seconds, in which humans decide if they can trust someone and in which they gather information. Therefore in such situations, statements should have read between 7and 9 sec (21–27 words with a maximum of 30 words). The message itself should contain only three items (Covello 2001).

On the other hand, sometimes, risk communication has to be carried out over centuries and even over millenniums, for example radioactive material storage. Here, other communication rules have to be considered. For example, to remember hazards like tsunamis (which might happen only once per century), memorials have been used.

4.8 Conclusion

In many discussions, subjective risk judgement is considered a model with low relation to objective conditions. Therefore, there is a general trend to substitute subjective judgement with objective risk measures. However, such risk measures depend heavily on the system borders of the models chosen. Therefore, subjective risk judgement can be seen as another model with different system borders and different weighting values. Keeping this in mind, subjective judgement might not be considered wrong anymore, but as a different viewpoint. Mortality statistics might show that cardiovascular diseases are the main cause for death in developed countries, and subjective judgement does not consider this in an appropriate way. However, this disease might also be caused by some social conditions like working pressure. Interestingly, subjective judgement focusses very strongly on such indicators. Therefore, the subjective judgement model has different start points of the causal chains. Which model is better remains to be shown. In general, it is recommended not to simply refuse subjective models. The involvement of all aspects of risks as shown in Fig. 4-43 is mainly not answered by pure scientific models since they focus so strongly on details. One should keep in mind the results from chapter "Indetermination and risk". On the other hand, this makes it so difficult to state all the different variables that influence subjective judgement.

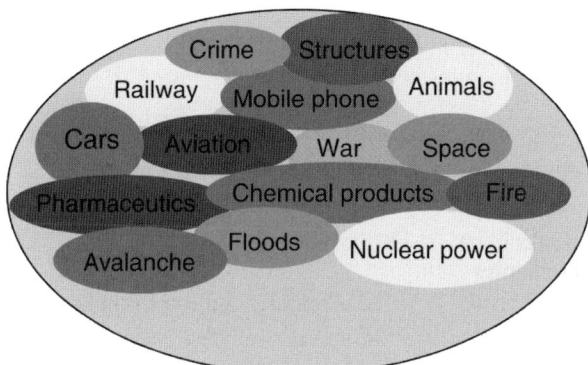

Fig. 4-43. Desideration of safety of the entire field of actions

References

Adams J (1995) Risk. University College London Press

Ajzen I (1988) Attitudes, Personality and Behavior. Open University Press, Milton Keynes

Axelrod R (1984) The Evolution of Cooperation. Basic Books, New York

Balderjahn I & Wiedemann PM (1999) Bedeutung von Risikokriterien bei der Bewertung von Umweltrisiken. Universität Potsdam

Becker GS (1976) The Economic Approach to Human Behavior. University of Chicago Press, Chicago, London

Beinder E (2007) Fetalzeit und spätere Gesundheit. Deutsches Ärzteblatt, Jg. 104, Heft 10, 9th March 2007, B567-B572

British Department for Transport (2004) 58924 Optimism Bias Guidance Document 10 06 04

Coleman J (1990) Foundations of Social Theory. Harvard University Press, Cambridge

Cook KS & Levi M (1990) The Limits of Rationality. University of Chicago Press, Chicago, London

Covello VT (2001) Canada Information Office workshop, January 2001, Canada, http://www.centreforliteracy.qc.ca/publications

Covello VT, Peters RG, Wojtecki JG & Hyde RC (2001) Risk Communication, the West Nile Virus Epidemic, and Bioterrorism: Responding to the Communication Challenges Posed by the Intentional or Unintentional Release of a Pathogen in an Urban Setting. Journal of Urban Health: Bulletin of the New York Academy of Medicine, Vol. 78, No. 2, June 2001, pp 382–391

Dake K (1991) Orientating dispositions in the perceptions of risk: an analysis of contemporary worldviews and cultural biases, Journal of Cross-Cultural Psychology 22. pp 61–82

Davis MD (1973) Game Theory: A Nontechnical Introduction. Basic Books, New York

Dawes RM (1988) Rational Choice in a Uncertain World. Harcourt Brace Janovich, San Diego

Dehaene S (2003) The neural basis of the Weber–Fechner law: a logarithmic mental number line. TRENDS in Cognitive Sciences, Vol. 7, No. 4, April 2003, pp 145–147

Dombrowski WR (2006) Die letzten Meter – Kein Kommunikations- sondern ein Interaktionsproblem. 7. Forum und Gefahrentag. DKKV & GTZ, Eschborn 2006, Presentation

Douglas M & Wildavsky A (1982) Risk and culture. Berkely

Easterlin R (1974) Does Economic Growth Improve the Human Lot. Nations and Households in Economic Growth: Essays in Honour of Moses Abramovitz, New York Academic Press, Inc., N.Y., USA

Ebbinghaus H (1913) Memory. Teachers College Press, New York

Ehlers H (1996) Lernen statt Pauken. Ein Trainingsprogramm für Erwachsene. Augustus Verlag

Eisenführ F & Weber M (1993) Rationales Entscheiden. Springer, Berlin

Erben R & Romeike F (2003) Allein auf stürmischer See – Risikomanagement für Einsteiger. Wiley-VCH GmbH & Co., Weinheim.

Feather NT (1990) Bridging the Gap between Values and Actions. Recent Applications of the Expectancy-Value Model. In: ET Higgins & RM Sorrentino (Eds), Handbook of Motivation and Cognition. Foundations of Social Behavior, Vol. 2, Guilford Press, New York

Fischhoff B, Lichtenstein S, Slovic P, Derby SL & Keeney RL (1981) Acceptable Risk, Cambridge University Press, Cambridge

Flood MM (1952) Some experimental games. Management Science 5, pp. 5–26.

Flyvbjerg B (2004) Procedures for Dealing with Optimism Bias in Transport Planning: Guidance Document (London: UK Department for Transport, June 2004)

Frank HG & Lobin G (1998) Ein bildungswissenschaftlicher Beitrag zur interlinguistischen Sprachkybernetik. München

Fromm E (1968) The revolution of hope: Towards a humanized technology. Harper & Row

Fromm E (2005) Haben oder Sein. Die seelischen Grundlagen einer neuen Gesellschaft. DTV

Geller S (2001) Sustaining participation in a safety improvement process: 10 relevant principles from behavioral science. American Society of Safety Engineers. September 2001, pp 24–29

Grams T (2007) KoopEgo. www.hs-fulda.de/~grams: Berücksichtigung von Nachbarschaft

Gray PCR & Wiedemann PM (1997) Risk and Sustainability: Mutual lessons from approaches to the use of indicators (Second Edition). Arbeiten zur Risiko-Kommunikation. Heft 61 Jülich, September 1996 (Revised and extended August 1997), Programmgruppe Mensch, Umwelt, Technik (MUT), Forschungszentrum Jülich

GTZ (2004) – German Technical Co-operation. Linking Poverty Reduction and Disaster Risk Management. Eschborn

Haney C et al. (1973) Interpersonel dynamics in a simulated prison. International Journal of Criminology & Penology 1(1), pp 69–97

Hedström P (2005) Dissection the Social. On the Principles of Analytical Sociology. Cambridge University Press, Cambridge

Huber PM (2004) Die Verantwortung für den Schutz vor terroristischen Angriffen. 12. Deutsches Atomrechtssymposium. Forum Energierecht 8, Nomos Verlagsgesellschaft, pp 195–215

IKSR – Internationale Kommission zum Schutz des Rheins (2002). Hochwasservorsorge – Maßnahmen und ihre Wirksamkeit. Koblenz

ILO (2007) The International Labour Organization, http://www.ilo.org/

Job RFS (1999) Human capacity of absolute and relative judgments and optimism bias. Application of Statistics and Probability (ICASP 8), Sydney, 1999, Vol. 1, pp 19–23

Johnson BB, Sandman PM & Miller P (1992) Testing the Role of Technical Information in Public Risk Perception http://www.piercelaw.edu/Risk/Vol3/fall/Johnson.htm

Jungermann H, Pfister HR & Fischer K (1998) Die Psychologie des Entscheidens. Spektrum Akademischer Verlag, Heidelberg

Keim DA (2002) Datenvisualisierung und Data Mining. Datenbank Spektrum, Vol. 1, No. 2, January 2002

Kienholz H, Krummenacher B, Kipfer A & Perret S (2004) Aspect of Integral Risk Management in Practice – Considerations. Österreichische Wasser- und Abfallwirtschaft. Vol. 56, Heft 3-4, March-April 2004, pp 43–50

Klinke A & Renn O (1999) Prometheus Unbound – Challenges of Risk Evaluation, Risk Classification and Risk Management. Arbeitbericht 153. Akademie für Technikfolgenabschätzung in Baden-Württemberg, Stuttgart

Knack S & Keefer P (1997) Does Social Capital Have an Economic Payoff? A Cross-Country Investigation. Quarterly Journal of Economics, 112(4), pp 1251–88.

Knack S & Zack P (2001) Trust and Growth. Economic Journal, April 2001

Koch K, McLean J, Segev R, Freed MA, Berry MJ, Balasubramanian V & Sterling P (2006). How Much the Eye Tells the Brain. Current Biology, Vol. 16, 14, pp 1428–1434

Koren G & Klein N (1991) Bias Against Negative Studies in Newspaper Reports of Medical Research, JAMA - Journal of American Medical Association, 266, pp 1824–1826

Koso M. & Hansen S (2006) Executive function and memory in posttraumatic stress disorder: a study of Bosnian war veterans. European Psychiatry. April 2006, 21(3), pp 167–173

Krimsky S & Golding D (1992) Social Theories of Risk. Praeger Publisher, Westport

Krueger LE (1989) Reconciling Fechner and Stevens: Toward a unified psychophysical law. Behavioral Brain Sciences 12, pp 251–267

Laney C & Loftus EF (2005) Traumatic memories are not necessarily accurate memories. Canadian Journal of Psychiatry. November 2005. 50(13), pp 823–828

Lübbe H (1989) Akzeptanzprobleme. Unsicherheitserfahrung in der modernen Gesellschaft. In: G Hohlneicher & E Raschke (Hrsg) Leben ohne Risiko. Verlag TÜV Rheinland, Köln 1989, pp 211–226

Luhmann N (1997) Die Moral des Risikos und das Risiko der Moral. In: G Bechmann (Hrsg) Risiko und Gesellschaft. Westdeutscher Verlag, Opladen, pp 327–338

Lustig T (1996) Sustainable Management of Natural Disasters in Developing Countries. In: V Molak (Ed), Fundamentals of Risk Analysis and Risk Management, CRC Press, Inc. 1996 pp 355–375

McNally RJ (2005) Debunking myth about trauma and memory. Canadian Journal of Psychiatry. November 2005. 50(13), pp 817–822

Metzner A (2002) Die Tücken der Objekte – Über die Risiken der Gesellschaft und ihre Wirklichkeit, Campus Verlag GmbH, Frankfurt Main

Milgram S (1974) Obedience to authority. An experiment view. Harper & Row

Milgram S (1997) Das Milgram-Experiment. Rowohlt

Moravec H (1998) Robot: mere machine to transcendent mind. Oxford University Press

Naftulin DH et al. (1973) The doctor fox lecture: A paradigm of educational seduction. Journal of medical education 48(7), pp 630–635

Newman MEJ, Girvan M & Farmer D (2002) Optimal design, robustness and risk aversion. Physical Review Letters 89, Issue 2, DOI: 10.1103/PhysRevLett.89.028301

Peters RG, Covello VT & McCallum DB (1997) The Determinants of Trust and Credibility in Environmental Risk Communication: An Empirical Study. Risk Analysis 17(1), pp 43–54

Pidgeon N, Kasperson RE & Slovic P (2003) The Social Amplification of Risk. Cambridge University Press, Cambridge

Planck W (2007) Unser Gehirn – besser als jeder Computer. www.netschool.de

Plapp T & Werner U (2002) Hochwasser, Stürme, Erdbeben und Vulkanausbrüche: Ergebnisse der Befragung zur Wahrnehmung von Risiken aus extremen Naturereignissen, Sommerakademie der Studienstiftung des Deutschen Volkes, Rot an der Rot, Lehrstuhl für Versicherungswissenschaft,

Graduiertenkolleg Naturkatastrophen, Universität Karlsruhe (TH), August 2002

Pliefke T & Peil U (2007) On the integration of equality considerations into the Life Quality Index concept for managing disaster risk. In: L Taerwe & D Proske (Eds), Proceedings of the 5th International Probabilistic Workshop, pp 267–281

Renn, O (1992) Concepts of risk: A classification. In: Krimsky S & Golding D (Hrsg), Social theories of risk. Praeger, London pp 53–79

Riemann F (1961) Grundformen der Angst – eine tiefenpsychologische Studie. Ernst Reinhardt GmbH & Co Verlag, München

Rode C, Cosmides L, Hell W & Tooby J (1999) When and why do people avoid unknown probabilities in decisions under uncertainty? Testing some predictions from optimal foraging theory. Cognition 72 (1999), pp 269–304

Rousseau DM, Sitkin SB, Burt RS & Camerer C (1998) Not so different after all: A cross-discipline view of trust. Academy of Management Review 23. pp 393–404

Ruggles RL (1996) Knowledge Management Tools. Butterworth-Heinemann Ltd

Sandman PM (1987) Risk communication – Facing Public Outrage. EPA Journal, November 1987, pp 21–22

Sandman PM (1994) Mass Media and Environmental Risk: Seven Principles.

Schacter DL (1999) Wir sind Erinnerung, Gedächtnis und Persönlichkeit, Rowohlt, Hamburg

Schütz H & Wiedemann PM (2005) Risikowahrnehmung – Forschungsansätze und Ergebnisse. Abschätzung, Bewertung und Management von Risiken: Klausurtagung des Ausschusses "Strahlenrisiko" der Strahlenschutzkommission am 27./28. Januar 2005. – München, Urban und Fischer

Schütz H, Wiedemann P, Hennings W, Mertens J & Clauberg M (2003) Vergleichende Risikobewertung: Konzepte, Probleme und Anwendungsmöglichkeiten. Abschlussbericht zum BfS-Projckt StSch 4217. Forschungszentrum Jülich GmbH, Programmgruppe "Mensch, Umwelt, Technik"

Schütz H, Wiedemann PM & Gray PCR (2000) Risk Perception Beyond the Psychometric Paradigm. Heft 78 Jülich, Februar 2000, Programmgruppe Mensch, Umwelt, Technik (MUT), Forschungszentrum Jülich

Shepard RN, Kilpatric DW & Cunningham JP (1975) The internal representation of numbers. Cognitive Psychology 7, pp 82–138

Shigley JE & Mischke CR (2001) Mechanical Engineering Design. 6th ed. McGraw Hill. Inc., New York, 2001

Shors TJ (2006) Significant life events and the shape of memories to come: A hypothesis. Neurobiology of Learning and Memory March 2006, 85(2), pp 103–115. Epub.

Shubik M (1971) The Dollar Action Game: A Paradox in Noncooperative behaviour and escalation. Journal of Conflict Resolution. 15, pp 109–111

Simon LA, Robertson JT & Doerfert DL (2003) The inclusion of Risk Communication in the Agricultural communication Curriculum: A preassessment of need. In: Thompson G. & Warnick B. (Eds), Proceedings of the 22nd Annual

Western Region Agricultural Education Research Conference, Vol. 22, Portland (Troutdale), Oregon 2003

Slaby M & Urban D (2002) Risikoakzeptanz als individuelle Entscheidung – Zur Integration der Risikoanalyse in die nutzentheoretische Entscheidungs- und Einstellungsforschung. Schriftenreihe des Institutes für Sozialwissenschaften der Universität Stuttgart, No. 1/2002 Stuttgart

Slovic P (1993) Perceived risk, trust and democracy, Risk Analysis, 13 No. 6, pp 675–682

Slovic P (1996) Risk Perception and Trust, Chapter III, 1, In: V Molak (Ed), Fundamentals of Risk Analysis and Risk Management, CRC Press, Inc., pp 233–245

Slovic P (1999) Trust, Emotion, Sex, Politics, and Science: Surveying the Risk-Assessment Battlefield, Risk Analysis, 19, No. 4, pp 689–701

Slovic P, Fischhoff B & Lichtenstein S (1980) Facts and fears. Understanding perceived risk. In: RC Schwing & WA Albers (Eds), Societal Risk Assessment: How Safe is Safe Enough? Plenum Press, New York, pp 181–214

Smith K (1996). Environmental Hazards – Assessing Risk and Reducing Disaster. Routledge, London

Starr C (1969) Social Benefit versus Technological Risk. Science, Vol. 165, September 1969, pp 1232–1238

Teger AI (1980) Too Much Invested to Quit. Pergamon. New York

Tversky A & Kahneman D (1974) Judgment under Uncertainty. Heuristics and Biases. Science, 85, pp 1124–1131

Tversky A & Kahneman D (1981) The Framing of Decisions and the Psychology of Choice. Science, 211, pp 453–489

Viscusi W (1995) Risk, Regulation and Responsibility: Principle for Australian Risk Policy. Risk. Regulation and Responsibility Promoting reason in workplace and product safety regulation. Proceedings of a conference held by the Institute of Public Affairs and the Centre for Applied Economics, Sydney, 13 July 1995. http://www.ipa.org.au/ Conferences/viscusi.html

Von Förster H & Pörksen B (1999) Die Wahrheit ist die Erfindung eines Lügners. Heidelberg

Wahle P, Patz S, Grabert J. & Wirth MJ (2005) In dreißig Tagen für das ganze Leben lernen. Rubin 2/2005, Ruhr-Universität Bochum, pp 49–55

Watzlawik P (Edr) (1985) Die erfundene Wirklichkeit – Wie wir wissen, was wir zu wissen glauben? Piper GmbH & Co. KG, München

WBGU (1998) – Wissenschaftlicher Beitrag der Bundesregierung Globale Umweltveränderungen. Jahresgutachten 1998. Welt im Wandel – Strategien zur Bewältigung globaler Umweltrisiken. Springer Heidelberg

WBGU (1999) – Wissenschaftlicher Beirat der Bundesregierung Globale Umweltveränderungen. Welt im Wandel – Umwelt und Ethik. Sondergutachten, Metropolis-Verlag, Marburg

Welter F (2004) Vertrauen und Unternehmertum im Ost-West-Vergleich. In: Vertrauen und Marktwirtschaft – Die Bedeutung von Vertrauen beim Aufbau marktwirtschaftlicher Strukturen in Osteuropa. Jörg Maier (Hrsg.)

Forschungsverband Ost- und Südosteuropa (forost) Arbeitspapier Nr. 22, Mai 2004, München, pp 7–18

Wettig J (2006) Kindheit bestimmt das Leben. Deutsches Ärzteblatt, Jahrgang 103, Heft 36, 8. September, pp B 1992–B1994

Wiedemann PM (1999) Risikokommunikation: Ansätze, Probleme und Verbesserungsmöglichkeiten. Arbeiten zur Risikokommunikation, Heft 70 Jülich, Februar 1999

Wiig S & Lindoe PH (2007) Patient safety in the interface between hospitals and risk regulators. ESRA 2007, pp 219–226

Wiltshire A & Amlang S (2006) Early Warning – From Concept to Action. The conclusion of the Third International Conference on Early Warning. 27–29 March 2006, Secretariat of the international Strategy for Disaster Reduction. German Committee for Disaster Reduction. Bonn

Wojtecki JG Jr & Peter RG (2000) Communication organizational change: Information technology meet the carbon-based employee unit. The 2000 Annual, Volume 2, Consulting. Jossey-Bass/Pfeiffer, San Francisco

Zwick MM & Renn O (2002) Wahrnehmung und Bewertung von Risiken: Ergebnisse des Risikosurvey Baden-Württemberg 2001. Nr. 202, Mai 2002, Arbeitsbericht der Akademie für Technikfolgenabschätzung und der Universität Stuttgart, Lehrstuhl für Technik- und Umweltsoziologie

5 Quality of Life – the Ultimate Risk Measure

5.1 Introduction

If indeed, as discussed in the previous chapter, people mainly compare risks and benefits instead of comparing only risks, it becomes clear that risk considerations alone are insufficient. Therefore, pure risk assessment should be exchanged by basing risks on possible benefits or advantages.

Since quality of life is described in its simplest form as access to advantages (Korsgaard 1993), it is not surprising that one of the great movements in many fields of science over the last decades has been the development and application of quality-of-life parameters. On one hand, the parallel development of these parameters in different fields can be seen as an affirmation of the conclusions. On the other hand, however, this parallel development has yielded different terms and understandings mainly caused by the different systems in the different fields of science. Therefore, it becomes important to first discuss what the construct "quality of life" describes, after which different quality-of-life parameters in different fields of science can be introduced.

5.2 Term

As suggested earlier, there are different stimuli for the actions of humans, such as possible benefits and possible risks. Such stimuli will have to be identified to understand the behaviour of humans. Here, happiness as an ultimate goal of human actions will be considered. Other stimuli concerning risk judgement have just been discussed in the preceding chapter. The definition of happiness, however, reveals the same problems as observed with the terms safety and risk. As declared by Aristotle (Joyce 1991), "When it comes to saying in what happiness consists, opinions differ, and the account given by the generality of mankind is not at all like that of the wise. The former take it to be something obvious and familiar, like pleasure or money or eminence… and often the same person actually changes his opinion. When he falls ill he says that happiness is his health, and when he is hard up he has it that it is money." As seen later in this chapter, in one definition, happiness can be translated into quality of life. Therefore, it is assumed here that the term happiness can be substituted by quality of life. The Roman writer Seneca used the term quality of life in his book, "Qualitas Vitae" (Schwarz et al. 1991). However, in antique times, the term "amenities of life" was mainly used (Amery 1975). It is not known whether happiness and quality of life were not of much interest during medieval times or were simply not documented. Different religions such as Catholicism, Judaism and Buddhism have tried to deal with the issue of happiness after they were introduced (Pasadika 2002, Barilan 2002).

Only during the 19th century did philosophers like Arthur Schopenhauer start to define terms such as "life satisfaction", which to a certain extent are comparable to quality of life: "Life satisfaction is the proportion between demands and property." (Schwarz et al. 1991) The use of terms like satisfaction already indicates the application of subjective judgement in the assessment process. That such a process also might include some cognitive errors as shown in Table 5-1, where the same external conditions might be differently interpreted by different individuals.

The term quality of life was probably introduced into scientific discussions by economist A.C. Pigou in 1920 (Pigou 1920, Noll 1999). Especially after World War II, however, with the new definition of health by the WHO (1948) and the increasingly visible limitations of the growth of material goods, other measures for the performance of life, society and mitigation action were required. This measure has increasingly become quality of life. Table 5-2 is a collection of some definitions of quality of life and gives quite a good impression about the diversity, including some core content of the construct. The definitions are in chronological order beginning with the earliest definition.

Table 5-1. Interpretation of different living conditions according to the welfare concept (Zapf 1984)

	Subjective living conditions	
Objective living conditions	Good	Bad
Good	Wellbeing	Dissonance
Bad	Adoption	Deprivation

Table 5-2. Examples of different definitions of quality of life mainly according to Noll (1999), Frei (2003), Joyce (1991)

Author	Definition
Welfare Research (Noll 1999)	"Measure of the congruence between the conditions of a certain objective life standard and the subjective evaluation of the thereby marked group of population."
Schweizer Vereinigung für Zukunftsforschung (Holzhey 1976)	"...the value in which the needs to survive, to develop and the well-being for every human according to their importance are fulfilled."
Zapf 1976 (Noll 1999)	"A new social balance by improvement of the public services, more participation of the particular in the national and private bureaucracies, more solidarity with the disorganized groups, humanization of the work with secured jobs, the human school with extended chances of education, fairer income and property distribution with constant growth, prevention instead of late repair, foresighted planning instead of short-sighted waste – these are some closer regulations for modern welfare politics, whose goals one can summarize... with the formula quality of life."
Campbell, et al. 1976 (Noll 1999)	"Quality of life typically involves a sense of achievement in one's work, an appreciation of beauty in nature and the arts, a feeling of identification with one's community, a sense of fulfilment of one's potential."
Milbrath 1978 (Noll 1999)	"I have come to the conclusion that the only defensible definition of quality of life is a general feeling of happiness."
Kirshner & Guyatt 1985 (Joyce 1991)	"The way a person feels and how he or she functions in daily activities."
Bombardier 1986	"Recent concern about cost-effectiveness of alternative treatments has reinforced the need for a comprehensive assessment of the patients' overall health, frequently referred to as quality of life."
Walker & Rosser 1988 (Joyce 1991)	"A concept encompassing a broad range of physical and psychological characteristics and limitations which describe an individual's ability to function and to derive satisfaction from doing so."
Patrick & Erickson 1988 (Joyce 1991)	"It can be defined as the value assigned to the duration of life as modified by the social opportunities, perceptions, functional states, and impairments that are influenced by disease, injuries, treatment or policy."
Küchler & Schreiber 1989	"Quality of life is a philosophical, political, economical, social science and medicinal term."
UNO 1990	"The sum of all possibilities to which are provided to a human over his/her lifetime."

(Continued)

Table 5-2. (Continued)

Author	Definition
Hasford 1991 (Frei 2003)	"The operability, illness and treatment-caused symptoms, the psychological condition as well as the measure of the social relations are substantial determinants and partially at the same time components of the quality of life."
Glatzer 1992 (Noll 1999)	"Quality of life is the goal formula of the post industrial affluent society, which has reached to the borders of growth and sees its ecological basis of existence threatened."
WHOQOL Group 1994 (Frei 2003)	"Quality of life is 'individuals' perceptions of their position in life in the context of the culture and value systems in which they live and in relation to their goals, expectations, standards and concerns."
Bullinger (1991)	"Health-referred quality of life describes a psychological construct as the physical, psychological, mental, social and functional aspects of the condition and the operability of the patients from their view."
Steinmeyer (1996) (Frei 2003)	"A high quality of life exists in the fulfilment of an internally felt or externally specified standard for internal experiencing, the observable behaviour and the environmental condition in physical, psychological, social and everyday areas of life."
Lane (1996) (Noll 1999)	"Quality of life ... defined as subjective well-being and personal growth in a healthy and prosperous environment."
Lane (1996)	"Quality of life is properly defined by the relation between two subjective or person-based elements and a set of objective circumstances. The subjective elements of a high quality of life comprise: (1) a sense of well-being and (2) personal developments, learning growth ... The objective element is conceived as quality of conditions representing opportunities for exploitation by the person living a life."
Lehmann (1996)	"…patients' perspectives on what they have, how they are doing, and how they feel about their life circumstances. At a minimum, quality of life covers persons' sense of well-being; often it also includes how they are doing (functional status), and what they have (access to resources and opportunities)"
Lexikon für Soziolgie (1997) (Noll 1999)	"Quality of life is a synonym for the being able to access all the benefits provided by a functioning economy in an industrialised country. This includes equal chances in training and occupation, material supply of the population with goods and services, a fair income distribution, the humanization of the working sphere and more equality and justice,…"
Noll (1997) (Noll 1999)	"Quality of life includes all important areas of life and covers not only the material and individual well-being being issued, but also immaterial and collective values, like liberty, justice, the security of the natural bases of life and the responsibility in relation to future generations."

Table 5-2. (Continued)

Author	Definition
Orley et al. (1998)	"…in that the latter concerns itself primarily with affective states, positive and negative. A quality of life scale is a much broader assessment and although affect-laden, it represents a subjective evaluation of oneself and of one's social and material world. The facets (of quality of life) are largely explored, either implicitly or explicitly, by determining the extent to which the subject is satisfied with them or is bothered by problems in those areas. … quality of life is thus an internal experience. It is influenced by what is happening 'out there', but it is colored by the subjects' earlier experiences, their mental state, their personality and their expectations."
Frei (2003)	"Quality of life is the result of an individual, multi-dimensional evaluation process of the interaction between person and environment. Social standards, individual value conceptions and affective factors are criteria for the evaluation of the quality of life"
Steffen 1974 (Amery 1975)	"Quality of life means deliberate creation of development of economy and society considering …safety of freedom for all and protection …of rights and laws."
Erikson 1974 (Noll 1999)	"Individual command over, under given determinants mobilizable resources, with whose help she/he can control and consciously direct his/her living conditions."

The overall impression about the term quality of life given by Table 5-2, however, can only be considered as a first step. First, it can be seen that the term includes a high degree of vagueness. On the other hand, it can be used as a basis for making decisions, and therefore not only a qualitative but, also a quantitative expression of quality of life is required. Here, the difficulties about the definition of the term become even more visible during the identification of the most important input variables for the quality-of-life parameter. Figure 5-1 divides the possible input variables into three dimensions: time, social space and space of experience. Another approach is shown in Table 5-3. Here, the possible input variables are separated into objective, subjective and social variables. Obviously the subjective variables represent a more individual approach in contrast to the societal variables, which are heavily used by politicians for decision-making. Nevertheless, subjective measures have to be included, as already shown in chapter "Subjective risk judgement"; also decision-making of politicians is strongly affected by social perceptions (Veenhofen 2001, Noll 2004, Layard 2005). This becomes visible not only in the field of welfare research but also in medicine, where it has been found that doctors are not particularly good in assessing the quality of life of the patients based on some numerical data about the health status. This is especially true for old and seriously ill people (Mueller 2002). Also, the desire to survive even with a low "quality of life" in old or seriously ill patients is systematically underestimated (Mueller 2002, Rose et al. 2000). Even people with fatal diseases and chronic handicaps point to their good quality of life under some conditions (Koch 2000a, Koch 2000b). At this point already, limitations of the numerical based quality-of-life parameters becomes visible. Here, in particular, objective measures fail to consider changes of the fundamental assumptions. For example, such changes can be the reframing of people. This term describes the positive adaptation of individuals or societies to impairing conditions. Surveys with people injured with paraplegia and people with a lotto win one year after the event showed in average the same level of life satisfaction (Brickmann et al. 1978).

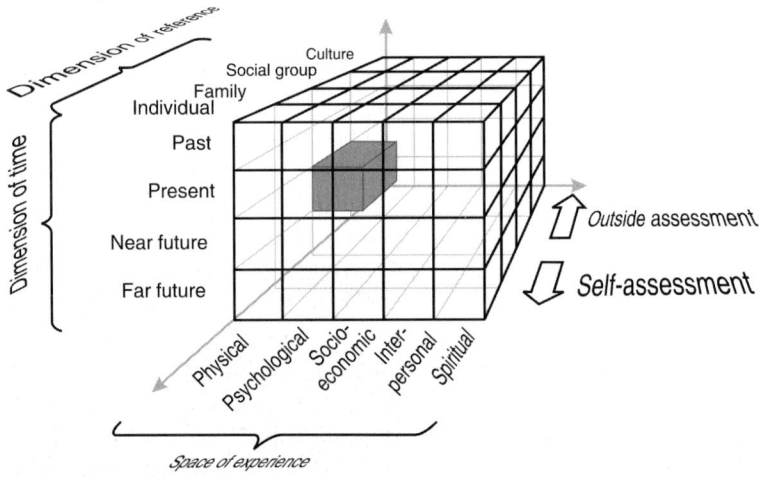

Fig. 5-1. Dimensions of quality of life according to Küchler & Schreiber (1989)

Table 5-3. Dimensions of quality of life (Delhey et al. 2002)

Objective living variables	Subjective wellbeing	Quality of society
Housing	Domain satisfaction	Social conflicts
Household composition	General life satisfaction	Trust in other people
Social relations	Happiness	Security, freedom
Participation in social life	Anxieties	Justice
Life standard	Subjective class membership	Social integrity
Income	Optimism/pessimism about	
Health	future developments	
Education and work	Judgement of the personal living conditions	

A strong theory about satisfaction, individual needs and quality of life was developed by Maslow (1970). In this theory, a classification of needs is introduced (Fig. 5-2). The classification starts with the supply of air, food or drinking water, which are essential for the survival of humans, but are weakly linked to a high quality of life. Here, further measures like the need for belonging of humans has to be fulfilled. This fits very well to some social theories, which consider freedom, safety, equity and trust as major components of quality of life (Bulmahn 1999, Knack & Zack 2001).

This consideration fits also in the concept based on the fulfilment of the three items: being, loving and having (Allardt 1975). A further concept is shown in Fig. 5-3, which is more concerned with social influences on quality of life (Noll 1999). There exist many further theories about the human needs, for example from Thomas (1923), Krech & Chrutchfield (1958), Etzioni (1968) or Alderfer (1972) (taken from Kern 1981), which will not be discussed here.

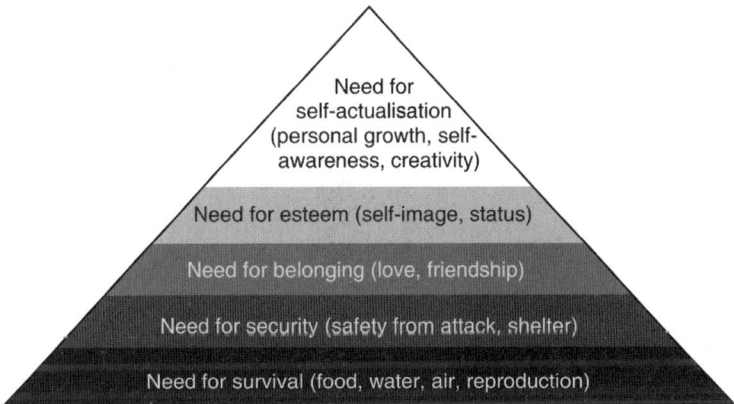

Fig. 5-2. Maslow's hierarchy of needs (Maslow 1970)

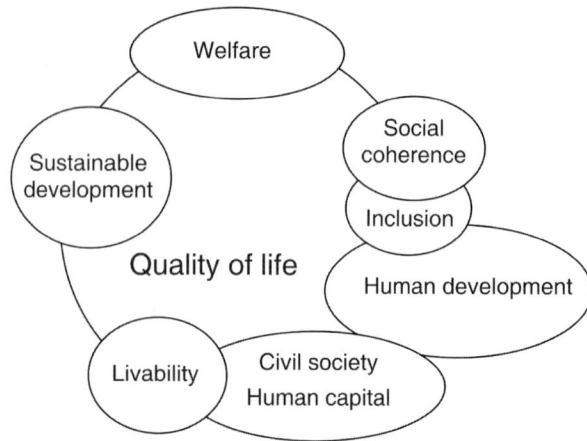

Fig. 5-3. Concept of quality of life according to Noll (1999)

In general, it becomes quite clear, not only from Fig. 5-1 and Table 5-3, but also from the definitions provided in Table 5-2, that there does not yet exist a final and clear understanding of the term (Noll 2004). Considering the interpretations about language in Chapter 1, it is doubtful that a final definition will ever be developed.

Safety, risk and quality of life are such complex constructions of real circumstances, that a simple *and* complete explanation cannot be done. As known from physics, such dualism can be found with many objects like light, which is described both as wave and as quantum.

5.3 Requirements for Quality-of-Life Measures

Although the concept will be limited, it might still be useful. The development of such measure depends on the field of application. According to Herschenbach & Henrich (1991), the following has to be considered if a quality-of-life measure is to be developed:

- Subjective versus objective data
- Self-assessment versus external assessment
- One-dimensional or multidimensional description
- Weighting functions
- Item formulation if items are kept
- Specific or general
- Duration of evolution

While Hagerty et al. (2001) have published an extended list of requirements of social quality-of-life parameters, here only some basic requirements are listed. According to Prescott-Allen (2001) and Herschenbach & Henrich (1991), quality-of-life parameters should fulfill the following properties. They should be

- Representative
- Reliable
- Feasible

A measure is representative if it considers the most important aspects and is able to identify spatial and temporal differences and developments. The latter property is also called discriminative. A measure is reliable if it reflects the objectivity, has a sound foundation, is accurate and the input data is measurable with standardised procedures. The feasibility includes the availability of data. It does not make sense to introduce indeterminable parameters.

Based on the different understanding and the different goals of application, a great variety of quality-of-life measures have been introduced. Some of the measures will be discussed in the subsequent chapters based on their field of development and application.

5.4 Medical Quality-of-Life Measures

In 1946, the WHO gave a new definition of health that extended the challenges for physicians: "Health is a state of complete physical, mental and social well-being and not merely the absence of disease or infirmity." In 1948, the Karnofsky Scale Index was introduced as a first tool to consider such a definition of health (Karnofsky et al. 1948). It classified patients based on their functional impairment using a three-level model. Table 5-4 shows the classification system. Still, the Karnofsky Index never mentioned the word quality of life, instead it spoke about "useful life" (Basu 2004).

Sir Robert Platt stated, at the Linacre Lecture in 1960, "…How often, indeed, do we physicians omit to enquire about the facts of happiness and unhappiness in our patients' lives." The term quality of life, however, was still not yet used. In an editorial published in the *Annals of Internal Medicine*, perhaps the term was introduced into the medical field: "This is nothing less than a humanistic biology that is concerned, not with material mechanisms alone, but with the wholeness of human life, with the spiritual quality of life that is unique to man." That sentence was probably partially borrowed from a quotation of Francis Bacon (1561–1625), who delegated the duty of medicine "to tune…and reduce it to harmony"; this editorial defined quality of life as "the harmony within a man, and between a man and his world" (Basu 2004).

Table 5-4. Karnofsky index classification system (Karnofsky et al. 1948)

Able to carry on nor-mal activity and to work; no special care needed	100	Normal; no complaints; no evidence of disease
	90	Able to carry on normal activity; minor signs or symptoms of disease
	80	Normal activity with effort; some signs or symptoms of disease
Unable to work; able to live at home and care for most personal needs; varying amount of assistance needed	70	Cares for self; unable to carry on normal activity or to do active work
	60	Requires occasional assistance, but is able to care for most of his personal needs
	50	Requires considerable assistance and frequent medical care
Unable to care for self; requires equivalent of institutional or hospital care; disease may be progressing rapidly	40	Disabled; requires special care and assistance
	30	Severely disabled; hospital admission is indicated although death not imminent
	20	Very sick; hospital admission necessary; active supportive treatment necessary
	10	Moribund; fatal processes progressing rapidly
	0	Dead

Since then, quality-of-life parameters have virtually experienced inflation in the field of medicine. While in the years 1966–1974, only 40 papers were published using the term quality of life, from 1986 to 1994, over 10,000 papers using the term quality of life are known (Wood-Dauphinee 1999). About 61,000 references from the years 1960 to 2004 are mentioned in the Medline search system. According to different publications, the current number of developed and applied health-related quality-of-life parameters lie somewhere around 159 (Gill & Feinstein 1994), 300 (Spilker et al. 1990), 800 (Bullinger 1997, Ahrens & LeININger 2003, Frei 2003) or possibly more than 1,000 (Porzsolt & Rist 1997) and 1,500 (Kaspar 2004).

This set of health-related quality-of-life parameters can be subdivided into groups according to different properties. One possible classification is the separation into illness-comprehensive and illness-specific quality-of-life parameters (Table 5-5). For example, the SF-36, which will be explained later, is a parameter applied for different diseases. To bypass the limitation of such illness-comprehensive measures, more and more specific quality-of-life measures were developed. The increase of differentiation of such measures is shown in Fig. 5-4. Table 5-6 lists different quality-of-life parameters for mental illness including the number of items or parameters.

The value differs between 15 and more than 300. Quality-of-life measures in other fields might also include up to 40 input variables (Estes 2003, Hudler & Richter 2002). Even if some control questions are included in such surveys, the large model differences become clearly visible. And even if different intentions are considered (Schwarz et al. 1991), some differences remain.

Table 5-5. Examples of illness-comprehensive and illness-specific health-related quality-of-life measures

Illness-comprehensive quality-of-life parameters	Illness-specific quality-of-life parameters
Nottingham Health Profile	Quality of Life Index – Cardia Version III (QLI)
Sickness Impact Profile	Seattle Angina Questionnaire (SAQ)
SF-36 (SF-12)	Angina Pectoris Quality of Life Questionnaire
WHOQoL	Minnesota Living with Heart Failure Questionnaire
EuroQol	Asthma Quality of Life Questionnaire (AQLQ)
McMaster Health Index	Questionnaire for Quality of Life with Asthma (FLA)
Questionnaire	Questionnaire for Asthma Patients (FAP)
MIMIC-Index	Asthma Questionnaire (AQ20/AQ30)
Visick-Skala	Osteoporosis Quality of Life Questionnaire
Karnofsky-Index	Quality of Life Questionnaire for Osteoporosis
Activities-of-Daily-Living I.	Osteoporosis Assessment Questionnaire (OPAQ)
Health-Status-Index	QOL Questionnaire of the European Foundation for Osteoporosis (QualEFFO)
Index-of-Well-being	Juvenile Arthritis QOL-Questionnaire (JAQQ)
Rosser-Matrix	Pain Sensitivity Scale (SES)
Rosser & Kind Index	Pain Disability Index (PDI)
Quality of Well Being Scale	

Table 5-6. Quality-of-life measure instruments for psychiatric patients (Frei 2003)

Quality-of-life measure instruments	Number of parameters
Social Interview Schedule (SIS)	48
Community Adjustment Form (CAF)	140
Satisfaction of Life Domain Scale (SLDS)	15
Oregon Quality of Life Questionnaire (OQoLQ)	246
Quality of Life Interview (QoLI)	143
Client Quality of Life Interview (CQLI)	65
California Well-Being Project Client Interview (CWBPCI)	304
Quality of Life Questionnaire (QoLQ)	63
Lancashire Quality of Life Profile (LQoLP)	100
Quality of Life Index for Mental Health (QLI-MH)	113

(Continued)

Table 5-6. (Continued)

Quality-of-life measure instruments	Number of parameters
Berlin Quality of Life Profile (BeLP)	66
Quality of Life in Depression Scale (QLDS)	35
Smith-Kline Beecham Quality of Life Scale (SBQoL)	28
Quality of Life Enjoyment and Satisfaction Questionnaire (Q-LES-Q)	93

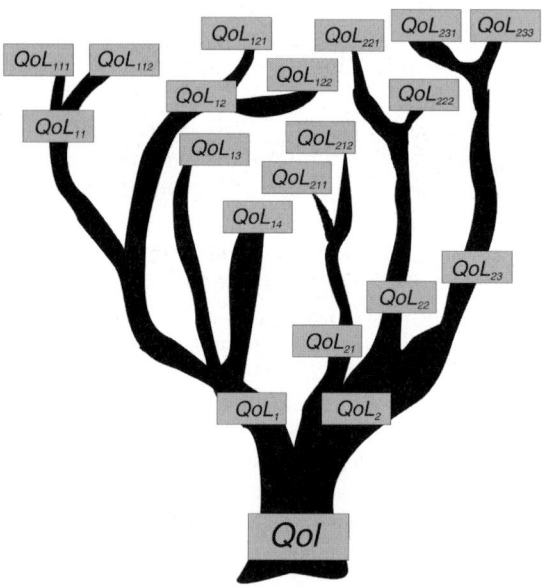

Fig. 5-4. Splitting up into increasingly specialised quality-of-life parameters (QoL – quality-of-life measure)

Another possible distinction in the design of quality-of-life measures is the differentiation between profile and index quality-of-life parameters. Profile parameters include various introductory figures or groups of introductory figures that cannot be summed up, whereas index parameters include all introductory figures in one indicator. Examples for those profile parameters from the field of medicine are the SF-36, the Sickness Impact Profile (SIP) and the Nottingham Health Profile (NHP). Examples for index parameters, also from the field of medicine, are the Karnofsky Index, the EuroQol and the Quality-of-Well-being Scale. Figure 5-5 visualises SF-36 data. The different dimensions, which have not been further combined, are visible at the axis of the diagram. Since the visualisation of high-dimensional data is a problem, there exists many different ways to present such health-related quality-of-life parameter profiles. Figure 5-6 shows a rather unusual visualisation of the same data as shown in Fig. 5-5. But, in contrast, the so-called Chernoff faces (Chernoff 1973) are used for visualisation. In Fig. 5-5, the values of the single items are clearly visible on the axes of the diagram, whereas in Fig. 5-6 the values are presented in some geometric properties of faces. The advantage of such types of diagrams is the presentation of many different dimensions; the disadvantage is the rather unscientific look.

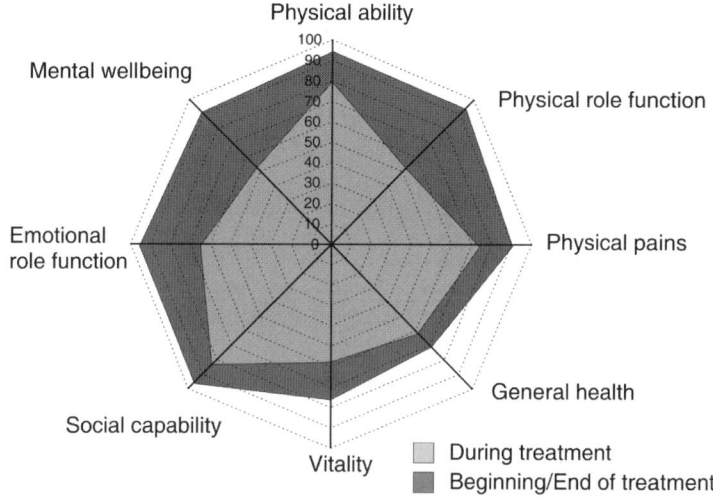

Fig. 5-5. Visualisation of SF-36 (Köhler & Proske 2007)

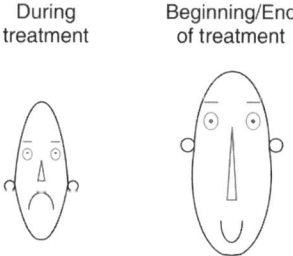

Fig. 5-6. Visualisation of SF-36 using Chernoff faces

5.4.1 SF-36

The SF-36 belongs to the medical quality-of-life surveys. SF stands for short form. Originally when the survey was developed, it included more items. However, it was found that the use of only 36 questions can yield to reliable results. Therefore, the number 36 is related to the 36 questions that are included in this measure. There has also been an even shorter form developed: the SF-12 including only 12 questions. But still, the SF-36 can be completed in less than 10 min. Tables 5-7–5-16 list the questions. Also, the questions are simple to understand, and influences from the questions to possible outcomes have been intensively investigated. Especially, translations of the SF-36 were developed very carefully. Additionally, the SF-36 also turned out to be able to show slight changes in aspects of health and wellbeing and, therefore, enabling doctors to realise the effectiveness of their treatment over time. Therefore, the SF-36 can be used over some time period to identify changes of the quality of life (Ravens-Sieberer et al. 1999, Jenkinson et al. 1993).

Table 5-7. Survey questions of the SF-36, Part 1 (Ravens-Sieberer et al. 1999)

1. In general, would you say your health is:	Excellent	Very good	Good	Fair	Poor
2. Compared to one year ago, how would you rate your health in general now?	Much better now than a year ago	Somewhat better now than a year ago	About the same as one year ago	Somewhat worse now than one year ago	Much worse now than one year ago

Table 5-8. Survey questions of the SF-36, Part 2

3.	The following items are about activities you might do during a typical day. Does your health now limit you in these activities? If so, how much?	Yes, limited a lot	Yes, limited a little	No, not limited at all
3.a	Vigorous activities, such as running, lifting heavy objects, participating in strenuous sports	1	2	3
3.b	Moderate activities, such as moving a table, pushing a vacuum cleaner, bowling or playing golf	1	2	3
3.c	Lifting or carrying groceries	1	2	3
3.d	Climbing several flights of stairs	1	2	3
3.e	Climbing one flight of stairs	1	2	3
3.f	Bending, kneeling or stooping	1	2	3
3.g	Walking more than one mile	1	2	3
3.h	Walking several blocks	1	2	3
3.i	Walking one block	1	2	3
3.j	Bathing or dressing yourself	1	2	3

Table 5-9. Survey questions of the SF-36, Part 3

		Yes	No
4.	During the past 4 weeks, have you had any of the following problems with your work or other regular daily activities as a result of your physical health?		
4.a	Cut down the amount of time you spent on work or other activities	1	2
4.b	Accomplished less than you would like	1	2
4.c	Were limited in the kind of work or other activities	1	2
4.d	Had difficulty performing the work or other activities (for example, it took extra time)	1	2

Table 5-10. Survey questions of the SF-36, Part 4

		Yes	No
5.	During the past 4 weeks, have you had any of the following problems with your work or other regular daily activities as a result of any emotional problems (such as feeling depressed or anxious)?		
5.a	Cut down the amount of time you spent on work or other activities	1	2
5.b	Accomplished less than you would like	1	2
5.c	Didn't do work or other activities as carefully as usual	1	2

Table 5-11. Survey questions of the SF-36, Part 5

		Not at all	Slightly	Moderately	Quite a bit	Extremely
6.	During the past 4 weeks, to what extent has your physical health or emotional problems interfered with your normal social activities with family, friends, neighbours or groups?	1	2	3	4	5

Table 5-12. Survey questions of the SF-36, Part 6

		Not at all	Slightly	Moderately	Quite a bit	Extremely
7.	How much bodily pain have you had during the past 4 weeks?	1	2	3	4	5

Table 5-13. Survey questions of the SF-36, Part 7

		Not at all	Slightly	Moderately	Quite a bit	Extremely
8.	During the past 4 weeks, how much did pain interfere with your normal work (including both work outside the home and housework)?	1	2	3	4	5

Table 5-14. Survey questions of the SF-36, Part 8

		All of the time	Most of the time	A good bit of the time	Some of the time	A little of the time	None of the time
9.	These questions are about how you feel and how things have been with you during the past 4 weeks. For each question, please give the one answer that comes closest to the way you have been feeling. How much of the time during the past 4 weeks.						
9.a	Did you feel full of pep?	1	2	3	4	5	6
9.b	Have you been a very nervous person?	1	2	3	4	5	6
9.c	Have you felt so down in the dumps nothing could cheer you up?	1	2	3	4	5	6
9.d	Have you felt calm and peaceful?	1	2	3	4	5	6
9.e	Did you have a lot of energy?	1	2	3	4	5	6
9.f	Have you felt down-hearted and blue?	1	2	3	4	5	6
9.g	Did you feel worn out?	1	2	3	4	5	6
9.h	Have you been a happy person?	1	2	3	4	5	6
9.i	Did you feel tired?	1	2	3	4	5	6

Table 5-15. Survey questions of the SF-36, Part 9

		All of the time	Most of the time	Some of the time	A little of the time	None of the time
10.	During the past 4 weeks, how much of the time has your physical health or emotional problems interfered with your social activities (like visiting friends, relatives, etc.)?	1	2	3	4	5

Table 5-16. Survey questions of the SF-36, Part 10

11.	How true or false is each of the following statements for you?	Definitely true	Mostly true	Don't know	Mostly false	Definitely false
11.a	I seem to get sick a little easier than other people	1	2	3	4	5
11.b	I am as healthy as anybody I know	1	2	3	4	5
11.c	I expect my health to get worse	1	2	3	4	5
11.d	My health is excellent	1	2	3	4	5

The answers to the questions will be transferred into natural numbers according to the tables. Of course, sometimes the numbers have to be calibrated or re-sorted. Because the SF-36 belongs to the class of profile indexes, the single answers are summed up according to some categories and shown in Table 5-17. As seen the computation of the SF-36 is rather simple. There are more complicated procedures to deal with missing data (Bullinger & Kirchberger 1998).

Table 5-17. Computation of the different items of the SF-36 (Bullinger & Kirchberger 1998)

Category	Sum of all item values after re-sorting	Min and max value	Max range
Physical functioning	3a+3b+3c+3d+3e+3f+3g+3h+3i+3j	10, 30	20
Role limitations	4a+4b+4c+4d	4, 8	4
Bodily pain	7+8	2, 12	10
General health perceptions	1+11a+11b+11c+11d	5, 25	20
Vitality, energy or fatigue	9a+9e+9g+9i	4, 24	20
Social functioning	6+10	2, 10	8
Role limitations due to Emotional problems	5a+5b+5c	3, 6	3
General mental health	9a+9c+9d+9f+9h	5, 30	25

Afterwards, the values are normalised according to formula (5.1):

$$\frac{\text{Raw value} - \text{Theoretical minimum value}}{\text{Theoretical maximum range}} \times \text{Range} \qquad (5\text{-}1)$$

The results can be used not only for the evaluation of treatment of single patients but also to give more general results for different patient groups compared to the entire population. Therefore, a so-called standard population has been introduced. The category values of the standard population were acquired during a representative survey. For example, in Germany about 4,500 people were consulted, of which about 3,000 answered survey sheets. Some general information about the standard population: average age 48 years, 56% were females and 78% live in couples (Bullinger & Kirchberger 1998).

Figure 5-7 shows the results of this survey for different ages and for the different categories. Obviously young persons experience a much better health-related quality of life compared to aged people. However, the decrease in the categories differs. While the reduction is rather high in the categories "physical functioning", "role limitations" and "bodily pain", the loss is comparably low in the categories "social functioning", "role limitations" due to emotional problems and "general mental health". Here, experience enables the aged people to develop alternative strategies. If one compares the sexes, it seems to be that females experience a lesser health-related quality of life. This is surprising since in many countries females experience a significantly higher life expectancy. But, based on the original goal of the application of this measure, the SF-36 should show a decreased level of health related quality of life for ill persons. Indeed this is shown in Figs. 5-8–5-10. For comparison reasons, the categories for the standard population with an age between 14 and 20 is included in the diagrams as well (Bullinger & Kirchberger 1998).

Also, as mentioned earlier, the survey can be used to describe the change in quality of life over the time of treatment. In Fig. 5-11, the increase of quality of life based on that measure is shown for patients after insertion of a mechanical heart valve. Of course, the quality of life of young and healthy people cannot be reached again, but a significant improvement is visible (Bullinger & Kirchberger 1998).

The capability of separation of different groups is a very important property of such measures. This property is also sometimes called "discriminate validity". To achieve such validity, usually a rather high amount of questions have to be asked. The SF-36 balances in a very elegant way the number of questions and the validity (Bullinger & Kirchberger 1998).

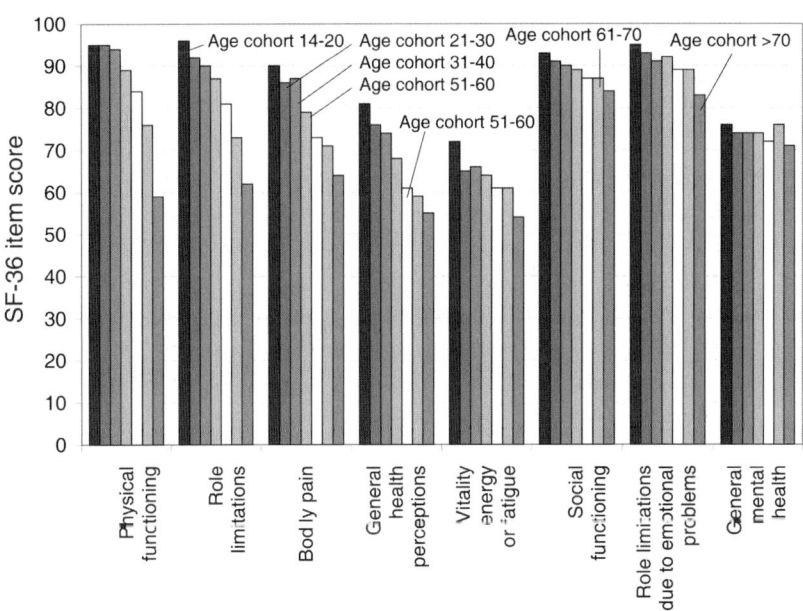

Fig. 5-7. SF-36 results for different age cohorts in Germany (Bullinger & Kirchberger 1998)

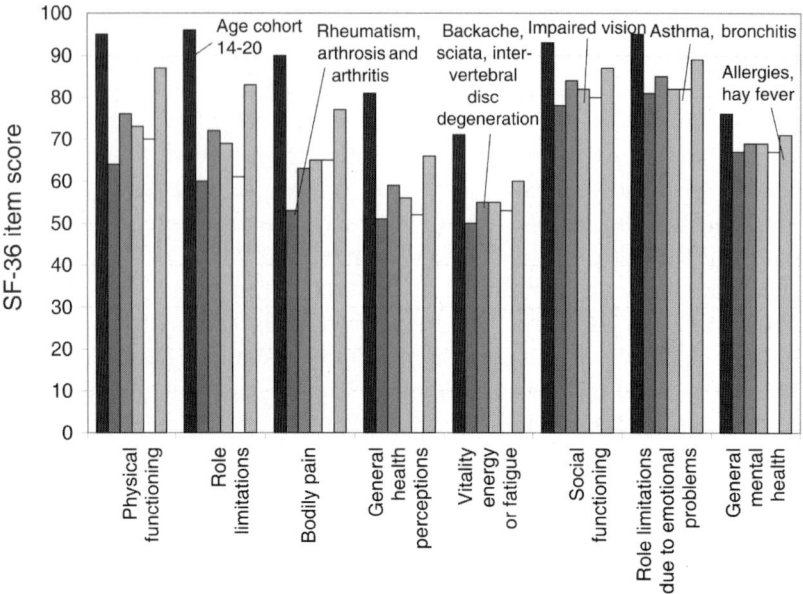

Fig. 5-8. SF-36 for different patients and one comparison cohort (Bullinger & Kirchberger 1998)

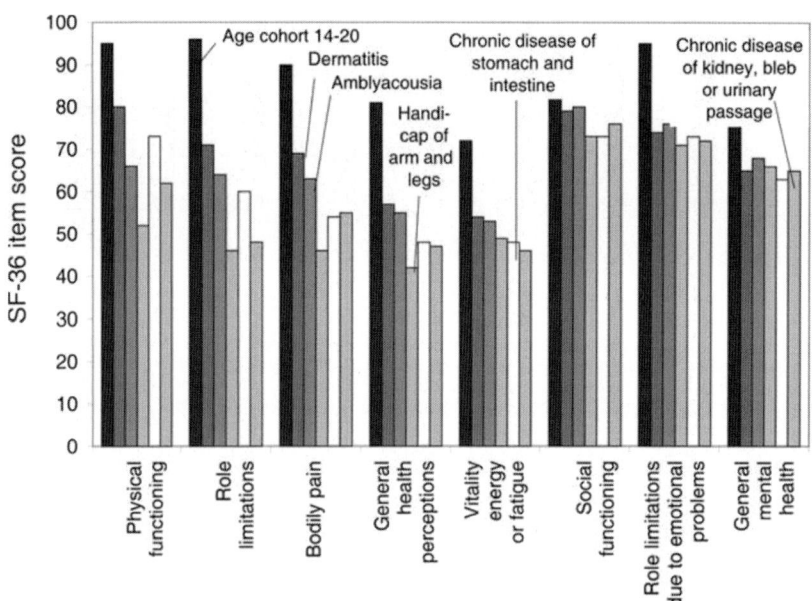

Fig. 5-9. SF-36 for different patients and one comparison cohort (Bullinger & Kirchberger 1998)

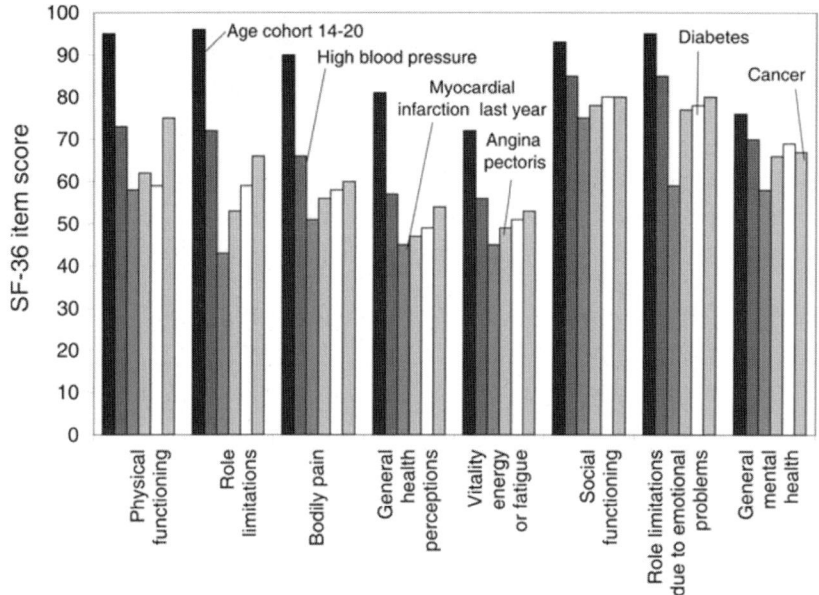

Fig. 5-10. SF-36 for different times of treatment for patients (Bullinger & Kirchberger 1998)

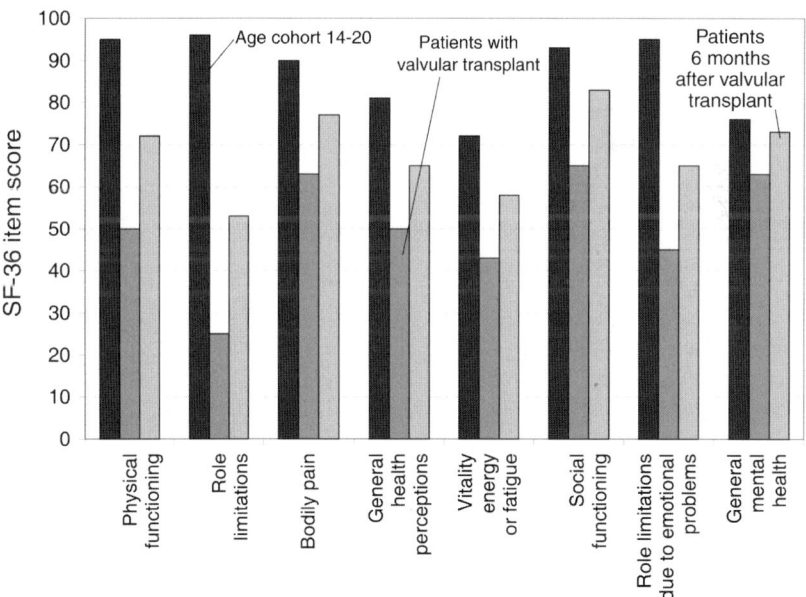

Fig. 5-11. SF-36 for patients with valvular transplant and age cohort 14–20 (Bullinger & Kirchberger 1998)

But, of course, other measures have been introduced as well. In the following, the questions from the EORTC QLQ-C30 Version 2.0 are listed in Tables 5-18–5-21 only for comparison reasons (Ravens-Sieberer et al. 1999).

The EORTC QLQ-C30 has been mainly developed for the evaluation of the quality of life for cancer patients, but some disease-specific versions have also been introduced. Here also, the number of studies worldwide lies in the range of several thousands (Ravens-Sieberer et al. 1999).

Table 5-18. Survey questions of the EORTC QLQ-C30, Part 1

		Not at all	A little	Quite a bit	Very much
1.	Do you have any trouble doing strenuous activities, like carrying a heavy shopping bag or a suitcase?	1	2	3	4
2.	Do you have any trouble taking a long walk?	1	2	3	4
3.	Do you have any trouble taking a short walk outside of the house?	1	2	3	4
4.	Do you need to stay in bed or a chair during the day?	1	2	3	4
5.	Do you need help with eating, dressing, washing yourself or using the toilet?	1	2	3	4

Table 5-19. Survey questions of the EORTC QLQ-C30, Part 2

During the last week		Not at all	A little	Quite a bit	Very much
6.	Were you limited in doing either your work or other daily activities?	1	2	3	4
7.	Were you limited in pursuing your hobbies or other leisure time activities?	1	2	3	4
8.	Were you short of breath?	1	2	3	4
9.	Have you had pain?	1	2	3	4
10.	Did you need to rest?	1	2	3	4
11.	Have you had trouble sleeping?	1	2	3	4
12.	Have you felt weak?	1	2	3	4
13.	Have you lacked appetite?	1	2	3	4
14.	Have you felt nauseated?	1	2	3	4
15.	Have you vomited?	1	2	3	4
16.	Have you been constipated?	1	2	3	4
17.	Have you had diarrhoea?	1	2	3	4
18.	Were you tired?	1	2	3	4
19.	Did pain interfere with your daily activities?	1	2	3	4

Table 5-19. (Continued)

During the last week	Not at all	A little	Quite a bit	Very much
20. Have you had difficulty in concentrating on things, like reading a newspaper or watching television?	1	2	3	4
21. Did you feel tense?	1	2	3	4
22. Did you worry?	1	2	3	4
23. Did you feel irritable?	1	2	3	4
24. Did you feel depressed?	1	2	3	4
25. Have you had difficulty remembering things?	1	2	3	4
26. Has your physical condition or medical treatment interfered with your family life?	1	2	3	4
27. Has your physical condition or medical treatment interfered with your social activities?	1	2	3	4
28. Has your physical condition or medical treatment caused you financial difficulties?	1	2	3	4

Table 5-20. Survey questions of the EORTC QLQ-C30, Part 3

29. How would you rate your overall health during the past week?						
1	2	3	4	5	6	7
Very poor						Excellent

Table 5-21. Survey questions of the EORTC QLQ-C30, Part 4

30. How would you rate your overall quality of life during the past week?						
1	2	3	4	5	6	7
Very poor						Excellent

Visible quality-of-life surveys differ in the formulation of the questions and in the assembly of the survey. Such differences can also be found in the weighting of the input data and in summarising the data. These differences, which partly can be explained by some high specialisation of some health-related quality-of-life parameters can also be found in other fields where quality-of-life parameters are used to evaluate certain actions.

5.4.2 Mental Quality-of-Life Measures

While in the SF-36 the mental state was only one sub-item, it can also be used as a specialised health-related quality-of-life parameter. Such parameters can, for example, describe mental intrusive thoughts and worries (Fehm 2000). As a simple technique, only the number and duration of worries per day has been counted. Healthy people without psychological diseases spend approximately 20% of their day worrying. Further studies show a greater diversity in results: from less than 5% up to 60% of the day spent with worries. This group of people was divided into "worry-people" and "non-worry-people". People with a general fear disturbance worry more than 60% of the day.

It is interesting to note that if people were asked directly after a worry to write down the duration of the worry, the differences between "normal" and those with a general fear disturbance disappeared and became closer. This yields to the question whether the actual duration of worries is different or the memory of worries is different. It might simply be the case that the non-worry-people just delete the memory of the time of worry and continue (Fehm 2000). Many questionnaires were developed for the investigation of worries (Table 5-22).

Table 5-22. Examples of quality-of-life measures concerning either intrusive thoughts or worries (Fehm 2000)

Quality-of-life measures	
Intrusive thoughts	Worries
Intrusive Thoughts Questionnaire	Penn State Worry Questionnaire
Distressing Thoughts Questionnaire	Worry Domains Questionnaire
Cognitive Instructions Questionnaire	Student Worry Scale
Obsessive Instructions Inventory	Worry Scale
Thought Control Questionnaire	Anxious Thoughts Inventory
White Bear Suppression Inventory	Interviews
Thought Intrusion Questionnaire	Diaries
Interviews	
Diaries	

Usually psychological burdens and diseases are correlated. A good measure for that is the so-called life change unit (LCU). Sick people usually show, in general, a much higher LCU than a standard group. Even higher mortality rates can be found in the group with higher LCU. For example, the mortality rate of women where their husband has just died is 12% as compared to 1% for the normal population (Kasten 2006).

Table 5-23 shows a list of events and their LCU values. If several events happen simultaneously, usually LCU values are added up, but values higher than 300 are not considered.

But, it should not be concluded that humans do not need any stress. As shown in the Yerkes-Dodson curve (1908) or by Hüther (2006), some stress is required to support the development of individual humans and their brain.

Table 5-23. List of LCU for cumbering events taken from Kasten (2006) and Heim (1976)

Event	Points
Minor violation of law	11
Christmas celebration	12
Vacation	13
Change in eating habits	15
Change in sleeping habits	16
Small hypothec	17
Change in social activities	18
Change in recreation	19
Relocation	20
Change in school	20
Change in working hours	20
Trouble with boss	23
Revision of personal habits	24
Change in living conditions	25
Begin or end school	26
Wife starts working	26
Outstanding personal achievement	28
Son or daughter leaving home	29
Professional change	29
Change of employment	36
Death of friend	36
Complete change of profession	39
Sex difficulties	39
Pregnancy	40
Change in health of family members	44
Retirement	45
Fired at work	47
Marriage	50
Serious disease	53
Imprisonment	63
Death of close family member	63
Marital separation	65
Divorce	73
Death of spouse	100

5.5 Social and Socio-Economical Quality-of-Life Measures

Historically, the meaning of quality of life has extended from mainly economical considerations to a much broader understanding. However, the economic influence still remains very strong. Not only the media reports regularly about the development of economical indicators but also such economic indicators still highly influence social indicators. Nearly all social indicators consider the gross domestic product per capita. The huge influence of the parameter is based on some historical development during the World War II when the growth of the weapon production was of utter importance. Later, further parameters were introduced in social indicators, like the life expectancy, the size of people (Komlos 2003), some health-related measures or some education measures. Some of such parameters will be introduced here in short.

5.5.1 Human Development Index

The Human Development Index (HDI) of the United Nations has also reached media attention. The indicator considers the average life expectancy, the per capita income and the capability to read and write (literacy). Some of the numbers are transformed, for example the income is logarithmised. Important for the HDI is the fact that not only average values can be considered but also the distribution of the values. To conduct such a consideration, usually minimal and maximal values are considered. An example of the calculation of a related value looks like that:

$$I_{ij} = \frac{\max_j X_{ij} - X_{ij}}{\max_j X_{ij} - \min_j X_{ij}} \tag{5-2}$$

The j is the counter for a country and the i is the counter for the input parameter. Adjacently, the average value from the three input variables is calculated:

$$I_j = \frac{\sum_{i=1}^{3} I_{ij}}{3} \tag{5-3}$$

The HDI is defined as

$$HDI_j = 1 - I_j \tag{5-4}$$

The computation will be shown on an example. First, the input data is required: the average life expectancy on earth is assumed as 78.4 years and the minimum average life expectancy in a country is assumed with 41.8 years. The maximum literacy in a country on the planet is assumed with 100% and the lowest value reaches only 12.3%. The logarithm of the maximum per capita income is 3.68 and the lowest value reaches 2.34. In the next step, the data from the country is required; here, it is assumed as

59.4 years life expectancy, literacy of 60 % and logarithm of per capita income reaches 2.9 (UNDP 1990a, 1990b).

According to Eq. (5-2), the following parameters are computed:

$$I_{1j} = \frac{78.4 - 59.4}{78.4 - 41.8} = 0.591 \tag{5-5}$$

$$I_{2j} = \frac{100.0 - 60.0}{100.0 - 12.3} = 0.456$$

$$I_{3j} = \frac{3.68 - 2.90}{3.68 - 2.34} = 0.582$$

Merging the three parameters, one achieves:

$$I_X = \frac{0.591 + 0.456 + 0.582}{3} = 0.519 \tag{5-6}$$

The HDI is thus

$$\text{HDI}_X = 1 - 0.519 = 0.481 \tag{5-7}$$

Based on some critics after the introduction of the measure in 1990, the HDI has been changed. Now the parameter includes not only the literacy but also the years of schooling. Also, the logarithm of the per capita income was substituted by Atkinsons formulae.

The HDI can be given as a series for many countries over time. This permits the comparison of the performance of the countries. In Fig. 5-12, the development is shown over 25 years for more than 100 nations. Additionally, Fig. 5-13 shows the frequency of the HDI for the years 1975 and 2000 worldwide. Of course, the figures can only give a rough overview, but based on these, there seems to be a constant growth of quality of life in most countries worldwide. But, not only is the increase angle different but also some nations failed to increase the quality of life at all. While in developing countries the HDI increased by up to 0.2, in developed countries the growth reached about 0.1, and some countries like Zambia or Iraq showed a declining HDI. On average, the HDI has increased worldwide by 0.1.

Although the HDI has only been introduced in 1990, the parameter has also been applied to some historical data. For example, there is HDI data available for Germany from 1920 to 1960 (Wagner 2003).

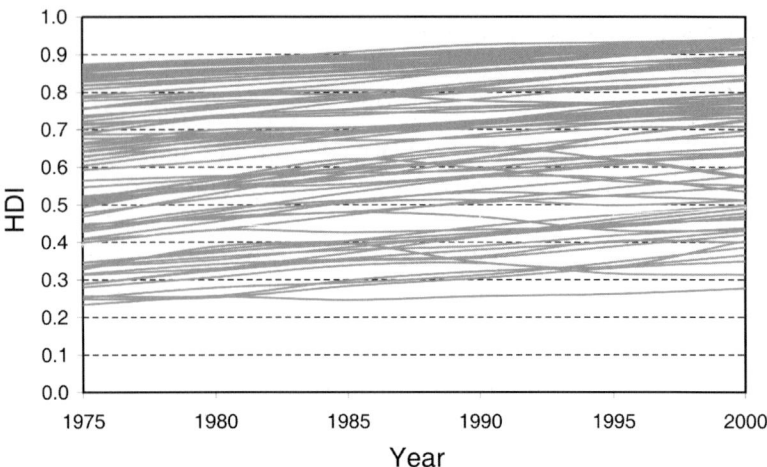

Fig. 5-12. Development of the HDI for most countries worldwide between 1975 and 2000 (UNO)

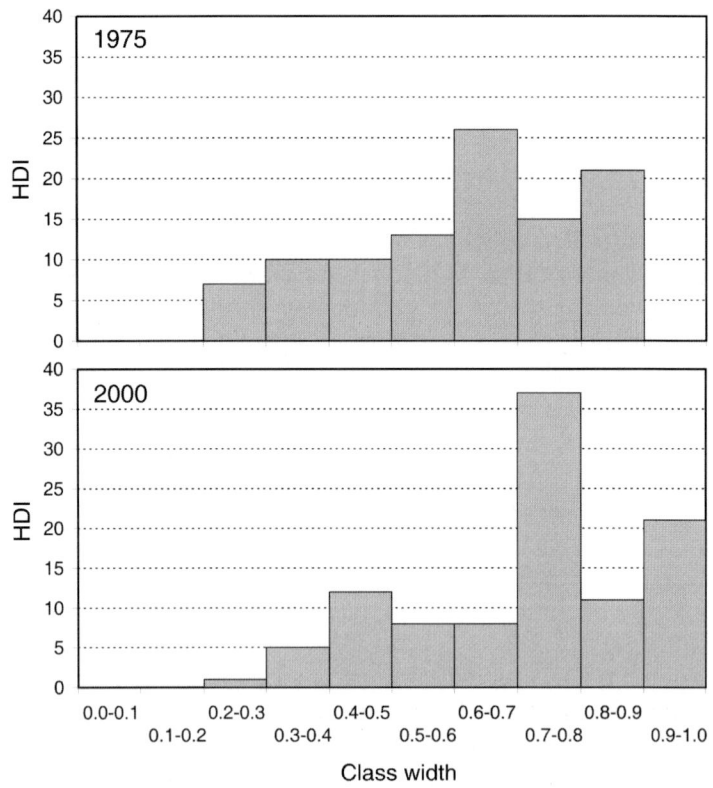

Fig. 5-13. Distribution of the HDI between 1975 and 2000 (UNO)

5.5.2 Index of Economic Well-Being

The Index of Economic Well-Being (IEWB) mainly describes the economical situation of a society as a substitute for quality of life. The measure includes four categories, which are the "consumption flow per capita",
"accumulation of productive stocks", "poverty and inequality" and the
"uncertainty in the anticipation of future incomes". These four categories
consist of further sub-parameters, which are shown in Table 5-24. The four
categories are merged using different weighting functions. For example,
consumption flow is weighted with 0.4, wealth stocks with 0.1, income
distribution with 0.25 and economic security with 0.25. But, of course,
such a weighting function is under discussion, and equal weighting of the
different categories might also be useful (Hagerty et al. 2001, CSLS 2007).

Table 5-24. Input data for the computation of the IEWB

IEWB	Consumption flow	Market consumption per capita adjusted for variation in household size
		Unpaid work per capita
		Government spending per capita
		Value of variation in work time per capita
		Regrettable expenditures per capita
		Value of variation in life expectancy time per capita
	Wealth stocks	Capital stock per capita
		Research and development per capita
		Natural resources per capita
		Human capital per capita
		Net foreign debt per capita
		Cost of environmental degradation per capita
	Income distribution	Poverty rate and poverty intensity
		Gini coefficient
	Economic security	Risk imposed by unemployment
		Financial risk from illness
		Risk from single parent poverty
		Risk from poverty at old age

The IEWB has a strong theoretical background and considers a wide
range of input data. It can be used on country level as well as on region
level to describe development. Although it has only been introduced in
1998, still times series for the parameter are available for many countries
from mainly the 1970s and 1980s, and for the US even from the beginning
of the 1960s. Taking the time period for Canada, the IEWB shows some

correlation to the development of the gross domestic product (Fig. 5-14). However, there seems to be a major difference since 1988 depending on the weighting function. If equal weights are assumed, the IEWB has declined since 1988 for some years and has only recently reached the value of 1988 again. In contrast, the development of the gross domestic product experienced a drop around 1990 but has recovered quickly. In general the IEWB shows a smoother development compared to the gross domestic product (Hagerty et al. 2001).

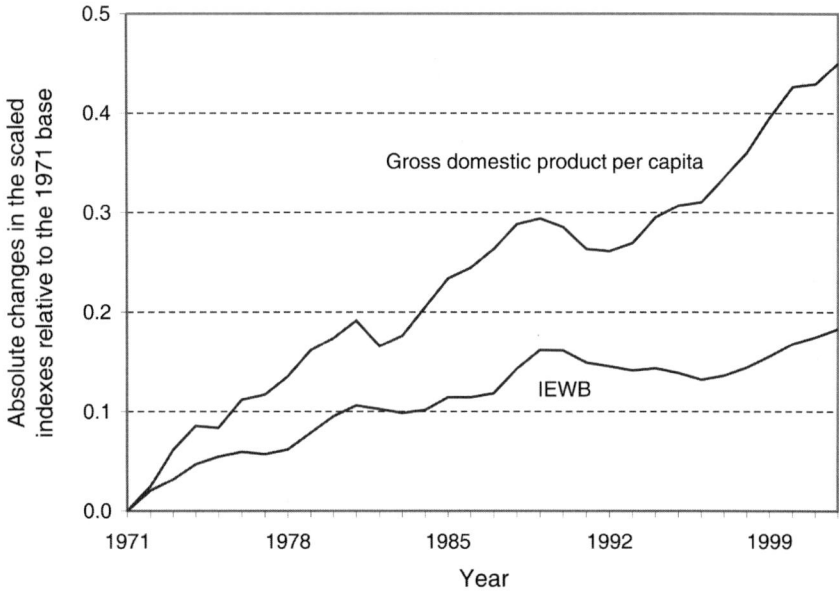

Fig. 5-14. The IEWB versus gross domestic product per capita in Canada from 1971 to 2002 (CSLS 2007)

5.5.3 Genuine Progress Indicator

The Genuine Progress Indicator (GPI) is another measure with a strong basis in economy. This becomes easily visible since it is given in dollars. However, in contrast to the gross domestic product which it heavily critics, it tries to include further conditions and activities by putting a monetary value to them. Furthermore, it subtracts some financial items included in the gross domestic product. For example, reconstruction activities after a disaster are actually increasing the gross domestic product, whereas from a common understanding, a disaster itself yields to a decline of quality of life and the restructuring is actually only the return to the wellbeing level before the disaster happened (Cobb et al. 2000).

In detail, the GPI starts with personal consumption expenditures, which are adjusted for income distribution based on the Gini coefficient. To the reached value, the following items are monetised and added:

- values for time spent at home doing house work, parenting or voluntary work
- values for the services of consumer durables like refrigerators
- values for long-lasting investigations like highways
  Also, the following items are monetised, if required, and subtracted:
- values for maintaining the level of comfort such as security to fight crime, dealing with the consequences of car accidents or air pollution
- values for social events, like divorces or crimes
- values for the decline of environmental resources like loss of farmland, wetland, forests or reduction of fossil fuels (Cobb et al. 2000)

Although there are some heavy critics against the GPI by the monetisation of some items, the results of the GPI show alarming signs. According to Venetoluis & Cobb (2004), the quality of life reached a maximum value in the middle of the 1970s and is, although not consistently, falling (Fig. 5-15). Up until 2002, the maximum value had not been reached again.

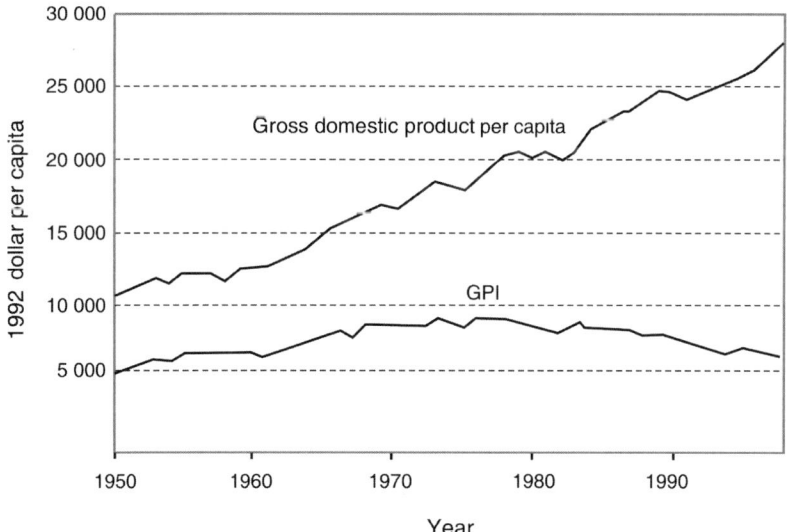

Fig. 5-15. Development of the GPI and the gross domestic product per capita in the US from 1950 to 1998 (Cobb et al. 1999)

5.5.4 American Demographics Index of Well-Being

The American Demographics Index of Well-Being is another economically based quality-of-life measure consisting of five categories, which are again split into different components (Table 5-25). The weights of the different components and categories were adjusted based on historical data. As a summary, the measure showed an overall growth of wellbeing in the US from 1990 to 1998 of about 4% (Hagerty et al. 2001).

Table 5-25. Input data for the computation of the American Demographics Index of Well-Being (Hagerty et al. 2001)

Categories	Weight	Components of categories	Weight
Consumer attitudes	1%	Consumer confidence index	47%
		Consumer expectations index	53%
Income and employment opportunity	21%	Real disposable income per capita	39%
		Employment rate	61%
Social and physical environment	10%	Number of endangered species	32%
		Crime rate	43%
		Divorce rate	25%
Leisure	50%	168 minus weekly hours worked	90%
		Real spending on recreation per capita	10%
Productivity and technology	18%	Industrial production per unit labour	69%
		Industrial production per unit labour	31%

5.5.5 Veenhoven's Happy Life-Expectancy Scale

Veenhofen generally criticises whether substitute measures about economy or other actions inside a society yield to a causal reaction in humans. Therefore, substitute measures are unable to describe the quality of life whether they are merged together or not. Therefore, Veenhofen assumes to measure the quality of life directly from the people and considering additionally the life expectancy. Both parts combined yield to "happy life expectancy" (HLE), which is seen as the years happily lived. Such surveys have been carried out for many countries. The maximum is reached in the Scandinavian countries with more than 60 happy life years, whereas minimum values are reached in Africa with values below 35 years (Fig. 5-16) (Hagerty et al. 2001).

The HLE is usually higher in nations that offer people freedom, education, affluence and harmony. These conditions are able to explain about 70% of the variation of the HLE measure. However, some very common measures like unemployment rate, state welfare, religiousness and trust in institutions do not seem to have stronger impact on the HLE. Veenhofen states that the measure is easy to understand, has sound theoretical background and is easy to use (Hagerty et al. 2001, Veenhoven 1996). A further work about happiness is that by Layard (2005).

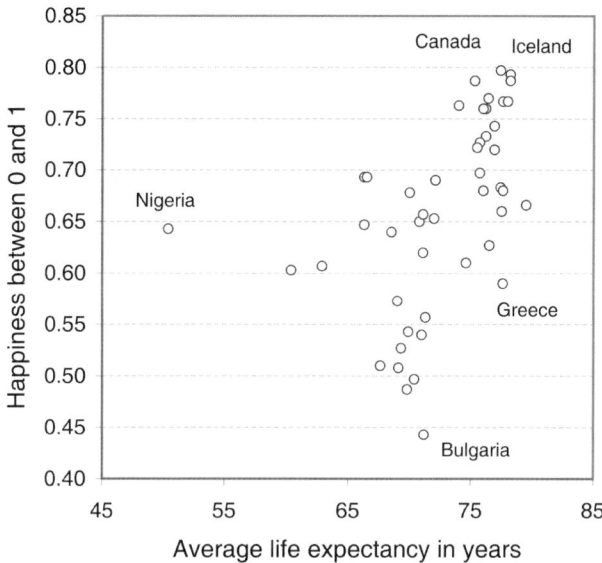

Fig. 5-16. Life expectancy versus happiness for about 50 countries at the beginning of the 1990s (Veenhoven 1996)

5.5.6 Johnston's Quality of Life Index

This parameter is entirely based on objective measures, which are used for time series analysis. The input values include health, public safety, education, employment, earnings and income, housing, family and equality. Overall, there are 21 variables. The index is dominated by some economical indicators (Hagerty et al. 2001).

5.5.7 Miringoffs Index of Social Health or the Fordham Index

This quality-of-life measure was developed by Miringoff and Miringoff in 1996 at the Fordham Institute of Innovation and Social Policy. The measure includes 16 or 17 objective and reliable components from several domains. The components have changed over time and were:

- Infant mortality
- Child abuse

- Children in poverty
- Teenage suicide
- Drug abuse
- High school drop-out
- Teenage birth
- Unemployment
- Average weekly earnings
- Health insurance coverage
- Poverty among people over 65
- Life expectancy at age 65
- Out-of-pocket health costs among those aged 65 and over
- Violent crime rate
- Alcohol-related traffic fatalities
- Food stamp coverage
- Access to affordable housing
- Income inequality

In contrast to the other quality-of-life parameters, here parameters are related to age groups. For example, the first three parameters are for children, whereas others are only for young people, adults or elderly. A few components are used for all age groups. The components are somehow normalised and then merged together by equal weighting of the components. Figure 5-17 shows the development of this parameter in comparison to the gross domestic product per capita for Canada for a time period of about 30 years (Carrie 2004, HRSDC 1997, Hagerty et al. 2001).

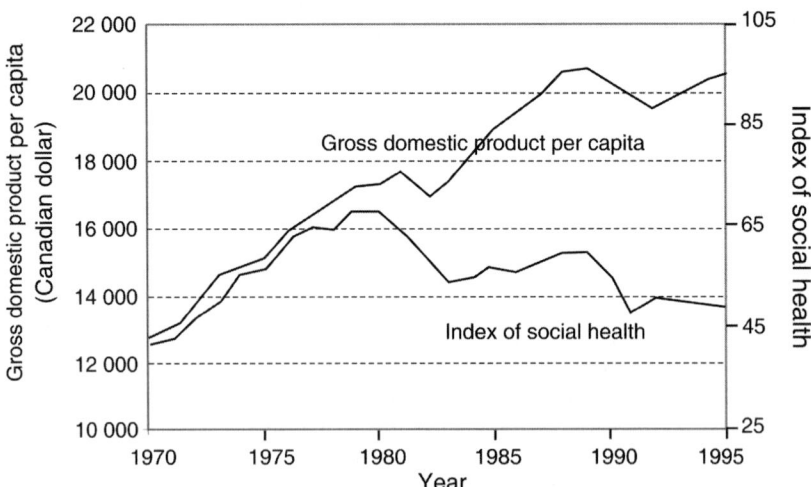

Fig. 5-17. Index of social health and gross domestic product per capita for Canada from 1970 to 1995 (FIISP 1995)

5.5.8 Estes Index of Social Progress

Estes has introduced in 1984 this Index of Social Progress. There exist variations from the original index, such as the weighted Index of Social Progress. This one includes 10 categories and 40 components, but the latter one includes 46 parameters. Table 5-26 lists input data for the previous one. Figure 5-18 gives an overview about the diversity of the development of different countries based on Estes Index of Social Progress (Hagerty et al. 2001, Estes 1992, 1988, 1990).

Table 5-26. Input data for the computation of Estes Index of Social Progress

Categories	Component
Education	Adult literacy rate
	Primary school completion rate
	Average years of schooling
Health status	Life expectation at birth
	Infant mortality rate
	Under-five child mortality rate
	Physician per 100,000 population
	Percent of children immunised against DPT at age 1
	Percentage of population using proved water sources
	Percent of population undernourished
Women status	Female adult literacy as % of males
	Contraceptive prevalence among married women
	Maternal mortality ratio
	Female secondary school enrollment as % of males
	Seats in Parliament held by women as % of total
Defense effort	Military expenditures as % of gross domestic product
Economy	Per capita gross national income (as measured by PPP)
	Percent growth in gross domestic product
	Unemployment rate
	Total external debt service as % of exports of goods and services
	GINI Index score, varied
Demography	Average annual population growth rate
	Percent of population aged 14 years and younger
	Percent of population aged 65 years and older
Environment	Nationally protected areas
	Average annual disaster-related deaths per million population
	Per capita metric tonnes of carbon dioxide emissions
Social chaos	Strength of political rights
	Strength of civil liberties
	Total deaths in major armed conflicts since inception
	Number of externally displaced persons per 100,000 population

(Continued)

Table 5-26. (Continued)

Categories	Component
	Perceived corruption index
Cultural diversity	Largest percentage of population sharing the same or similar racial/ethnic origins
	Largest percentage of population sharing the same or similar religious beliefs
	Largest share of population sharing the same mother tongue
Welfare effort	Age first national law – old age, invalidity and death
	Age first national law – sickness and maternity
	Age first national law – work injury
	Age first national law – unemployment
	Age first national law – family allowance

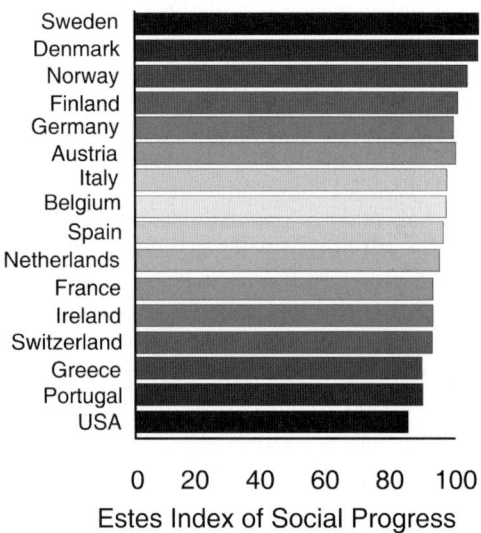

Fig. 5-18. Estes Index of Social Progress for the year 2001 (Burke 2004)

5.5.9 Diener's Basic and Advanced Quality of Life Index

Diener has complained about the limitations of different quality-of-life parameters, such as no sound theoretical work, the choice of weighting is rather arbitrary, rank order statistics might not visualise differences in a proper way and scale values are partially unable to reflect some universal values. Based on some work by Schwarz and Maslow, two parameters were developed by Diener: the Basic Index and the Advance Index. Both indexes incorporate seven variables, which also might be further built up by other variables (Hagerty et al. 2001).

5.5.10 Michalos' North America Social Report

The Michalos North America Social Report compared living conditions in the US and Canada between 1964 and 1974. The parameter has used up to 126 input variables (Hagerty et al. 2001).

5.5.11 Eurobarometer

The Eurobarometer is a public opinion survey in the European Union. The questions are asked to a representative sample of the population of the member states. The survey is carried out twice a year. The survey includes questions about the satisfaction with life as a whole and with the democratic conditions in the country (Hagerty 2001).

5.5.12 ZUMA Index

The ZUMA Index has been developed for the comparison of the development among the European Union member states. The parameter was developed at the Centre for Survey Research and Methodology (ZUMA) in Germany. The measure is based on 25 indicators, for example considering the per capita gross domestic product, expenditures on education as percent of gross domestic product, rate of female employment, unemployment rate, infant mortality, size of railway system, emissions of carbon dioxide and sulfur dioxide, percentage of personal income spent on food, average life span, number of physicians per 1,000 inhabitants, ready access to water, noise and air pollution or other environmental damages (Hagerty 2001).

5.5.13 Human Poverty Index

The Human Poverty Index (HPI) is generally considered a better parameter for the description and change of life conditions in poor countries. It is based on

- the probability at birth of not reaching the age of 40
- percentage of adults lacking functional literacy skills
- weighted measure considering the percentage of children underweight and percentage of population without access to safe water

5.5.14 Further Measures

The collection of social measures is by no means complete. Some examples are the International Living Index, State Level Quality of Life Survey,

Philippines Weather Station, Netherlands Living Conditions Index (LCI) and the Swedish ULF System (Hagerty 2001).

Additionally, in many journals, life satisfaction is investigated through, for example, maps of future development in Germany (Handelsblatt 2007) or the Life Satisfaction Map of Germany (DNN 2005). Such procedures are also applied internationally like the worldwide quality of living survey by Mercer (MHRC 2007) or Money's Best Places to Live.

5.6 Economical Quality-of-Life Measures

5.6.1 Gross Domestic Product

The gross domestic product (GDP) is the total value of goods and services produced by labour and property within a country during a specific period. In 1991, GDP became the US government's primary measure of economic activity in the nation, replacing gross national product (GNP), which is the total value of goods and services produced by labour and property supplied by US residents (but not necessarily located within the country). The value itself is not so much of interest, but mainly the growth is often quoted as an indicator for improvements.

Figure 5-19 shows the development of growth of the GNP over the last 150 years for Germany. The diagram does not only show the yearly values but also a moving average giving a better impression about longer development of this indicator.

Before 1850, about 50–60% of the German population were working in the agricultural sector. And only from 1871, Germany was unified. Since about 1850, there was a constant average growth rate of 2%, reaching in some years up to 8%. After the unification of Germany, the average value reached up to 4% only to drop then for some time. This period was called industrial expansion period crisis. From about 1883 to the start of the World War I, the average growth rate reached nearly 3%. The population working in the industry increased from 20% in 1883 to 40% in 1914. During the World War I, production fell and the pre-war level was only reached in 1928. The growth was again stopped by the world economy crisis. While in the times before the World War I, the unemployment rate was between 1 and 6% and after the war, it was between 1 and 2%; this value increased up to 30% during the major economical crisis. Additionally, Germany was strongly hit by the heavy inflation, and many citizens lost their entire savings. The great independence of the German Federal Bank is one of the results of the events in the 1920s (Pätzold 2004).

Since the beginnings of the 1930s, the economy recovered in Germany and reached up to 10% growth. However, already in 1934, the Nazis started to change the German industry towards a war industry. This was visible by the much lower growth of consumption compared to the overall growth of the industry, a loss of income for employees and an absorption of income by saving programmes of the government. Therefore, this time was called "deformed growth".

After World War II, in the 1950s, West Germany reached an extremely high economical growth of up to 10%. This growth slowed down to 2–4% in the 1960s, which still was a time of full employment. Approximately since the middle of the 1970s, Germany reached only a restrained growth of 1–2%. This value corresponded very well to the average growth over the last 200 years. During that time some 5–8-year economical cycles were visible, but also longer periods were known (Kontradieff cycles).

For comparison reasons, Figs. 5-20–5-25 show the developments of the GDP per capita for most regions of the world over the last 50 years in 1990 US dollars converted at Geary Khamis purchasing power parity (GGDC 2007).

In general, the development in Germany was comparable to many other countries. However, some countries showed some special effects, for example the discovery of oil in Norway or the crisis in some Asian countries at the end of the 1990s. Also, the crisis in the eastern European countries after the collapse of the communist systems is visible. It took for these countries several years to reach the pre-collapse level. Finally, wars are visible as a major cause of collapse of economical systems, for example in Iraq.

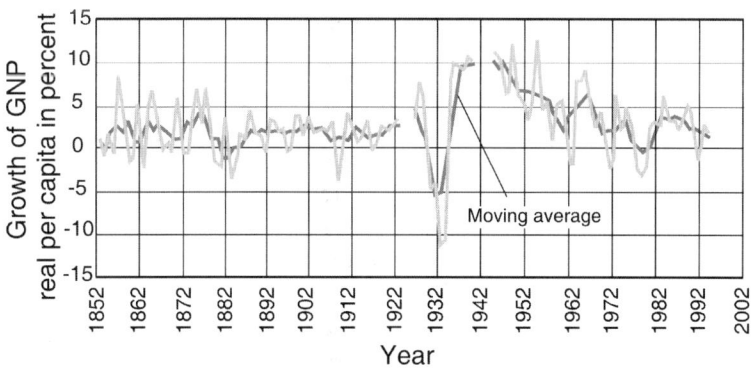

Fig. 5-19. Development of German GNP real per capita in percent (Pätzold 2004)

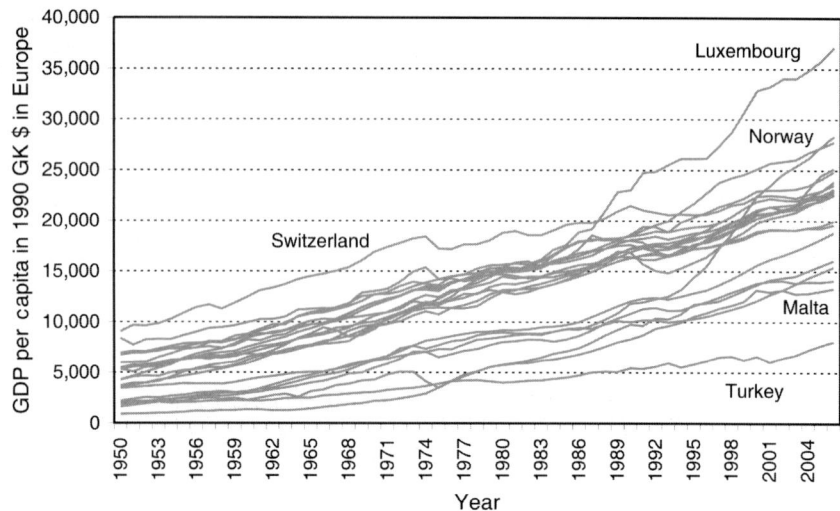

Fig. 5-20. Development of the GDP per capita in dollars in the last years in Europe (GGDC 2007)

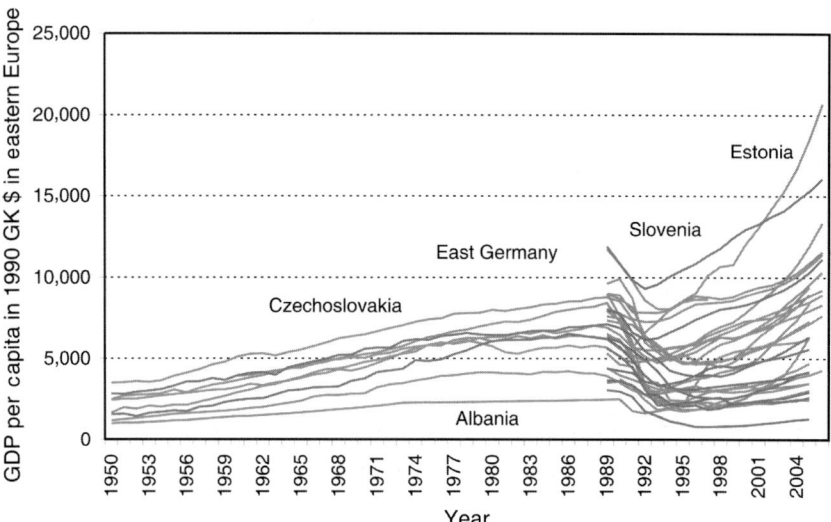

Fig. 5-21. Development of the GDP per capita in dollars for eastern Europe (GGDC 2007)

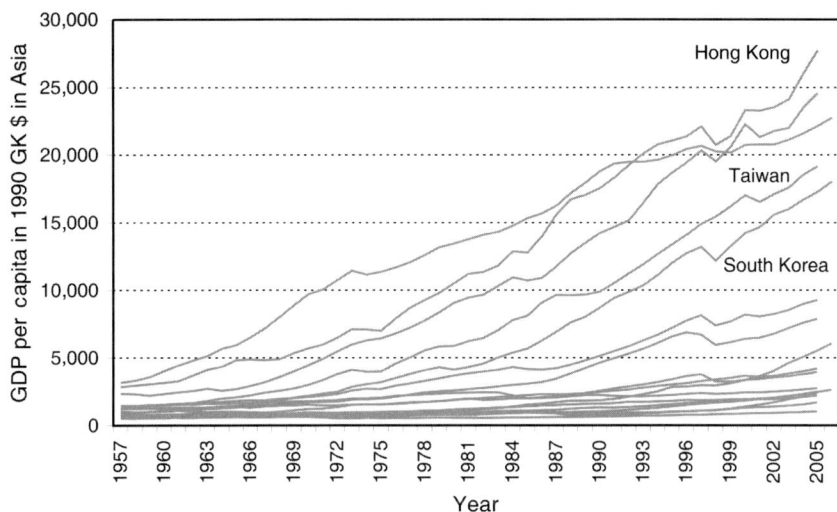

Fig. 5-22. Development of the GDP per capita in dollars for Asia (GGDC 2007)

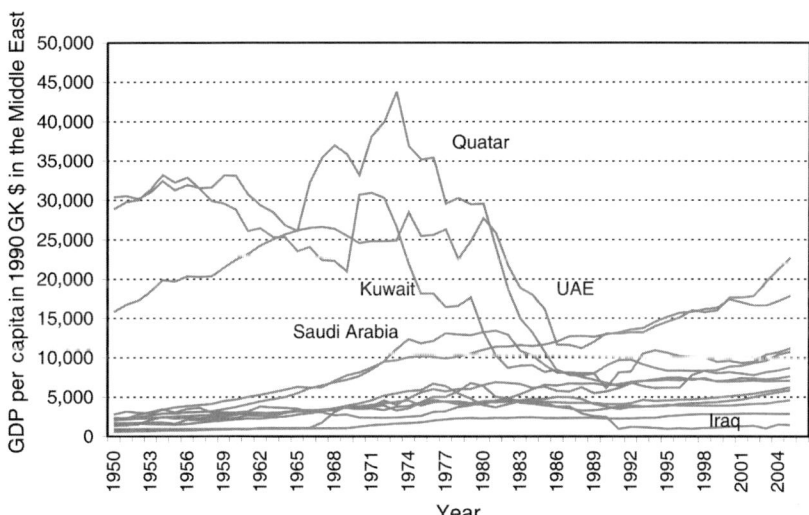

Fig. 5-23. Development of the GDP per capita in dollars for the Middle East (GGDC 2007)

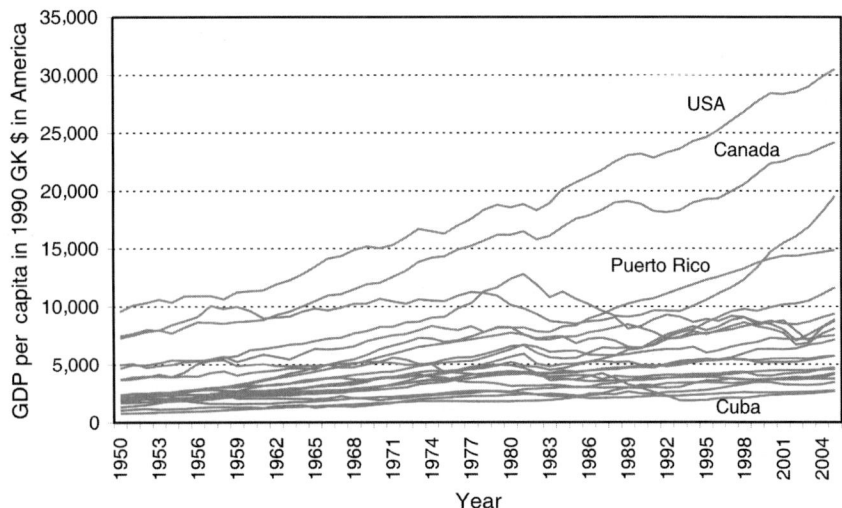

Fig. 5-24. Development of the GDP per capita in dollars for America (GGDC 2007)

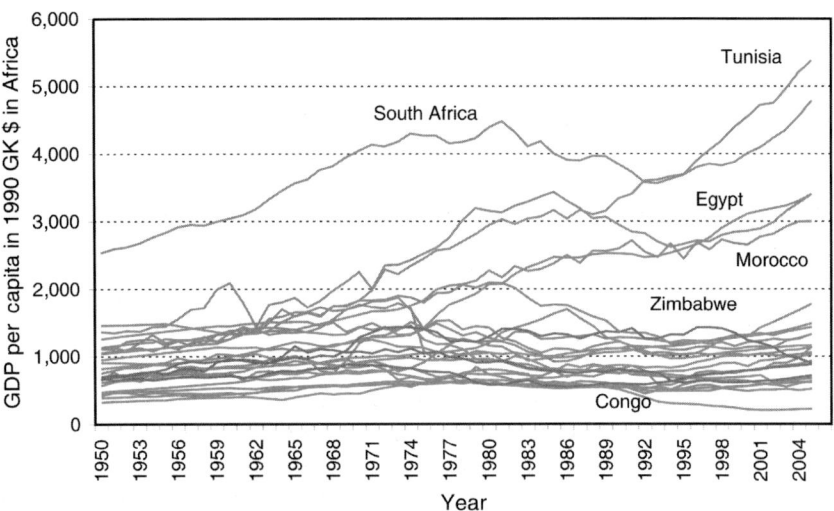

Fig. 5-25. Development of the GDP per capita in dollars for Africa (GGDC 2007)

5.6.2 Unemployment Rate

As already mentioned, serious economical crises are joined by high unemployment rates in a region or a country. This parameter can also be used for quality-of-life estimations for a society.

Figure 5-26 shows the development of the unemployment rate in Germany since about 120 years. Highest values were reached during the economical crisis in the early 20th century. However, in the last 30 years, the unemployment rate shows a growth, only interrupted by some economical cycles. In some east German regions, already values close to the unemployment rate of the 1920s were reached (Metz 2000).

In contrast to the unemployment rate, which follows the economical growth, some alternative measures like the net investment rate can be used. Nevertheless, such values are more difficult to relate to quality of life.

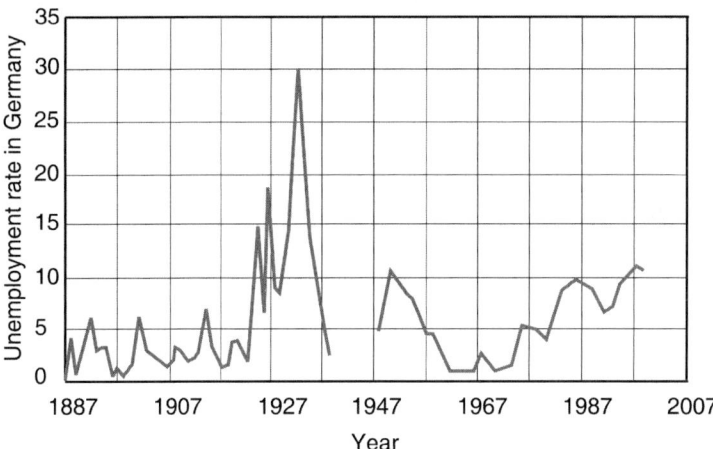

Fig. 5-26. Development of unemployment rate in Germany over the last 120 years (Metz 2000)

5.6.3 Ownership of Goods

Results of economical power in terms of distribution and number of consumer products might be more appropriate for quality-of-life investigations based on economical numbers. Indeed, the ownership of goods can be used as an easy measure, but still some requirements have to be fulfilled. For example, such products will have to be in use for some time to permit comparison. Also, they should be helpful for most members of the public. Such technical products are cars, telephones, refrigerators, television or internet access. Figure 5-27 shows the spatial and temporal distribution of the average ownership of telephones. Figure 5-28 gives data for the geographic distribution of telephones, traffic density and internet hosts. However, the limitations have already become visible since mobile phones have completely changed the conditions of the use of telephones in the last decade.

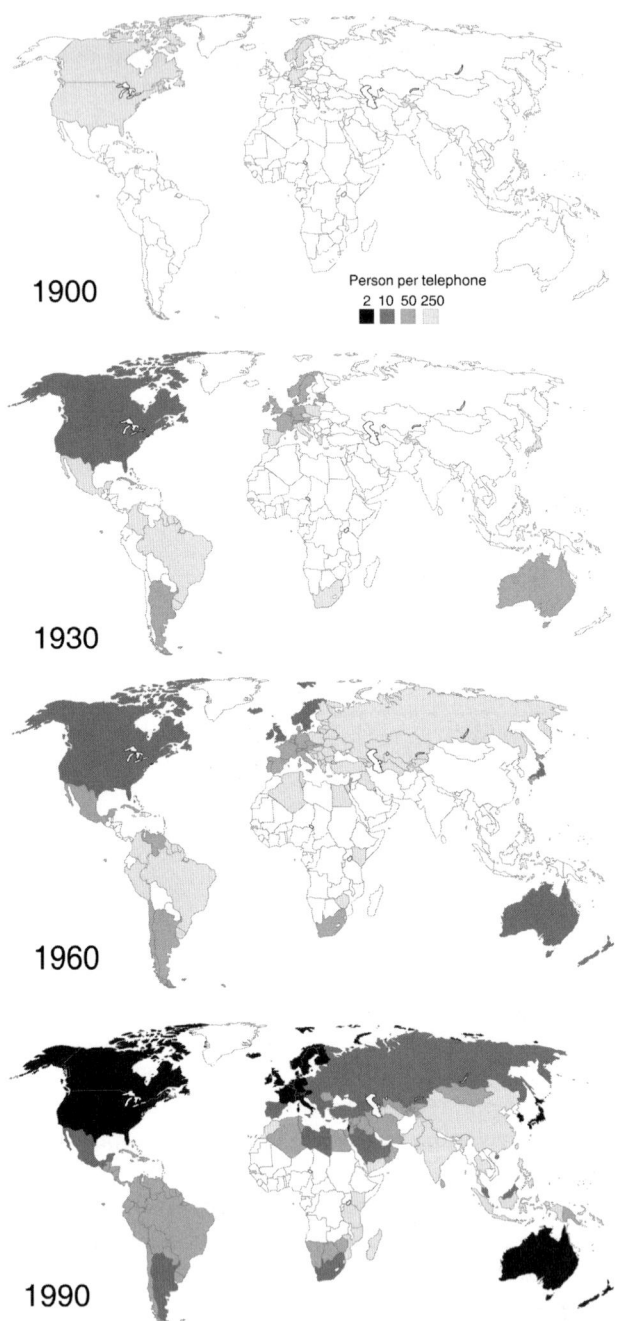

Fig. 5-27. Spatial distribution and development of the telephone (White 2007)

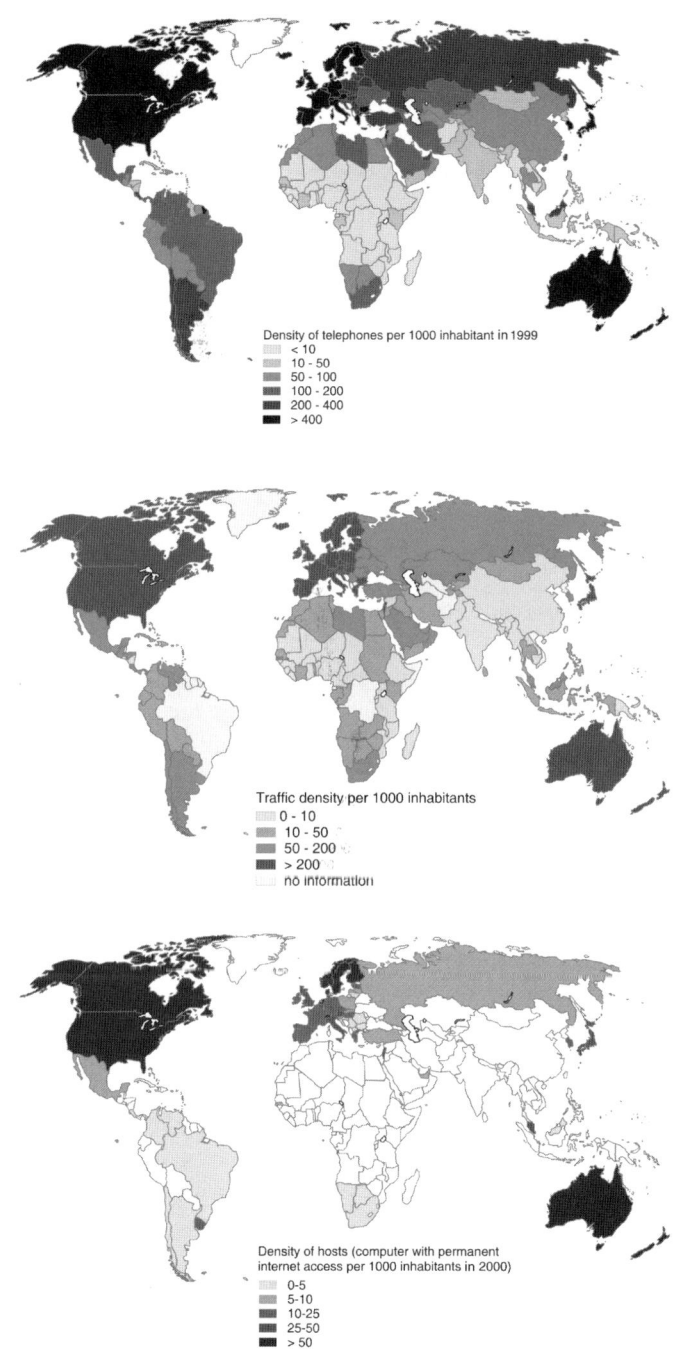

Fig. 5-28. Spatial distribution of the ownership of telephones, traffic density worldwide and internet hosts (Diercke Weltatlas 2002)

5.7 Environmental Measures and Quality of Life

Economical growth has been seen for a long time as a major indicator for human wellbeing. Nevertheless, the truth of this consideration is limited since some major assumptions about market behaviour are often not fulfilled and market failure is well known. For example, the upbringing of children is usually not considered as a performance in an economical sense. However, if this is not performed, the economy will experience a total collapse. Also, the performance of the biosphere in terms of providing oxygen in the air or the performance of the sun as major energy supplier are not part of the economical systems. However, the permanent extension of the concept of economical descriptions can be easily seen in the financial values of land. While probably several thousand years ago, land was considered as boundary condition, but now the value is expressed in market prices. Constanza et al. (1997) have tried to identify the performance of the ecosystem of earth in financial terms to include it in economical considerations.

The decline of environmental resources might have direct impacts not only on the economical systems but also on the quality of life. Therefore, in the last years, some combined quality-of-life parameters have been developed, not only considering the wellbeing of humans but also of the environment. One such parameter was already introduced: the Genuine Progress Indicator.

But, the same problem occurs during the development of some human quality-of-life parameters: the choice of the input variables for the wellbeing of nature. Based on a work by Mannis (1996) and Winograd (1993), some environmental parameters are listed in Tables 5-27 and 5-28.

Table 5-27. Matrix of environmental indicators under consideration by UNEP (World Resources Institute 1995)

Issues	Parameters
Climate change	Green house gas emissions
Ozone depletion	Halocarbon emissions and production
Eutrophication	N, P emissions
Acidification	SO_x, NO_x emissions
Toxic contamination	POC, heavy metal emissions
Urban environmental quality	VOC, NO_x and SO_x emissions
Biodiversity	Land conversion, land fragmentation
Waste	Waste generation in municipality and agriculture
Water resources	Demand in private, agriculture and industry
Forest resources	Use intensity
Fish resources	Fish catches

Table 5-27. (Continued)

Issues	Parameters
Soil degradation	Land use changes
Oceans/coastal zones	Emissions, oil spills, depositions
Environmental index	Pressure index

Table 5-28. Selected indicators of sustainable land and natural resources use (Winograd 1993)

Variable	Example
Population	Total population density
	% urban and rural
Socioeconomic	Unemployment
development	External debt and debt service
Agriculture and food	Food production calories per capita
	Annual fertilizer and pesticide use
	Agricultural land per capita
	Percent of grain consumed by livestock
	Percent of agricultural lands
	Percent of soil with limitation potential
Energy and materials	Firewood and coal per capita
	Traditional fuels as a percent of total requirements
	Bioenergetic potential
	Per capita materials consumption
Ecosystems and land	Current and natural primary production
use	Percent change jobs per hectare
	Annual production and value net emissions
	Species used
	Using fossil fuels
Forests and pastures	Surface area of dense and open forests
	Annual deforestation
	Annual reforestation
	Annual deforestation rate
	Ratio of deforestation and reforestation
	Wood production per capita
	Wood reserves per capita and by hectare
	Ratio of production/reserves
	Percent change in pastures
	Percent change in livestock index of load capacity
	Percent change in meat production dollars per hectare
Biological diversity	Percent of threatened animals species
	Threatened plants per 1,000 km
	Percent of protected areas
	Index of vegetation use
	Index of species disappearance risk
	Current Net Value

(Continued)

Table 5-28. (Continued)

Variable	Example
Atmosphere and climate	Emission of carbon dioxide – Total, per capita and per GNP emission
Information and participation	Number of environmental profiles and inventories
	Number of NGOs per area of activity
	Perception of environmental problems
Treaties and agreements	Signing and ratification of international treaties
	Funds generated for conservation
Land use projections	Potential productive land per capita
	Agricultural land necessary in 2030
	Index of land use deforestation
	Rate and ratio of re/deforestation
	Average annual investment cost and benefit of rehabilitation
Agroforestry	Carbon absorption through reforestation

5.7.2 Happy Planet Index

The Happy Planet Index (HPI) is a measure combining human and environmental wellbeing. Actually, it describes the efficient usage of environmental resources to provide a long and happy life for individuals, regions or countries. The index is a survey asking questions dealing with health, wellbeing of persons, conditions of the lifestyle as well as the consumption of environment due to travel behaviour. Example questions are shown in Table 5-29 (Marks et al. 2006, NEF 2006).

Table 5-29. Some questions from the HPI survey

Question	Possible answers
Which phrase best describes the area you live in?	A big city
	The suburb of a big city
	A town or small city
	A country village
	A farm or countryside home
And which best describes your home?	A detached house or bungalow
	A semi-detached house or large terraced
	A small terraced house
	A flat/apartment
	Any accommodation without running water

Table 5-29. (Continued)

Question	Possible answers
With whom do you live?	Alone
	1 person
	2 people
	3 people
	4 people
	5 or more people
Does that include a partner or spouse?	Yes
	No
Thinking about how you get about, which of the following do you do on a typical working day?	I walk over 20 min over the day
	I cycle
	I use public transport
	I drive (up to 20 miles each way)
	I drive (20 miles or more each way)
	None of the above
Roughly how many hours (in total) do you fly each year?	None
	< 5 h
	6–18 h
	19–50 h
	> 51 h
Which of the following best describes your diet?	Vegan
	Vegetarian
	Balanced diet of fruit, vegetables and pulses, with meat no more than twice a week
	Regular meat (every or every other day)
	Regular meat (every or every other day), including more than two hot dogs, sausages, slices of bacon or similar each week
	There have been days when I couldn't afford to get at least one full and balanced meal
Where does your food normally come from?	A mix of fresh and convenience
	Mostly fresh food
	Mostly convenience food
Do you smoke?	No, never
	No, but I'm often with smokers
	Ex-smoker or social smoker
	1–9 filtered cigarettes/day
	10–19/day
	20–29/day
	>30/day

(Continued)

Table 5-29. (Continued)

Question	Possible answers
In an average week, on how many days do you take at least 30 min of moderate physical exercise (including brisk walking)? The 30 min need not be all at once.	0 1–2 3–4 5–7
In the past 12 months, how often did you help with or attend activities organised in your local area?	At least once a week At least once a month At least once every three months At least once every six months Less often Never

The parameter is built upon three sub-parameters, which are combined according to the following formula:

$$\text{HPI} = \frac{\text{Life satisfaction} \times \text{Life expectancy}}{\text{Ecological footprint}} \qquad (5\text{-}8)$$

The three parameters are introduced in short. Life satisfaction is understood as being, in contrast to the concept of daily change in happiness, a more general and more stable measure. It is often investigated by a simple question like "How satisfied are you with your life?" Since it is only one question, the parameter undoubtedly has its limitation. On the other hand, the developers persist that the measure behaves rather well in comparison to other national statistics parameters (Marks et al. 2006, NEF 2006).

The next item considers the life expectancy at birth. However, the life expectancy itself is multiplied with the average life satisfaction to give happy life years. This parameter experiences some drawbacks, for example the suggestion that happy life-years are constant over a population or over time. Possible solutions to this problem have been seen in a correlation between disability adjusted life-years and happy life-years. However, such a correlation is only partly valid. Even worse, there are now special questions in the survey for children, and it is simply assumed that children are as happy as adults. Here, further extensions of the survey are required (Marks et al. 2006, NEF 2006).

The ecological footprint describes the size of area on earth required to provide a certain level of consumption, technological development and resources for a given population. Such area is summed up by land use for the growth of plants for nutrition as well as for the trees to absorb carbon dioxide and for the ocean to provide fish. Examples for the computation of the footprint are given below. In general, the ecological footprint can also be understood as the impact of a human society to the worldwide environment (Greenpeace 2007).

In 2001, the earth provided about 11.2 billion hectares for mankind or 1.8 global hectares per person. However, the consumption and living style of mankind utilised resources that can only be provided by more than 13.7 billion hectares or 2.2 hectares per person. That means that the current capacity of the earth's biosphere is exceeded by more than 20%, not considering the resources required by the non-human species at all. This exceedance is not equally distributed worldwide. While the average Indian requires only 0.8 hectare per person, the average American citizen requires 9 hectares per person; on the other hand, an Austrian citizen needs about 4.9 hectares per person (Greenpeace 2007).

Table 5-30 shows the major contributions for an average Austrian citizen to the footprint. Obviously, the biggest single contributor is animal nutrition, such as meat. Here, a simple reduction of the meat supply could have tremendous impacts on the footprint. The next major contributor is heating of flats. Advanced heating technologies could provide future savings here. The subsequent single contributor is motorised traffic and air traffic. But, what is quite surprising is the strong contribution by paper consumption.

Table 5-30. Major contributors to the ecological footprint. The two right columns are sub-items of the left groups

Contribution to the ecological footprint	Group	Contribution to the ecological footprint	Major contributor to the group
0.33	Overall nutrition	0.23	Animal nutrition
0.25	Living	0.225	Heating and electricity
0.2	Mobility	0.18	Motorised individual traffic and airplanes
0.167	Consumption, goods, services	0.05	Paper

The example of the consumption of paper should be used to show the computation of the footprint. In western Europe, about 250 kg of paper are consumed per capita per year. Depending on the amount of recycled paper and some technological conditions in Germany, for about 1 tonne of paper about 0.8 m^3 of wood are needed. Using the world wide average for wood production with 1.48 tonne per hectare per year, one can compute:

$$\frac{250 \text{ kg per year per person} \times 0.8 \text{ m}^3 \text{ per tonne}}{1,000 \text{ kg} \times 1.48 \text{ m}^3 \text{ per hectare per year}} \qquad (5\text{-}9)$$

$= 0.14$ hectare per person per year

Checking the value of $0.05 = 5\%$ of the overall ecological footprint from Table 5-30 gives

$0.05 \times 4.9 = 0.245$ hectare per person per year

Of course, such a simple procedure has also limitations. For example, no areas for the wild flora and fauna are considered in the model, but at least about 20% of the global area is needed to permit at least a minimum wild biosphere. Also, double effects are not considered, for example the carbon dioxide assimilation by farming is not considered, but only the area for the supply of food is considered.

In general, it seems that the HPI is biased by preferring small countries and countries on coastlines. This yields to the question whether people are indeed happier on the coast and in small countries.

5.7.3 Wellbeing of Nations

Another indicator has been introduced by Prescott-Allen (2001). Again here, the GDP per capita has been heavily criticised as a performance measure. Major drawbacks of the GDP are:

- weak indicator for economic development
- weak indicator for welfare or wellbeing
- GDP has no consideration of income distribution
- depreciation of buildings or machines is added
- GDP does not consider roles of families or communities
- GDP does not consider depletion of natural resources
- GDP can count natural degradation as a benefit
- GDP cannot distinguish costs and benefits

Therefore, a wide range of parameters has been chosen and combined to give an overall measure of human and environmental wellbeing. An overview about the different input parameters is given in Figs. 5-29; in Fig. 5-30, arbitrarily some root of the parameters has been followed. For the grey watermarked fields, there are no parameters chosen yet.

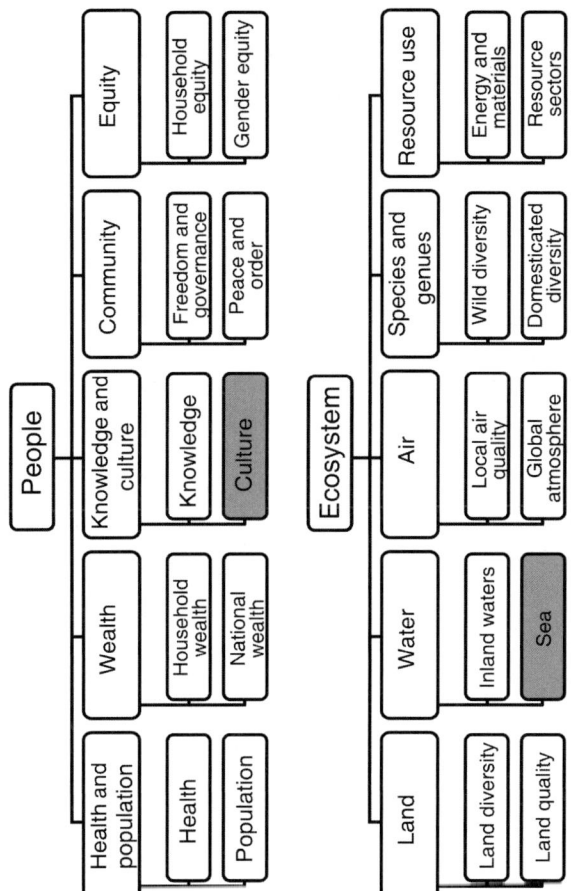

Fig. 5-29. Structure of the computation of the wellbeing index according to Prescott-Allen (2001). The *grey* fields are not numerically included

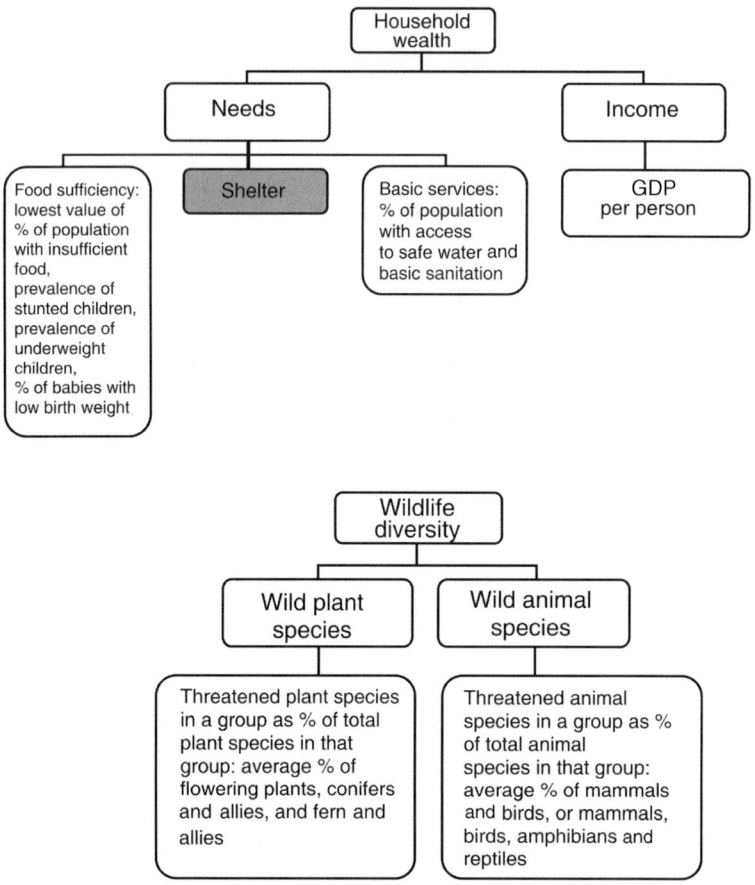

Fig. 5-30. Examples of sub-items for the computation of the wellbeing index: top shows the parameters for household wealth and bottom shows the development of the wildlife diversity measure according to Prescott-Allen (2001)

Using the statistics, diagrams as shown in Fig. 5-31 can be developed. Here, human and environmental wellbeing is shown in percent. There seems to be a tendency that nations start at the bottom right, climb upward to an area on the left and start to return at a higher level again to the right side. But, no country at present shows sustainable behaviour. The leaders like Sweden, Finland, Norway, Iceland and Austria score well on the human wellbeing index, but only score marginally on the ecosystem wellbeing index. Some countries have low demands on the environment but are desperately poor. In 141 countries, the environment is highly stressed without increasing the wellbeing for humans (Prescott-Allen 2001).

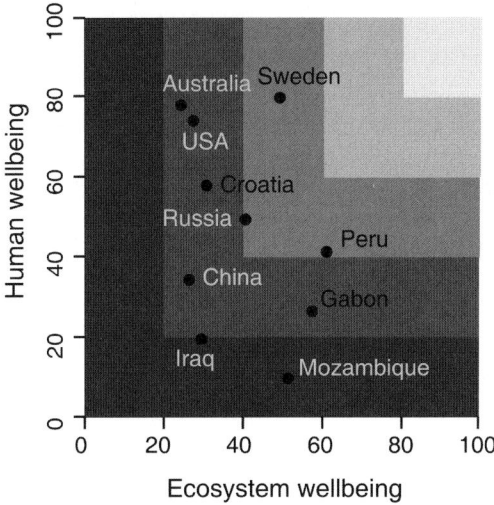

Fig. 5-31. Human and ecosystem wellbeing index for several nations according to Prescott-Allen (2001)

5.8 Engineering Quality-of-Life Measures

A very important contribution to the discussion of quality-of-life parameters for the efficient evaluation of certain mitigation measures had been introduced by Canadian researchers at the end of 1990s (Nathwani et al. 1997). This so-called Life Quality Index (LQI) has found wide application in different fields of engineering, reaching from maritime engineering to civil engineering.

During the development of this parameter, it was generally assumed that quality of life can be expressed in a function based on some social indicators:

$$L = f(a,b,c,...,e,...)$$ (5-10)

To simplify the entire procedure, only two social indicators were chosen. The two indicators are per capita cross domestic product g and the average life expectancy e. The per capita cross domestic product is used here as an indicator for available resources, whereas the life expectancy is used as a measure for the useable time t. Both indicators are multiplied giving a two-dimensional quality-of-life parameter

$$L = f(g) \cdot h(t)$$ (5-11)

Furthermore, the usable lifetime is limited by the procurement of resources expressed in the per capita GDP. This interrelation between both indicators is described here using the ratio w of the entire working time versus the average life expectancy. The entire lifetime can then be separated in a working part $w \times e$ and in a free part $(1-w) \times e$. As already stated in the assumptions, the quality-of-life parameter should be based on functions, and this function should be differentiable. Additionally, it can be assumed that only changes of the parameters are of interest:

$$\frac{dL}{L} = c_1 \cdot \frac{dg}{g} + c_2 \cdot \frac{dt}{t} . \qquad (5\text{-}12)$$

Furthermore, it should be assumed:

$$\frac{c_1}{c_2} = \text{constant} \qquad (5\text{-}13)$$

The solution of a pair of differential equations yields to

$$f(g) = g^{c_1} \text{ and } h(t) = t^{c_2} = ((1-w) \cdot e)^{c_2} . \qquad (5\text{-}14)$$

Assuming that the per capita income and the working lifetime can be related, the LQI is then suggested as

$$L = (c \cdot w \cdot e)^{c_1} \cdot ((1-w) \cdot e)^{c_2} \qquad (5\text{-}15)$$

Further, it should be assumed that if the working time can be used to control the quality of life, then a maximum quality of life is reached by

$$\frac{dL}{dw} = 0 \qquad (5\text{-}16)$$

The differentiation of the function yields to

$$0 = c_1 - c_2 \cdot \frac{w}{1-w} . \qquad (5\text{-}17)$$

A further constant defined as

$$\bar{c} = c_1 + c_2 \qquad (5\text{-}18)$$

yields to $c_1 = \bar{c} \cdot w$ and $c_2 = \bar{c} \cdot (1-w)$ and giving

$$L = g^{\bar{c} \cdot w} \cdot e^{\bar{c} \cdot (1-w)} \cdot \bar{c} \cdot (1-w)^{\bar{c} \cdot (1-w)} \qquad (5\text{-}19)$$

Further, it is assumed that $\bar{c} \approx 1$ and for values of w between 0.1 and 0.2, $(1-w)^{(1-w)} \approx 1$ is valid, which is a rather bold assumption:

$$L \approx g^w \cdot e^{(1-w)} \text{ or } L \approx g^{\frac{w}{1-w}} \cdot e = g^q \cdot e \qquad (5\text{-}20)$$

Since the parameter is mainly designed for the evaluation of safety measures, the change of life quality is of interest rather than the absolute quality of life. The quality of life should improve by some actions, when it gives

$$\frac{dL}{L} = \frac{de}{e} + \frac{w}{1-w}\frac{dg}{g} \geq 0 \qquad (5\text{-}21)$$

This net benefit criterion describes a positive change of life. The change can either be reached by an increase of the average life expectancy e or by an increase of income g. Usually the increase of life expectancy is brought. These yields to a loss of income, and a trade-off can be made according to the equation. But, some further simplifications are required to permit computation under practical circumstances. First, the decrease of income should be approximated to

$$-\frac{dg}{g} \approx -\frac{\Delta g}{g} = 1 - \left(1 + \frac{\Delta e}{e}\right)^{1-\frac{1}{w}} \qquad (5\text{-}22)$$

Also, the change of life expectancy can be simplified by

$$-\frac{de}{e} \approx -C_F \cdot \frac{dM}{M} \qquad (5\text{-}23)$$

and the change of quality of life can be written as

$$\frac{dL}{L} = -C_F \cdot \frac{dM}{M} + \frac{w}{1-w} \cdot \left[1 - \left(1 + \frac{\Delta e}{e}\right)^{1-\frac{1}{w}}\right] \geq 0 \qquad (5\text{-}24)$$

The change of mortality is computed as the ratio of the number of possible fatalities versus the population size:

$$dM = \frac{N_F}{N} \qquad (5\text{-}25)$$

The variables represent
dM – change of mortality
M – general mortality

de – change of average life expectation
e – average life expectation
N_F – number of fatalities caused by an accident
N – population size
C_F – shape of the age pyramid

The factor C_F describes the curvature of the age pyramid for a chosen nation. If all people in a nation would die at the same age, the pyramid would actually be rectangular, and the value of C_F would be zero. A real pyramid would have C_F value of 0.5 and corresponds with the age linear decrease of the population. An age pyramid with the shape of exponential function would have the value of 1 and represents a constant mortality over the age. A negative pyramid with a wider head than foot could develop in an overaged population. The different types are shown in Fig. 3-32.

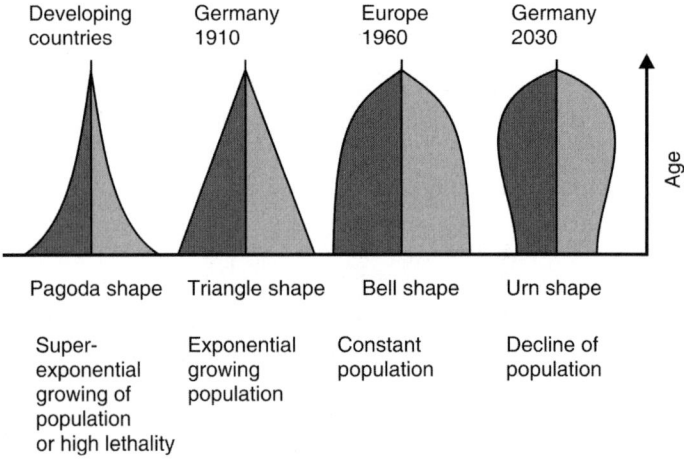

Fig. 5-32. Different types of population pyramids and examples according to Pätzold (2004)

Historically, most developed countries went from values around 0.5 to values as close to 0.1. As an example, the population pyramid for Germany and India are shown in Fig. 5-32. In 1910, the age pyramid was nearly completely developed in Germany. In 1999, the age pyramid at least partially shows a rectangle. However, in some age cohorts, there are still effects from the world wars and from times with higher birth rates. In 2050, it is predicted that a negative pyramid will have developed. Most parts of the population will be older than 50 years. Nearly 40% of the population will have an age higher than 60 years. The portion of the young people of age less than 20 will decline from 20% to 15%. Over the long run, there are also such changes possible for India, but currently India shows the same shape of the age pyramid of Germany in 1910, as shown in Fig. 5-33.

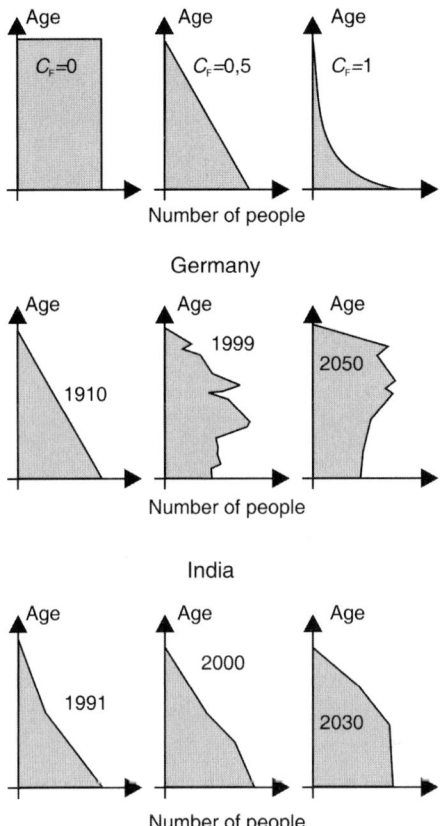

Fig. 5-33. Population pyramids for India and Germany for different times

Whether a population grows or declines depends on the ratio of mortality and natality. Natality is the birth rate and mortality is the death rate. Both parameters have heavily changed over the last centuries as shown in Fig. 5-34. The parameters depend on some socioeconomic conditions and some cultural preferences of a country. It should be mentioned that both influences are correlated with each other. While both values remained high in feudal societies, but since the beginning of industrialisation first the mortality and then the natality have fallen. Since the natality decreased significantly later, there was a strong population growth. The difference between both parameters during that time is called the demographic overhang.

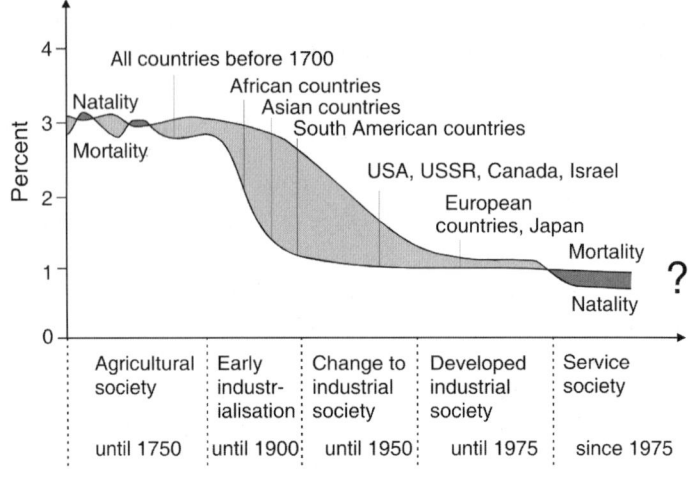

Timeline for Europe

Fig. 5-34. Development of natality and mortality over the last centuries according to Pätzold (2004)

But, changes of population are a constant appearance, not in the last century alone. Data for the last 2,000 years for west Europe are shown in Table 5-31. In particular, the sound conditions during the prime of the Roman empire yielded to a growth of the population, whereas the collapse caused a decrease of the population. Only during medieval ages, especially in the 12th century, population could recover since agriculture and trade improved. The next setback occurred mainly due to the heavy plague pandemic in the 14th century. Since then, population grew constantly, but only the 17th century shows a drop due to the 30-year war. With the beginning of industrialisation, growth increased rapidly.

Table 5-31. Average of the west European population growth since 2,000 years (Streb 2004)

Year	Population growth in %
0–200	0.06
201–600	–0.10
601–1000	0.08
1001–1300	0.28
1301–1400	–0.34
1401–1500	0.32
1501–1600	0.24
1601–1700	0.08
1701–1820	0.41
1821–1998	0.60

While the development of the GNP has been discussed in section "Economical quality-of-life parameters" and the development of the average life expectancy discussed just here, the third parameter of this quality-of-life parameter model has to be investigated. This parameter is the working-to-lifetime ratio w.

In indigene societies, the working time was about 20 h per week (Kummer 1976) and about 1,100 h per year, and in the year 1700, an employee worked about 2,300 h per year. One hundred years later, the value had already increased to about 2,500 h, and a maximum was reached in 1850 with about 3,000 h per year (Tables 5-32, 5-33 and Fig. 5-35). At that time, working 60 h per week was an average value. Since then, the working time has more or less constantly fallen and now reaches about 1,700 h per year (Table 5-34). Additionally, the average life expectancy has also increased since 1850, and therefore humans can enjoy a much brighter life (Table 5-33). Unfortunately, with globalisation working time has started to rise again in developed countries (Streb 2004).

Table 5-32. Average working time per year (Streb 2004)

Year	Days per year out of	365	Working days	Working hours per		
	Sundays per year	52		Day	Week	Year
1700	Holidays	72	189	12	43.6	2,268
	Blue Mondays	52				
1800	Holidays	53	208	12	48	2,496
	Blue Mondays	52				
1850	Holidays	53	260	12	60	3,120
1975	Holidays	15	225	9	38.9	2,025
	Vacation	21				
	Free Saturday	52				
1985	Holidays	12	219	8	33.7	1,752
	Vacation	30				
	Free Saturday	52				

Table 5-33. Proportion between working and non-working years (Streb 2004)

Year	Start of work at age	End of work at age	Average life expectancy in years	Working–living years	Non-working–living years	Ratio of working to non-working–living years
1700	8	30	30	22	8	2.8:1
1800	7	35	35	28	7	4:1
1900	14	45	45	31	14	2.2:1
1975	16	64	71	48	23	2.1:1

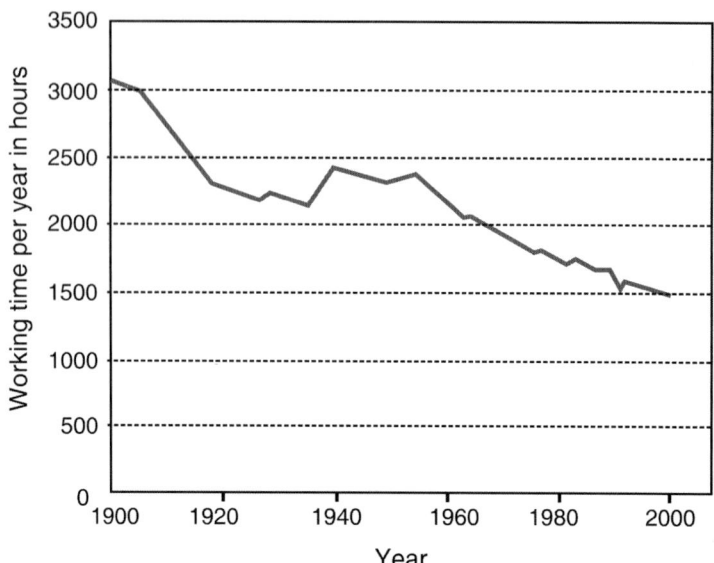

Fig. 5-35. Development of working time per year in hours in Germany during the last century (Miegel 2001)

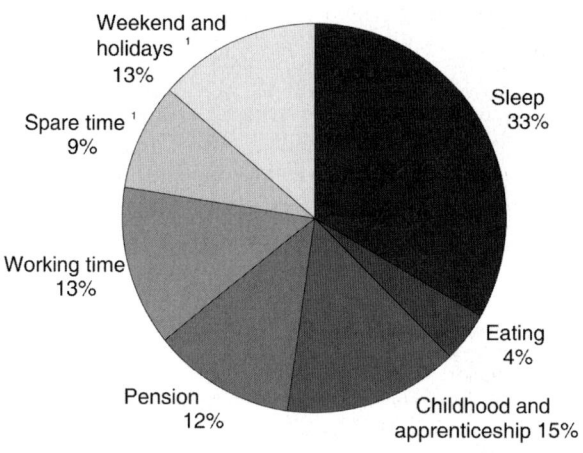

[1]during working years

Fig. 5-36. Distribution of activities over the entire lifetime in percent

Table 5-34. International year working time in 2000 in hours (OECD 2002)

Country	Year working time in hours
USA	1835
Japan	1821
New Zealand	1817
Spain	1814
Canada	1794
Finland	1721
Great Britain	1708
Ireland	1690
Sweden	1624
Italy	1622
France	1590
Switzerland	1568
Denmark	1504
Germany	1481
Netherlands	1381
Norway	1375

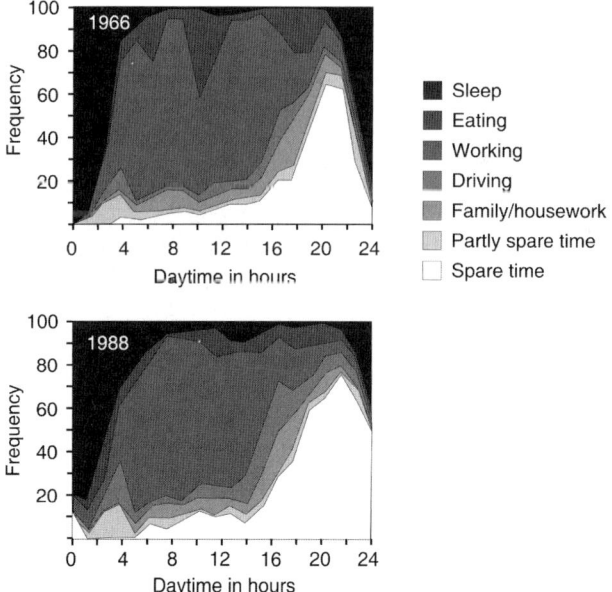

Fig. 5-37. Distribution of day-time activities in Germany in 1966 and 1988 by Lüdtke (2000)

An example of how the ratio of working per lifetime has changed is not only shown in Table 5-33 but will also be discussed based on Fig. 5-36. In Fig. 5-36, a human with a lifetime of about 80 years is assumed. The working time includes 45 years; there are 15 years of pension time and 20 years of childhood, school and education. The weekly working time is assumed with 45 h, but also 30 days of holidays and 5 bank holidays per year are expected. This yields to a working lifetime value of about 13%. An increase of the working time at 1 h per day increases the value by 1%. Working 12 h per day over 45 years already gives a value of over 20% working time per lifetime. But, on the other hand, if the life expectancy increases by 1 year, then the working lifetime value decreases by about 0.25%. If the life expectancy decreases to 75 years, then w reaches 14%. Also, if the start of working is put forward, then w increases by 0.3% per year.

Figure 5-37 shows the change of the distribution of working time over daytime. While in the 1960s, the working time was rather determined, but in the last decades, working time is spreading over the daytime.

After the discussion of single input parameters, it seems to be useful to study the behaviour of the quality-of-life model outcome. This is visualised in Fig. 5-38. The diagram clearly shows that small financial resources can improve lifetime substantially. This property seems to be valid for an average income up of 2,000–4,000 Euros or dollars. Higher income marks a region of change, where only a rather slow increase of life expectancy can be reached. At about that value, a functioning economy is establishing. At about 15,000 Euros or dollars, a third region can be seen marking developed countries.

Considering also the date information, it seems that countries follow a main curve during development. This is also shown in Fig. 5-39. However, even under some time, limited stagnation, countries keep increasing the life expectation. This is called lateral effects.

If one transforms the axis of this diagram to logarithmic scale, the points build up a line as shown in Fig. 5-40 (b). Furthermore, the ratio of working to lifetime can be changed according to the average per capita income (Fig. 5-40 (c)). And finally, one can neglect the absolute values of income and life expectancy, since the interesting idea of this concept is to decide whether some action is improving the quality of life or not. This is shown in Fig. 5-40 (d). Figure 5-41 continues Fig. 5-40.

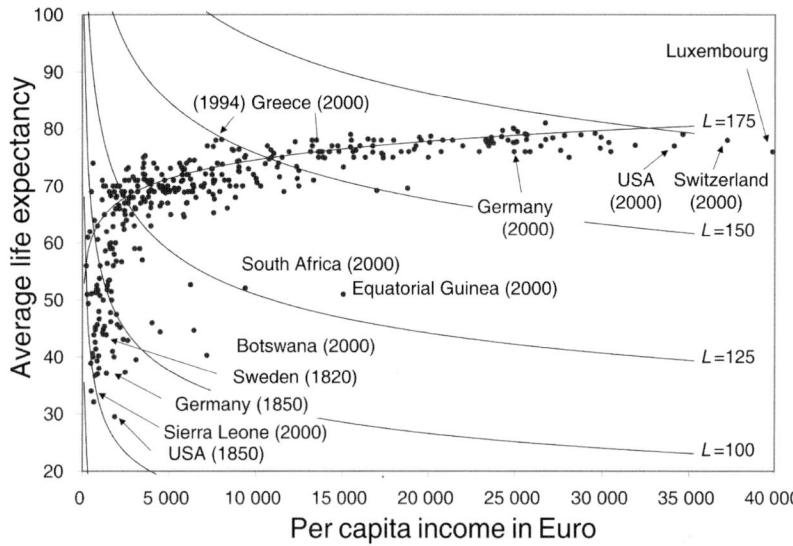

Fig. 5-38. Per capita income versus average life expectancy including lines of constant quality of life using data from Easterlin (2000), Cohen (1991), NCHS (2001), Rackwitz & Streicher (2002), Skjong & Ronold (1998), Becker et al. (2002), IE (2002), Statistics Finland (2001)

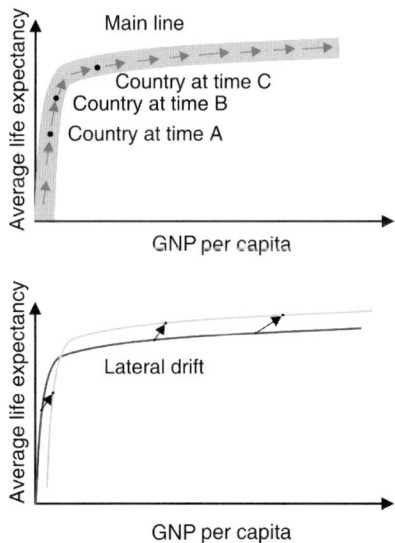

Fig. 5-39. The development of a nation is not only described by the average life expectancy and the GDP. There are additional effects yielding to some lateral drifts

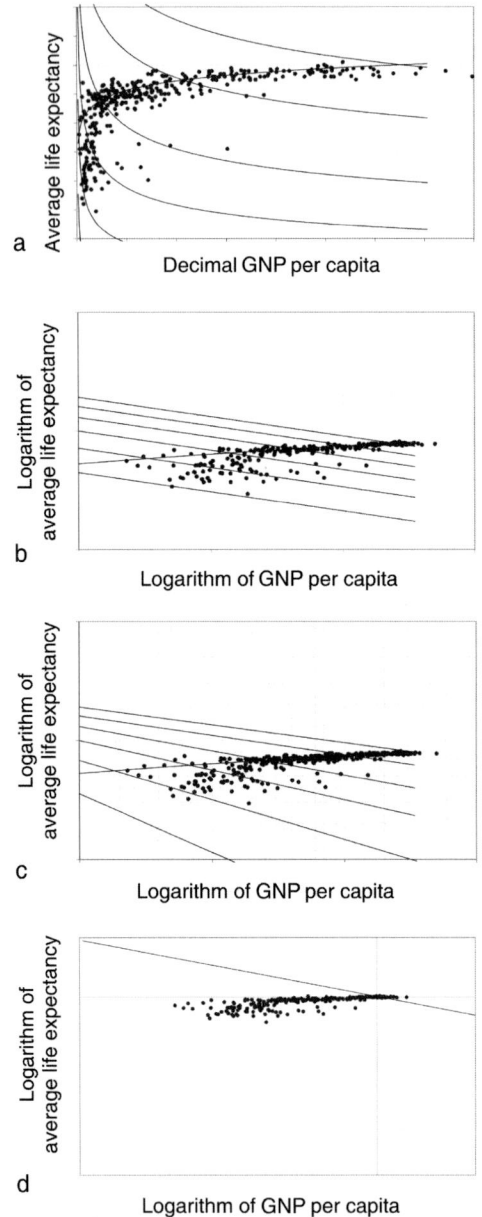

Fig. 5-40. Per capita income versus average life expectancy diagram (**a**) is transformed from decimal type to the logarithm type (**b**), then the working time is adapted to income conditions (**c**) and finally the origin of the diagram is shifted to the current state of quality of life in a certain nation (**d**)

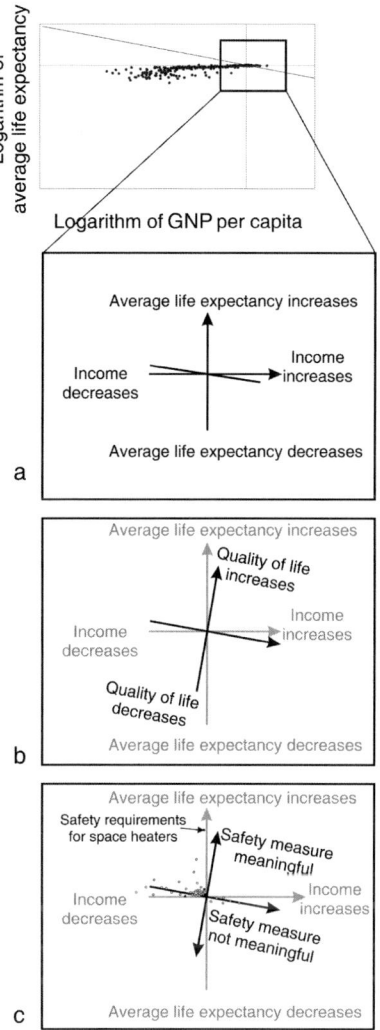

Fig. 5-41. Based on Fig. 5-40 (**d**), the new origin is scaled up (**a**), then the coordinate system is rotated according to the line of working time per lifetime (**b**), and finally some examples are put into the diagram (**c**)

This concept can further be used to estimate a financial value to save a single human life. Figure 5-42 gives such values in terms of ICAF (Implied Cost Averting a Fatality), WTP (Willingness to Pay) or VSL (Value of a Statistical Life). The detailed references about the data can be found in Proske (2004). Such values can also be age-adapted as shown in Fig. 5-43.

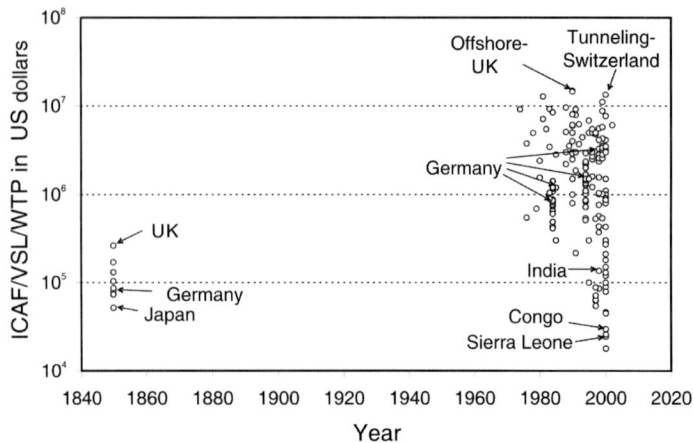

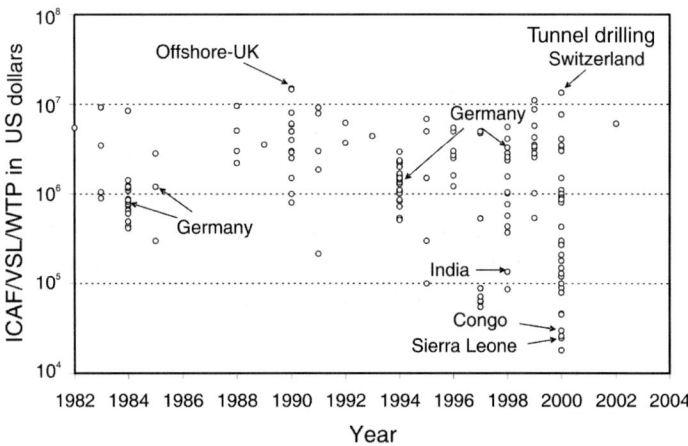

Fig. 5-42. Example values for ICAF, VSL and WTP. The bottom diagram is simply a part of the top diagram. The list of references for the data can be found in Proske (2004)

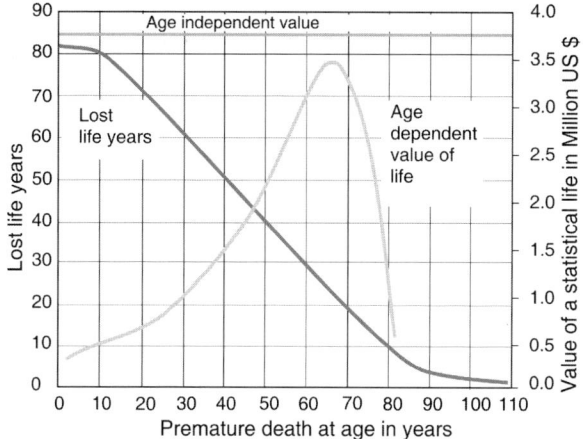

Fig. 5-43. Development of statistical life over lifetime. For comparison, there is the curvature of lost life-years and an age-independent statistical life value included according to Hofstetter & Hammitt (2001)

Furthermore, instead of relating such values to the number of humans, it might also be related to the number of gained years, as done by Tengs et al. (1995) and Ramberg & Sjoberg (1997) and shown in Fig. 5-44. However, assuming putting more money into historical effective measures might not yield to success under all conditions. As shown in Fig. 5-45, the single values are part of functions, and increased overall invested money into one mitigation measure might yield to a reduction in efficiency.

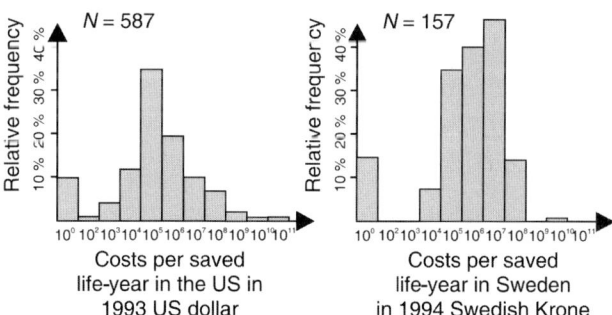

Fig. 5-44. Distribution of costs per saved life-year in the US (Tengs et al. 1995) and in Sweden (Ramberg & Sjoberg 1997)

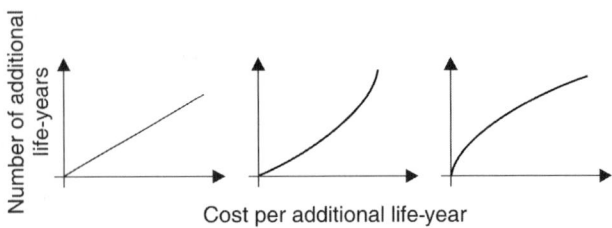

Fig. 5-45. Possible functions for cost development per additional life-year

Although the idea of the LQI is impressive, it also includes some major assumptions, which might influence further success. Therefore, further work is currently being carried out to improve the parameter (Pliefke & Peil 2007). This work will be introduced in short here.

In general, it is stated that the LQI is an anonymous individual utility function. It does not consider current quality-of-life status and assumes an ordinal utility function. Furthermore, as pointed out here, the optimisation of quality of life subject to the working time is a major assumption of the theory. Such assumptions include also, for example, an unequal part of longevity and consumption data. But, referring to the subjective judgement of risks, humans do not consider the risks actually according to their statistical value, but rather on other items, such as the equity of the gains. While the LQI considers the efficiency of possible mitigation measures according to the works of Pareto and the Kaldor-Hicks compensation test, it lacks the capability to consider equity considerations. The application of such efficiency measures to reach maximum wealth in a society is now widely accepted. Nevertheless, other factors might limit the possible application of the pure efficiency measures.

This can be easily seen in a system trying to retain its integrity even with short-term non-efficient behaviour. One of the major drawbacks of the application of efficiency measures is a strong concern about the future behaviour of systems.

Therefore, equity considerations can be seen as efficient support of the major economic frames, mainly not included in the efficiency models (Jongejan et al. 2007). Pandey & Nathwani (1997) already had tried to improve the measure by incorporating quality-adjusted income. Pliefke & Peil (2007) have adapted the LQI in the form of

$$EAL = L \cdot \alpha(1 - \text{Gini}(QAI)) \tag{5-26}$$

by using the Gini index and applying:

$$EAL_{\text{final}} > EAL_{\text{final}} \tag{5-27}$$

Furthermore, according to Pliefke & Peil (2007), a second step should be used for the assessment of mitigation measures by life quality measures, which is defined as

$$\text{Gini}_{\text{inital}} - \text{Gini}_{\text{final}} \geq a \qquad (5\text{-}28)$$

Therefore, a mitigation action might have an extremely low net-benefit, but it might increase equity over a society. This fits very well to the example of a state budget including strong equity actions in terms of welfare. However, there seems to be a theoretical maximum equity since the Yerkes-Dodson curve shows that complete equity limits the efficiency of a society.

The definition of the Gini index is strongly related to the Lorenz curve (Fig. 5-46). The Lorenz curve describes the concentration of some properties over the certain population since statistical measures of dispersion might fail to describe concentration properly. The Lorenz curve indicates equity if the curve is a constant 45° line, whereas if the line is lower than the constant 45° line, then the curve indicates inequity. The ratio of the area under the 45° line and the Lorenz curve gives the Gini index, which is an overall measure for equity. In perfect equity, the Gini index reaches 0, whereas in perfect inequity, the values reaches 1. For example, on a worldwide level, the Gini index for the GDP has risen from 0.44 in 1960 to 0.55 in 1989 (Dahm 2002).

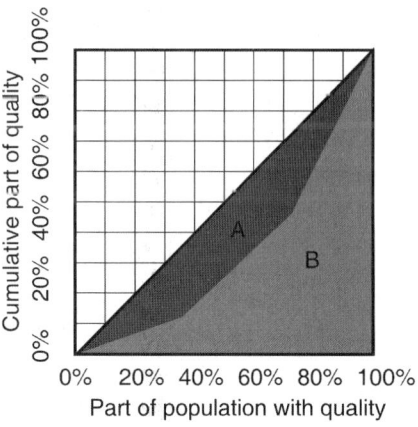

Fig. 5-46. Lorenz curve and Gini index

5.9 Quality of Life in Religions

Quality-of-life considerations have a long history in religions. One could easily assume religions as a collection of rules to reach a good quality of life.

On the other hand, current developments in some religious organisations have considered the limits of the concept of quality of life. For example, in Barragán (2005), the main goal of quality of life concept is seen in the distinction between high and low quality of life expressed in some measures. Furthermore, such concepts can be applied for the identification of "not living worthwhile lives". Here, measures such as intelligent coefficients less than 20 as an indicator for minimum intellectual capacity, minimum self-awareness, self-control, capacity to communicate or the capacity to control one's own life are mentioned. Barragán's critics about the possible application of quality-of-life parameters for the distinction between living worthwhile and not living worthwhile lives is right and has been found in other fields as well. However, quality-of-life parameters do not necessarily have to be applied only for that distinction. They are also applicable for other questions.

5.10 Quality of Life in Politics

In 1776, the American Declaration of Independence argued for "certain inalienable rights that among these are Life, Liberty and the Pursuit of Happiness". David Eisenhower (1890–1969) has used quality of life in his election campaign, and President Dwight Eisenhower's 1960 Commission on National Goals has used the term. In 1964, President Lyndon Johnson gave a speech about: "The task of the great society is to ensure our people the environment, the capacities, and the social structures which will give them a meaningful chance to pursue their individual happiness. Thus the great society is concerned not with how much, but with how good–not with the quantity of goods but with the quality of their lives." (Noll 1999)

German Chancellor Willy Brandt used quality of life as an election campaign in 1972: "…more production, profit and consumption does not automatically mean an increase in satisfaction, luck and development possibilities for the particular. Quality of life represents more than a higher standard of living. Quality of life presupposes liberty, also liberty of fear. It is security by human solidarity, the chance for the self-determination and self implementation… Quality of life means enriching our life beyond the material consumption." (Noll 1999, Amery 1975)

Especially, the oil crisis in the 1970s yielded to considerations about the final goal of economic growth and wealth (Schuessler & Fisher 1985). At that time, many conferences, for example, the one organised by the IG Metal in Germany, were discussing further goals of societies (Eppler 1974).

Although the political use of the term of "quality of life" has decreased since then, still some parties, like the Green Party of Aotearoa in New Zealand, use "quality of life" as a slogan. Also, many politicians summarise social indicators to describe the quality of life in a nation – for example US president Clinton's State of the Union speech on 19 January 1999. In addition to this development, many major cities describe their development in terms of quality of life, such as the quality of life reports for several cities in New Zealand (QoLNZ 2003) or the quality of life report for Southern Tyrol (LfS 2003).

5.11 Limitation of the Concept

Reaching an optimal life in an optimal society is the major goal of the application of quality-of-life parameters. The more efficiently the resources are spent, the better off are the members of the society. But, as mentioned in chapter "Indetermination and risk", the question remains whether the tools used for such optimisation and efficiency assessment are suitable.

First of all, a critical look at the success of quality-of-life measures should be carried out. Gilbody et al. (2002) stated that regular quality-of-life investigations were costly and did not give robust results for the improvement of psychological outcomes of patients. In addition, the UK700 Group (1999) concluded that clinical and social variables are only able to cover 30% of the variance of an individual's quality of life. One should keep those limitations in mind when carrying out optimisation procedures using quality-of-life measures.

An optimisation is a mathematical technique to identify those combinations of input factors that yield to an extreme value of the investigated function. The questions arising here are: "Does one know the function and does one know the initial conditions?" If one looks at many fields, optimisation and efficiency assessment are of utmost importance. For example, in car manufacturing, much effort had been spent to develop optimal car elements. However, under real-world conditions, quite often such solutions failed. If the initial assumptions of the investigations were not valid or only valid to a limited degree, the results of optimisation investigation would not behave optimally under real-world initial conditions (Marczyk 2003,

Buxmann 1998). Indeed, for real-world conditions, one never knows either the function or the data. One can only investigate the effectiveness and apply the optimisation procedure backwards.

Indeed, the last years have shown the change of direction away from optimisation towards more robustness and fitness in many fields. Such solutions include a high malfunction resistance and remain functioning under unexpected situations. Mainly such systems are indetermined and include redundant elements. These elements might appear to be abandonable, however only in certain situations. Considering a traffic system of a city, where many parallel roads exist, one might decide to use only one main road. But, in the case of an accident or some maintenance work, the traffic there might be restricted and drivers will try to avoid such a road for a certain time. So, in general, additional roads may not be required, but under some conditions they are essential for the functioning of the entire system. Another example is small disturbances at workplaces, such as a decoration on a table. This may increase efficiency over the long run: a sparkling idea might be caused by a disturbance as psychological research has shown (Piecha 1999, Richter 2003). Such techniques are already known from mathematical stochastic optimisation search techniques. And a third example is the general commitments of societies. Such commitments are, for example, found in the constitution of many states, but also in some organisations. For example, the military organisations very often promise to free prisoners in war situations. Over the short term, this shows non-efficient behaviour: many soldiers are endangered to free one or two soldiers. However, over the long term, soldiers are motivated, and people are more willing to join the organisation. The consideration of time horizons on economical considerations is well known (Münch 2005). Therefore, the promise eventhough not efficient under simple models turns out to be of utmost importance for the shear existence of the organisation. This can also be seen in the promises in several constitutions to accept the dignity and life of humans under all situations. There was a strong discussion in Germany about the possible shooting-down of passenger airplanes in case of possible terrorist use of the particular airplane. Here, an efficiency consideration (saving more people by preventing the terrorist attack) might yield severe damage of the entire social system by destroying trust in the social system. According to Knack & Zack (2001) and Knack & Keefer (1997), trust in social systems is the major motivation for people to produce wealth, whereas countries with a lack of trust in the social system motivate the people to simply take wealth from other people instead of producing on their own.

Therefore, for example, social forms such as dictatorship may give easier conditions to perform an efficiency assessment and optimisation procedures, which is rather impossible under democratic rules, according to Arrow (1951). However, such systems might not behave robustly. This has become clearly visible in the current development of Iraq or former Yugoslavia. Such situations can only be changed very slowly.

To summarise the comments: the application of quality-of-life indicators for efficiency consideration of mitigation measures might decrease risks to humans. However, it introduces a new risk: the possibility of unintended consequences is infiltrated by initial assumptions in the model. And furthermore, general promises by the establishment of societies, like respect to humans or trust into societies, might be heavily damaged. This can be most clearly seen in the field of risk and justice where clear definitions of risk and safety are refused due to such concerns.

References

Ahrens A & Leininger N (2003) Psychometrische Lebensqualitätsmessung anhand von Fragebögen. Christian-Albrechts-Universität zu Kiel, Vorlesungsunterlagen 7.7.2003

Allardt E (1975) Dimensions of welfare in a comparative Scandinavian study. University of Helsinki

Amery C (1975) Lebensqualität – Leerformel oder konkrete Utopie? In: Uwe Schultz (Hrsg) Lebensqualität – Konkrete Vorschläge für einen abstrakten Begriff. Aspekte Verlag, Frankfurt am Main

Arrow KJ (1951) Social Choice and Individual Values, Wiley, New York

Barilan YM (2002) Health and its Significance in Life in Buddhism. Gimmler A, Lenk C & Aumüller G (Eds), Health and Quality of Life. Philosophical, Medical and Cultural Aspects. LIT Verlag, Münster, pp 157–172

Barragán JL (2005) Ten years after "Evangelium Vitae". The quality of life. Pontifica Academia pro Vita. Quality of life and the ethics of health. Proceedings of the XI assembly of the pontifical academy for life. Vatican City, 21–23 February 2005, Elio Segreccia & Ignacio Carrasco de Paula (Eds)

Basu D (2004) Quality-Of-Life Issues in Mental Health Care: Past, Present, and Future. German Journal of Psychiatry, June 2004, pp 35–43

Becker GS, Philipson TJ & Soares RR (2002) The Quantity and Quality of Life and the Evolution of World Inequality. May 10 2003

Brickman P, Coates D & Janoff-Bulman R (1978) Lottery winners and accident victims: is happiness relative? Journal of Personality and Social Psychology, 36, pp 917–927

Bullinger M & Kirchberger I (1998) SF-36 – Fragebogen zum Gesundheitszustand. Handanweisung Hogrefe – Verlag für Psychologie: Göttingen

Bullinger M (1991) Quality of Life-Definition. Conceptualization and Implications: A Methodologists View. Theoretical Surgery, 6, pp 143–148

Bullinger M (1997) Entwicklung und Anwendung von Instrumenten zur Erfassung der Lebensqualität. Bullinger (Ed), Lebensqualitätsforschung. Bedeutung - Anforderung - Akzeptanz. Stuttgart, Schattauer, New York, pp 1–6

Bulmahn T (1999) Attribute einer lebenswerten Gesellschaft: Freiheit, Wohlstand, Sicherheit und Gerechtigkeit. Veröffentlichungen der Abteilung Sozialstruktur und Sozialberichterstattung des Forschungsschwerpunktes Sozialer Wandel, Institutionen und Vermittlungsprozesse des Wissenschaftszentrums Berlin für Sozialforschung. Abteilung "Sozialstruktur und Sozialberichterstattung" im Forschungsschwerpunkt III Wissenschaftszentrum Berlin für Sozialforschung (WZB), Dezember 1999, Berlin

Burke A (2004) Better than money. Nordic News Network: www.nnn.se

Buxmann O (1998) Erwiderungsvortrag von Herrn Prof. Dr.-Ing. Otto Buxmann zur Ehrenpromotion: Ursachen für das mechanische Versagen von Konstruktionen bei Unfällen. Wissenschaftliche Zeitschrift der Technischen Universität Dresden, 47, 5/6, pp 145–147

Carrie A (2004) Lack of long-run data prevents us tracking Ireland's social health, Feasta Review, Number 2, pp 44–47

Chernoff H (1973) The use of faces to represent points in k-dimensional space graphically. Journal of the American Statistical Association, 68, pp 361–368

Cobb C, Goodman GS & May Kliejunas JC (2000) Blazing Sun Overhead and Clouds on the Horizon: The Genuine Progress Report for 1999. Oakland: Re-defining Progress

Cobb C, Goodman GS & Wackernagel M (1999) Why bigger is'nt better – The Genuine Progress Indicator – 1999 Update. Redefining Progress. San Francisco

Cohen BL (1991) Catalog of Risks extendet and updated. Health Pysics, Vol. 61, September 1991, pp 317–335

Constanza R, d'Arge R, de Groot R, Farberk S, Grasso M, Hannon B, Limburg K, Naeem S, O'Neill R, Paruelo R, Raskin RG, Sutton P & van den Belt M (1997) The value of the world's ecosystem services and natural capital. Nature, Vol. 287, pp 253–260

CSLS (2007) Centre for the Study of Living Standards: Index of Economic Well-being. http://www.csls.ca/iwb/oecd.asp

Dahm JD (2002) Zukunftsfähige Lebensstile – Städtische Subsistenz für mehr Lebensqualität. Universität Köln, Köln

Delhey J, Böhnke P, Habich R & Zapf W (2002) Quality of life in a European perspective: The Euromodule as a new instrument for comparative welfare research. Social Indicators Research. Kluwer Academic Publishers, 58, pp 163–176

Diercke Weltatlas (2002) Westermann Schulbuchverlag GmbH, Braunschweig 1988, 5. Auflage

DNN (2005) Der Osten verlässt das Jammertal. 28th April 2005, p 3

Easterlin RA (2000) The Worldwide Standard of Living Since 1800. Journal of Economic Perspectives, Vol. 14, No. 1, pp 7–26

Eppler E (1974) Maßstäbe für eine humane Gesellschaft: Lebensstandard oder Lebensqualität? Verlag W. Kohlhammer. Stuttgart

Estes R (1988) Trends in World Social Development, New York. Praeger Publishers

Estes R (1990) Development Under Different Political and Economic Systems, Social Development Issues, 13(1). pp 5–19

Estes R (1992) At the Crossroads: Dilemmas in Social Development toward the Year 2000 and Beyond. New York and London: Praedger Publishers

Estes R (2003) Global Trends of Quality of Life and Future Challanges. Challenges for Quality of Life in the Contemporary World. Fifth Conference of the International Society for Quality of Life Studies. 20.-23. Juli 2003, Johann Wolfgang Goethe Universität Frankfurt/Main

Fehm B (2000) Unerwünschte Gedanken bei Angststörungen: Diagnostik und experimentelle Befunde. PhD thesis, University of Technology Dresden

FIISP (1995) Fordham Institute for Innovation in Social Policy: Index of Social Health: Monitoring the Social Well-Being of the Nation, 1995. GDP and population figures are from the Statistical Abstract of the United States, U.S. Department of Commerce

Frei A (2003) Auswirkungen von depressiven Störungen auf objektive Lebensqualitätsbereiche. Dissertation. Psychiatrische Universitätsklinik Zürich. August

GGDC (2007) Groningen Growth and Development Centre and the Conference Board, Total Economy Database, January 2007, http://www.ggdc.net

Gilbody SM, House AO, Sheldon T (2002) Routine administration of Health related Quality of Life (HRQoL) and needs assessment instruments to improve psychological out-come – a systematic review. Psychological Medicine; 32. pp 1345–1356

Gill TM & Feinstein AR (1994). A critical appraisal of the quality-of-life measurements. Journal of the American Medical Association, 272, pp 619–626

Greenpeace (2007) Footprint – Footprint und Gerechtigkeit: http://www.einefueralle.at/index.php?id–2989 (08.10.2007)

Hagerty MR, Cummins RA, Ferriss AL, Land K, Michalos AC, Peterson M, Sharpe A, Sirgy J & Vogel J (2001) Quality of Life Indexes for national policy: Review and Agenda for Research. Social Indicators Research, 55, pp 1–96

Handelsblatt (2007) Zukunftsatlas 2007 für Deutschland. 26 March 2007, Nr. 60 und PROGNOS

Heim E (1976) Stress und Lebensqualität. Bättig & Ermertz (Eds), Lebensqualität: ein Gespräch zwischen den Wissenschaften. Birkhäuser Verlag Basel, pp 85–94

Herschenbach P & Henrich G. (1991) Der Fragebogen als methodischer Zugang zur Erfassung von "Lebensqualität" in der Onkologie. Lebensqualität in der Onkologie. Serie Aktuelle Onkologie. W. Zuckschwerdt Verlag München, pp 34–46

Hofstetter P & Hammitt JK (2001) Human Health Metrics for Environmental Decision Support Tools: Lessons from Health Economics and Decision Analysis. National Risk Management Research Laboratory, Office of Research and Development, US EPA, Cincinnati, Ohio, September 2001

Holzhey H (1976) Evolution und Verhalten: Bemerkungen zum Thema Lebensqualität aus der Sicht eines Ethologen. Bättig & Ermertz (Eds), Lebensqualität: ein Gespräch zwischen den Wissenschaften. Birkhäuser Verlag Basel, pp 193–205

HRSDC (1997) Human Resources and Social Development Canada: How Do We Know that Times Are Improving in Canada? Applied Research Bulletin – Vol. 3, No. 2 (Summer-Fall 1997)

Hudler M & Richter R (2002) Cross-National Comparison of the Quality of Life in Europe: Inventory of Surveys and Methods. Social Indicators Research, 58, pp 217–228

Hüther G.(2006) Manual for the human brain (in German). Vandenhoeck & Ruprecht, Göttingen

IE (2002) http://www.internationaleconomics.net/research-development.html

Jenkinson C, Wright L & Coulter A (1993) Quality of Life Measurement in Health Care. Health Service Research Unit, Department of Public Health and Primary Care, University of Oxford, Oxford

Jongejan RB, Jonkman SN & Vrijling JK (2007) An overview and discussion of methods for risk evaluation. Risk, Reliability and Societal Safety – Aven & Vinnem (Eds), Taylor & Francis Group, London, pp 1391–1398

Joyce CRB (1991) Entwicklung der Lebensqualität in der Medizin. Lebensqualität in der Onkologie. Serie Aktuelle Onkologie. W. Zuckschwerdt Verlag München, pp 3–10

Karnofsky DA, Abelmann WH, Craver LF & Burchenal JH (1948) The use of the nitrogen mustards in the palliative treatment of carcinoma." Cancer 1: pp 634–656

Kaspar T (2004) Klinisch-somatische Parameter in der Therapie schwerer Herzinsuffizienz unter besonderer Berücksichtigung der gesundheitsbezogenen Lebensqualität Verlaufsanalyse des Einflusses klinisch-somatischer Parameter auf die Lebensqualität, Ludwig-Maximilians-Universität, München

Kasten E (2006) Somapsychologie, Vorlesungsmaterial, Magdeburg

Kern R (1981) Soziale Indikatoren der Lebensqualität. Verlag: Verband der wissenschaftlichen Gesellschaften, Wien

Knack S & Keefer P (1997) Does Social Capital Have an Economic Payoff? A Cross-Country Investigation. Quarterly Journal of Economics, 112(4), pp 1251–88

Knack S & Zack P (2001). Trust and Growth. Economic Journal, April 2001

Koch T (2000a) Life quality versus the 'quality of life': Assumptions underlying prospective quality of life instruments in health care planning. Social Science & Medicine, 51, pp 419–427

Koch T (2000b) The illusion of paradox. Social Science & Medicine, 50, 2000, pp 757–759

Köhler U & Proske D (2007) Interdisciplinary quality-of-life parameters as a universal risk measure. Structure and Infrastructure Engineering. 25 May 2007

Komlos J (2003) Warum reich nicht gleich gesund ist. "Economics and Human Biology", http://www.uni-protokolle.de/nachrichten/id/10996

Korsgaard CH (1993). Commentary on GA Cohen: Equality of What? On Welfare, Goods and Capabilities; Amartya Sen: Capability and Well-Being. MC Nussbaum & Amartya Sen. The quality of life. Calrendon Press, Oxford pp 54–61

Küchler T & Schreiber HW (1989) Lebensqualität in der Allgemeinchirurgie – Konzepte und praktische Möglichkeiten der Messung, Hamburger Ärzteblatt 43, pp 246–250

Kummer H (1976) Evolution und Verhalten: Bemerkungen zum Thema Lebensqualität aus der Sicht eines Ethologen. Bättig & Ermertz (Eds), Lebensqualität: ein Gespräch zwischen den Wissenschaften. Birkhäuser Verlag Basel, pp 29–39

Layard R (2005) Die glückliche Gesellschaft. Frankfurt, New York

LfS (2003) Landesinstitut für Statistik. Indikatoren für die Lebensqualität in Südtirol. Autonome Provinz Bozen-Südtirol. Bozen

Lüdtke H (2000) Zeitverwendung und Lebensstile – Empirische Analysen zu Freizeitverhalten, expressiver Ungleichheit und Lebensqualität in Westdeutschland, LIT, Marbuger Beiträge zur Sozialwissenschaftliche Forschung 5

Mannis A (1996) Indicators of Sustainable Development, University of Ulster. Ulster

Marczyk, J (2003) Does optimal mean best? In: NAFEMS-Seminar: Einsatz der Stochastik in FEM-Berechnungen, 7.-8. Mai 2003, Wiesbaden

Marks N, Abdallah S, Simms A & Thompson S (2006) The Happy Planet Index. New Economics Foundation, London

Maslow A (1970) Motivation and Personality. 2nd Edition, Harper & Row, New York, pp 35–58

Metz R (2000) Säkuläre Trends in der deutschen Wirtschaft. (der) Michael North: Deutsche Wirtschaftsgeschichte. München, p 456

MHRC (2007) – Mercer Human Resource Consulting: 2007 Worldwide Quality of Living Survey. http://www.mercer.com

Miegel M (2001) Arbeitslosigkeit in Deutschland – Folge unzureichender Anpassung an sich ändernde wirtschaftliche und gesellschaftliche Bedingungen. Institut für Gesellschaft und Wirtschaft, Bonn, Februar 2001

Mueller U (2002) Quality of Life at its End. Assessment by Patients and Doctors. Gimmler A, Lenk C & Aumüller G (Ed), Health and Quality of Life. Philosophical, Medical and Cultural Aspects. Münster, LIT Verlag, pp 103–112

Münch E (2005) Medizinische Ethik und Ökonomie – Widerspruch oder Bedingung. Vortrag am 8. Dezember 2005 zum Symposium: Das Gute – das Mögliche – das Machbare, Symposium am Klinikum der Universität Regensburg

Nathwani JS, Lind NC & Pandey MD (1997) Affordable Safety by Choice: The Life Quality Method. Institute for Risk Research, University of Waterloo Press, Waterloo, Canada

NCHS (2001) – National Centre for Health Statistics: National Vital Statistics Report, Vol. 48, No. 18, 7th February, 2001

NEF (2006): The new economics foundation http://www.happyplanetindex.org/about.htm

Noll H-H (1999) Konzepte der Wohlfahrtsentwicklung. Lebensqualität und "neue" Wohlfahrtskonzepte. Centre for Survey Research and Methodology (ZUMA), Mannheim

Noll H-H (2004) Social Indicators and Quality of Life Research: Background, Achievements and Current Trends. Genov, Nicolai (Ed), Advances in Sociological Knowledge Over Half a Century. Wiesbaden: VS Verlag für Sozialwissenschaften

OECD (2002) OECD Statistics online: Annual Labour Force Statistics. Paris

Pandey MD & Nathwani JS (1997) Measurement of Socio-Economic Inequality Using the Life Quality Index. Social Indicators Research 39, Kluwer Academic Publishers, pp 187–202

Pasadika B (2002) Health and its Significance in Life in Buddism. Gimmler A, Lenk C & Aumüller G (Ed), Health and Quality of Life. Philosophical, Medical and Cultural Aspects. Münster, LIT Verlag, pp 147–156

Pätzold J (2004). Kurzscript zum Teil Wirtschaftsgeschichte im Rahmen der Vorlesung „Problemorientierte Einführung in die Wirtschaftswissenschaften – Teil Volkswirtschaftslehre". Universität Hohenheim

Piecha A (1999) Die Begründbarkeit ästhetischer Werturteile. Fachbereich: Kultur- u. Geowissenschaften, Universität Osnabrück, Dissertation

Pigou AC (1920) The Economics of Welfare. London: Mac-Millan

Pliefke T & Peil U (2007) On the integration of equality considerations into the Life Quality Index concept for managing disaster risk. Luc Taerwe & Dirk Proske (Eds), Proceedings of the 5th International Probabilistic Workshop, pp. 267–281

Porzsolt F & Rist C (1997) Lebensqualitätsforschung in der Onkologie: Instrumente und Anwendung. Bullinger (Ed), Lebensqualitätsforschung. Stuttgart, Schattauer, New York. pp 19–21

Prescott-Allen R (2001) The Wellbeing of Nations. A Country-by-Country Index of Quality of Life and the Environment. Island Press. Washington, Covelo, London

Proske D (2004) Katalog der Risiken. 1st Edition. Dirk Proske Verlag, Dresden

QoLNZ (2003) Quality of Life in New Zealand's Eight Largest Cities 2003

Rackwitz R & Streicher H (2002) Optimization and Target Reliabilities. JCSS Workshop on Reliability Bades Code Calibration. Zürich, Swiss Federal Institute of Technology, ETH Zürich, Switzerland, March 21–22 2002

Ramberg JAL & Sjoberg L (1997) The Cost-Effectiveness of Lifesaving Interventions in Sweden. Risk Analysis, Vol. 17, No. 4

Ravens-Sieberer MPH, Cieza A & Bullinger M (1999) Gesundheitsbezogene Lebensqualität – Hintergrund und Konzepte. MSD Sharp & Dohme GmbH, Haar

Richter UR (2003) Redundanz bleibt überflüssig? LACER (Leipzig Annual Civil Engineering Report) 8, 2003, Universität Leipzig, pp 609–617

Rose M, Fliege H, Hildebrandt M, Bronner E, Scholler G, Danzer G & Klapp BF (2000) Gesundheitsbezogene Lebensqualität, ein Teil der allgemeinen Lebensqualität. Lebensqualitätsforschung aus medizinischer und sozialer Perspektive. Jahrbuch der Medizinischen Psychologie Band 18. Hogrefe – Verlag für Psychologie, Göttingen, pp 206–221

Schuessler KF & Fisher GA (1985) Quality of life research and sociology. Annual Review of Sociology, 11, pp 129–149

Schwarz P, Bernhard J, Flechtner H, Küchler T & Hürny C (1991). Lebensqualität in der Onkologie. Serie Aktuelle Onkologie. W. Zuckschwerdt Verlag München, pp 3–10

Skjong R & Ronold K (1998) Societal Indicators and Risk acceptance. 17th International Conference on Offshore Mechanics and Arctic Engineering, 1998 by ASME, OMAE98-1488

Spilker B, Molinek FR, Johnston KA, Simpson RL & Tilson HH (1990) Quality of Life, Bibliography and Indexes, Medical Care, 28(12), pp D51–D77

Statistics Finland (2001) http://tilastokeskus.fi/index_en.html

Streb J (2004) Problemorientierte Einführung in die Volkswirtschaftslehre. Vorlesungsbegleitmaterial Teil III: Wirtschaftsgeschichte, Wintersemester 2003/2004

Tengs TO, Adams ME, Pliskin JS, Safran DG, Siegel JE, Weinstein MC & Graham JD (1995) Five-Hundred Life-Saving Interventions and Their Cost-Effectiveness. Risk Analysis, Vol. 15, No. 3 pp 369–390

UK700 Group (1999) Predictors of quality of life in people with severe mental illness. British Journal of Psychiatry, 175, pp 426–432

UNDP (1990a) United Nations Development Program: Human Development Report. Oxford University Press

UNDP (1990b) United Nations Development Program: Human Development Report 1990: Concept Measurement of human development. http://hdr.undp.org/rcports

Veenhofen R (2001) Why Social Policy Needs Subjective Indicators. Veröffentlichungen der Abteilung Sozialstruktur und Sozialberichterstattung des Forschungsschwerpunktes Sozialer Wandel, Institutionen und Vermittlungsprozesse des Wissenschaftszentrums Berlin für Sozialforschung. ISSN 1615 – 7540, July 2001, Research Unit "Social Structure and Social Reporting" Social Science Research Center Berlin (WZB), Berlin

Veenhoven R (1996) Happy Life Expectancy – A comprehensive measure of quality-of-life in nations, Social Indicators Research, vol. 39, pp 1–58

Venetoluis J & Cobb C (2004) The Genuine Progress Indicator 1950–2002 (2004 Update). Redefining progress for people, nature and environment. San Francisco

Wagner A (2003) Human Development Index für Deutschland: Die Entwickung des Lebensstandards von 1920 bis 1960. Jahrbuch für Wirtschaftsgeschichte 2, pp 171–199

White M (2007) Twentieth Century Atlas – Worldwide Statistics of Death Tolls. http://users.erols.com/mwhite28 (access 2nd November 2007)

WHO (1948) World Health Organization. WHO Constitution. Geneva

Winograd M (1993) Environmental Indicators for Latin America and the Caribbean: Towards Land Use Sustainability, Organisation of American States, and World Resources Institute, Washington, DC.

Wood-Dauphinee S (1999) Assessing Quality of Life in Clinical Research: From Where have we come and where are we going? Journal of Clin Epidemiology, Vol. 52, No. 4, pp 355–363

World Resources Institute (1995) Environmental Indicators: A Systematic Approach to Measuring & Reporting on Environmental Policy Performance in the Context of Sustainable Development", World Resources Institute, Washington, DC.

Yerkes RM & Dodson JD (1908) The Relation of Strength of Stimulus to Rapidity of Habit-Formation. Journal of Comparative Neurology and Psychology, 18, pp 459–482

Zapf W (1984) Individuelle Wohlfart: Lebensbedingungen und wahrgenommene Lebensqualität. W Glatzer & W Zapf (Eds), Lebensqualität in der Bundesrepublik. Objektive Lebensbedingungen und subjektives Wohlbefinden. Frankfurt/Main and New York, Campus, pp 13–26

6 Law and Risk

6.1 Introduction

Social, health, natural and technical risks affect an individual's life often without much choice. In many cases, the public is not informed about hazards and risks. For example, there are no signs about the probability of failure in front of bridges to inform users. The same is also true for food products. However if one buys food products usually only some rough indications of the food ingredients are given on the food package. It remains unknown to the consumer what risk that represents. One just simply assumes that the safety of the product is ensured.

This assumption of safety is mainly based on the idea of a functioning state (Kaufmann 1973). Such states promise a net safety benefit through their existence. If one considers the budgets of developed countries as shown in Figs. 6-1 and 6-2, it becomes clear that, indeed, the major spending is on safety measures. As shown in chapter "Objective risk measures", social risks can be considered as highest risks. Figure 6-1 shows that in Germany, probably more than 50% of the budget are spent to mitigate such risks. Defense can be considered as mitigation measure. However, the distribution of resources to different tasks indicates differences among the countries, as shown in Figs. 6-1 and 6.2.

Actually, even simple societies found in the fauna are built on the assumption of the improvement of safety. The alternative would be complete individual responsibility for safety. For example, people might have to wear a weapon all the time to defend themselves. In the fauna, some great predators live this way. In general, however, building a society can create a safer and less vulnerable environment for individual members. On the other hand, a society or a state cannot fulfill this general requirement on its own. It needs the support from individuals. Therefore, the responsibility for safety previously transferred from the citizens to the state will simply be transferred back to the individuals again. This transfer, of course, has to include some pressure to motivate the individual to take this burden.

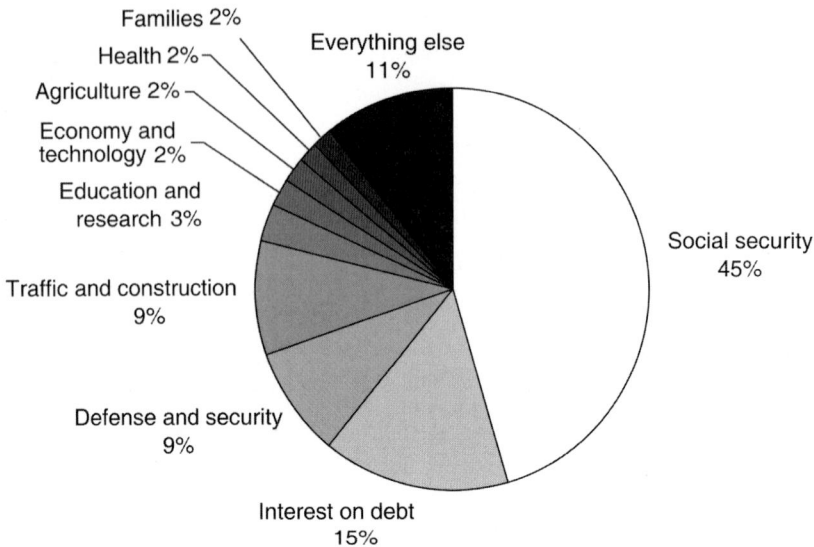

Fig. 6-1. Federal German budget in 2006

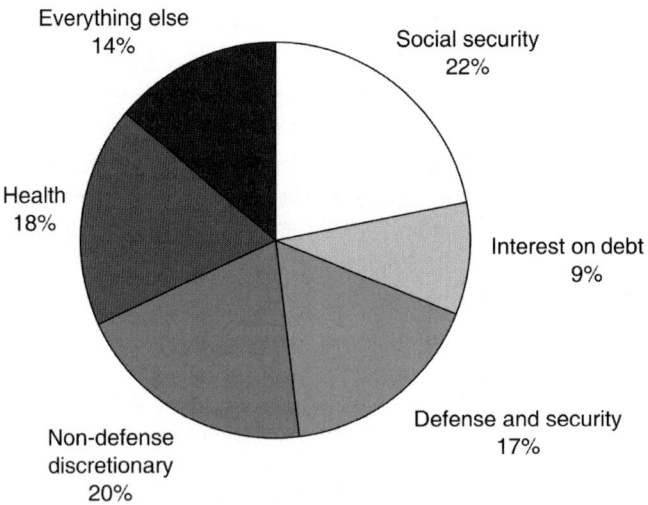

Fig. 6-2. Federal US budget in 2003

This simple function of a state can be clearly seen in the codex of Hammurabi created approximately 3,700 years ago. The codes of Hammurabi were not the first corpus juris, but because it was so excellent preserved, this early collection of laws is often referred to. The codex of King Hammurabi contains, depending on the counting, up to 280 laws. The law were concerned with different professions. In a very special way, the duties of each profession were described: "If a builder builds a house for someone, and does not construct it properly, and the house which he built falls in and kills the owner, then that builder shall be put to death. If the house kills the son of the owner then the son of that builder shall be put to death." (Murzewski 1974, Mann 1991).

This regulation was concerned with the safety of the public by transferring the issue of safety to the individual builder. The builder lived safely only if he provided proper buildings. Therefore, the public need to be concerned about the safety of buildings, since the builder is concerned about his own safety. Of course, the laws also dealt with a payment from the builder a penalty, and killing as a penalty was mainly caused by the insufficient penal system at that time. There were, for example, no prisons then.

Also interesting to note is the differentiation inside a profession about possible mitigation measures and impossible mitigation measures. For example, the codex of Hammurabi also stated: "If a building for livestock is hit by lightning or a lion kills some livestock, then the shepherd cleans them for God and hands back the killed livestock to the owner. But when the shepherd is responsible and causes the damage, then he should compensate the owner."

The transferring of responsibility for safety is, therefore, strongly connected to the definition of responsibility and the possibility of such a definition at all. As mentioned in chapter "Indetermination and risk", for some systems, a development of causal chains still remains impossible. While here further work has to be done, much work has been carried out about the transfer of responsibility of safety from the individual to the state. Nowadays, citizens in developed countries simply assume that safety is provided by the state and that acceptable levels of risk are achieved under all circumstances. The author believes that the duty of safety is the one and only goal of a state. In exchange, the state gets a monopoly of power. That also means the state has the right to punish people. This has been very clearly stated in the ideas about states in the 16th and 17th century: "Contracts without the mere sword are mere words and do not possess the power to prove even the lowest levels of safety to humans. In the case of a missing compulsory force or a too weak force which can not provide safety, everyone has the right to use and will use his own force and skills to protect himself" (Huber 2004).

The theoretical foundation of the monopoly of violence has influenced the development of modern constitutions, for example the "Bill of Rights" in 1689, the constitution of the USA in 1787 or the "Code civil des Français" in France in 1804 (Huber 2004). Since then, the right of life has been anchored in the United Nations Charter of human rights created in 1948. At this point, it should be noted that the extension of the monopoly of violence to international organisations seems the next logical step in war prevention. Indeed some authors recommend it for the United Nations (Ipsen 2005). This development, however, might require a few more decades; therefore, one should focus more on a national level. For example, in Germany, the right to life and to physical intactness can be found in the constitution under article 2, paragraph 2. In the beginning, it should be assumed that such a goal can be reached by providing a homogenous level of safety over all fields of a society. All special risks in all thinkable areas of daily life should have a comparable size. Under such standardised conditions, no warning or alarming is required. Only if the risk is substantially increased, risk measures and warnings are required. This assumption permits the citizens in a country to busy themselves with everyday life problems rather than with problems concerning their own existence. However, if the general requirement of the statement has to be fulfilled in all areas of societies, the regulations and laws have to be more specific. These laws should clearly state when mitigation measures and warnings are required and when acceptable levels of risks are reached.

Before we consider this, some terms widely used in jurisprudence are introduced. Even though jurists use the same words as engineers, for example, risk, acceptable risk and probabilities, it does not mean that the terms have the same meaning here.

Since the times of Bismarck's Prussia, danger is a situation that yields to damage when active prevention measures are not taken (Leisner 2002, HVerwG 1997). The possibility of damages is described by a sufficiently large probability. No absolute certainty is required. However, the possibility of a damage is actually a risk in engineering terms, whereas it is considered a danger in jurisprudence.

The difference in the meanings of terms depends on the likelihood of a damage event (Di Fabio 1994). Dangers from a lawyer's point of view are considered risks with a high likelihood, but from an engineer's point of view, risks with median or low likelihood are called risks by both and risks with very low likelihood are called acceptable risks by both. It should be mentioned here that the acceptable risk for engineers does not only depend on the likelihood.

Independent of this classification, the courts or the legislator demand measures for the avoidance of risks. First, dangers have to be removed. Danger prevention in terms of security requirements can be found in German civil law (damage substitute Section 823 paragraph 1 civil code, in product liability law Section 1, paragraph 1 or in administrative order Section 123, 80 paragraph 5).

While danger has to be eliminated completely, risks are juridically tolerated in a limited frame. This is obvious since 100% risk exclusion is impossible (Rossnagel 1986, Risch 2003). The partial acceptance of risks leads again to the concept of acceptable risk. Independent from the origin of the risk, either from technologies or from natural processes, it has to be carried by all citizens as a social load according to the German federal constitutional court (Risch 2003): "Uncertainty beyond this threshold is caused by the limitation of the human cognitive faculty. They are therefore inescapable and insofar are a socially adequate burden for all citizens." (Böwing 2004).

Therefore, in general, courts agree that there is no risk-free life in a society. This is not a theoretical fact, but has had consequences in justice. For example, in the field of natural hazards, some cases about acceptable risk have drawn major attention from the media and the public. One example was an accident on 13 March 1996 at Berglasferner in the Stubaier Alps, Austria. A mountain guide led a tour to the "Wilde Hinterbergl". On the way, the group passed a glacier, which was known to be rich in fissures. The guide informed himself about the ice conditions and decided that the entire group could pass the glacier without roping. One of the participants fell into a fissure and died. The guide was sued. The question was whether the group should have used ropes or not. This would have been a mitigation measure to drop the risk to an acceptable risk. The first court decided that the guide was innocent, whereas in the second court he was held guilty. In the latest decision, it was decided that he was innocent. In the reasoning the court stated: "... based on the statements and the assumptions about the circumstances, especially the missing of hazard indicators in the accident area, a descent without a rope at the time of the accident ... was not an infringement of the duty to take care and still in the region of an acceptable risk." Historically, it was assumed that guided tours are completely risk free, which as illustrated above cannot be the case (Ermacora 2006).

A second example is an accident that occurred in Jamtal on 28 December 1999. Here, three guides decided to climb the "Russkopf" with a group of 43 people. During the descent, the group passed a slope of about 40°, which they had also used for the ascent. There was a possibility to use another way, which would have resulted in a detour of only a few minutes (mitigation measure). During the descent, an avalanche killed nine people from the group. A court in Innsbruck, Austria, decided the guides were innocent (acceptable risk), whereas a court in Germany (most of the members of the group came from Germany) decided compensation may be given. (Ermacora 2006).

A third example was the killing of two Germans on the Gotthard highway in Switzerland by a rockfall in 2006. The Swiss government stated that the use of the Gotthard highway means the acceptance of a certain risk. Historically, there were occasional heavy rockfalls in the area, sometimes smashing the Gotthard highway.

Such risks are also called de minimis risks in juristic language. The term comes from the Latin expression "De minimis non curat lex", which means "the law does not concern with trifles". These risks are, therefore, acceptable. It does not mean, however, that such events cannot happen. The German federal constitutional court during the Kalkar case tried to define acceptable risk in terms of being "practically unimaginable" and "irrelevant"; however, an exact value was not given. The Kalkar case dealt with the safety of nuclear power plants in Germany.

As seen from example two, a clear definition of danger, risk and acceptable risk would be helpful. The explicit definitions of danger, risk and acceptable risk according to the mathematical definition of likelihood have not occurred up to now through the legislator. In contrast to some examples, such as DNA testing or blood alcohol values, legislators and courts oppose the definition of clear mathematical values: "It can not be explained rationally whether a danger begins with the likelihood of a certain damage event of 10^{-5}, 10^{-6} or 10^{-7} per year" (Becker 2004, De Fabio 1994). In contrast, courts declare risk as ones that significantly rise the general life risk (Lübbe-Wolff 1989).

Fortunately, in some court cases, numbers were given. While the Kalkar decision just stated "risk beyond practical rationality", the administrative appeals tribunal in Münster, 1975, gave a frequency value of 10^{-7} per year for a nuclear accident as an acceptable risk (OVB Münster 1975). This value was approved by the administrative tribunal in Freiburg, 1977 (VG Freiburg 1977). In that year, the administrative tribunal in Würzburg discussed that value (VG Würzburg 1977). In 1997, the Hessian administrative tribunal, however, refused to give a number about the frequency of an acceptable risk (HVerwG 1997). It should be mentioned that, in some cases, the courts decide only about the increase or decrease of risks. For example, in 2000, the administrative appeals tribunal Rhineland-Palatinate decided that the increase of ship size on the river Main does not increase the risk for bridge failure on this river (OVG RP 2000). For further investigation about the issue, please refer to chapter "Example – Ship impacts against bridges" (Luke 1980).

In the American judicature, there are also some examples dealing with the definition of acceptable risk: "If, for example, the odds are one in a billion that a person will die from cancer by taking a drink of chlorinated water, the risk clearly could not be considered significant (10^{-9}). On the other hand, if the odds are one in a thousand (10^{-3}) that regular inhalation of gasoline vapors that are 2% benzene will be fatal a reasonable person might well consider the risk significant and take the appropriate steps to decrease or eliminate it." (US Supreme Court 1980).

In England, already in 1949 lawyers discussed the comparison of risks: "Reasonably practicable is a narrower term than physically possible and seems to imply that a computation must be made by the owner in which the quantum of risk is placed on one scale and the sacrifice involved in the measures necessary for averting the risk (whether in money, time or trouble) is placed in the other. If it be shown that there is a gross disproportion between them – the risk being insignificant in relation to the sacrifice – the defendants discharge the onus on them." (Asquith 1949) Based on that idea the so-called ALARP-Principal (As Low As Reasonable Practicable) has been developed, which is widely used in the English-speaking countries, but it actually is also connected to the Kalkar decision.

In general, in laws, one mainly finds the requirement of safety. Thus, one sees in the Saxon building code (SächsBO 2004) under Section 3 general requirements: "(1) physical structures as well as other arrangements and facilities in terms of Section 1 paragraph 1 sentence 2 are to be ordered in such a way, to establish, to change, to repair and to keep in good condition so that the public security and order, in particular life or health or the natural resources are not endangered. ..."

This example already shows the conflict in determining the requirements of safety for the legislator. On one side, the legislator would like to permit the fundamental rights to occupational freedom, proprietary freedom and freedom of economic activity, and on the other side, the fundamental right to life and health has to be guaranteed. Physical structures are necessary to protect people in winter against cold, but these physical structures may also become a menace for the people who use these buildings. Both points of view are taken into consideration in Germany in the constitution: as already mentioned, article 2 paragraph 2 includes the right to life and health, article 12 includes the right to occupational freedom and article 14 describes the right of proprietary freedom. The solution to this tension between "freedom of technology" and the "protection of the citizens" according to the protective duty is incumbent upon the state (Risch 2003).

The state can use different instruments to comply with its commitments. Such instruments are also used for the evaluation of risks. The higher the risk, the higher the legitimisation of the state. The highest legitimised government body is usually the Parliament. All decisions concerning basic rights of citizens have to be dealt with by that body. This is especially true for decisions concerning possible risks to the entire population. For example, the German Federal Emission Control Act has a strong influence over the production permit of genetic technology facilities. One instance of this influence was the Hessian administrative court refusing to grant such permission in advance. The court declared that only the state could decide whether the hazard is acceptable for the population and, therefore, should develop such an act. The legislator has followed and has introduced the mentioned act, along with the Genetic Technology Act and the Embryos Protection Act.

In the case of the safety of structures, the legislators have refused to give detailed indications about the required safety. The only hint was given in terms of the permitted material. For example, carcinogenic materials such as asbestos are not permitted. If no further safety requirements are given by the legislatives, then the decisions have to be done by the administration. Administrations as executives are usually better prepared to evaluate risks. Major tools for administrations are administrative acts and administrative fiats. Such tools are faster and more flexible to apply as compared to the Parliamentarian instruments. In addition, the administrative regulations can include much more details, and in some cases the constitution requires the committal of the development of further rules from the Parliament to the administration.

It is interesting to note in many cases that the administration uses an accurate technical set of regulations provided by standardisation organisations. The administration employs private technical expertise to develop the regulations. This is again an example of how the administration transfers the responsibility of safety back to individuals and professional associations. The standardisation codes are usually given the status of administrative fiats, which helps the user of standardisation by giving a legal framework. But, of course, there are also conditions where the situation is not so clear. For example, civil engineering professionals are required to keep up to date with the current state-of-the-art. That means if some important developments have occurred and have been published in some journals, the professionals are required to learn that and consider that in their daily work. Additionally, in some cases, the so-called "permission" in individual case is given by the authorities. For example, the author has verified the static computation of the first textile reinforced bridge worldwide. In such a case, there was no standard available, and therefore special treatment of the case was required. Under such conditions, the evaluation of safety is given back to the administration, which has to check the requirements.

So far, only theoretical considerations were expressed. On the other hand, such complex issues as safety and risk have to be permanently assessed not only by the individuals but also by the authorities. This becomes even more visible if one looks at different court decisions. Therefore, the justice itself has to be included in the decisions since it influences both the individual's decision about acceptable risk and the legislators, forcing them to introduce a juristic frame. Many such court decisions can be found during the time when nuclear power plants were introduced in Germany. Such technology has yielded to intensive worldwide juristic argumentations about possible and acceptable risks. Here, the limitations of safety requirements in the basic law become clearly visible. In the Atomic Energy Act, the permission of this technology is given, but detailed information is missing. In the Atomic Energy Act Section 7, a precaution based on the current state-of-the-art is required. This general requirement has then been rarefied by further laws, for example the Radiation Protection Act. The administration has heavily focussed on the safety of humans and human health. For example, the permitted radioactive radiations are less than the natural radiation variance. On the other hand, still many details have not been ruled on yet. This forces the administration tribunals and federal constitutional courts to state their interpretation of terms such as danger, risk and acceptable risk. Prior to the concept of risk, courts, of course, dealt with the problem of safety using other techniques. For example, authorities used deterministic safety concepts such as declaring a state either safe or not safe. For nuclear power plants, more advanced safety concepts were introduced like the probabilistic safety concept and the risk concept. This makes sense especially for nuclear power plants since the damage potential is so high. The discussion about safety concepts has sometimes yielded to rather strange situations. In some cases, courts had to decide which safety evaluation concept should be used, but, of course, courts have limited technical knowledge and are unable to decide, for example, with which proceeding the safety of nuclear power plants during an earthquake should be considered. The German federal constitutional court has, therefore, declared that it does not consider scientific disputes as their task.

As an example, the safety concepts in structural engineering should be mentioned. The safety requirements according to building regulations have already been discussed. Also, some safety concepts were discussed in chapter "Indetermination and risk". But here, the development from some general legislation requirements to some details in technical regulations should be shown. While normal building codes use either simple deterministic or semi-probabilistic safety concepts, under some conditions, the codes, such as the German code E DIN 1055-9 (2000) or the Eurocode 1 (1998), permit risk assessment: "The exclusion of risks can in most cases not be achieved. Therefore the acceptance of certain risks is required ... For the definition of acceptable risks a comparison with other risks in commensurable situations which are accepted by the society should be carried out."

6.2 Examples

Consideration of an acceptable risk in the design process can give rather unintended results. Two cases are presented to illustrate this point.

In the middle of the 1990s, Patricia Anderson sued General Motors due to an explosion in the tank of the car during a rear-end collision. The automotive company knew the problem but considered the risk an acceptable. The estimated number of casualties was about 500 per year with 41 million cars from the company being used. The compensation claims were assumed to be approximately 100 million US dollars per fatality per year based on the prevailing case law at that time. The assumption, however, of an acceptable risk without any further risk mitigation measures being implemented convinced the court to increase the compensation claim up to 4.9 billion US dollars (Stern 1999).

This was actually only a second instance of an earlier case. In the 1970s, the car manufacturer Ford was sued due to the same problem. The tank of the Ford Pinto was prone to explosion during collisions. Ford assumed, at that time, the compensation claims would be approximately 200,000 US dollars for one fatality and 67,000 US dollars for the seriously injured. About 11 million cars of this type were sold per year, and Ford assumed about 2,100 cars were deflagrable. However, in 1978, a court in California approved a compensation of 128 million US dollars for an injured person.

The threat of punishment either in terms of money or in terms of personal punishment can be found, as already mentioned, historically. The requirement of the safety of structures can be found not only in the Roman law collection "Corpus Juris" but also in some theological books, for example The Bible (Hof 1991). Probably very early on, the safety of houses was of concern since houses were one of the first technical products. Some publications mention that the earliest fixed structures existed 20,000 years ago (Mann 1991). Since then, the variety of technical products has greatly expanded. With the expansion of such products, the legal net had to become denser to describe the safety requirements for all technical products.

Such a development of legislations can only follow technical development. A few examples should show this. Probably fire protection legislations are nearly as old as house safety legislations. A primary law of Hittites Empire from 2000 B.C. described incendiary and compensation claims (Schneider & Lebeda 2000). In Hammurabi's codex, statements about theft during a fire can been found. Also, there were fire brigades in the old Egyptian empire. Greeks, in 293 B.C., used mobile fire ladders (Werner 2004). In 23 B.C., the Roman emperor Augustus released a regulation introducing a fire brigade in Rome. After some conflagrations in

Rome, there were rules about minimum road widths to prevent the spreading of fires (Werner 2001) since many major cities experienced heavy conflagrations. Already in medieval times, the first fire protection law was imposed. The law stated that, in case of a fire all craft guild members have to aid in extinguishing the fire. The second oldest fire protection law might be the one in London enacted in 1189. According to this regulation, every owner of a house had to have a barrel of water in front of the house in summer. In 1268, the permanent fire watch was introduced, which was carried by the night watchman. The validity of this law was extended in 1285 to all English cities. In Vienna, the first fire protection law was introduced very early. In 1221, the town privilege given by Herzog Leopold IV included juridical consequences for the house owner, in whose house a fire started. Since then, the introduction of the Fire Protection Act can be found in many European cities, such as in 1276 in Augsburg, 1299 in Metz, 1351 in Erfurt, 1370 in Munich, 1371 in Paris, 1403 in Cologne, 1433 in Bremen and 1439 in Frankfurt (Werner 2004, Schneider & Lebeda 2000). Further examples are listed in Table 6-1.

The communal laws were only substituted by federal state laws after the Napoleon wars. Currently, large portions of the fire protection requirements are rooted in the federal building regulations. These general requirements are then fulfilled either through special fire protection regulations or on buildings regulations. Such special regulations are, for example, included in the Public Assembly Act, Commercial Building Regulations or Garage Regulations.

Another example of the regulations considering safety is rules for the production of explosive materials. The explosion of a ship carrying black powder caused 151 casualties in the Dutch city of Leiden in 1807, which is already mentioned in chapter "Risks and disasters". Napoleon visited the site and decided to impose an imperial act concerning the location of manufactories. In the act, it was stated that manufactories should be divided into different risk groups. Manufactories dealing with hazardous material were not permitted in the neighbourhood of houses. Manufactories dealing with less hazardous materials were permitted in the neighbourhood of houses but not in the city. Safe productions were permitted inside the cities (Ale 2003).

Table 6-1. Development of regulations concerning fire safety in cities (Werner 2004, Schneider & Lebeda 2000)

Year	City/Region	Act concerning
1342	Munich	Roof cladding only with bricks
1366	Esslingen	Minimum road width
1388	Flensburg	Roof cladding only with bricks
1427	Ulm	Checking of building plans by regulators
1558	Hessian	Smokestack requirements for brew houses
1680	Lübeck	Requirement of fire walls
1715	Karlsruhe	Escape routes considered in city planning
1735	Wuppertal	No wooden chimneys permitted
1814	Cologne	Introduction of building inspection department
1844	Prussia	Limitation of building heights
1868	Baden	Escape routes act
1890	Hessian	Fire extinguishing acts
1909	Prussia	Theater building acts
1918	Prussia	Domicile building regulations
1919	Prussia	City planning regulations
1931	Germany	City planning regulations
1931	Prussia	Regulations for commercial buildings
1942	Germany	German building law (not imposed)

In 1814, a further regulation dealing with damages, hazards and disturbances by production was introduced in the Netherlands. In that same year, a regulation concerning explosive materials was imposed. In 1876, a law for dealing, storage and transport of poisonous materials was introduced, and in 1875, a law for production was imposed. In 1896, the first occupational safety and health laws were developed. In 1886, a report pointed to the difference between risk to workers and risk to the public. Since then, the laws concerning working conditions, working risks or the production of explosive materials have been extended and have become more detailed. However, as seen in some disasters in the last years, for example, in Enschede, the Netherlands, there is still some work to do (Ale 2003).

In contrast, the development of legislation requirements in some fields had just started in the last decades. For example, the first laws concerning debris (for example, the Solid Waste Disposal Act) were introduced in 1965 in the US.

In the field of genetic and biological technologies, the history is even shorter. First steps to developing new laws concerning the issue can be traced to beginning of the 1970s. In 1974/1975, the first voluntary moratorium was developed (Berg et al. 1974). It should be mentioned that a first call for a moratorium was published already in the beginning of the 1960s.

In 1976, guidelines were introduced to replace the moratorium. In 1978, the European Commission developed a proposal (Lohninger 2007).

6.3 Codes

As already mentioned, codes of practice can be considered an administrative fiat. Therefore, the listing of codes also represents certain government approaches in dealing with risks. Such codes can be found in many fields, including civil engineering, food safety, medicine, military actions, mining and so on. In Table 6.2, a few codes are mentioned to give the reader an idea about the diversity of codes not only in different fields but also across different states.

Table 6-2. Examples of codes of practice dealing with risk assessment

S.No.	Norm
1	AS/NZS 4360: Risk Management. 1999
2	CAN/CSA-Q850-97: Risk Management: Guideline for Decision-Makers. 1997
3	DIN 40041: Zuverlässigkeit – Begriffe. Beuth Verlag Berlin 1990
4	DIN EN 14738: Raumfahrtproduktsicherung – Gefahrenanalyse; Deutsche Fassung EN 14738: 2004, August 2004
5	DIN EN ISO 14971: Medizinprodukte- Anwendung des Risikomanagements auf Medizinprodukte. Beuth Verlag Berlin 2001
6	DIN EN 61508-4: Funktionale Sicherheit elektrischer/elektronischer/ programmierbar elektronischer sicherheitsbezogener Systeme – Begriffe und Abkürzungen. Beuth Verlag Berlin 2002
7	DIN VDE 31000–2: Allgemeine Leitsätze für das sicherheitsgerechte Gestalten technischer Erzeugnisse – Begriffe der Sicherheitstechnik – Grundbegriffe. December 1984
8	E DIN 1055-100: Einwirkungen auf Tragwerke, Teil 100: Grundlagen der Tragwerksplanung, Sicherheitskonzept und Bemessungsregeln, Juli 1999
9	E DIN 1055-9: Einwirkungen auf Tragwerke Teil 9: Außergewöhnliche Einwirkungen. März 2000
10	Emergency Risk Management: Applications Guide (Emergency Management Australia)
11	Eurocode 1 – ENV 1991–1: Basis of Design and Action on Structures, Part 1: Basis of Design. CEN/CS, August 1994
12	EPA/630/R-95/002F: Guidelines for Ecological Risk Assessment. May 14, 1998
13	FM 100-14 US Army: Risk Management
14	HB 231: Information Security Risk Management Guidelines. 2000
15	HB 240: Guidelines for Managing Risk in Outsourcing. 2000

(Continued)

Table 6-2. (Continued)

S.No.	Norm
16	HB 250: Organisational Experiences in Implementing Risk Management practices. 2000
17	ISO 2394: General Principles on Reliability for Structures. 1996
18	ISO 13232-5: Motorcycles–Test and analysis procedures for research evaluation of rider crash protective devices fitted to motorcycles - Part 5: Injury indices and risk/benefit analysis. 1996
19	ISO 14001: Environmental Management Systems, Specification with Guidance for Use. 1996
21	ISO 14004 Environmental Management Systems, General Guidelines on Principles, Systems and Supporting Techniques. 1996
22	ISO 14121: Safety of Machinery – Principles of Risk Assessment. 1999
23	ISO 14971: Medical devices – Application of Risk Management to Medical Devices. 2000
24	ISO 17666: Space Systems – Risk Management. 2003
25	ISO 17776: Petroleum and natural gas industries – Offshore production installations – Guidelines on tools and techniques for hazard identification and risk assessment. 2000
26	ISO 8930: Allgemeine Grundregeln über die Zuverlässigkeit von Tragwerken; Liste äquivalenter Begriffe, Dezember 1987
27	ISO/IEC Guide 51: Safety Aspects – Guidelines for Their Inclusion in Standards. 1999
28	ISO/IEC Guide73: Risk Management Terminology. 2002
29	ISO/TS 14798: Lifts (Elevators), Escalators and Passenger Conveyors – Risk Analysis Methodology. 2000
30	ISO/TS 16312-1: Guidance for assessing the validity of physical fire models for obtaining fire effluent toxicity data for fire hazard and risk assessment –Part 1: Criteria. 2004
31	MG3 – A Guide to Risk Assessment and Safeguard Selection for Information Technology Systems and its Related Framework.
32	New Seveso Directive 82/96/EC (1997)
33	New Seveso Directive 96/82/EC
34	NIST 800-30: Risk Management Guide
35	NORSOK Standard Z-013: Risk and Emergency Preparedness Analysis. Norwegian Technology Standards Institution, March 1998
36	Norwegian Maritime Directorate: Regulation of 22. December 1993 No. 1239 concerning risk analyses for mobile offshore units
37	Norwegian Petroleum Directorate: Guidelines for safety evaluation of platform conceptual design 1.9.1981
38	Norwegian Petroleum Directorate: Regulations relating to implementation and use of risk analyses in the petroleum activities, 12.7.1990.
39	Norwegian Standards: NS 5814: Requirements for risk analyses. Norwegian Standard Organisation, August 1991
40	OSHA Standards: Occupational Exposure to Cadmium, Section: 6: Quantitative Risk Assessment

Table 6-2. (Continued)

S.No.	Norm
41	SAA HB 141: Risk Financing. 1999
42	SAA HB 142: A Basic Introduction to Managing Risk. 1999
43	SAA HB 203: Environmental Risk Management: Principles and Processes. 2000
44	SAA/NZS HB 143: Guidelines for Managing Risk in the Australian/New Zealand Public Sector. 1999
45	SAA/SNZ HB 228: Guidelines for Managing Risk in the Healthcare Sector. 2001
46	SNZ 2000: Risk Management for Local Government (SNZ 2000)

Subsequent to this discussion of risk parameters, subjective judgement and juridical background, the next chapter describes an example of the application of risk assessment.

References

Ale BJM (2000) Risk Assessment Practices in the Netherlands. Proceedings – Part 1/2 of Promotion of Technical Harmonization on Risk-Based Decision-Making, Workshop, May 2000, Stresa, Italy

Ale BJM (2003) Keynote lecture: Living with risk: a management question. (Eds) T Bedford &, PHAJM van Gelder: Safety & Reliablity – (ESREL) European Safety and Reliability Conference 2003, Maastricht, Netherlande, Balkema Publishers, Lisse 2003, Volume 1, pp 1–10

Asquith E (1949) National Coal Board. All England Law Reports, Vol. 1, p. 747

Becker P (2004) Schadensvorsorge aus Sicht der Betroffenen. 12. Deutsches Atomrechtssymposium. Forum Energierecht Band 8, Nomos Verlagsgesellschaft 2004 Baden-Baden, pp 133–148

Berg P, Baltimore D, Boyer HW, Cohen SN, Davis RW, Hogness DS, Nathans D, Roblin R, Watson JD, Weissman S & Zinder ND (1974) Potential Biohazards of Recombinant DNA Molecules. Proceedings of the National Academy of Sciences of the USA. Vol. 71, No. 7, pp 2593–2594

Böwing A (2004) Die Vorsorge gegen äußerst seltene, auslegungsüberschreitende Vorfälle im Atomrecht aus Sicht der Betreiber. 12. Deutsches Atomrechtssymposium. Forum Energierecht 8, Nomos Verlagsgesellschaft, pp 149–157

BVergGE (1978) – Bundesverfassungsgericht. BVergGE 49, 89. Kalkar I, 8. August 1978, pp 89 –147

Di Fabio U (1994) Risikoentscheidung im Rechtsstaat. Tübingen; Mohr Siebeck

E DIN 1055-9 (2000) Einwirkungen auf Tragwerke Teil 9: Außergewöhnliche Einwirkungen. März 2000

Ermacora A (2006) Restrisiko und der Umgang mit dem Strafrecht. In: Fuchs, Khakzadeh & Weber (Eds): Recht im Naturgefahrenmanagement. Studien-Verlag, Innsbruck, pp 197–207

Eurocode 1 (1998) ENV 1991-2-7: Grundlage der Tragwerksplanung und Einwirkungen auf Tragwerke – Teil 2-7: Einwirkungen auf Tragwerke – Außergewöhnliche Einwirkungen. Deutsche Fassung, August 1998

Hof W (1991) Zum Begriff Sicherheit. Beton- und Stahlbetonbau 86, Heft 12, pp 286–289

Huber PM (2004) Die Verantwortung für den Schutz vor terroristischen Angriffen. 12. Deutsches Atomrechtssymposium. Forum Energierecht 8, Nomos Verlagsgesellschaft, pp 195–215

HVerwG (1997) – Hessischer Verwaltungsgerichtshof: Urteil vom 25.3.1997: Az. 14 A 3083/89

Ipsen K (2005) The eternal war – capitulation of the international law for the reality? (in German). In: Rubin 2/2005. Ruhr-University Bochum, pp 26–31

Kaufmann F (1973) Sicherheit als soziologisches und sozialpolitisches Problem – Untersuchungen zu einer Wertidee hochdifferenzierter Gesellschaften. Enke: Stuttgart

Leisner A (2002) Die polizeiliche Gefahr zwischen Eintrittswahrscheinlichkeit und Schadenshöhe; DÖV 2002, pp 326–334

Lohninger AC (2007) Grundlagen der Gen- und Biotechnologie. Nomos, Wien

Luke R (1980) Gefahren und Gefahrenbeurteilungen im Recht. Teil II. Risikoforschung in der Technik Länderberichte Bundesrepublik Deutschland und Frankreich, Köln, Berlin, Bonn, München: Heymann

Lübbe-Wolff G (1989) Die Grundrechte als Eingriffsabwehrrechte, Baden-Baden, Nordrhein-Westfälische Verwaltungsblätter, Heft 9

Mann G (Edr) (1991) Propyläen Weltgeschichte – Eine Universalgeschichte. Propyläen Verlag Berlin – Frankfurt am Main

Murzewski J (1974) Sicherheit der Baukonstruktionen. VEB Verlag für Bauwesen, Berlin, DDR

OVG Münster (1975) Urteil vom 20.2.1975, Az.: VII A 911/69

OVG RP (2000) Urteil vom 27. Juli 2000, Az.: 1 C 11201/99, Juris Nr.: MWRE109040000

Risch BM (2003) Juristische Grundlagen von Risikountersuchungen - Recht und Risiko. 1. Dresdner Probabilistik-Symposium – Risiko und Sicherheit im Bauwesen, Technische Universität Dresden, Fakultät Bauingenieurwesen, Dresden, 14. November 2003, Tagungsband, pp 67–76

Rossnagel A (1986) Die rechtliche Fassung technischer Risiken; UPR 1986, pp 46 –56

SächsBO (2004) Sächsische Bauordnung. Fassung vom 28. Mai 2004

Schneider S & Lebeda C (2000) Baulicher Brandschutz. W. Kohlhammer, Stuttgart

Schröder R (2003) Verfassungsrechtliche Grundlagen des Technikrechts. In: Schulte M (Hrsg): Handbuch des Technikrechts, Berlin, Springer, pp 185–208

Stern (1999) Eiskalte Rechnung. 22.7.1999

US Supreme Court (1980) Industrial Union Department. vs. American Petrol Institut, 448 U.S. 607, 448 U.S. 607, No. 78-911. Argued Oct. 10, 1979. Decided July 2, 1980

VG Freiburg (1977) Urteil vom 14.3.1977, Az.: VS II 27/75 = NJW 1977, 1645 ff

VG Würzburg (1977) Urteil vom 25.3.1977, Az.: W 115 II 74 = NJW 1977, 1649 ff

Werner U-J (2004) Bautechnischer Brandschutz. Birkhäuser Verlag, Basel

7 Example – Ship Impacts Against Bridges

7.1 Introduction

The concepts introduced in this book so far should now be applied to a real-world problem. This is done in this chapter through an investigation on bridges. Bridges are technical products. Comparing to other technical products, however, bridges feature some special properties. Bridges, for example, might change the view of a landscape. Also, bridges have an exceptionally long and useful lifespan. Usually the lifespan of bridges exceeds the lifetime of humans. There exist bridges with a lifespan of more than two thousand years.

One can well imagine that the load conditions have changed considerably since then due to progress in other technical areas, such as the development of road vehicles. Vehicles were only invented 100 years ago. Nowadays, regular vehicles with an overall mass of up to 44 tonnes can use the roads in Germany. Vehicles with a mass of up to 70 tonnes can use the roads in Germany with special permission (Hannawald et al. 2003), and the request for this permission has grown exponentially over the last few decades (Naumann 2002). The same development pattern can be found for railways (Beyer 2001).

Many historical bridges were built between the mid-19th century and the beginning of the 20th century. This time frame is introduced in this chapter as a definition for the erection period of historical bridges. It was during this time that the Industrial Revolution occurred in Germany and many other European countries. Germany experienced a remarkable increase in infrastructure investment during the Industrial Revolution. For example, the length of the railway system in Germany grew from roughly 600 km in 1840 to approximately 62,000 km in 1910 (Mann 1991). This led to a noticeable rise in the number of bridges erected.

In many cases, these historical bridges, which were predominantly made of stone and steel, remain in service even today. For example, in the area controlled by the Zwickau Department of Highways and Bridges (Saxony,

Germany), one-third of all road bridges are historical, concrete or natural stone arch bridges (Bothe et al. 2004).

Natural stone arch bridges, in particular, are able to bear increasing vertical live loads due to their simple but effective static arch system and very high grade of natural stone used. These bridges also show a very large resistance to horizontal loads because of their large dead weight. In addition to the explained altered vertical loads, a new hazard has emerged for historical bridges that span inland waterways. Within the last century, due to the dramatic increase of dead weight loads and the speed of inland ships, the possibility of ship impacts resulting in bridge failure has increased substantially. In Germany, the mean mass value of motor ships sailing on inland waterways has doubled in the last 30 years (Proske 2003). Today, inland waterway ships reach a length of up to 200 m, a breadth of up to 12 m and a weight of up to several thousand tonnes. For comparison, it should mentioned that the pier width of the bridges mentioned is between 3 and 6 m and the length can be up to 12 m. These numbers alone illustrate the possibility of a bridge failure due to ship impact.

7.2 Problem

It sounds very unlikely that ship impacts against bridges happen. Nevertheless, history shows that this type of accident has happened quite regularly. Scheer (2000) lists bridge failures caused by ship impacts since 1850 (Fig. 7-1). Larsen (1993) gives a list of serious ship accidents including bridge failures due to ship impact (Fig. 7-2). Additional information on slip collisions with bridges are given by Van Manen & Frandsen (1998). The fact that bridge failures due to ship impact can lead to serious consequences, including massive fatalities, is illustrated in Table 7-1. Two examples of recent ship impacts are given: a ship impacted against a highway bridge in China in 2007, yielding to the collapse of the bridge; and a ship impact in Krems, Austria, in 2005 resulted in the pier moving more than 2 m (Simandl et al. 2006). Table 7-2 lists the probabilities of ship impacts against bridges for several European waterways.

The data show a peak in the number of impacts at the end of the 1970s and the beginning of the 1980s. After that, the number of impacts declined. Unfortunately, since the beginning of the 1990s, the number of ship impacts against bridges has risen again, especially in some areas of Germany. The development of the number of accidents, both for Germany (Stede 1997, Kunz 1998) and for the area controlled by the Würzburg Department of Highways and Bridges, is shown in Fig. 7-3.

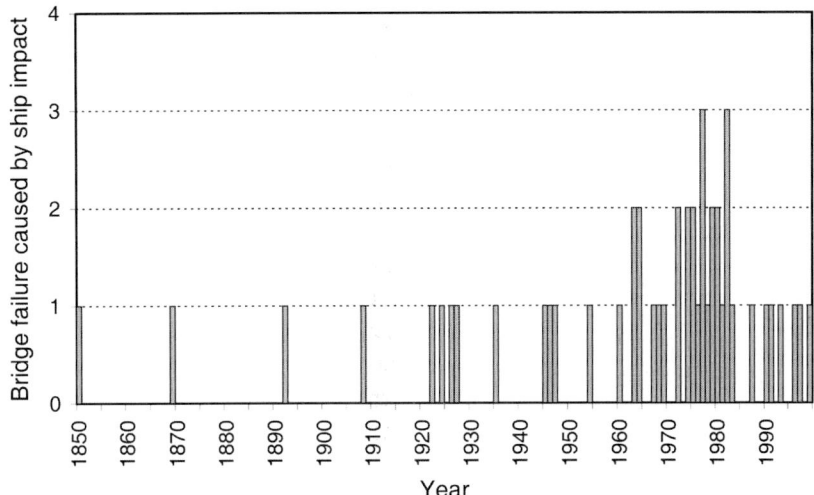

Fig. 7-1. Number of bridge failures caused by ship impacts (Scheer 2000)

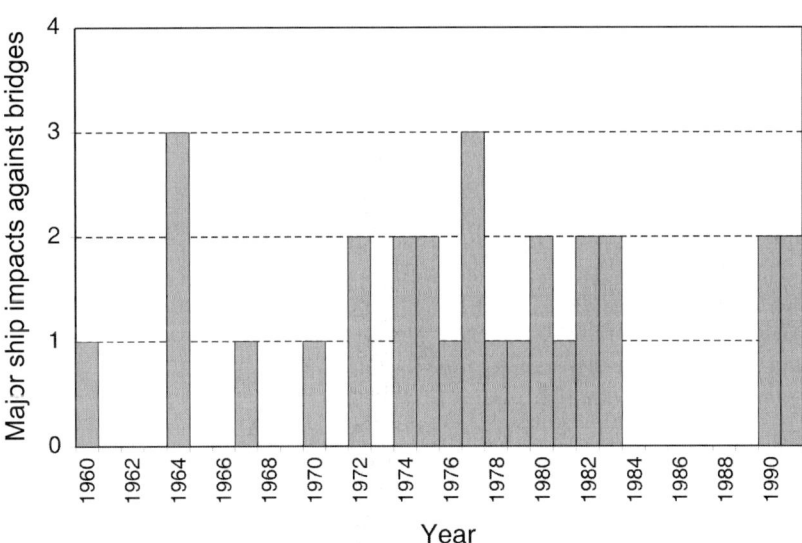

Fig. 7-2. Major ship impacts against bridges (Larsen 1993)

Table 7-1. Examples of bridge failure caused by ship impact (extended from Mastaglio 1997)

Bridge and location	Year	Fatalities
Severn River Railway Bridge, UK	1960	5
Lake Ponchartain, USA	1964	6
Sidney Lanier Bridge, USA	1972	10
Lake Ponchartain Bridge, USA	1974	3
Tasman Bridge, Australia	1975	15
Pass Manchac Bridge, USA	1976	1
Tjorn Bridge, Sweden	1980	8
Sunshine Skyway Bridge, USA	1980	35
Lorraine Pipeline Bridge, France	1982	7
Sentosa Aerial Tramway, China	1983	7
Volga River Railroad Bridge, Russia	1983	176
Claiborn Avenue Bridge, USA	1993	1
CSX/Amtrak Railroad Bridge, USA	1993	47
Port Isabel, USA	2001	8
Webber-Falls, USA	2002	12
Highway bridge between Foshan and Heshan, China	2007	>9

Table 7-2. Impact frequencies for German and European waterways

River	Impact per year per bridge	Impact per ship passage	Reference
Thames (UK)	0.2300	$10.7 \cdot 10^{-6}$	PTC 1 (1999)
Seine (France)	0.0313		PTC 1 (1999)
Seine (France)	0.0556	$15.7 \cdot 10^{-6}$	PTC 1 (1999)
Drogden Channel (DK/S)	1.7561	$59.0 \cdot 10^{-6}$	PTC 1 (1999)
Main (Germany)	0.0088	$0.7 \cdot 10^{-6}$	PTC 1 (1999)
Main (Germany)	0.0160	$61.0 \cdot 10^{-6}$	Proske (2003)
Main (Lohr, Germany)	0.0351	$21.0 \cdot 10^{-6}$	Kunz (1998)
Mosel (Germany)	0.0370	$0.7 \cdot 10^{-6}$	PTC 1 (1999)
Danube (Vilshofen, Germany)	0.1580		Kunz (1996)
General, Germany	0.0210		Stede (1997)
General, Germany	0.0095	$0.5 \cdot 10^{-6}$	Lohrberg & Keitel (1990)
Elbe (Dresden, Germany)	0.0380		Proske (2003)

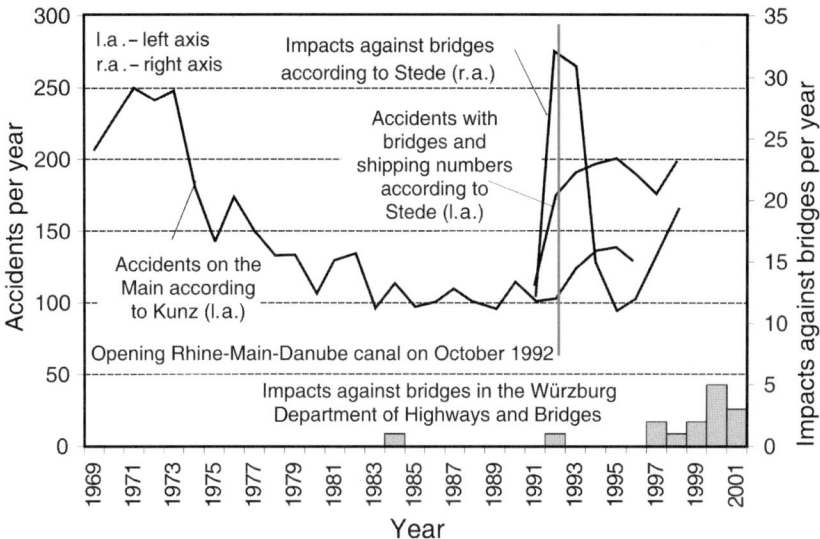

Fig. 7-3. Number of ship accidents and ship impacts against bridges

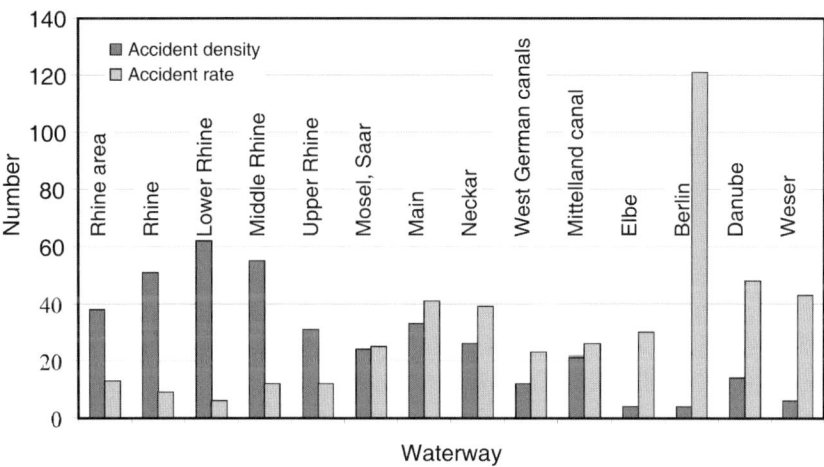

Fig. 7-4. Ship accident rate per billion tonne per kilometre and density for different waterways in Germany (Stede 1997)

Practically, all the available data indicate an increase in accidents at the beginning of the 1990s. The increase started around the year 1992, the maximal value occurred between 1992 and 1995, which is followed by a decrease in accidents. Since 1997, an increase has been recorded again, with the collision rate in the area of the Würzburg Department of Highways and Bridges reaching a dramatic value. Unfortunately, no data for the rest of the German inland shipping network are available since 2000. Since the Main River presumably does not have an over-proportionately high accident rate (Fig 7-4), one must conclude that the total collision rate in Germany has also increased. This worrying development was the reason why the Würzburg Department of Highways and Bridges required a numerical estimation of the safety of historical bridges with regard to collisions with ships.

Safety is an essential requirement for every technical product. The German state underlines this requirement through many legal regulations (German Civil Code Section 823, Product Liability Law Section 1 or the Constitution). The construction industry must comply with these regulations.

In civil engineering, safety is defined as the qualitative capacity of structures to resist loads (E DIN 1055-100 1999). A building cannot obviously withstand all theoretically possible loads, but it is essential that it resists the majority of loads to a satisfactory degree. The basis of this decision can only be quantitative. The reliability of a building is one quantitative measure of the capacity of a structure to resist loads. Reliability is interpreted in the current building regulations through the probability of failure. In addition, the concept of risk is introduced as a safety measure for accidental loads such as impacts (Eurocode 1 1994, E DIN 1055-9 2000). This safety measure is based on the probability of failure and additionally through considering the consequences of structural failure. Therefore, it becomes possible to classify the potential danger of buildings and bridges and compare them with other risks (E DIN 1055-9, para 5.1 (2) 2000).

This estimation of the safety of historical bridges with respect to ship impact has been carried out in terms of the operational failure probability and using risk parameters in accordance with the current regulations. Within the framework of a pilot project, this value has been determined for two historical bridges over the river Main (Proske 2003).

7.3 Investigated Bridges

Bridge 1 is situated near Würzburg in the south of Germany. The bridge is a six-arch bridge with a span of approximately 25 m. It was constructed between 1872 and 1875 with regular coursed ashlar stone work. The material used was red Main-sandstone, a high-quality coloured sandstone with a uniaxial compression strength of up to 140 MPa. Towards the end of World War II, one pier was blasted. This pier was rebuilt in 1945 and 1946 using concrete. During the reconstruction phase, the so-called explosion chambers were built inside the pier. Figure 7-5 shows a view of the site to-day. A detailed description of the bridge can be found in Proske (2003). The bridge was chosen to represent a typical historical German arch bridge over inland waterways. Several other arch bridges, such as the Main bridge in Marktheidenfeldt (1846), the Albert bridge in Dresden (1875), the Marien bridge in Dresden (1846), the Augustus bridge in Dresden (1910) and the old arch bridge in Pirna (1875), are to a certain extent comparable.

Bridge 2 was built in 1893 and consists of a steel frame superstructure with natural stone piers. Parts of the bridge were destroyed during the World War II. The statical system of Bridge 2 is a four-field beam with a span of approximately 39 m. Due to the different statical systems of the two bridges, both bridges will show different behaviour under impact. In contrast to the excellent natural stone material of Bridge 1, the material in Bridge 2 has lesser strength. Figure 7-6 shows a view of the site today. A detailed description of the bridge can be found in Proske (2003). This bridge was chosen to represent a typical historical German bridge with steel superstructures over inland waterways.

Fig. 7-5. Picture of Bridge 1

Fig. 7-6. Picture of Bridge 2

Attention should be drawn to Bridge 2 as well as Bridge 1 because both were hit by ships in the last years. Bridge 2 was in particular danger when a ship ran into one pier in 2000 (Fig. 7-7). Bridge 1 experienced an impact in 1999 (Fig. 7-8). The damages from these impacts on both the bridge and the ship have been used to calibrate the computation models. Results from the computation model for the impact against Bridge 2 are shown in Figs. 7-9 and 7-10 in terms of stress and cracks.

Fig. 7-7. Ship impact against Bridge 2 in 2000

Fig. 7-8. Ship impact against Bridge 1 in 1999

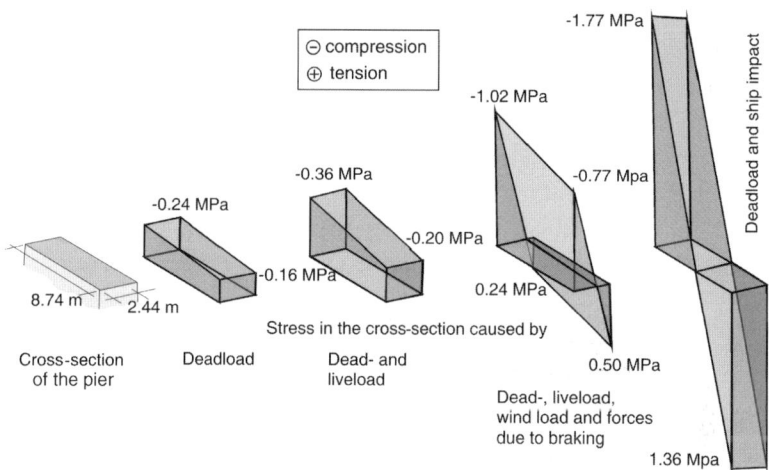

Fig. 7-9. Simple stress evaluation in the pier of Bridge 2

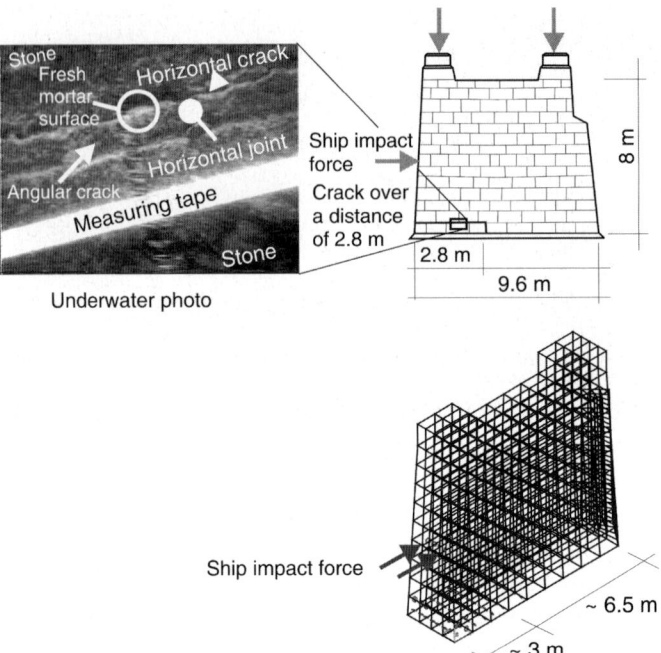

Fig. 7-10. The figure on the top shows location of a crack found at the hit pier of Bridge 2, and the figure at the bottom shows the results of the computation model for the pier using the finite element program ATENA. A 3-m-long crack was found in the finite element model at the same location as the real crack

7.4 Investigation

The investigation for both bridges was split into three steps. First, the bridge regions most severely stressed during the impact were detected by simple numerical calculations and then selected for drilling. Since Bridge 1 consisted of different materials due to the partial replacement of blasted piers at the end of World War II, the transfer of results from one structural element to the same element in another position was not possible. Moreover, Bridge 1 has a unique foundation. With due regard to these specific factors, 26 drillings of a length of up to 15 m were planned and carried out on this bridge. The drillings had an overall length of 150 m: 90 m in masonry and 60 m in concrete. Figure 7-11 shows the drilling cores from Bridge 1.

Fig. 7-11. Drilling cores from Bridge 1

The drilling produced a sufficiently large amount of bridge material for material testing. More than 500 material tests were carried out including compressive and tensile strength tests of the sandstone, mortar and concrete, Young's modulus tests, height and width measurements of the sandstone, density measurements and measurements of the shear strength of the masonry. With the data from the tests, it was possible to describe the material input parameters in terms of random distributions. The choice of the distribution type is discussed intensively in Proske (2003). Several statistical techniques have been used to determine the type of statistical distribution for the investigated material properties, such as:

- relating the coefficient of variance and the type of distribution
- relating skewness and kurtosis and the type of distribution
- the minimum sum square error based on histograms
- the χ^2 test and $n\omega^2$ test
- the Kolmogoroff–Smirnoff test
- the Shapiro–Wilk or Shapiro–Francia test
- probability plots
- quantile–correlation values

In addition, the strengths of the concrete compressive strength samples were multimodal. Therefore, this distribution was decomposed into original distributions. This statistical effect could also be identified visually on the testing specimen and historically with documents of reconstruction after the war.

Material tests were also carried out for Bridge 2. In contrast to Bridge 1, the number of material tests was much lower. The properties of the random variables for both bridges are shown in Table 7-3

(Proske 2003). Additionally, random properties for the ship impact forces are shown in this table. The ship impact forces have been calculated using Meier–Dörnberg's deterministic model (Meier-Dörnberg 1984). The model from Meier–Dörnberg was developed especially for inland water-way ships and has, therefore, been used in favour of the Woisin model (1977), which has been successfully applied for maritime ships and is the basis for regulations in many countries. Figure 7-12 shows the results of both formulas based on the input data. To evaluate the impact forces as random variables, some input data must also be random variables. These variables are the speed of the ship, the mass of the ship, the angle of the ship when going off-course and the exact location where the ship went off-course. Based on the work of Kunz (1998), the probability of hitting the bridge as well as the statistical properties of the ship impact force could be estimated.

Table 7-3. Statistical properties of input variables (statical properties)

Bridge	Parameter	Distribution	x_m[a]	s[b]	Unit
1	Sandstone compressive strength	Lognormal	75.40	21.30	MPa
	Concrete compressive strength	Lognormal	47.90	22.28	MPa
	Sandstone splitting strength	Lognormal	4.72	1.30	MPa
	Concrete tensile stress	Lognormal	1.15	0.69	MPa
	Young's modulus sandstone	Lognormal	28,534.	7,079.6	MPa
	Young's modulus concrete	Lognormal	22,552.	8,682.1	MPa
	Density sandstone	Normal	2.27	0.15	kg/dm^3
	Density concrete	Normal	2.26	0.10	kg/dm^3
	Mortar compressive strength	Lognormal	11.00	7.25	MPa
	Ship impact force (frontal)	Lognormal	2.04	1.5	MN
	Ship impact force (lateral)	Lognormal	0.61	0.385	MN
	Sandstone height	Normal	0.7	0.13	m
	Sandstone width	Lognormal	0.8	0.08	m
	Mortar joint height	Lognormal	0.037	0.048	m
	Impact height	Normal	3.0	0.5	m
2	Natural stone compressive strength	Normal	21.2	2.4	MPa
	Natural stone splitting strength	Normal (Lognormal)	0.38	0.094	MPa
	Mortar compressive strength	Normal	15.5	3.58	MPa
	Ship impact force (frontal)	Lognormal	2.04	1.5	MN
	Ship impact force (protection)	Lognormal	0.046	0.8368	MN
	Ship impact force (lateral)	Lognormal	0.61	0.385	MN
	Impact height	Normal	3.0	0.5	m
	Normal load	Normal	0.242	0.0242	MPa

[a] x_m – empirical mean, [b] s – empirical standard deviation.

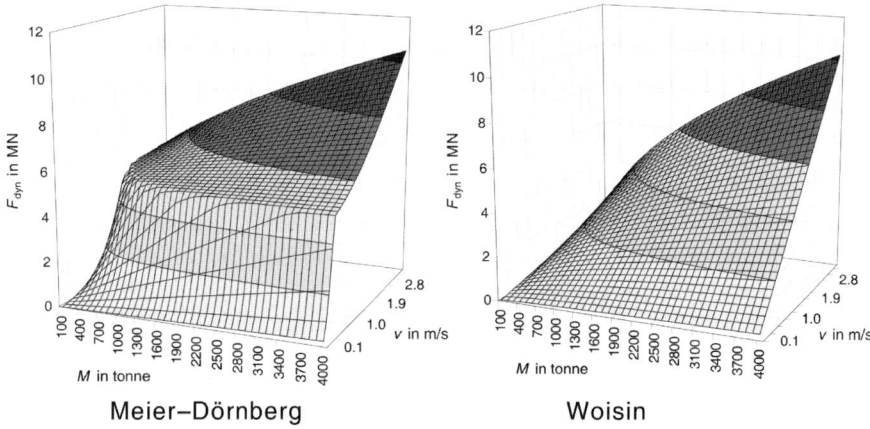

Meier–Dörnberg   Woisin

Fig. 7-12. Ship impact forces according to Meier–Dörnberg and Woisin models

In the second step, the mechanical behaviour of the bridges under ship impact has been modelled with finite element programs (ANSYS and ATENA). The model of Bridge 1 was particularly complex due to the in-homogeneous structural system. Figure 7-13 permits a view inside the bridge. Figure 7-14 shows the finite element model using area and volumes (unmeshed), and Fig. 7-15 shows the principal compression stress inside a hit pier. The typical element size was about 0.5 m, but smaller elements were used in some regions. Cracks observed on both bridges caused by dead- and liveloads could be approved with the used models.

To achieve such results, a realistic numerical model for the description of the load-bearing behaviour of natural stone masonry had to be incorporated into the finite element program. There exist several different techniques to describe the load-bearing behaviour of natural stone structures under normal and shear forces. The models from Mann, Hilsdorf, Berndt and Sabha for one layer walls and the models from Warnecke and Eger-mann for multi-layer walls (Warnecke et al. 1995) have been investigated (Proske 2003). The model of Berndt (1996) was chosen after intensive numerical investigations to describe the load-bearing behaviour of the natural stone piers of the bridges. This model is valid for normal forces and shear forces, and therefore is able to describe the load conditions under impact. Also, this model has proved to reach acceptable results in the comparison of the load-bearing behaviour with a wide range of experimental data. In addition, the implementation into the finite element program was convenient.

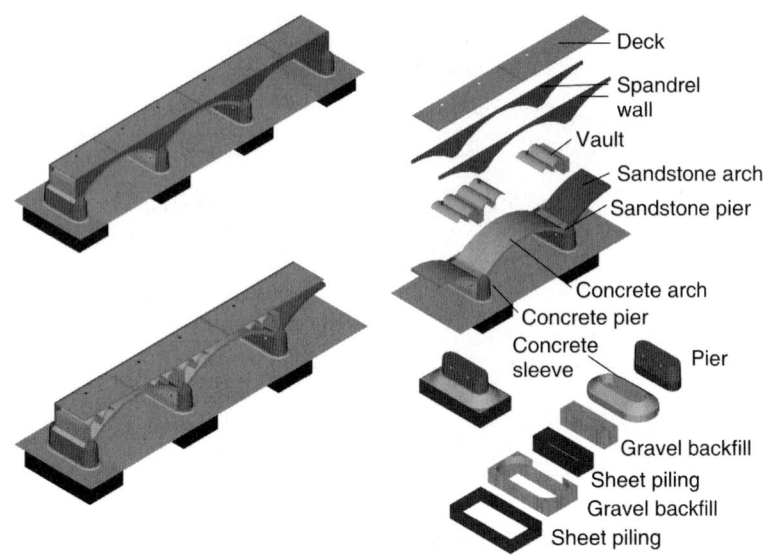

Fig. 7-13. Assembly of Bridge 1

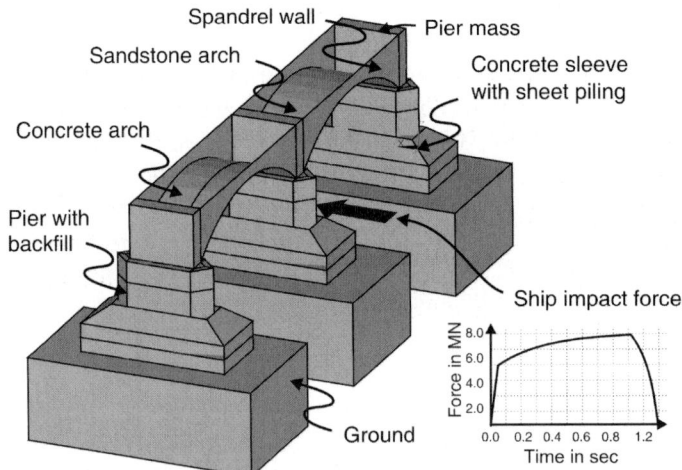

Fig. 7-14. Example of a finite element model of Bridge 1 (not meshed)

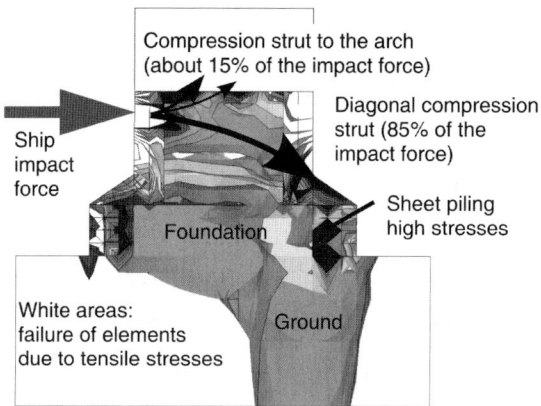

Fig. 7-15. Principal compression stress in the longitudinal section of frontal hit pier of Bridge 1

Using this model, the calculation of one deterministic dynamic impact against Bridge 1 on an IBM workstation with a power II processor took about 1h. Of course, simple, linear, elastic static and dynamic models for the bridges have also been used to check the results of the sophisticated models.

To incorporate the random variables in the third step, a probabilistic calculation was done using first-order (FORM) and second-order reliability methods (SORM). After the FORM (Spaethe 1992) calculation, different SORM methods were applied to improve the quality of the first. The best-known and most widespread SORM method is probably Breitung's formula (Spaethe 1992). In addition to that, the methods of Köylüoglu & Nielsen (1994) and Cai & Elishakoff (1994) were used. In addition, importance sampling has been used to back the results (Spaethe 1992). All the methods used gave results that were comparable from an engineering point of view.

The criteria for probabilistic calculation included results of the dynamic finite element calculation. Therefore, the so-called limit state function was not available in an analytically closed form. One way to obtain results with the probabilistic calculation with a known limit state function, which is not analytically closed, is the application of the response-surface methodology, for example Rajashekhar & Ellingwood (1993). This procedure was included into the finite element program ANSYS (Curbach & Proske 1998). The FORM and SORM techniques were also incorporated into the finite element program ANSYS using the customised capabilities of the program. For that purpose, the techniques had to be provided as FORTRAN subroutines, which could then be compiled and linked into the program.

Based on both, the intensive investigated input data and the complex mechanical model, the evaluation of the safety of both bridges could be carried out.

7.5 Results

First, the maximum possible impact forces have been investigated. In the next evaluation step, the probabilistic investigation has been accomplished. The results of the probabilistic investigation are shown in Table 7-4. Several structural solutions to increase the load-bearing capacity of the bridges under ship impact were also investigated. They are visualised in Fig. 7-16. The results are shown as the probability of failure either per impact $P(V|A)$ or per year $P(V \cap A)$. The value per year also includes the probability of the ship impact event. To show that the models of the bridges are comparable, the probability of failure under dead- and liveload conditions are also evaluated. Lines 6 and 14 from Table 7-4 show approximately the same value. Only these two lines refer to the failure of the piers under normal stress; all other lines refer to shear failure of the piers or the arch.

To allow a better comparison of both bridges, Table 7-5 summarises the major properties of the bridges. Also, Table 7-6 compares the results with other probabilistic calculations of historical bridges based on references or calculated by different authors. The author is well aware of the influence of different mechanical models or different input data. However, these historical bridges have existed under real-world load conditions for a long time, and the results of a probabilistic calculation must reflect this fact in terms of comparability. Further details about the probabilistic investigations of historical arch bridges can be found in Proske et al. (2006).

The maximum permitted probability of failure per year is about 1.3×10^{-6} (E DIN 1055-100 1999). Due to unsatisfactory results (see Table 7-4), the description of safety in terms of risk has been extended.

Table 7-4. Probability of failure for different structural versions. For both bridges, there exist different probabilities of impact – row VIII and IX for Bridge 2 and row X and XI for Bridge 1. To compare both bridges, Bridge 1 has in row VIII and IX the same probabilities of impact applied as for Bridge 2. (Note: The probabilities are given as multiples of 10^{-6}.)

I #	II Bridge	III Load	IV Element	V Version	VI $P(V|A) \cdot 10^{-6}$ per impact	VII $\beta(V|A)$	VIII $P(V \cap A) \cdot 10^{-6}$ per year	IX $\beta(V \cap A)$	X $P(V \cap A) \cdot 10^{-6}$ per year	XI $\beta(V \cap A)$
1	2		Pier 2	Damage (crack)[a]	313,667.7	0.4854	5,018.7	2.5745		
2		Frontal impact	Pier 2	No damage (no crack)[b]	154,256.0	1.0183	2,468.1	2.8111		
3			Pier 2	Impact fender system	1,540.5	2.9595	24.6	4.0605		
4			Pier 2	Pier increased ×2.3[c]	11,843.4	2.2621	189.5	3.5541		
5			Pier 2	Tension strength ×2[d]	43,179.2	1.7149	690.9	3.1985		
6		Dead and liveload		Normal stress[f]	240,0	3.4919	4.8	4.4259		
7		Lateral	Pier 2	No damage[b]	328,986.4	0.4427	5,263.8	2.5580		
8		Impact	Pier 2	Impact fender system	84,539.3	1.3751	1,352.6	2.9994		
9	1		Pier II	Explosion chamber[e]	80,760.0	1.4002	1,292.2	3.0132	596.0	3.2410
10		Frontal	Pier II	No explosion chamber[f]	23,300.0	1.9904	372.8	3.3722	172.0	3.5799
11		impact	Pier II	Prestressed[g]	340.0	3.3977	5.4	4.3958	2.5	4.5639
12		Impact	Pier II	Reinforced concrete[h]	32.0	3.9976	0.5	5.0370	0.2	5.0370
13			Pier II	GEWI-elements[i]	28.0	4.0290	0.4	5.0370	0.2	5.0370
14		Dead and liveload		Normal stress[f]	203.0	3.5363	4.1	4.4619	4.1	4.4619
15		Frontal	Pier III	Explosion chamber[e]	35,930.0	1.8004	578.8	3.2511	265.2	3.4651
16		Impact	Pier III	No explosion chamber[f]	28,720.0	1.9004	459.5	3.3143	212.0	3.5249
17			Pier III	Prestressed[g]	30.0	4.0128	0.5	5.0493	0.2	5.0493
18			Arch		1,500.0	2.9681	24.0	4.0652	11.1	4.2421
19		Lateral	Pier III	No explosion chamber[f]	25,670.0	1.9491	410.7	3.3454	54.9	3.8678
20		Impact	Pier III	No explosion chamber[f]	10,720.0	2.3006	171.5	3.5809	36.1	3.9688

[a] The pier has been found with a 3 m crack.
[b] Assumption of closing the crack.
[c] Size of the pier increased by factor 2.3.
[d] Hypothetical material with higher tensile strength.
[e] Explosion chamber was found inside the piers.
[f] Closing of the explosion chamber.
[g] Prestressing of the pier with no-bond tendons (2 × 2 MN and 2 × 4 MN, respectively).
[h] Reinforced concrete replacement type piles (2 × 3 Ø 1.5 m) inside the pier and closing the explosion chamber.
[i] Use of threaded rods (GEWI) inside the piers (2 × 4) and closing explosion chamber.
[f] Not considering an impact.

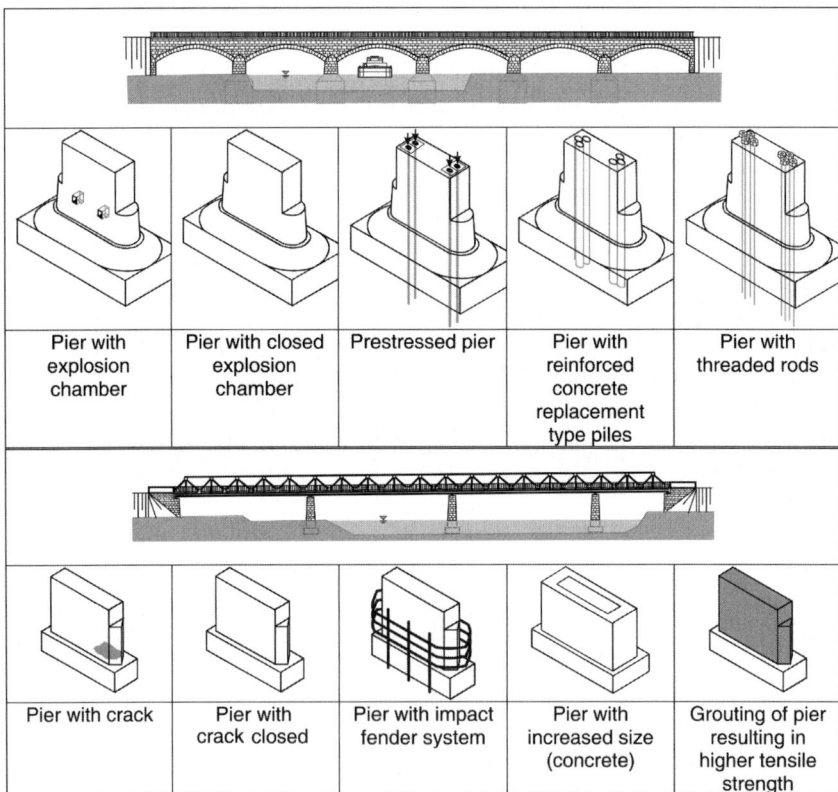

Fig. 7-16. Visualisation of the different strengthening technologies investigated for Bridge 1 (top) and Bridge 2 (bottom)

Table 7-5. Comparison of the properties of Bridge 1 and Bridge 2

	Bridge 1		Bridge 2	
Probability of failure per impact	0.023	$(1.0)^a$	0.15	$(6.5)^a$
Probability of failure under dead- and liveloads	$2.030·10^{-4}$	$(1.0)^a$	$2.400·10^{-4}$	$(1.2)^a$
Area of the pier in m^2	48	$(1.9)^a$	25	$(1.0)^a$
Deadload in MN	37	$(7.4)^a$	5	$(1.0)^a$
Existing normal stress in MPa	0.84	$(3.2)^a$	0.26	$(1.0)^a$
Acceptable normal stress in MPa	25	$(2.5)^a$	10	$(1.0)^a$
Acceptable shear stress in MPa	0.8	$(2.7)^a$	0.3	$(1.0)^a$
Maximal dynamic impact force in MN	13.0	$(2.9)^a$	4.5	$(1.0)^a$
Quantile value of the impact force in %	99.99		97.00	

[a] Numbers in parentheses give the ratio between the two bridges.

Table 7-6. Probabilities of failure for different historical arch bridges

Proof under dead- and liveload	Chosen value	per	Reference
Muldenbridge Podelwitz 1888[a]	$591.50·10^{-6}$	Year	Möller et al. (1998)
Flöhabridge Olbernhau	$0.04·10^{-6}$	Year	Möller et al. (2002)
Syraltalbridge Plauen 1905	$360.00·10^{-6}$	Year	Möller et al. (2002)
Bridge 1 (Bernd model) 1875	$4.10·10^{-6}$	Year	Proske (2003)
Bridge 2 1893	$4.80·10^{-6}$	Year	Proske (2003)
Marienbridge Dresden 1846	$1,279.0·10^{-6}$	Load	Busch (1998)
Bridge 1 (Mann model) 1875	$33,430.·10^{-6}$	Load	Proske (2003)
Bridge 1 (Berndt model) 1875	$248.0·10^{-6}$	Load	Proske (2003)
Bridge 2 1893	$203.0·10^{-6}$	Load	Proske (2003)
Foreshore Bridge Rendsburg 1913	$343.0·10^{-6}$	Load	
Bridge Schwarzenberg 1928	$159.0·10^{-6}$	Load	
Artificial bridge	$130.0·10^{-6}$	Year	Schueremans et al. 2001
Magarola bridge	$5.48·10^{-2}–10^{-12}$	Year	Casas (1999)
Jerte bridge	$2.33·10^{-4}–10^{-10}$	Year	Casas (1999)
San Rafael bridge	$1.90·10^{-8}–10^{-12}$	Year	Casas (1999)
Duenas bridge	$4.05·10^{-1}–10^{-12}$	Year	Casas (1999)
Quintana bridge	$1.28·10^{-12}$	Year	Casas (1999)
Torquemada bridge	$3.40·10^{-6}$	Year	Casas (1999)

[a] Limit state of serviceability.

One possibility to prove the safety of the bridges is through the comparison of individual risks in terms of mortality per year per person for a certain activity or in a certain situation. The author has done this with many different circumstances in chapter "Objective risk measures". For example, about one million fatalities per year due to car accidents, a few thousand fatalities per year due to aircraft accidents and about ten fatalities per year due to bridge failure caused by ship impact are listed worldwide as average over the last decade (Proske 2003). However, this representation does not take into consideration the severity of a single accident, which is an important issue for the public. For this reason, the so-called *F-N* diagrams or Farmer-diagrams were used for the representation of risks. In these diagrams, the consequences of failure or an accident were plotted against the frequency. The consequences were predominantly given in the number of fatalities and occasionally in monetary units. The data basis for the estimation of fatalities caused by bridge failure due to ship impact is shown in Fig. 7-17. The results are shown in Fig. 7-18. Furthermore, Fig. 7-18 includes several goal curvatures for the proof of safety in these diagrams. In these diagrams, both bridges, but especially Bridge 2, show a safety level that is insufficient. Therefore, the bridges have to be strengthened.

In conclusion, it is necessary to find the most effective method of strengthening bridges to enable them to bear ship impact loads. To find this, the so-called "Life Quality Index" (Voortman et al. 2001) introduced in chapter "Quality of life–the ultimate risk measure" has been used. This index considers the change of the probabilities of failure from the original bridge to the strengthened bridge. It also includes further data such as the mean life expectancy in Germany, mean mortality rate, ratio of working to leisure time and the per capita income. The results are shown in Table 7-7, assuming a fatality number of either 10 people (based on the history of only road bridge failures through ship impact) or 22 people (based on the history of all bridge failures through ship impact) (Proske 2003).

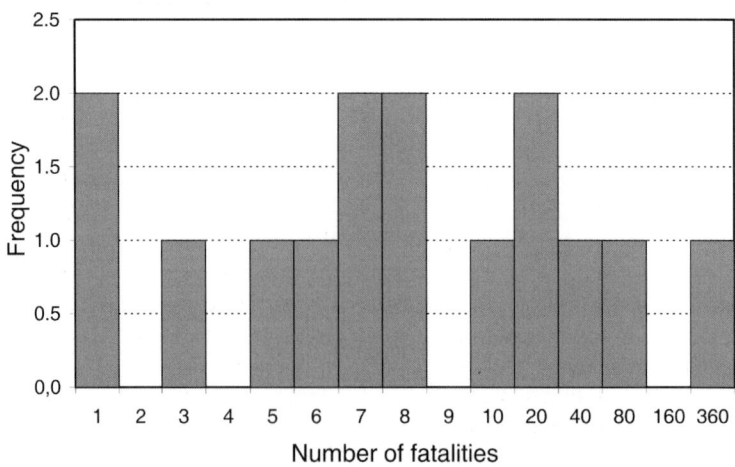

Fig. 7-17. Number of fatalities during bridge failures caused by ship impacts

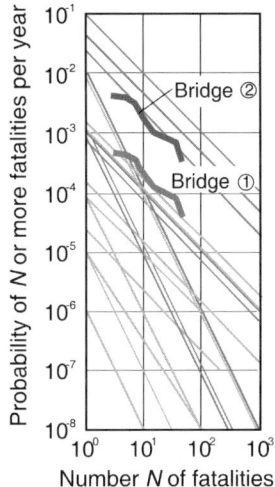

Fig. 7-18. *F-N* diagram for the two bridges

Table 7-7. Investment to increase safety through different structural strengthening solutions using the life quality index (B.-Bridge, ✞ - Number of fatalities for indices see Table 7-4)

B.	Pier	Element	As-it-is	Strengthening	✞	€
2	2		Damage (crack)[a]	No Damage (no crack)[b]	10	52,885
	2	Front	Damage (crack)[a]	Impact fender system	10	103,548
	2	Impact	Damage (crack)[a]	Pier increased × 2.3[c]	10	100,130
	2		Damage (crack)[a]	Tension strength × 2[d]	10	89,735
	2	Lateral	No Damage (no crack)[b]	Impact fender system	10	81,095
1	II		Explosion chamber[e]	No explosion chamber[f]	10	19,062
	II	Front	Explosion chamber[e]	Prestressed[g]	10	26,679
	II	Impact	Explosion chamber[e]	Reinforced concrete[h]	10	26,782
	II		Explosion chamber[e]	GEWI-Elements[i]	10	20,986
	III		Explosion chamber[e]	No explosion chamber[f]	10	2,392
	III		Explosion chamber[e]	Prestressed[g]	10	11,910
2	2		Damage (crack)[a]	No damage	22	116,347
	2	Front	Damage (crack)[a]	Impact fender system	22	227,806
	2	Impact	Damage (crack)[a]	Pier increased × 2.3[c]	22	220,287
	2		Damage (crack)[a]	Tension strength × 2[d]	22	197,416
	2	Lateral	No Damage (no crack)[b]	Impact fender system	22	178,410
1	II		Explosion chamber[e]	No explosion chamber[f]	22	41,937
	II	Front	Explosion chamber[e]	Prestressed[g]	22	58,695
	II	Impact	Explosion chamber[e]	Reinforced concrete[h]	22	58,919
	II		Explosion chamber[e]	GEWI-Elements[i]	22	46,107
	III		Explosion chamber[e]	No explosion chamber[f]	22	5,262
	III		Explosion chamber[e]	Prestressed[g]	22	26,202

The application of threaded rods (GEWI-Elements) and closing the explosion chambers has turned out to be the most effective and realisable structural solutions for Bridge 1, and the application of a fender system was recommended for Bridge 2. Both systems have now been installed and are protecting the bridges.

Acknowledgement

Many thanks to OBR W.-M. Nitzsche from the Würzburg Department of Highways and Bridges (Straßenbauamt Würzburg) and also the Federal Waterways Engineering and Research Institute (Bundesanstalt für Wasserbau) in Karlsruhe for the support they gave to the ship impact research work.

References

Berndt E (1996) Zur Druck- und Schubfestigkeit von Mauerwerk – experimentell nachgewiesen an Strukturen aus Elbsandstein. Bautechnik 73 (1996), Heft 4, pp 222–234

Beyer P (2001) 150 Jahre Göltzschtal- und Elstertalbrücke im sächsischen Vogtland 1851–2001. Vogtland Verlag Plauen

Bothe E, Henning J, Curbach M, Proske D & Bösche T (2004) Nichtlineare Berechnung alter Bogenbrücken auf der Grundlage neuer Vorschriften. Beton- und Stahlbetonbau 99 (2004) 4, pp 289–294

Busch P (1998) Probabilistische Analyse und Bewertung der Tragsicherheit historischer Steinbogenbrücken: Ein Beitrag zur angewandten Zuverlässigkeitstheorie. Dissertation. Technische Universität Dresden

Cai GQ & Elishakoff I (1994) Refined second-order reliability analysis. Structural Safety, 14 (1994), pp 267–276

Casas, JR (1999) Assessment of masonry arch bridges. Experiences from recent case studies. Current and future trends in bridge design, construction and maintenance in Bridge design construction and maintenance. (Eds) PC Das, DM Frangopol & AS Nowak. Thomas Telford, London 1999, pp 415–423

Curbach M & Proske D (1998) Module zur Ermittlung der Versagenswahrscheinlichkeit. 16. CAD-FEM Nutzermeeting. Grafing

E DIN 1055-100 (1999) Einwirkungen auf Tragwerke, Teil 100: Grundlagen der Tragwerksplanung, Sicherheitskonzept und Bemessungsregeln, Juli 1999

E DIN 1055-9 (2000) Einwirkungen auf Tragwerke, Teil 9: Außergewöhnliche Einwirkungen. März 2000

Eurocode 1 (1994) ENV 1991–1: Basis of Design and Action on Structures, Part 1: Basis of Design. CEN/CS, August 1994

Eurocode 1 (1994) ENV 1991–1: Basis of Design and Action on Structures, Part 1: Basis of Design. CEN/CS, August 1994

Hannawald F, Reintjes K-H & Graße W (2003) Messwertgestützte Beurteilung des Gebrauchsverhaltens einer Stahlverbund-Autobahnbrücke. Stahlbau 72, 7, pp 507–516

Köylüoglu HU & Nielsen SRK (1994) New approximations for SORM integrals. Structural Safety, 13, pp 235–246

Kunz C (1996) Risikoabschätzung für Schiffsstoß an der Straßenbrücke Vilshofen, Donau-km 2249, Bundesanstalt für Wasserbau, Karlsruhe, 10. Juni 1996

Kunz C (1998) Probabilistische Stoßlast-Ermittlung für Schiffsstoß an der Alten Mainbrücke Lohr, Main-km 197,9. Bundesanstalt für Wasserbau (BAW), Auftragsnummer: BAW-Nr. 97 116 407, Karlsruhe

Larsen OD (1993) Ship Collision with Bridges, The Interaction between Vessel Traffic and Bridge Structures. IABSE (International Association for Bridge and Structural Engineering), Zürich

Lohrberg K & Keitel V (1990) Zur Frage der Eintrittswahrscheinlichkeit des Schiffsstoßes auf Brücken. Bw/ZfB – Zeitschrift für Binnenschifffahrt und Wasserstraßen – Nr. 1 – Februar 1990, S. 16–21

Mann G (Edr) (1991) Propyläen Weltgeschichte – Eine Universalgeschichte. Propyläen Verlag Berlin – Frankfurt am Main

Mastaglio L (1997) Bridge Bashing. Civil Engineering. April 1997, pp 38–40

Meier-Dörnberg K-E (1984) Entwurfsgrundsätze zum Ausbau der Saar – theoretische und experimentelle Untersuchungen zur Ermittlungen statischer Ersatzlasten für den Anprall von Schiffen an Bauwerken. TH Darmstadt

Möller B, Beer M, Graf W, Schneider R & Stransky W (1998) Zur Beurteilung der Sicherheitsaussage stochastischer Methoden. Sicherheitsrisiken in der Tragwerksmodellierung, 2. Dresdner Baustatik-Seminar. 9. Oktober 1998, Technische Universität Dresden, pp 19–41

Möller B, Graf W, Beer M & Bartzsch M (2002) Sicherheitsbeurteilung von Naturstein-Bogenbrücken. 6. Dresdner Baustatik-Seminar. Rekonstruktion und Revitalisierung aus statisch-konstruktiver Sicht. 18. Oktober 2002. Lehrstuhl für Statik + Landesvereinigung der Prüfingenieure für Bautechnik Sachsen + Ingenieurkammer Sachsen. Technische Universität Dresden, pp 69–91

Naumann J (2002) Aktuelle Schwerpunkte im Brückenbau. 12. Dresdner Brückenbausymposium. Lehrstuhl für Massivbau und Freunde des Bauingenieurwesens e.V., pp 43–52

Proske D (2003) Ein Beitrag zur Risikobeurteilung von alten Brücken unter Schiffsanprall. Ph.D. Thesis, Technische Universität Dresden

Proske D, Lieberwirth P & van Gelder P (2006) Sicherheitsbeurteilung historischer Steinbogenbrücken. Dirk Proske Verlag Dresden. Second edition will be printed by Springer in 2008/2009

PTC 1 (1999) – Working Group Nr. 19, Permanent International Association of Navigation Congresses: "Problems of Collision due to the Presence of bridges". Konzeptpapier vom 17.12.1999.

Rajashekhar MR & Ellingwood BR (1993) A new look at the response surface approach for reliability analysis. Structural Safety, 12 (1993), pp 205–220

Scheer J (2000) Versagen von Bauwerken. Band I: Brücken. Ernst & Sohn Verlag für Architektur und technische Wissenschaften GmbH, Berlin

Schueremans L, Smars, P & Van Gemert D (2001) Safety of arches – a probabilistic approach. 9th Canadian masonry symposium

Simandl T, Glatzl J, Schweighofer B & Blovsky S (2006) Schiffsanprall an die Eisenbahnbrücke in Krems. Erstmalige Anwendung des neuen Eurocode EN 1991-1-7 für Außergewöhnliche Einwirkungen. Beton- und Stahlbetonbau, Vol. 101, Issue 9, pp 722–728

Spaethe G (1992) Die Sicherheit tragender Baukonstruktionen, 2. Neubearbeitete Auflage, Wien, Springer Verlag

Stede J (1997) Binnenschifffahrtsunfälle 1991 bis 1996. StBA. Wirtschaft und Statistik 12

Van Manen, SE & Frandsen AG (1998) Ship collision with bridges, review of accidents. In: Ship Collision Analysis. Proceedings of the International Symposium on Advances in Ship Collision Analysis, Copenhagen/Denmark, H Gluver & D Olsen (eds), A.A. Balkema, Rotterdam, pp 3–12

Voortman HG, Van Gelder PHAJM, Vrijling JK & Pandey MD (2001) Definition of acceptable risk in flood-prone areas. Structural Safety and Reliablitity (ICOSSAR 01). (Eds) RB Corotis, GI Schuëller & M Shinozuka. Rotterdam: Balkema, on CD

Warnecke P, Rostasy FS & Budelmann H (1995) Tragverhalten und Konsolidierung von Wänden und Stützen aus historischem Natursteinmauerwerk. Mauerwerkskalender 1995, Ernst & Sohn, Berlin, pp 623–685

Woisin G (1977) Die Kollisionsversuche der GKSS. Jahrbuch der Schiffbautechnischen Gesellschaft, 70. Band, Springer Verlag, Berlin

Index

Printing: Krips bv, Meppel, The Netherlands
Binding: Stürtz, Würzburg, Germany